19.95
80U

The Continuous Wave

Hugh G. J. Aitken

The Continuous Wave:
Technology and American Radio,
1900-1932

PRINCETON UNIVERSITY PRESS
PRINCETON, NEW JERSEY

Contents

Plates

Figures

Acknowledgments

One of the pleasures of finishing a book is the opportunity it furnishes to thank those who have helped you along the way. This particular book has been a long time in the making and I have incurred heavy obligations to individuals who gave me the benefit of their advice and criticism and to institutions that provided support and made their resources available.

In the latter category I should express my thanks especially to the National Museum of American History at the Smithsonian Institution, the Manuscript Division of the Library of Congress, the Schaffer Library at Union College, the Sterling Library at Yale University, the Widener Library at Harvard University as well as the Mackay Engineering Library at that institution and the Baker Library of the Harvard University Graduate School of Business Administration. In California I received indispensable assistance from the Bancroft Library at the University of California, Berkeley, the Special Collection Department of the Stanford University Library, and the archives of the Foothill College Electronics Museum. While working with the Fessenden Papers I received excellent cooperation from the staff of the North Carolina State Archives. Wendy Chu, manager of library services at the David Sarnoff Library in Princeton, New Jersey, was most helpful in response to several requests for information. Special mention must be made of the Owen D. Young Library at Van Hornesville, New York, where the hospitality of Josephine and Everett Case transformed an otherwise laborious week of research into the kind of idyllic experience that lingers in the memory.

I am deeply indebted to the personnel of the Frost Library at my home institution, Amherst College, and particularly to the unfailingly efficient and cheerful staff of the Reference Department. Time and again I thought, when submitting to them an apparently impossible request, that this time I had surely tried their patience to the limit. But that never happened.

Anyone carrying out an extended piece of historical research at a small college somewhat distant from major urban libraries becomes acutely aware of how much he depends on the knowledge of skilled librarians and on the efficient functioning of interlibrary cooperation. It is a pleasure to acknowledge my debt in this case. The staff of Frost Library performed miracles for me, not once but regularly.

I must also record my obligation to the Academic Computer Center at Amherst College, and especially to its director, Betty Steele, and her talented staff of student assistants. I tried their patience many times.

A number of individuals were good enough to read drafts of the chapters in this book, and although I am of course solely responsible for any errors or omissions it may still contain, I am very conscious of how much more imperfect it would have been without their assistance. David Noble took time from his own research to write an extended critique of an early outline that I referred to often in later years. Susan Douglas, though deeply involved in preparing her own work for publication, gave me the benefit of her keen critical sense and her knowledge of early radio history. My friends Robert S. Lurie and Russell Janis helped to orient me to recent literature in the field of industrial organization; Margaret Graham shared her knowledge of the later history of RCA; and Bernard Finn saved me from several errors in connection with the technology of submarine cables. George Wise, historian in the R&D Communications Operation of the General Electric Company, was always ready to make available to me his detailed knowledge of the history of that company, and Thorn Mayes and Leonard Fuller generously shared their knowledge of the Federal Telegraph Company. Kaye Weedon read every chapter as it was written and provided detailed criticisms and suggestions; my information on the involvement of the Swedish government in American radio policy and on de Groot's arc station in Java comes almost entirely from him. Alan Douglas was particularly helpful in connection with Fessenden and the National Electric Signaling Company, and Robert Rosenberg graciously shared with me his knowledge of the history of engineering education.

There are a few people who played such a special role in connection with the writing of this book that they must be mentioned separately. One of them is James Brittain, who made available to me without constraint the typescript of his forthcoming biography of Ernst Alexanderson, along with several fine illustrations. Such generosity is not always to be counted on when scholars are working in closely related areas. Second, I must thank my friend James H. Clark, who took time from his busy schedule to read every chapter in draft and strengthen my determination to continue. Somehow or other the chapters did get written, and for that much of the credit must go to the support of a good friend. And lastly, I owe a special debt of gratitude to my wife Janice and to my daughter Ellen, who have lived with the book for as long as I have. When my confidence failed, theirs did not. When I was ready to give up, they had courage. Surely every author goes through phases when, as he sees it, the only sensible thing for him to do is burn his manuscript. That I never quite reached the point of acting on that impulse is largely due to Janice and Ellen.

H.G.J.A.
Summer Harbor, Maine
August, 1983

This material is based upon work supported by the National Science Foundation under Grant No. SOC 79-07208 and Award No. SES 82-03870. Any opinions, findings, and conclusions or recommendations expressed in this publication are those of the author and do not necessarily reflect the views of the National Science Foundation.

Chronology of Events in
The Continuous Wave

1887 Hertz's experiments at Karlsruhe
1893 Fessenden joins Western University of Pennsylvania
1894 Marconi's experiments at Bologna
      Lodge demonstrates wireless signaling at Oxford
1898 Tesla demonstrates radio guidance
1899 Marconi telegraphs across the English Channel
      De Forest completes doctoral dissertation at Yale
1900 Fessenden approaches GE about a radiofrequency alternator
1901 Fessenden discovers heterodyne principle
      Marconi reports transatlantic reception
      Duddell announces the "musical arc"
1902 Fessenden forms NESCO
      Alexanderson joins GE
      Elwell arrives in San Francisco
      American De Forest Wireless Telegraph Company formed
1903 Poulsen arc patented
      De Forest invents the "flame detector"
1904 Alexanderson designs 150 KHz alternator for Fessenden
      Fleming invents the "oscillation valve"
1905 De Forest invents the four-electrode flame detector
1906 Fessenden reports successful radiotelephony
      De Forest invents the three-electrode audion detector
      United Wireless Company formed
1907 Elwell graduates from Stanford
      American De Forest Wireless Telephone Company formed

1908   Elwell examines Poulsen system in Denmark
       GE completes 100 KHz 2 kw. alternator

1909   Elwell brings first Poulsen arc to U.S.
       Poulsen Wireless Company formed
       Navy invites bids for Arlington station

1911   Federal Telegraph Company formed
       Fessenden leaves NESCO
       De Forest joins Federal Telegraph
       Lowenstein invents "ion controller" amplifier

1912   De Forest invents audion amplifier
       Lowenstein and de Forest demonstrate amplifiers to AT&T

1913   AT&T buys rights to de Forest's amplifier
       Armstrong perfects the feedback circuit
       Navy commissions Arlington station
       Federal Telegraph wins contract for Darien station

1914   World War I breaks out in Europe
       Navy assumes control of Tuckerton station
       American Marconi announces plans for New Brunswick station
       Alexanderson urges GE to develop a 50 kw. alternator
       British Marconi acquires rights to Poulsen arc

1915   Navy assumes control of Sayville station
       Marconi inspects GE's 50 kw. alternator
       First GE-Marconi contract

1917   U.S. enters World War I
       Navy assumes control of Marconi and Federal stations
       50 kw. alternator installed at New Brunswick

1918   Navy purchases Marconi coastal and ship stations
       Navy purchases Federal Company's patents
       200 kw. alternator installed at New Brunswick
       Armistice declared in Europe

1919   Peace Conference convenes
       Marconi and GE resume negotiations
       Bullard and Hooper intervene in Marconi-GE negotiations
       RCA incorporated

1920   Navy relinquishes control of Marconi stations; RCA takes over
       Navy calls for cooperation on vacuum tube patents
       AT&T and GE sign cross-licensing agreements
       Station KDKA begins broadcasting

1921   RCA's "Radio Central" in service
       Westinghouse signs cross-licensing agreement with AT&T, GE,
           and RCA
       Federal Trade Commission begins investigation of radio industry

1922   AT&T divests itself of investments in RCA

1923   RCA-AT&T cross-licensing dispute goes to arbitration

1926   AT&T sells station WEAF to RCA
       New cross-licensing agreement signed between "Radio Group"
           and "Telephone Group"

1929   RCA-Victor Company formed

1930   Sarnoff becomes president of RCA
       "Unification" agreement signed by RCA, GE, and Westinghouse
       RCA indicted under Sherman Antitrust Act

1932   Consent decree separates RCA from GE and Westinghouse

The Continuous Wave

ONE

Prologue

THIS book, although designed to be read independently, is in one sense a continuation of an earlier work, *Syntony and Spark: The Origins of Radio*, published in 1976.[1] That book dealt with the very earliest phase of radio technology, when the scientific work of James Clerk Maxwell and Heinrich Hertz was being transformed into a technology of communication by men like Oliver Lodge and Guglielmo Marconi and the first attempts were being made to base commercial enterprises on that technology. The present volume picks up the story in the closing décades of the nineteenth century and carries it through the 1920s, when the advent of popular broadcasting transformed radio from a means of point-to-point communication, competing with the wired telegraph, into the agency of mass communication it is today. I discuss the origins of broadcasting only briefly in this book. My interest is in the origins of the technology that made broadcasting possible. This was the technology of the continuous wave.

In the earliest days of "signalling without wires" the only known method of generating radio waves was by means of sparks. An induction coil, or sometimes a bank of capacitors, was used to place a high voltage across a spark gap; when a spark jumped the gap it created an electromagnetic disturbance that could be detected at a distance. A series of sparks following each other in rapid succession gave rise to a chain of such disturbances—a radio wave, in short—that could be interrupted to form the dots and dashes of the Morse code and thereby convey information. Such waves travelled at a constant velocity: the speed of light. Each wave had a specific wavelength—the distance between succeeding

[1] Hugh G. J. Aitken, *Syntony and Spark: The Origins of Radio* (1976; Princeton, 1985).

peaks or troughs—usually measured in meters; and therefore, given the constant velocity, it had a specific frequency (the number of cycles per second).[2] Each wave, that is to say, had a particular "place" on the electromagnetic spectrum, defined by its wavelength or frequency.[3] If it was to be detected, the apparatus used for receiving had to be capable of responding to waves of that frequency—that is, it had to find and react to a signal at that "place" and, if possible, reject all others. Today we do this by a process we call tuning. In the earliest days of radio it was more common to speak of "syntony." Receiving and transmitting circuits were said to be in syntony when they resonated at the same frequency.

Syntony and spark were the characteristics that gave unity to that first phase of radio history. Technological development consisted of devising more effective spark transmitters, receivers that could detect and respond to spark-generated waves, and syntonic circuits that made it possible for transmitters and receivers to "find" each other in the radio spectrum. Important elements in this process were the development of antennas that could radiate and pick up signals efficiently, and the trial-and-error discovery of which wavelengths were most suitable for transmission over long distances.

The radio wave generated by a spark transmitter was a wave of a particular type. Each spark discharge generated a series of oscillations that diminished rapidly in amplitude as its energy was radiated into space and absorbed by the internal resistance of the components. A common simile, and an appropriate one, was to compare the antenna to a bell struck by a clapper. The bell, when struck, would sound a note, radiating energy in the form of sound waves. But the strength of the note would diminish more or less rapidly, as the vibrations of the bell diminished in amplitude. If the vibrations were "damped," as for instance if one placed a hand on the bell's surface, the sound would die away very quickly. So it was with a spark discharge: it had a degree of damping, depending on the internal resistance of the circuit and the rate at which it radiated energy into space. The radio wave generated by a succession of spark discharges consisted of a series of these damped oscillations. In that sense,

[2] The term "Hertz" is now commonly used for the older "cycles per second," and that convention will be generally but not slavishly followed in this book. The abbreviations kHz and MHz refer to one thousand and one million cycles per second respectively.

[3] The radiofrequency spectrum is generally taken today as lying between 20 kHz and 30,000 MHz (30 GHz).

a spark transmitter, although it might radiate continuously, did not generate a true continuous wave. (See Fig. 1.1)

It can be shown mathematically, by a technique known as Fourier analysis, that a damped oscillation such as that generated by a spark transmitter (or indeed any other complex waveform) can be decomposed into a large number of other oscillations, each with a frequency and wavelength of its own.[4] These constituent oscillations are sine waves, in which the signal changes in an exactly prescribed way through a full cycle, going first positive, then negative, following the sine function in trigonometry. (See Fig. 1.2) This is no mere mathematical transformation: if such a train of damped oscillations were radiated from an antenna, its constituent sine waves would appear on the electromagnetic spectrum and (unless filtered out by tuned circuits) affect any receiver with enough sensitivity to detect them. From this fact certain important practical implications followed. A spark radio transmitter generated not one radio wave, but a very large number of them. Its signal was not at a single "place" on the electromagnetic spectrum but at a very large number of places. A true unmodulated continuous sine wave, in contrast, if one could have been generated, would have had one frequency only; it would have appeared at one place in the spectrum and only at that place.

This is the reason why, today, spark transmitters are universally outlawed. The radiofrequency spectrum is, to be sure, a unique resource, in that it can never be used up. But it can be overused; it can be overcrowded. Congestion of the radio spectrum creates a form of pollution in which transmitters interfere with each other and receivers are unable to select the signal that conveys the desired information from among interfering signals that are essentially noise. The danger always exists of a new "tragedy of the commons," in which overuse of a resource freely available to all creates a situation in which it is available to no one.[5] To prevent such a situation, governments and international supervisory agencies resort to frequency allocation—essentially, the rationing of scarce space on the spectrum.

A spark transmitter is inevitably a dirty transmitter. It pollutes the

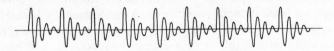

Fig. 1.1: A train of damped oscillations.

[4] For further discussion, see Aitken, *Syntony*, pp. 70-75.

[5] See Garrett Hardin, "The Tragedy of the Commons," in *Economic Foundations of Property Law*, ed. Bruce A. Ackerman (Boston, 1975).

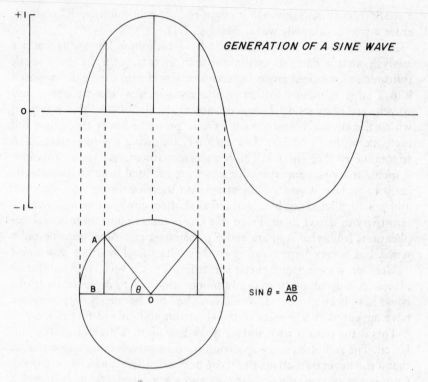

Fig. 1.2: Geometric generation of a sine wave.

spectrum by contaminating frequencies far removed from those nominally being used to carry the message. Its undesired effects can be minimized by reducing the degree of damping, which is to say by approximating more and more closely to a continuous wave. But there is a point, with spark, beyond which amelioration cannot go. Furthermore, spark transmissions make selective tuning much harder to achieve. The spark gap itself is a high-resistance element; its presence unavoidably lowers what engineers call the "Q" (quality) of the transmitting circuit and introduces a large damping coefficient. A spark-generated wave, therefore, is necessarily "broad." Quite apart from the harmonics it generates, it occupies an undesirably large space on the spectrum.

In the closing years of the nineteenth century and the first years of the twentieth, a few radio experimenters and scientists arrived at the conviction that spark radio transmission would have to be abandoned. They were, at first, a very small minority and they had difficulty making their case. On the one hand, the full potentials of spark transmitters had by no means been exhausted. Each year more powerful and more sophisticated spark transmitters appeared on the scene. Higher spark rates and lower damping coefficients approximated ever more closely to a continuous wave. Spark was a familiar and proven technology and there seemed no need to abandon it. On the other hand, devices and circuits that could generate true continuous waves were not at hand. A few experiments had been made with alternators, similar in principle to those that generated alternating current electricity for homes and factories, but they were low-frequency devices and delivered little output power. Some interesting work had been done with oscillating arcs that could be made to generate sound waves. Oscillating triode vacuum tubes, later to be the almost universal method of generating continuous waves, were unknown. A continuous wave transmitter that could generate substantial amounts of power at radio frequencies did not exist in 1900. In the circumstances, to believe that continuous wave radio could and should replace spark called for an act of faith.

Historians of technology have learned to recognize situations of this type and to attach special importance to them. Edward Constant, for example, writing about the introduction of the turbojet engine, points out that the men who, in the 1930s, built the first turbojets were not responding to any current failure of the conventional piston engine, which at that time had by no means reached the limits of its development.[6] They were responding, rather, to insights that left them convinced that the conventional aircraft propulsion system would inevitably run into difficulties at some point in the future, and that a new and radically different system could and should be built. Such a situation sets the stage for discontinuous change, for what Constant, borrowing a term from Thomas Kuhn, calls a shift in technological paradigms.[7] In Kuhn's model, designed originally to explain major discontinuities in the history of scientific theory, attempts to develop a new paradigm begin not when the explanatory power of its predecessor is fully exhausted but when anomalies begin to accumulate: phenomena that accepted theory cannot explain, or that it manages to explain only by successive ad hoc adjustments and extensions. Similarly, for Constant, attempts to develop a new

[6] Edward W. Constant, Jr., "A Model for Technological Change Applied to the Turbojet Revolution," *Technology and Culture* 14 (October 1973), 553-72. For a more extended discussion see Constant, *The Origins of the Turbojet Revolution* (Baltimore, 1980).

[7] Thomas S. Kuhn, *The Structure of Scientific Revolutions*, 2nd ed. (Chicago, 1970). See also Kuhn, *The Essential Tension: Selected Studies in Scientific Tradition and Change* (Chicago, 1977).

technological paradigm begin not when conventionally accepted practice has failed in any absolute sense, but when a minority has become convinced that at some point in the future it will fail.[8] At the time, however, there is still much potential for development in the conventional system; the anomaly is presumptive, not presently existing. It is visible to some but not to others. And those who see it and become convinced of its reality require, as part of their motivation, a kind of dedicated determination that to outsiders often seems unreasonable. To more commonsensical people, what they seek is probably unattainable and there is no real need for it anyway.

For Constant, the source of the insights that convince some individuals of the existence of a presumptive anomaly in technological practice is always science—in his case, advances in aerodynamics, the branch of physics that deals with gas flows. Presumptive anomaly occurs when scientific insight, or assumptions derived from science, indicate either that, under some future conditions, the conventional technological system will fail or will function badly, or that a radically different system will do a much better job or do something entirely novel.[9] Science, according to Constant, provides the rational component that balances the nonrational "fanaticism" of the technological innovators—the "*provocateurs*," as he calls them.

In this book, although we shall use Kuhn's overworked term sparingly, we shall indeed be dealing with a major paradigm shift in radio technology—a shift that created the technical base for what we recognize as radio today and that was in its time as radical as the shift from piston engine to turbojet of which Constant writes. But we shall not take it for granted that what powered that shift were new insights derived from science, if by science is meant a body of articulated theory and a set of repeatable experimental observations. To assume that information generated by science is the only possible source for the detection of presumptive anomalies comes close to assuming that advances in science are the only possible source of major technological change. And that in turn comes near to defining technology as applied science—an identification that few scholars today are willing to make.[10]

[8] Constant defines a technological paradigm as "an accepted mode of technical operation, the usual means of accomplishing a technical task . . . the conventional system as defined and accepted by a relevant community of technological practitioners" ("A Model for Technological Change," p. 554).

[9] Ibid., p. 555.

[10] See, for example, Edwin T. Layton, "Mirror-Image Twins: The Communities of Science and Technology in 19th-Century America," *Technology and Culture* 12 (October 1971), 562-80.

What role science played in the technological shift that is the central concern of this book is a question to be asked, not an assumption to be made. The men we shall be dealing with were not without scientific training; they had a keen sense of scientific literature and scientific personnel as resources on which they could draw; and they understood the importance of clothing their judgments in the prestigious language of science. But this is not to say that their commitment to the continuous wave was founded on scientific insight, nor that their perception of the presumptive anomaly facing spark technology was deduced from any body of scientific theory. On the contrary, if we are to judge by their own statements, their dissatisfaction with spark was based on more pragmatic grounds. The test of performance was whether or not radio communication could be reliably maintained over considerable distances. By this standard, spark was in their judgment a poor prospect, primarily because, with its broad signal and multiple harmonics, a spark transmitter dissipated its power. Concentrate the available energy on a single frequency and your chances of achieving distance, of cutting through interference, fading, and atmospherics, were likely to be very much better. That kind of thinking called for no sophisticated knowledge of Fourier analysis. It was a strictly practical matter. Could spark do the job that its apologists claimed? Advocates of the continuous wave believed that it could not. There had to be a better way.

And there had to be a better way of detecting radio signals than by the device—the coherer—that typically accompanied spark transmission. Even the best of coherers was a temperamental device, hard to keep in adjustment; the need for "tapping-back" to restore sensitivity after a signal had been received, meant slow transmission speeds; and, since the coherer responded to voltage impulses, it was incapable of discriminating between signals and atmospheric noise. Escaping from spark technology called, therefore, not only for developing transmitters capable of generating continuous waves but also for finding receivers capable of detecting them. This was to require new circuits as well as new devices.

Those who decided to abandon spark and find a better way were responding not to new insights derived from science but to their sense that spark was a technological dead end and that continued reliance on spark would jeopardize radio's economic viability. The issue was whether radio could find for itself an economic niche in which it could grow and develop. Was this likely as long as spark reigned supreme? Some thought not. The presumptive anomaly that these individuals saw on the horizon appeared, not where technology and science met, but at the hazy boundary where radio stopped being a matter for visionary experimenters and started to become a hardheaded business capable of gaining and holding a commercial market. The criteria that the new technology would have to meet were economic criteria: could capital invested in radio earn the going rate of return? The expectation that spark radio would fail this market test was the rational ground for turning to the continuous wave.

Breaking away from spark, however, was not easy. Spark was the technology through which radio had come into existence; it had provided the only dramatic successes that the new means of communication could claim; and it was the technique to which the major operating organization of the day—the Marconi Company—seemed irrevocably committed. What was involved in the shift to the continuous wave was not just an incremental improvement in radio technique; it was a change in the way you thought about radio, the way you conceptualized it and visualized it. If you took it seriously it had some of the elements of religious conversion: it affected everything you did in the field thereafter. This had economic implications. To insist, as for example Reginald Fessenden did, that spark radio was following a fundamentally wrong track piled a new uncertainty on top of serious economic uncertainties already present. That did not make it easier to find and keep financial backers. And it had personal implications. Those who advocated such unconventional and visionary ways of thought and action paid a price in terms of the personalities they developed and the style of life they followed.

One element in the vision that these individuals followed was wireless telephony: the transmission by radio of voice and music. Without this element—that is, if radio communications had continued to be thought of exclusively as the dots and dashes of Morse code telegraphy—it is questionable whether continuous wave radio would have seemed a gamble worth taking. Without exception, the early devotees of continuous wave radio had in mind the transmission of the human voice and not merely a marginal improvement in radio telegraphy. This had implications for the standards of performance that continuous wave radio was expected to meet. A slight competitive advantage over spark telegraphy was not enough to attract venture capital: the lure was wireless telephony—initially not for broadcast entertainment but to provide the kind of point-to-point communication that the American Telephone and Telegraph Company provided, only without the fixed costs of a wired network.

It is often the case, when a radically new and different technological system appears on the horizon, that it is at first judged to be less efficient than the system it eventually replaces. There are two main reasons for this. First, the conventional system has had the benefit of considerable developmental improvement since it was first introduced and it is familiar to users. The system that challenges it is imperfect, incomplete, risky, and often disconcerting. The second reason is more subtle: the standards of performance by which the new system is appraised have been worked out in terms of the jobs that the old system has done and the criteria especially relevant to those jobs. Two examples will make the point clear. In the early eighteenth century an overshot water wheel, with centuries of development behind it, was undoubtedly a more efficient power source for conventional uses than the novel atmospheric engine. A millwright comparing the two would not have hesitated to choose the water wheel. It was only for particular uses, such as for pumping water out of deep mine shafts, and in particular contexts, such as locations where cheap coal was available, that the Newcomen engine had the advantage. These particular functions and contexts, however, gave the new technology of steam power the toehold it needed to set out on its own course of refinement and development, a course that eventually made it capable of performing functions (such as overland transportation) that no water wheel could ever perform. Similarly, as Constant points out, for conventional uses and with conventional airframes, the first turbojet engines—"volatile contrivances," he calls them—offered little if any advantage over the best piston engines of their day.[11] Their true superiority began to show only as aircraft approached and then exceeded the speed of sound and as new airframe construction techniques were developed. Changes in these parameters altered the performance criteria that an aircraft power plant was expected to meet.

Examples could be multiplied, but the point is essentially simple. A radically new technology does more than merely perform old functions better; it makes it possible to perform functions that the technology it replaces could not perform at all. It not only solves a problem; it "oversolves" it. It literally creates its own future. One implication is that, like one of Kuhn's scientific paradigms, it has to be appraised not in terms of the old functions, but in terms of the new possibilities that it opens up. Judged by the criteria of the system being displaced, its differential advantage may at first seem trivial—just enough, perhaps, to give it a point of entry into commercial acceptance. Further development depends on finding new markets. So it was with continuous wave radio. For radio telegraphy its superiority was arguable—in the judgment of respected authorities, nonexistent. But for voice transmission it was a different matter. Many people tried to transmit the human voice by modulating the output of a spark transmitter. They all failed. Even at very high spark rates and with low damping coefficients, spark transmitters were too

[11] Constant, "A Model for Technological Change," pp. 570-71.

noisy to transmit speech without intolerable distortion. Only a transmitter that generated true continuous sine waves could do the job.

Anyone who could devise such a transmitter, therefore, had at his disposal a technology with potentials that transcended those of spark. Implicit was the possibility not only of point-to-point communication in plain spoken language but also of "broadcasting" information and entertainment to anyone with a suitable receiver. In 1900, to be sure, no one was thinking of broadcasting in those terms. The fact that radio communications *were* "broadcast"—that they could not be kept secret unless coded or encrypted—was generally thought of as a serious limitation of the new technology, as compared with the relative privacy and security offered by wired systems. When the U.S. Navy used the term "broadcasting," it referred to a practice whereby ships receiving radio traffic were exempted from the usual requirement to acknowledge receipt. When Woodrow Wilson's adviser on communications, briefing the president for the Paris Peace Conference, discussed the use of radio for "broadcasting," he was referring to the possibility of disseminating American news and propaganda to foreign news agencies without depending on the submarine cables. These forms of "broadcasting" still exist today, but they are not what we usually mean by the word. We are creatures of the age of continuous wave radio and our language reflects the fact.

The development in the 1920s and 1930s of broadcasting in the popular sense—that is, the transmission of news, entertainment, and advertising to the general public by radio—would have been impossible without previous advances in continuous wave radio technology that had originally been made with quite different objectives in view. One can, of course, find a few farsighted individuals who thought that some such development was possible. All of the early experimenters with the continuous wave tried their hands at transmitting speech and music to anyone who would listen; and the legend of David Sarnoff and his "radio music box" in the home is well known. With these qualifications, it remains true that the rise of radio broadcasting is a classic example of the unanticipated consequences of technological change. Continuous wave radio opened a Pandora's box of consequences for life in the modern world—consequences completely out of proportion to the limited objectives of those who developed and sponsored the new techniques. Those who write of "autonomous technology," of "technology out of control," have precisely such episodes in mind.[12]

[12] See, for example, Langdon Winner, *Autonomous Technology: Technics-out-of-Control as a Theme in Political Thought* (Cambridge, Mass. and London: MIT Press, 1977), esp. pp. 91-98.

The irony is that continuous wave radio, once its technical feasibility had been proved, was a highly supervised and "managed"—even a politicized—innovation. It attracted the attention of both governments and major corporations. Attempts to manage the innovation, however, had nothing to do with what turned out to be its most socially disruptive use: public broadcasting. They were directed, rather, to its implications for what we would now call telecommunications policy, and specifically toward the new possibility it opened up for creating a corporate entity— a "chosen instrument," as it were—to advance and protect American national interests in world communications. One outcome of those efforts was the Radio Corporation of America, an organization whose original function and reason for being were to manage the deployment of continuous wave technology in what was then its most advanced form: the General Electric Company's radiofrequency alternator. External relations—communications between the United States and the rest of the world—were originally the corporation's primary concern, not domestic communications within the United States. Very quickly, however, RCA also became an instrument for the consolidation of domestic interests in radio and for the allocation of exclusive fields of corporate activity in domestic communications. The outcome was an elaborate network of intercorporate treaties intended to control the new technology and minimize the probability of conflict. This structure was not designed with public broadcasting in mind; it was, however, the structure that existed when broadcasting began, and the way broadcasting developed was profoundly affected by that fact. Broadcasting itself was an unplanned social innovation, but the corporate context in which it appeared was the result of a great deal of planning.

The invention of continuous wave radio technology provides the substance for approximately half of this book; the attempts to manage the innovation are the substance of the second half. As a "bridge," to illustrate the political and ideological context in which postwar policies were formulated, I have included a short chapter on telecommunications issues at the Paris Peace Conference, focusing on the acrimonious dispute over the disposition of the German submarine cables. In the chapters that make up the first half I have organized the presentation around certain of the individuals who played important roles in the process of invention: Reginald Fessenden; Cyril Elwell and his successor as chief engineer of Federal Telegraph, Leonard Fuller; and Lee de Forest. In the second half I have likewise given much attention to the individuals who were deeply involved in the organization of RCA, notably Owen D. Young and Lt. Comdr. Stanford C. Hooper, USN. This mode of presentation should not be taken as implying that I subscribe to a heroic theory of either technological or organizational change. It does reflect a belief that, although the development of technology and the rise of complex hierarchical organizations can be depicted as historical forces that transcend individual personalities, nevertheless they work themselves out through the decisions of individual men and women, and there is a certain gain in directness and immediacy when we show this process in operation.

* * *

Such, then, is the historical content of this book. And I hope that the chapters that follow are substantial enough in themselves to hold the attention of readers who have no interest in questions of historical method. Such questions inevitably arise, however, and in an academic discipline as new as the history of technology it may be useful to comment on them briefly.

Eugene Ferguson has warned against what he calls "the engineer's insistence that everything be defined as a first step."[13] This is, he suggests, a convenient device for avoiding any hard thought about implications that do not come immediately to mind, and he expresses the hope that historians will recognize that, when they have completely defined their problems, they have also solved them. With this caution in mind, I refrain from offering any of the standard and readily available definitions of technology, which normally emphasize either hardware or methods. Such definitions reduce technology either to machines or to techniques. Behind these definitions there lies an older and less limiting idea: the conception of technology as knowledge—Aristotle's "reasoned state of capacity to make." This is the conception that underlies this book.[14] I think of technology as one form of organized information—that which deals with man's capacity to manipulate the natural environment for human ends—and of the history of technology as one branch of intellectual history or of the history of ideas. Technological knowledge may indeed be given physical form in machines, and it may manifest itself in techniques. Questions about the origin and evolution of machines and techniques, however, are probably best tackled by examining the information that they embody and asking how that information came to be organized in that particular way.

The assumption is that, when the questions are phrased in these terms,

[13] Eugene S. Ferguson, "Toward a Discipline of the History of Technology," *Technology and Culture* 15 (January 1974), 21n.

[14] Compare the seminal article by Edwin T. Layton, "Technology as Knowledge," *Technology and Culture* 15 (January 1974), 33.

there is a reasonable probability that they can be answered from the historical record—that is, from surviving documents and other artifacts. An emphasis on the informational content of technology suggests that, when analyzing devices and processes that are new, we should ask to what extent they resulted from net increments to the stock of human knowledge—as might emerge, perhaps, from scientific discovery—and to what extent they represented novel combinations of knowledge already in existence. In either case the new device or process—the invention, if you will—can be thought of as a new configuration of information—a new gestalt, to use the psychologist's term. Some elements in that gestalt may be information that is new in an absolute sense; others will be drawn from the preexisting stock. What is new about the invention is the novelty of the combination—the way the elements are put together.[15]

If, then, we intend to analyze the process of invention, we need concepts that will enable us to come to grips with the ways in which flows of information come together to produce new combinations. We may take for granted the proposition that invention is a social process: the new combinations are indeed formed as ideas held by individuals, but those individuals function within formal and informal networks of communication. Such networks are social facts; they provide the channels over which information moves. Some of them are long-lived, serve as organized command and control systems, and are coterminous with formal organizations: the communications system of the General Electrical Company, perhaps, or of the U.S. Navy Department. But many are not. They may be highly informal and evanescent—a relaxed conversation between two engineers looking out over Gloucester harbor and chatting about yachts and vacuum tubes, to cite an example from one of the chapters that follow. Such networks serve not as the control systems of organizations but rather to interconnect organizations and individuals.

If we ask how inventions come into being, a sound operational rule is to examine the flows of information that converged at the point and at the time when the new combinations came into existence. A hypothesis worth testing is that the points of confluence of information flows define the social locations where there is a high probability of new combinations being made. And a derivative hypothesis is that the most interesting and striking of new combinations are likely to occur when information flows

[15] Judicial interpretations of patent law, in the United States, parallel this view: courts have consistently held that all elements in a patent claim may be old and yet the claim itself can still be proper subject matter for patent protection. See David A. Blumenthal, "Life-forms, Computer Programs, and the Pursuit of a Patent," *Technology Review* (February/March, 1983), 30.

meet, by chance or by design, that have not met before. In such a case a new node is formed and the possibility exists, at least as long as the interconnection lasts, that networks previously disjunct will be able to exchange information with each other.

This way of thinking about invention offers several advantages. It avoids the heroic view of invention that has afflicted so many popular histories. At the same time it calls attention to those individuals who find themselves, or place themselves, at the points of confluence of information flows and have the wit, the curiosity, and the imagination to put together items of information, and sometimes kinds of information, that have never been so synthesized before. Also, such an approach avoids determinism: it gives no warrant for asserting any kind of necessity in the process. But neither are we thrown back on blind chance. It is a matter of probabilities: the probability of new combinations being formed is higher at the points of confluence of information flows than it is elsewhere. And thirdly it enables us to take into account both demand and supply. Clearly the process of technological change has a certain logic of its own. Not all inventions, but only a certain set of them, are possible at any given time; it depends on the stocks of existing knowledge, the nature of the new knowledge being injected into the system, and the layout of the communications networks over which knowledge moves. For these reasons a "demand-pull" theory of invention and innovation, though it tells us much of what we need to know, can only be part of the story. The stock of scientific and technical knowledge is not a kind of putty out of which almost anything can be shaped.[16] There are important supply-side constraints on what is possible. But on the other hand, the way in which technological change tracks over time is not independent of demand-side factors, for information on social "need," in the form of market signals, legislation, government procurement policies, and so on, also influences decision-making. Some technological possibilities are exploited, others are not. Into some resources are poured generously, while others live on a starvation diet. Demand and supply interact to determine the outcome.

In *Syntony and Spark* I tried to grapple with these issues on a rather abstract level by identifying within any society three subsystems labelled science, technology, and the economy. Every society, I argued, has some form of scientific knowledge and some form of scientific activity, though

---

[16] The leading example of "demand-pull" theories of invention is Jacob Schmookler, *Invention and Economic Growth* (Cambridge, Mass., 1966). For critical comments, see Nathan Rosenberg, "Science, Invention, and Economic Growth," in *Perspectives on Technology* (Cambridge, 1976), 265.

possibly quite different from that typical of modern industrialized nations. It unquestionably has a technology of some sort. And it necessarily has an economic system, no matter how unspecialized and loosely articulated. These systems were depicted as linked to each other, more or less closely, by exchanges of information and resources; the historian's problem was to describe the content and volume of these exchanges in particular societies at particular times. One hypothesis suggested by the model was that, in determining the rate and direction of technological change in a society, a critical role was played by individuals who functioned at the interfaces between the three systems, translating the information generated in one into a form intelligible to participants in the others and organizing the movement of resources between them. These individuals I called "translators." In using that word I had in mind both the literal meaning of translation as the moving of something from place to place—social "place" in this case—and its more familiar meaning as the conveying of information from one language to another. It seemed important that science, technology, and the economy were usually organized around different values, rewarded different modes of behavior, and provided their members with different kinds of signals to guide their decisions. They literally did use different vocabularies, speak in different languages, and respond to different cues. There was, therefore, a general and continuing need for coordination and intercommunication, and these were the functions that the "translators" performed. How effectively they performed them, and what institutions they developed to help their performance, would presumably influence the rate and direction of technical change, which was my primary concern, but there would also be implications for the advance of science and for the rate of economic growth.

As a framework for organizing one's thoughts and marshalling the evidence, this simple model had its uses. It suffered, however, from at least three limitations. In the first place, try as one might to emphasize how complex were the interactions between science, technology, and the economy, the image left in the minds of most readers was of a process in which new knowledge was generated by science, converted into useful devices by technology, and then put to use by the economic system. The effect was to reinforce a stereotype that by the 1970s had been rejected as inadequate by most serious students of the history of technology, on the grounds that it left unexplained vast areas of technological development and seriously underestimated the extent to which scientific advance depended on technology rather than vice versa. To the degree that the model served to breathe new life into that moribund stereotype, it was a step backward in scholarship, not forward.

Secondly, the model tended to encourage what logicians used to call the "fallacy of misplaced concreteness." The impression left in the reader's mind was of science, technology, and the economy as in some sense real entities, rather than the highly abstract categories they were. In particular, the image presented was of the three systems as being distinct and clearly bounded. This may be a bias implicit in systems theory as it is often practiced.[17] One treats subsystems as "black boxes" with inputs and outputs, and the usual way of diagramming the system is to depict the several subsystems as neat rectangles with definite boundaries, connected by lines that indicate the exchanges going on between them. As an aid to analysis this technique is no doubt often useful. Inevitably, however, it encourages a two-dimensional view of reality and a linear theory of causation. The "afterimage" one carries away from any such presentation is of the subsystems as being clearly separated, distinct, and sharply bounded.

The more I learned about the history of technology, however, the less satisfactory that kind of image came to appear. Certainly there had been periods in history and there had been cultures in which science and technology lived in separate worlds, as it were, where their practitioners formed distinct communities with no common membership, where interaction between them was reduced to a kind of sporadic arm's length trading. But in the period in which I was working—the late nineteenth and early twentieth centuries—one of the obvious historical trends had been toward closer integration of the two systems. Edwin Layton, in a famous metaphor, had argued that by the end of the nineteenth century science and technology had become "mirror-image twins," as engineers acquired scientific training, applied the methods of science to engineering problems, and adopted the professional organization and (to a degree) the attitudes of scientists.[18] But also, it seems to me, science, at least in some of its branches, had become more technological, in the sense that the pace of scientific advance had come to be greatly influenced by the rate at which the necessary technology could be made available. Controlled nuclear fusion and genetic engineering, in the contemporary world, were obvious examples. The image that seemed appropriate was not one of discrete bounded systems but rather of communities that overlapped and intermingled.[19] There was a grey area where scientific technology and technological science shaded into each other.

[17] Compare Robert Lilienfeld, *The Rise of Systems Theory: An Ideological Analysis* (New York, 1978), esp. pp. 196-224.

[18] Layton, "Mirror-Image Twins."

[19] On the significance of interpenetration, see Talcott Parsons, *The System of Modern Societies* (Englewood Cliffs, N.J., 1970), pp. 5-7. Leonard Reich has

What was true of science and technology was even more true of the third element in the triad: the economic system. By the early decades of the twentieth century, the evidence seemed to suggest, the economic calculus had penetrated deeply into technological decision-making. Calculations of expected returns on capital, of costs and benefits, had become a major determinant of the direction in which technology moved. And there were indications that economic rationality and a market orientation had in some fields come to influence scientific decision-making, despite the fact that it ran counter to the traditional ethos. Among the obvious reasons were the increasing cost of scientific experimentation, the vastly enlarged role of government as the sponsor of scientific research, and the intimate connection between scientific advance and national security. Decisions on the funding of high-cost research projects depended on estimates of their probable net contribution at the margin to national welfare, preparedness, or prestige; and wherever that kind of calculation came into play the interpenetration of the scientific and the economic systems was clear.

The third major reason for dissatisfaction was that the model included no government sector. As long as the government could be regarded exclusively as a consumer of the outputs of science and technology, this introduced no serious error, for the government then was merely one component of the economic system, albeit an important one. But governments in the twentieth century had clearly come to play a major role in directing the course of scientific and technological discovery, and not merely through their leverage as consumers. They had become major sources of finance; they had taken the initiative in steering research and development in particular directions; they had intervened to block the transfer of scientific and technical information in certain cases and expedite it in others; and, through patent and copyright law, they had helped to structure the reward system that guided private endeavors. So pervasive, indeed, did the influence of government appear to have become, not merely on the rate of scientific and technological advance but also on its trajectory, that a clear case could be made for splitting off the government sector and analyzing separately its role in technological change.

---

pointed out that in an industrial research laboratory scientists often function as engineers and engineers as scientists, so that it becomes impossible for the historian to tell them apart except by their choice of projects and methods of research. In that environment the "mirror-image twins" model of distinct communities begins to break down. See Leonard S. Reich, "Irving Langmuir and the Pursuit of Science and Technology in the Corporate Environment," *Technology and Culture* 24 (April 1983), 199-221.

Here again, however, the danger of misplaced concreteness seemed very real. No government, in its relations to science, technology, and the economic system, was a monolith. Quite the contrary: it acted through a myriad of agencies and bureaus, pursued a host of different policies, not all of them consistent, and exercised its influence on the movement of information and resources through very many different channels. Nor was it a clearly bounded system. Between the government, the corporate sector, and the scientific and technological communities there was a constant interchange of personnel and a good deal of ambiguity as to where the boundary between public and private lay. As with the other sectors, overlap and interpenetration seemed the phenomena to be emphasized, rather than separateness and sharp boundaries.

The preliminary model, in short, had emphasized the characteristics that differentiated the world of technology from the world of science on the one hand and the world of economic activity on the other. One ended up with a scheme that was logically tidy and that helped to sort out the evidence—a kind of intellectual filing system. But a price was paid for the tidiness and for the emphasis on differentiation. One tended to oversimplify the interactions between the systems, to overemphasize their separateness and distinctness—the "black box" fallacy—and to reify what were in truth only mental categories.

Some ideas, however, were worth salvaging. There remained intact the conception of society as a complex system of communications networks; the hypothesis that social locations where the probability of invention was high could be identified as points of confluence, where information flows came together, whether fortuitously or by design; and the image of the critical role of the translators. In some respects, indeed, moving away from a model in which social subsystems were depicted as sharply bounded and highly differentiated promised to make this "communications approach" to technological change somewhat more attractive. For if we were to think now of science, technology, the economy, and government as systems with "soft edges," characterized by a good measure of overlap and interpenetration, it became easy to conceive of these areas of overlap as precisely the locations where information networks interconnected; and the individuals we had called translators were then those who monitored and managed the interconnections, converting information from one coded form into another and—a consideration of some importance—deciding what to translate and what to ignore, what information to pass on and what to block. The points of interconnection were, in that sense, switching points, and the question of who controlled the switches promised to be interesting.

The word "translator," if it is more than just a figure of speech, suggests that we are identifying particular individuals who perform these functions of communication and control. And indeed there are certain advantages to conducting the analysis at that level of disaggregation, when the surviving historical record permits it. In certain cases it is possible to identify the individuals who, in a particular context and at a particular moment, were responsible for the transfer of a critical piece of information from one network to another; the chapters that follow provide several examples. It is not only the richness and immediacy of the historical narrative that benefits when this is done. It also serves as a useful offset to the tendency referred to earlier, to reify the systems under discussion and oversimplify the interactions between them. It becomes evident, for example, that the individuals most critically engaged in transfers of information, and occasionally in the blocking of such transfers, are often not the nominal leaders of the organizations involved, the people at the top of the hierarchy, but rather individuals at the second or third level down, or in staff positions, who assemble, organize, and filter the information on which decisions are made. This is a fact of life familiar to anyone who has worked in a formal bureaucracy, but it is a fact that historians sometimes forget. Similarly, when discussing how networks interconnect, it is is easy to concentrate on formal and relatively long-lasting channels of communication, particularly those involving the exchange of written documents, and to overlook the vital role that is often played by more informal and transient interconnections, particularly face-to-face conversations and the telephone. Information moving in oral form over transient interconnections is inevitably hard for historians to track and evaluate, but anyone interested in how new ideas arise and are diffused can hardly afford to overlook its role.

For these reasons I have not hesitated to present the analysis at some points in highly particularistic form. More is involved than a fascination with detail. It is important to get a sense of the multiple levels at which organizations interact, of the complex circuits over which information moves, and of the role played by oral communications and often transitory interconnections between networks. Fortunately for the historian of continuous wave radio, the surviving evidence is at points rich enough to make this possible.

There is, however, another side to the story. Clearly evident in the events to be described is a trend toward greater formalization, toward the institutionalization of processes that earlier had been handled on a more personal and idiosyncratic basis. In *Syntony and Spark* it had been possible to use a biographical approach almost exclusively. Hertz, Lodge, Marconi—these were unambiguously the individuals responsible for the "translations" that made a commercial communications system out of

Maxwell's equations. Formal organizations played a certain role—Marconi's Wireless Telegraph Company, the British Post Office—but they remained largely in the background. As the story moved into the twentieth century, however, this would no longer do. Indeed, much of the interest lay precisely in the creation of institutions for the management of the new technology and in the intervention of formal organizations—the Navy Department, General Electric, AT&T—to protect and advance their interests. The formation of RCA was itself a classic example of the deliberate creation of a formal organization to oversee the deployment of a technological innovation.

Where these new institutions appeared was precisely in the areas of overlap where science, technology, business, and government met, and it was to manage the complex interactions taking place in those areas that they were created. That was where there was most at stake, where the need for mutual intelligibility and cooperative action was greatest, the task of translating different modes of thought most demanding. Institutionalization of the translator function that earlier had been handled on a highly personal basis was not easy: it still presents problems today. The first two decades of the twentieth century were the period when American industry and government began to learn which organizational forms and managerial techniques worked in that arena and which did not.

We can begin telling the story of continuous wave radio, therefore, by describing the achievements and problems of individuals. But we end by describing the strategy and tactics of large corporate bureaucracies and government departments. In organizational terms the world of continuous wave radio changed dramatically in two decades. How that change took place is the subject matter of this book. It is not another study of the "impact" of new technology; rather it is an inquiry into how technological change and organizational change influence each other. Interdependence and interaction, not one-way linear causation, is the theme. In particular we shall be interested in the emergence of new institutions in that "grey area" where technology, government, and the economy meet, overlap, and interpenetrate, where resources and information flow between the systems. Failure to resolve the problems arising in that area— as evidenced by inadequate financing, poor management, and a faulty sense of the market—was a characteristic feature of the early phase: the period of individual inventors and small, highly personalized firms. This was a highly creative period in the technical sense: the critical inventions and innovations were made then. With few exceptions, however, they were imperfectly integrated into the economy. The entry of large corporations into continuous wave radio, best symbolized by AT&T's purchase of de Forest's audion patents in 1912 and GE's creation of RCA in 1919, represented attempts to bring the new technology under control, to develop it in directions that would further corporate interests and control its use in areas where it might threaten those interests. Long-term capital could be assured; predictable markets could be guaranteed (or so it was thought); and the organized intelligence of corporate research laboratories could be mobilized to perfect and extend what individual inventors had created.

It is a complex story but a fascinating one. And it has implications that extend beyond radio and beyond the communications industry. The processes we see operative in this case have been at work more generally. Indeed, they have done much to shape the modern world. Since the closing decades of the nineteenth century the large, multiunit, hierarchically organized corporation, staffed by salaried managers, has emerged as the most powerful and characteristic of all private economic institutions. In the perspective of history it ranks as an organizational innovation of the first importance. The rate at which such giant corporate hierarchies appeared on the scene differed from country to country and, because of differences in legal codes, the structure of national markets, and other factors, the precise form differed. But the process itself transcended national boundaries: in one country after another large bureaucratically organized corporations since the late nineteenth century came to play an ever-larger role in the coordination of economic life, and other possible institutional arrangements, such as competitive markets, cartels, trade associations, and the like, an ever-smaller one.[20]

Theories intended to explain this historical process emphasize the market power that such large consolidations can exercise or the economies of production and distribution that large size often makes possible. These are of course complementary lines of explanation. It is a truism of economic theory that any set of competing firms can increase their joint revenue by suspending competition and combining to act jointly. And it is a well-verified empirical observation that, in some fields but not all, modern industrial technology makes it possible for the large firm to achieve, up to a point, significant economies of scale. The special con-

[20] Alfred D. Chandler, Jr., *Strategy and Structure* (New York, 1966); Chandler, *The Visible Hand: The Managerial Revolution in American Business* (Cambridge, Mass., 1977); and Alfred D. Chandler and Herman Daems, eds., *Managerial Hierarchies: Comparative Perspectives on the Rise of the Modern Industrial Enterprise* (Cambridge, Mass. and London, 1980). On the role of technology in the "Chandler thesis," see especially Oliver E. Williamson, "Emergence of the Visible Hand: Implications for Industrial Organization," in *Hierarchies*, ed. Chandler and Daems, pp. 182-202.

tribution of modern scholarship, at the hands of A. D. Chandler, Jr., in particular, has been to call attention to the organizational economies that can result from centralized management and control, in addition to the economies of production and distribution that are more conventionally emphasized.

Technology's role in the process has been recognized to some extent. Early analyses of the rise of the giant corporation in the United States tended to emphasize the appearance of a national market in the post-Civil War decades. This was, in effect, an explanation in terms of technology. The emergence of a national market in an economically significant sense—that is, an integrated market system in which prices and rates of return tended to equalize among regions—depended less on the growth of population and the westward movement of settlement than on specific developments in the technology of communications, the railroad and the telegraph in particular. Similarly, it had long been recognized that economies of scale in the conventional sense depended on industrial technology. But it was no less true that the organizational economies of the large corporation depended on prior technological change in the area of communications and information-processing. On several grounds, therefore, it was plausible to argue that technological change was at least a necessary condition for the rise of the modern large corporation, if not a sufficient one.

There may, however, be a little more to it than that. To generalize from one or a few case studies is of course dangerous, and the formation and expansion of each individual large firm takes place in a particular context that differentiates it from others. Nevertheless, the example of the radio industry raises a question whether the cautious "necessary but not sufficient condition" formula adequately describes the role that technology can play.

The problem with the conventional wisdom concerning the rise of the modern large corporation lies in its implicit assumption that the technology around which corporations are built is always under control, that it is "managed" from inception to maturation, and that at least the first-order effects are fully anticipated. Technology, in short, is depicted as passive, controlled, and predictable. This image of a docile technology may seem somewhat incongruous to a generation grown increasingly wary of the unanticipated consequences of technological change, but it underlies much conventional historiography. Technology, we are asked to believe, provides the context within which organizational innovation takes place; it presents the opportunities to which creative entrepreneurship responds. But it does not itself spring any surprises.

In some respects, we must admit, the early history of continuous wave radio confirms and reinforces the conventional interpretation. By the end of the First World War three feasible technologies for continuous wave radio transmission had emerged: the oscillating arc, the radiofrequency alternator, and (at a somewhat earlier stage of development) the oscillating triode vacuum tube. Arc transmitter technology had been pioneered in Denmark and brought to a high level of sophistication by the engineers of the Federal Telegraph Company of California. In 1918, to ensure continued American control of the technology, the key patents were purchased by the U.S. Navy. The radiofrequency alternator had been developed by the General Electric Company, originally on special order for Reginald Fessenden of the National Electric Signaling Company, later in the expectation of selling the machines to the Marconi organization. In 1919 the U.S. Navy intervened to block the sale of alternators to Marconi; after a complex series of moves, GE formed the Radio Corporation of America as an operating company that would place its alternators in service and function as the "chosen instrument" of American national policy in telecommunications.

The arc and the radiofrequency alternator represent "managed technology," and it is noteworthy that in both cases management required joint action by government and corporate interests. The oscillating triode vacuum tube is a quite different story. The device was invented (or so the courts would eventually hold) by Lee de Forest in 1912. Commercial rights to its use were purchased by the American Telephone and Telegraph Company. AT&T's interest in the vacuum tube had originally little to do with radio; it was looking for an amplifier for its long-distance wired telephone circuits. Aware of its probable use in radio, however, and of radio's potential threat to wired systems, the Telephone Company protected itself by acquiring radio as well as line amplifier rights. And these rights were its major contribution to the RCA patent pool when AT&T joined that organization in 1920.

RCA, at its inception, was designed to perform two functions: it was to represent the American national interest in telecommunications; and it was to arrange a consolidation of the key patents that controlled continuous wave radio in the United States. Vacuum tube transmitters had shown what they could do in short-range military service during World War I, and AT&T had carried out some long-distance radiotelephone tests in 1915. Engineers at GE, RCA, AT&T, and elsewhere knew very well that at some time in the future high-powered vacuum tubes would replace the alternators and the arcs that dominated long-distance radio in 1919. But no particular problem was foreseen in gradually phasing these new devices into RCA's operating system as they became available. The whole situation was, surely, under managerial control. RCA was a very carefully assembled corporate structure. The cross-licensing agreements that held it together were marvels of the lawyer's art. But the whole structure depended on the assumption that continuous wave technology could be controlled and in fact had been controlled.

Within two years that assumption was proved false. The explosive growth of popular broadcasting meant an exponential growth in the demand for vacuum tubes, to which RCA held exclusive rights, and RCA found itself transformed overnight from a gallant champion of American rights into a despised "radio trust" that threatened the American public's access to broadcast radio. Broadcasting itself, in its new meaning, made the careful lawyer's terminology of RCA's cross-licensing agreements meaningless or irrelevant. And, within a few years more, the art of long-distance radio was itself transformed by the discovery of short-wave ionospheric propagation and "beam" transmission. RCA's "state-of-the-art" long-wave system, based on the alternator, was threatened with instant obsolescence, and RCA itself, disillusioned with the profit potential of telecommunications, began turning its gaze to network broadcasting, recordings, and the movies.

All these changes were based on the vacuum tube, the innovation that had at first seemed so tamely under control but in fact was anything but that. No corporate executive ever designed a broadcast network around the alternator. No small-town entrepreneur ever set himself up in the broadcasting business by installing a Federal arc. RCA was established on the assumption that the true business of radio was telecommunications and that the appropriate technology required for that function involved very large, capital-intensive systems such as alternators and arcs. But there was more to continuous wave radio than that: the technology had some surprises to spring, and RCA found itself not controlling the trend of events, but adapting to it.

The point is not that the individuals responsible for the establishment and expansion of RCA should have shown more foresight than they did. That would indeed be an inane conclusion. The moral is, rather, that sometimes a technology seems to take charge of events and exercise what is almost a legislative power of its own. In situations of that type technology is more than a context, a passive environment within which decisions are made. It is a participant in the drama, a protagonist in the dialectical struggle between aspirations and constraints.

The great economist, Joseph Schumpeter, in analyzing modes of entrepreneurship, made a distinction between the adaptive and the creative response. In making an adaptive response to some change in circumstances, the entrepreneur would do something that was already within the realm of existing practice. This was the type of response that neoclassical economic theory, with its elegant apparatus of marginal adjustments in quantities supplied and demanded, explained very well. In contrast, when making a creative response, the entrepreneur would break away from normal practice and do something that could not have been predicted from knowledge of the preexisting market situation. It was to describe and explain this second type of behavior that Schumpeter devised his own theory of entrepreneurship and derived from that his own theory of capitalist economic development.[21]

Schumpeter's distinction has proved useful for many purposes but sometimes it fails us. There are situations in which the potentials of a new technology unfold so rapidly that merely to adapt to the new opportunities it presents and the threats it poses requires creative entrepreneurship of a high order. Technological management in such a context is management with a time lag. Decision-making becomes reactive. Events are always one step ahead of policy, entrepreneurship is a matter of grappling with changes that have already taken place, and the most creative response that can be made is to find a strategy that permits successful and continuous adaptation. In such a situation the distinction between the adaptive and the creative response, so clear in theory, becomes blurred in practice. Survival requires adaptation, and adaptation requires creativity.

Situations of this type can emerge when a technology that has been developed for one purpose is found to have ready applicability to other uses. If these other uses turn out to constitute a large, rapidly growing, and potentially lucrative market, there can ensue a phase of explosive growth that disrupts industrial structures and established managerial strategies. In such a situation, technology for a time becomes more than an instrumentality, a means for achieving ends already decided upon. It creates a world of new possibilities in which new goals have to be formulated. In that sense it acquires a determinative force to which organizational decision-making has to adapt. One such situation emerged when the technology of continuous wave radio, developed with other uses in mind, encountered the latent market for public broadcasting.

[21] Joseph A. Schumpeter, "The Creative Response in Economic History," *Journal of Economic History* 7 (November 1947), 149-59, and Schumpeter, "Economic Theory and Entrepreneurial History," in *Explorations in Enterprise*, ed. Hugh G. J. Aitken (Cambridge, Mass., 1965), pp. 45-64.

TWO

Fessenden
and the Alternator

ON the 22nd of November 1899 Professor Reginald A. Fessenden
of the Western University of Pennsylvania addressed the American Institute of Electrical Engineers, meeting in New York, on
"The Possibilities of Wireless Telegraphy."[1] The program for the evening
had listed this as a "Topical Discussion," and Fessenden started off on
a suitably informal note. There were many advantages, he said, to living
in a city—he was referring to Pittsburgh—where a widespread and intelligent interest was taken in scientific work. But there was one offsetting
disadvantage. When some striking new discovery was made, the professor's friends and the directors of the institution he was connected with
expected him to set aside his own work and plunge into "the novelty of
the hour." This had happened to him several years before, Fessenden
recalled, when he had been induced to take up X-ray work. He had little
to show for his labors in that field. Since then he had considered himself
proof against the seductions of such things as "liquid air and wireless
telegraphy." But recently once again, he said, he had allowed himself to
be diverted from his main line of work. In December of the previous year
he had been asked by the *New York Herald* to report the international
yacht races by wireless telegraphy. He had declined and they had contacted Marconi instead. But, thinking it over later, it had become clear
to him that there were serious scientific questions still unresolved in that
field. In none of the work done so far had any exact measurements been
made. The theory of electromagnetic waves was, he thought, reasonably
well understood, but if that theory was correct it was difficult to account
for some of the empirical results achieved by Marconi.

[1] R. A. Fessenden et al., "The Possibilities of Wireless Telegraphy," *Transactions* of the American Institute of Electrical Engineers, 16 (1900), 607-51.

This account of how Fessenden came to be involved in wireless telegraphy may not have convinced every member of his audience, which included his old friend and co-worker, Arthur Kennelly. Kennelly must have remembered that ten years earlier, when they were both working in Thomas Edison's laboratory, Fessenden had sought Edison's permission to begin work on the newly discovered Hertzian waves. And since that time, as professor of electrical engineering first at Purdue and then at the Western University of Pennsylvania, Fessenden had lectured on electromagnetic waves and encouraged his students to experiment with them. At Pittsburgh, too, he had the help of a talented assistant, S. M. Kintner, who was later to succeed him in the professorship, and the two of them had been working on Hertzian waves for several years. Fessenden's interest in wireless telegraphy, in short, was not as recent in origin as his remarks suggested; nor was his research in that field a reluctant concession to "the novelty of the hour."[2]

When Fessenden and his listeners thought of wireless telegraphy, they had in mind what had already become an accepted and conventional technology. In essence it differed little from the methods that Heinrich Hertz had used in his laboratory at Karlsruhe in 1886-1887. Electromagnetic waves were generated by the discharge of a capacitor across a spark gap and radiated from a dipole antenna. Instead of Hertz's "ring resonator"—a tricky form of detector even under laboratory conditions— Marconi, Branly, Lodge, and others had introduced the coherer, in its simplest form a glass tube filled with metal filings between two electrodes. Marconi had also introduced the tall vertical antenna and the connection to ground, in place of Hertz's horizontal dipole. And in the search for greater distance he had moved down from the very high frequencies at which Hertz had worked to longer and longer wavelengths. Oliver Lodge

[2] There is no good biography of Fessenden. That written by his wife, Helen M. Fessenden—*Fessenden: Builder of Tomorrows* (New York, 1940)—is useful but understandably uncritical. Ormond Raby's *Radio's First Voice: The Story of Reginald Fessenden* (Toronto, 1970) is a romanticized account in which it is impossible to distinguish between verifiable fact and imaginative reconstruction. Orrin E. Dunlap, Jr., offers a brief biographical sketch in his *Radio's 100 Men of Science* (New York and London, 1944). Certain early sections of an intended autobiography were published by Hugo Gernsback in *Radio News* 6 (January 1925) through 7 (November 1925); they include very little that is directly relevant to his work on radio and it is not hard to understand why Gernsback cut him off after eleven installments. Many of Fessenden's papers are preserved in the State Archives of North Carolina; others are in the Clark Radio Collection at the Smithsonian Institution. The papers of his Pittsburgh backers, T. H. Given and Hay Walker, Jr., have not been located.

had made explicit the importance of tuning, or what he called syntony. This had been implicit in earlier work; Hertz's transmitting and receiving antennas had a natural resonant frequency, like a taut violin or guitar string; so did Marconi's grounded verticals. But Lodge stressed that, for maximum transfer of energy between transmitter and receiver, both had to be tuned to the same wavelength; and the sharper the tuning, the less interference would be experienced from signals on other wavelengths. This required careful calculation and adjustment of resonant circuits: to rely on the natural resonance of the antenna was not good enough, particularly when a spark gap was part of the antenna circuit at the transmitter and a coherer part of the corresponding circuit at the receiver.[3] These and other innovations had converted Hertz's laboratory apparatus into a technological system that could transmit information by means of coded signals and that might even have some commercial value. But essentially the technique was Hertz's: you generated an electromagnetic disturbance by means of a spark, and you created that spark by the sudden discharge of the energy stored in a capacitor.

Most of the modifications in the original Hertzian apparatus between 1888 and 1899 had been made in a highly empirical, trial-and-error manner. This was particularly true of Marconi's work. Marconi himself had little formal scientific training: what he knew about Hertzian waves he had picked up from Augusto Righi's lectures at the University of Bologna, from private reading, and from his own experiments. He was interested in science primarily for its instrumental value: he wanted to construct a system of wireless telegraphy that could communicate reliably over long distances; and he wanted to make money from it. This single-ness of purpose had already begun to pay off, in public repute if not in cash: in the popular consciousness of the late 1890s, wireless telegraphy and Marconi were almost synonymous. But it also meant that many of the technical assumptions that underlay the Marconi system had escaped rigorous examination. There had been a lot of cut-and-try, a lot of im-provisation. There had not been much controlled experimentation or careful measurement or consideration of alternative approaches. And there had been little input from the scientific community. With notable exceptions, such as Oliver Lodge in England and Ferdinand Braun in Germany, academic scientists had shown little interest in the use of Her-tzian waves for signalling. Those who had followed in Hertz's footsteps took his work in a direction quite different from the one Marconi chose: up to the ultrahigh frequencies where radio waves began to behave like

[3] These matters are covered in more detail in the author's *Syntony and Spark: The Origins of Radio* (1976; Princeton, 1985).

light, rather than down to the lower frequencies that seemed to be more useful for signalling.

Fessenden's remarks to the AIEE in 1899 were not a frontal attack on the Marconi system: that would come later. He did point out, however, that the links between Marconi's work and the physical theory of electromagnetic radiation were tenuous; and he deplored the absence of measurements. This, he pointed out, was not just a regrettable oversight: the kind of detector used by Marconi, the coherer, was singularly ill-suited for scientific research. A coherer, when properly built and adjusted, was like a switch that was either on or off: normally nonconducting, it became conducting when the energy from an incoming wave reached its electrodes and created a difference in electrical potential between them. Then it had to be tapped back to its nonconducting state in readiness for the next incoming wave. This "trigger" action made it very difficult to measure the strength of received signals. In the absence of such measurements, scientific investigation of transmitter and receiver design, of wave propagation, of antennas, and of a host of other problems relevant to physics and to practical wireless telegraphy was not possible.

For Fessenden, the coherer was the most obvious weak link in the Marconi system. At its best it was an insensitive detector: it took almost one volt, Fessenden thought, to trigger it from the nonconducting to the conducting state, and that was a lot to expect from a receiving antenna.[4] What was needed was a detector that responded to the total energy of the received signal, not just to its peak voltage. And he wanted one that would make it possible to take quantitative measurements. He presented three types that he and Kintner and their students had been working on at Pittsburgh. True, they were laboratory instruments, not intended for commercial service; but any one of them was more sensitive than a coherer. And, more important, they opened the way to controlled experiments and measurement. Fessenden listed a dozen lines of research that were either already under way or planned for the immediate future.

The men who made up Fessenden's audience that night were a mixed lot. Some of them were highly trained academic physicists: Michael Pupin, for example, professor of mathematical physics at Columbia University, trained at Cambridge University and at the University of Berlin under Helmholtz. Others were scientist-engineers, like Charles Proteus Steinmetz, master of alternating current theory and a key man in General Electric's research laboratory at Schenectady. And others were down-to-earth experimenters and operators, like W. J. Clark, who followed Fes-

[4] Fessenden, "Possibilities," p. 617. The reader must bear in mind that in 1899 there was no means of amplifying the strength of received signals.

senden at the lectern and felt it necessary to explain that he was not "a theoretical man." What brought them together was a common interest in electricity. Wireless telegraphy was a new use for electricity; it posed intriguing problems both for the theorist and for the practical man—problems seemingly different from those encountered in the more familiar fields of heat, light, and power. What made the November session of the AIEE productive and exciting was that it provided a forum in which physicists and engineers could interact. None of them asked whether they were doing science or engineering; they were obviously doing both. Out of this interaction came a kind of rigorous scrutiny that the Marconi system had so far escaped, and an informed search for alternatives to the technology that it embodied.

Fessenden's detectors emerged from just such a critical scrutiny. His comments on the inherent limitations of the coherer foreshadowed his later rejection of spark technology as a whole. But Pupin's comments as official discussant gave an even clearer indication of what lay ahead. The topic for the evening, he reminded the gathering, was the *possibilities* of wireless telegraphy. The papers he had heard, though interesting, hardly addressed that subject. Two possibilities seemed to him important: better tuning and greater distance. In both of these respects the system currently in use—the system Marconi was using—had serious defects, and wireless telegraphy would never realize its full possibilities until these defects were overcome.

Take the question of tuning. Marconi's transmitters radiated waves by creating sparks across a spark gap. What was the length of the wave that resulted? No one could say: there was bound to be "an oscillation of all sorts of unrelated frequencies." And the reason for this was that a great deal of energy was dissipated right in the spark gap. The antenna therefore radiated a highly damped wave, one in which the oscillations diminished very quickly in amplitude. "That," said Pupin, "is the reason why they have not been able to tune their receiving apparatus in England." There was no single wave to which a receiver *could* be tuned. Even if the receiving and transmitting antennas were tuned to precisely the same frequency, as long as the transmitter emitted those highly damped waves "you will get no appreciable resonance. To produce strong resonance you must send forth oscillations which have little damping."[5]

[5] Ibid., p. 623. For a parallel discussion in terms of Fourier analysis, see Aitken, *Syntony*, pp. 71-72. Pupin expressed serious doubt that, in a typical Marconi vertical antenna, the waves emitted would bear any simple harmonic relation to each other, on the ground that the capacity of the antenna per unit length was not constant but varied with its distance from the ground.

The implication was clear: a way had to be found to transmit undamped waves—or at least waves with as little damping as possible. There would always be some damping, because the transmitting antenna was radiating energy. But, said Pupin, "if you could excite *oscillations in a wire without a spark-gap* [emphasis in original]," you would get very close to undamped waves. "They would be, so to speak, sonorous oscillations, as when you strike a bell made of fine bell-metal, it continues to ring for a long time after the stroke is delivered. But if you put a finger on the bell and strike it then, or put in some resistance, the sound dies out very rapidly. So that is the state of affairs in this transmitting wire with a spark gap . . . you have only a very few waves sent out after each spark. Now, when you have a train consisting of very few rapidly diminishing waves, they cannot produce much resonance."

How one might excite oscillations in an antenna without a spark gap Pupin did not say: he mentioned the possibility as a theoretical ideal, rather than something feasible. And his example of the bell makes it clear that he was still thinking in terms of sparks. The bell had to be struck with the clapper and then allowed to ring free, so that the resonant vibrations died out only slowly. That was the physical image he had in mind. But suppose a way could be found to make that bell ring continuously, without striking it at all, but supplying energy to it in some other way, to replace the energy it lost by radiating its sound waves? Then one would have a wave with no damping at all, a true continuous wave.

Pupin did not explore these possibilities but he was clearly unhappy with the highly damped waves and low spark frequencies of his day. The same theme recurred when he turned to the second of his two "possibilities": greater distance. A way had to be found to increase the effective radiating power of wireless transmitters. Think of the transmitting and receiving antennas as a matched pair of tuning forks, said Pupin. The problem was to get the tuning fork at the receiving end to resonate as strongly as possible when the tuning fork at the transmitting end was struck. One way of doing this was obviously to strike the transmitting tuning fork harder: this would be the equivalent of increasing the wireless transmitter's power—a longer spark gap, in the parlance of those days. But another way was to strike it more frequently—in other words, a more rapid train of sparks. Marconi used a spark frequency of about eight per second: that, said Pupin, was nothing at all—"the coarsest kind of dilettante work." He would like to see spark frequencies of at least a thousand a second. Set that beside the idea of undamped oscillations and you would really begin to get "the accumulated effect of resonance." The vibrations in the receiving tuning fork would have little chance to die out: ". . . before the energy which the second tuning fork has received, has decayed, if you strike the first tuning fork again and again, the resonance of the second tuning fork will continually increase until it reaches a maximum effect, and you cannot go beyond that."

Again Pupin was thinking and speaking in terms of sparks. Only as a theoretical possibility could he conceive of a true continuous wave radiator using no spark at all. That was an ideal to be approximated as closely as possible, by using resonant circuits and by increasing the spark frequency so that one train of sparks followed very closely after the preceding one. It was not something Pupin presented as attainable. In that respect he was still working within the same mind-set as Marconi: Hertzian waves could be generated only by a spark discharge. The limits that Pupin saw to that approach were purely practical ones: how much power could you feed into a spark gap before the electrodes fused together? There might, in short, be developmental difficulties as spark transmitters moved to higher power. But neither from Pupin nor from any member of the group was there an explicit assertion that the spark gap approach should be abandoned.

And yet, implicitly, the idea was there. The most significant statement Pupin made during his comments was thrown out almost as an aside. "There is nothing mysterious or even strange," he said, "about these waves employed in wireless telegraphy, they being perfectly simple waves like any other electrical waves and can be made to obey the same rules." Now, if this were really so, much followed. It followed, for example, that between power engineering and radio engineering there could be no unbridgeable chasm. It followed that the theory of alternating currents could be applied both to the transmission of power and to the transmission of information. And it followed that the methods and, moreover, the vision of men like Nikola Tesla and C. P. Steinmetz, pioneers in the design of alternating current machinery, could be used for wireless telegraphy. To think of wireless telegraphy as a special application of the general theory of alternating current electricity was an intellectual innovation of some importance. That was not how Marconi thought of it.

Nobody questioned Pupin about that suggestion. No one wondered, except perhaps in the privacy of his own imaginings, whether a machine could be built that would generate alternating current electricity, not at the 25, 50, or 60 cycles per second used for light and power systems but at the 20,000 or more needed for wireless signalling.[6] Alternating current itself was a new thing. The first European demonstration of an alternating current distribution system had taken place in 1884, a bare fifteen years

[6] The lower limit of the radiofrequency spectrum is generally taken as 20 kHz. See James M. Moore, *Radio Spectrum Handbook* (New York, 1970), pp. 9-10.

earlier. George Westinghouse, against the advice of his engineers and patent lawyers, opted for alternating rather than direct current in 1885. The Niagara Falls alternating current generators first went into service in August 1895.[7] All these systems had operated at low frequencies—the Niagara Falls alternators at a mere 25 Hertz—and for supplying power to street railways, urban lighting systems, and the electrochemical industry this was enough.[8] A radiofrequency alternator was a different proposition entirely.

Furthermore, the mental habit of thinking in terms of spark was hard to break. The system worked—sometimes, it is true, not very well and, in 1899, not very far. But no other system had been shown to work at all. No other method of generating Hertzian waves was known. Already the elements of the Marconi system—the high vertical antenna, the spark gap, the coherer, the faith in long waves for long distance—had crystallized into a technological paradigm, a standard, accepted approach, with its own assumptions, its own criteria of performance, its own group of adherents and skilled practitioners.[9] The mind-set that accompanied it was difficult to escape.

* * *

The Marconi system of wireless telegraphy, as we have seen, had developed in a highly empirical manner, once the initial scientific insights derived from Hertz were absorbed. It was a matter of improvisations and expedients. The components of the system—antennas, detectors, transmitters—were chosen, not as a result of scientific analysis, but because, for reasons imperfectly understood, they seemed to work. Sparks were used to generate Hertzian waves because no one knew any other

[7] See Harold C. Passer, *The Electrical Manufacturers, 1874-1900: A Study in Competition, Entrepreneurship, Technical Change, and Economic Growth* (Cambridge, Mass., 1953), esp. pp. 129-50 and 276-320. Westinghouse's interest in alternating current was first aroused by reading a report of a display of the Gaulard and Gibbs system at the Inventions Exhibition in London in 1884-1885. He was encouraged to proceed by a Budapest firm, Ganz and Company, the European pioneers in alternating current.

[8] The writer can recall, however, the perceptible flicker in incandescent bulbs in Toronto in 1947, when that city was still using 25 Hz current.

[9] Compare the concept of a "normal technology" as presented in Edward W. Constant II, *The Origins of the Turbojet Revolution* (Baltimore, 1980). Reese Jenkins, in his *Images and Enterprise: Technology and the American Photographic Industry, 1839 to 1925* (Baltimore, 1975), pp. 4-6, makes good use of the allied concept of a technological "mind-set."

way. And for a time, as Marconi and his emulators scrambled to find a place for their systems in the world of practical affairs, that was good enough. Before long, however, to undergird this body of empirical knowledge, to explain why the methods in use worked, there appeared a theoretical rationalization. One of its central propositions directly contradicted what Pupin had asserted. Hertzian waves were not "perfectly simple waves like any other electrical waves." They were waves of a special type, radiated in a special way, and the spark discharge was necessary for their creation. And this was presented, not as a mere empirical observation, but as a deduction from the laws of physics. Its author was an English scientist in the service of Marconi's Wireless Telegraph Company: John Ambrose Fleming.

Fleming, professor of electrical engineering at the University of London, became associated with the Marconi Company in 1900. A respected physicist, he brought to the company not only an intense interest in the new art but also a thorough knowledge of its foundations in physical theory. In 1906 he published the first edition of his monumental *Principles of Electric Wave Telegraphy*, destined to remain for many years and through several editions the leading treatise on the subject. The book was notable for its comprehensive coverage, its detailed descriptions of the equipment in use, and its careful and often mathematical presentation of the underlying theory. It was an authoritative statement of the physical principles on which the Marconi system was believed to rest.[10]

Fleming began his treatise with an explanation of what was meant by high frequency currents and the distinction between damped and undamped oscillations. This was followed by a description of machines that had been built to generate sustained high frequency currents. There were not many of these in 1906, and none of them so far had exceeded a frequency of 10 or 15 thousand cycles per second (10 or 15 kHz). How good were the prospects of being able to generate, by purely mechanical means, frequencies an order of magnitude higher than this—say 100 kHz? In Fleming's opinion, not good at all; and even if it could be done, the power output would be small. This meant that such machines, though their design might present interesting problems for the engineer, were unlikely to exhibit the phenomena that Fleming intended to discuss. Even if they could be built, it was doubtful that any appreciable radiation would result.[11]

[10] J. A. Fleming, *The Principles of Electric Wave Telegraphy*, 1st ed. (London, 1906). All page references are to this edition unless otherwise stated; all quotations are by permission of the copyright holder, Longman Group Limited.

[11] Ibid., pp. 4-14, 81. As Fleming is reported to have stated in court testimony,

The only method of generating Hertzian waves that had so far been found possible was by the oscillatory discharge of a condenser of some kind. There could be no Hertzian waves without the Hertzian spark. In order to create an electric wave, one had to create a state of strain in a dielectric (that is, in a normally nonconducting medium) and then release that constraint very suddenly. This was true of wave motion in general. You could, for example, move your hand through the air, or swing a bell in a church steeple, and all that would happen would be that the air would flow gently around the moving object, creating whirls or vortices that absorbed the energy. To create a sound wave a violent motion was necessary—clapping two hands together, for example, or striking the bell with a clapper. Similarly with electromagnetic waves: it was not enough merely to create electrical oscillations. There had to be a sudden discharge before waves could be radiated into space.

To explain how such radiation took place Fleming presented a physical model. Its essential feature was what he called, in 1906, "decussation."[12] Consider the kind of radiating antenna Hertz had used: essentially two metal rods placed end to end, but separated by a small spark gap. When these rods (the two elements of a dipole antenna) were conected to the secondary circuit of an induction coil, they became charged with electricity, one acquiring a positive charge, the other a negative one. When these charges reached a certain threshold value, the air insulation separating the two rods would break down and a spark would jump across the gap.

At that moment the two rods became in effect one conductor and an electrical oscillation took place as the positive and negative charges neutralized each other. If that oscillation were started sufficiently suddenly, Fleming argued, some of the energy would be thrown off in the form of an electromagnetic wave—the so-called displacement wave. And if the induction coil were kept going, you would have groups of these oscillatory discharges, and successive trains of waves would be thrown off, to travel or spread out through the surrounding medium.

What was happening in this process, according to Fleming's model, was that lines of electric strain were being formed in the ether, as the

---

"unless some form of condenser is discharged to cross the spark gap there cannot be any production of Hertzian waves—the disruptive discharge is the one essential condition for the production of Hertzian waves." (Quoted in R. A. Fessenden, "How Ether Waves Really Move," *Popular Radio* 4 [November 1923], 340.) Compare Fessenden Papers, 1140-27, memorandum, "More Important Fessenden Wireless Patents."

[12] To decussate is to divide crosswise, as in the Roman symbol X for the number ten. Decussation meant for Fleming the intersection of lines of strain in the ether.

two arms of the antenna acquired their opposing charges. When the spark jumped the gap, these lines of strain began to collapse inward. For radiation to take place, it was essential that they collapse rapidly. It was at this point that the theory of decussation entered. If the discharge was slow and gradual, the lines of strain would collapse inwards and then be re-created in an opposite direction, and that would be all. "If, however, the oscillations are sufficiently rapid, the lines of strain are unable to accommodate themselves quickly enough. Each line, or rather the medium in which it exists, possesses an inertia, and the lines of strain cannot instantly be annihilated or recreated in any place. Hence it follows that there is a decussation or crossing of some of the lines of strain during the discharge. . . . When this decussation takes place the line of electric strain is nipped off at the crossing point, and part of it is detached as a closed loop of electrical strain. This process is repeated at each alternation, and results in throwing off normally from the rod self-closed lines of electric strain."

As an image with which to visualize the radiation of Hertzian waves, Fleming's model was probably effective. One could vividly see, in the mind's eye, the lines of strain suddenly collapsing when the spark jumped the gap; as they collapsed they would intersect each other and be "nipped off" at the point where they crossed, so that they could be "thrown off" into space. And it was evident how, to anyone visualizing the process in this way, it would seem highly unlikely, if not impossible, for continuous oscillations to create and radiate Hertzian waves. The collapse of the lines of strain had to be sudden.

Fleming, however, was not entirely comfortable with his metaphor, and for good reason. It required him to think of the lines of strain as objective physical realities, not just aids to thought. It required him to accept the existence of the ether as the medium in which the lines of strain existed and through which they were propagated. And, perhaps most demanding of all to a thoughtful physicist, he would have to believe in an ether that had inertia, for otherwise there was no reason to hold that the lines of strain could not "instantly be annihilated or recreated in any place."

Partly for these reasons, partly because in the meantime it had been convincingly demonstrated that continuous wave generators could and did radiate, when the third edition of Fleming's treatise appeared in 1916 the theory of decussation had disappeared. In its place there was inserted a theory of "kinks." As Fleming expressed it, "If the end of a line of electric strain has a sudden movement given to one end . . . the result is to create in the line a *kink* [emphasis in original] which travels outwards, just as would a *kink* in a stretched rope if the end were given a jerk at right angles to the direction of the rope. If the end of a line of electric force terminates on a point-charge of electricity or so-called electron, then a sudden movement of this electron . . . will be accompanied by the outward propagation of *kinks* or places of sudden bend or flexion along the lines of electric strain."[13]

Whether the idea of kinks in the ether represented much of an advance on the earlier theory of decussation was perhaps arguable. On the really critical points Fleming did not yield. The later edition, just as the earlier one, contained the flat statement that ". . . in order to create an electric wave we have to create a state, called, for the sake of definiteness, electric displacement in a dielectric, and to release that constraint very suddenly." His skepticism about continuous waves was muted; a passage questioning whether a high frequency alternator could ever radiate appreciable power was deleted; and the book closed with a well-informed discussion of advances in radiotelephony, which assumed efficient continuous wave radiation. But the imagery of the spark discharge and the suddenly collapsing lines of strain in the ether remained intact.

It has often been observed that the Marconi companies, originally highly innovative, were by the second decade of the twentieth century followers rather than leaders in the introduction of new radio techniques. The innovations that made continuous wave telegraphy possible were made in the United States and in continental Europe, not in Britain. And the same is true of the early work on radiotelephony. Part of the explanation probably lies in technological conservatism in the narrowest sense: the tendency of any organization to cling to the formulas that first brought it success, the reluctance to shift to something new and untried when what is familiar and available seems to work well enough. But Marconi's persistent dedication to spark may also reflect, in part, a kind of intellectual failure. If Fleming's work represents the scientific theories—or, more precisely, the scientific images—on which Marconi practice was based, a serious question can be raised about their adequacy. The practice of searching for physical models that could be visualized had, it is true, paid handsome dividends in other fields, such as mechanical engineering. And even in electromagnetic theory, for men like Faraday and Maxwell, it had proved immensely useful. But it may be that, in carrying the technique into the realm of what we now call electronics, Fleming was taking it beyond its limits. His models were not wrong; they were just inadequate. In particular they were inadequate to describe modes of radio propagation other than by spark discharges. This element may well have added its weight to the technical conservatism that seems to have afflicted

---

[13] Fleming, *Principles*, 3rd ed. (London, 1916), p. 420.

the Marconi companies after 1910. Caught short in the transition to continuous wave radio, they had to acquire the necessary technology from others.

Fessenden, originally critical of particular components in the Marconi system, moved rapidly to a rejection of the system as a whole. He became convinced that Marconi and those who followed his example were on the wrong track. Marginal improvements on what Marconi was doing were not enough. It was necessary to start over again on a different basis and build a system that was not slightly better than Marconi's, but capable of doing things that Marconi's techniques could never do. The key to this alternative system was the continuous wave. Fessenden was willing to use spark if he had to—very high frequency spark, as Pupin had suggested—but only as a temporary expedient. What he was after was a device that would generate true continuous waves of constant frequency—waves that could be interrupted to send Morse code but that could also be modulated with speech and music. Such a system of transmission would make many other innovations necessary—new methods of reception, for example, for no coherer could ever detect radiotelephony. But above all it would necessitate a radical change in ways of thinking. Fessenden and those who followed him had to break away from the habits of thought characteristic of spark telegraphy. This was no small task, since at the turn of the century these habits were shared by almost everyone involved with wireless and underlay the only systems that were known to work. Accomplishing it exacted its costs.

* * *

Those who have written about Reginald Aubrey Fessenden typically use a common set of adjectives to describe his personality: vain, egotistic, arrogant, bombastic, irascible, combative, domineering—it becomes a familiar litany. And certainly a man who could tell one of his most valued employees—not once but often—"Don't try to think—you haven't the brain for it"—would seem to deserve some such characterization.[14] On the other hand Ernst Alexanderson of General Electric, who worked as closely with Fessenden as any man and differed from him on important issues, found him not at all difficult to deal with.[15] And there has been

[14] Raby, *Fessenden*, pp. 104-105.

[15] Ernst Alexanderson, "Reminiscences" (Columbia University Oral History Collection), p. 17. All quotations from the Ernst F. W. Alexanderson memoir in the Radio Pioneers Series of the Columbia University Oral History Collection, in this chapter and later, are copyright 1976 by The Trustees of Columbia University in the City of New York, and are used with their permission, which is hereby acknowledged.

preserved a letter from the assistant general manager of the Westinghouse Company, recommending Fessenden for the chair of electrical engineering at Purdue: "He is every inch a gentleman and an agreeable man to get along with."[16] This may confirm one's suspicions of letters of recommendation; on the other hand it may suggest that the personality usually attributed to Fessenden was not something he was born with.

He was born in 1866, in East Bolton, a small town in what is now the Province of Quebec (then, before Confederation, known as Canada East). East Bolton was part of the area called the Eastern Townships, at that time a mostly English-speaking Protestant enclave within the predominantly French-speaking Catholic culture. Fessenden's father was a minister in the Episcopal Church and the family was far from wealthy. They changed their place of residence several times while the boy was still young, as his father was transferred from one parish to another. By the time he was nine years old they were living at Niagara Falls in Ontario. Reginald Fessenden spent one year at a military college on the American side of the river, acquiring there the erect, rigid posture that characterized him for the rest of his life and, perhaps, something of his authoritarian manner. In 1877 he transferred to Trinity College School at Port Hope, Ontario, and from there he graduated at the age of fourteen. He seems to have been an excellent pupil.

Up to this point there is no evidence to suggest that Fessenden had any particular technical or scientific bent. His training at Trinity College School was in languages, particularly the classics, and in mathematics.[17] In 1881 he received an offer of a teaching position in mathematics at Bishop's College School, in Lennoxville, in the Province of Quebec. This institution was affiliated with Bishop's College, where his father had prepared for ordination, and it is entirely possible that Fessenden's abilities, qualifications, and needs for financial assistance had been made known through the informal communications network of the Episcopal Church in Canada. The terms of the offer were that he would teach mathematics and other subjects to the pupils in the school; in return he would receive a nominal salary, board and lodging, and the privilege of being credited with a year's work at the college without having to attend classes, provided he passed the final examinations. Fessenden accepted. The fact that, at the age of fifteen, with no previous teaching experience,

16 Fessenden Papers, 1140-94, William P. Zimmerman to President J. H. Smart, 8 July 1892.

17 Fessenden, "Autobiography," *Radio News* 6 (May 1925), 2055. "The secular studies were substantially confined to classics and mathematics, taught in the old-fashioned way." This reference to secular studies may suggest a curricular emphasis on preparation for holy orders.

he was immediately appointed senior mathematics master is certainly testimony to the confidence his employers placed in his talents; it may also suggest that there were not many competitors for the job.

Fessenden was never graduated from Bishop's College, although apparently he completed "substantially all work necessary for a degree."[18] Whether he may properly be termed "college educated" is, therefore, largely a question of how much importance one attaches to the diploma. More to the point, his service at Bishop's College gave him a chance to improve his mathematics in one of the best ways possible—by teaching it to others. And, for the first time, if his recollections on the point may be trusted, he became intrigued by matters technical and scientific.[19] These came to his attention, not in the classroom—there he studied mathematics, Greek and Latin, a little Hebrew and Arabic, and some history—but in the college library. There he found copies of *Nature* and *Scientific American* which particularly intrigued him, and to the latter publication he even submitted a formal communication—not, however, acknowledged or accepted by the editor.

The image we have at this point is still that of a bright, personable young man, moderately competent in mathematics and languages, heading for a respectable if undistinguished career as a schoolteacher or perhaps as a scholar. His next move probably resulted from the same kind of personal recommendation as had taken him to Lennoxville. This was in response to the offer of the principalship of the Whitney Institute in Bermuda. Having completed most of the required work at Bishop's College, he felt there was little to keep him there; he needed more income, partly for his own needs and partly to help pay for the schooling of his younger brothers; and besides, as he later put it, he was "restless with the feeling that there were fields more constructive than investigation of the Greek particles."[20]

Once again the title of the new position was grander than the reality. Fessenden was indeed principal of the institute: he was also its only teacher. Teaching everything that needed to be taught kept him busy, but not too busy to participate in the social life of the island. Tall, red-haired, with an irreproachable Canadian background, he must have been considered an eligible bachelor, albeit an impecunious one. He became engaged to Helen May Trott, daughter of one of Bermuda's better-known

[18] Ibid., *Radio News* 6 (June 1925), 2217.

[19] Ormond Raby, in his biography, depicts Fessenden as fascinated by these subjects from his earliest boyhood. There is nothing in Fessenden's autobiography, nor in the biography written by his widow, to support this interpretation.

[20] Fessenden, "Autobiography," *Radio News* 6 (June 1925), 2217.

merchants and produce growers—a gentleman blessed with nine daughters and one son. Marriage, however, required a larger income. And school-teaching by this time was coming to look like a dead end, at least in the absence of stronger academic credentials than he then possessed. In 1886 he left for New York City.

Fessenden's biographers make much of the fact that his determination at this point was to work for Thomas Edison in his laboratory. His own recollections are somewhat less positive: ". . . I decided I might as well learn my practical electricity under Edison as anywhere else."[21] However firm the intention, it marked a decided shift from the trajectory that Fessenden's life had followed previously. It was a move, perhaps symbolic, from the small town to the big city. It was a move out of teaching into engineering and research. It was a move away from the academic study of mathematics and the classics to the world of science and technology. It was also an audacious move—almost naively so. Why should Edison hire this unknown twenty-year-old schoolteacher? The audacity lay not so much in the fact that Fessenden had no formal scientific or engineering credentials: Edison was no uncritical admirer of diplomas and degrees, and institutions in North America in 1886 where one could get formal training in electrical engineering were few. More to the point, Fessenden could offer no evidence whatever of his ability to conduct scientific research or to work in an experimental laboratory. Nor were any of the personal recommendations he could muster from previous friends and acquaintances likely to help him in this venture.

One suspects, therefore, that there was more in Fessenden's mind than the hope of learning electricity by working with Edison, even though, in the light of later history, this is what it is tempting to emphasize. This may well have been his long-term objective; but in the short run he intended to support himself by journalism. One of his acquaintances in the boarding house where he had lived in Bermuda was a dedicated supporter of Henry George and the single tax movement. George was at that time running for election as mayor of New York City, and it was by joining his entourage as a writer that Fessenden hoped to support himself initially.[22] Recommendations had been provided; unfortunately, they got Fessenden nowhere. Neither did a proposal to write scientific articles for the *Tribune*. And neither did a direct approach to Edison.

Eventually he did get a job working for Edison. The Edison Machine Works was at this time laying down electric light mains in Manhattan between 14th and 52nd Streets and Fessenden, after many applications,

[21] Ibid., p. 2218.
[22] Ibid., p. 2217.

was hired as an assistant tester. This was at least a cut above common laborer: what it involved was scraping the insulation off the conductors where they emerged from the conduits, so that the tester could check for ground faults. It was, Fessenden later recalled, "harder work than it sounds, but I had got a start and was putting in my lunch hour in working at electrical theory and analytical mechanics, which we had not had at college."[23] Intelligence and hard work, in classic Horatio Alger style, paid off, and he was soon promoted to tester and then chief tester. By the time his section of the project was completed, in December 1886, he was inspecting engineer—the first time he had been able to claim that title. He was offered a choice between two new positions: either to continue with the Edison Machine Works at their Schenectady headquarters, or to become one of Edison's assistants at the new Llewellyn Park laboratory in West Orange, New Jersey, and work on dynamo development. Fessenden chose the latter.

Fessenden worked with Edison for a little over three years. His responsibilities were mostly in the area of industrial chemistry—new insulating materials for cables, new lacquers for dynamo windings—and despite the fact that he had no previous training in that field he appears to have given Edison the kind of help he needed. Three things impressed him about the experience. The first was the chance to observe Edison's methods at first hand. The second was access to the laboratory's library, containing (or so it seemed to Fessenden) "complete sets of every scientific transaction and proceeding and publication which had been printed up to that date."[24] And the third was the close working relations with other men in the laboratory. Particularly important was his friendship with Arthur Kennelly, then Edison's chief electrician. Kennelly, later to be professor of electrical engineering at Harvard and famous in radio history for his research on ionospheric propagation, was, like Fessenden, largely self-educated as far as science and engineering were concerned, and the two men proved highly congenial.[25] They read physics and mathematics together during their lunch hours and they collaborated on several proj-

[23] Ibid., p. 2274.

[24] Ibid., *Radio News* 7 (August 1925), 156.

[25] Arthur Edwin Kennelly was born in India, educated in England, Scotland, France, and Belgium, and emigrated to the United States to work under Edison in 1887—the same year that Fessenden joined the laboratory. He is best known for his demonstration in 1902 of the existence of what has come to be called the Kennelly-Heaviside layer, a belt of ionized air that reflects radio waves of certain frequencies back to earth and thus makes long-distance shortwave radio possible. I am indebted to Professor C. Stewart Gillmor for the information that Kennelly was self-educated in science and engineering.

ects. The friendship was important, for it is in this period that we get the first clear indication of Fessenden's emerging interest in high frequency alternating currents. This was not an interest likely to arise from any work he did for Edison, with his dedication to direct current. Its origin lay in joint study of electrical theory with Kennelly and exploitation of the resources of the Llewellyn Park library.

For Edison personally Fessenden developed deep admiration. This was based, not on the image of the ingenious empiricist that Edison liked to present to the world, but on observation of the man at work and respect for his leadership. There were some obvious lessons learned: the importance of patents, for example. And there were some less obvious ones, such as the importance of a systematic search of the literature, of not jumping at the first solution that seemed to work, of not quitting until the full range of possibilities had been tested. Probably the most important lesson was the element that later historians have emphasized as critical to Edison's success: his consistent practice of inventing whole systems, rather than separate components for the systems of others.[26] And he probably understood Edison's reasons for doing this: his drive to innovate, not just to invent, and his appreciation of the way in which imbalances within a system tended to stimulate further invention. These are characteristics clearly evident in Fessenden's later career. Like Edison he strove to integrate his inventions, not resting content with particular elements but driving always for a complete system, recognizably different from and independent of the systems of others. And, like Edison but with less success, he tried to maintain personal control of his inventions as they moved from laboratory to commercial use.

There were also, however, lessons that he could have learned but did not. Primary among these was Edison's respect for the market—his determination never to invent anything for which there was not a clear commercial demand, never to be too far ahead of, or too far behind, his time. There was a hardheaded sagacity in Edison's approach to the business of inventing that Fessenden never acquired. Both men were at their best working on the frontier of technology; but the signals that Edison followed when choosing new fields for exploration were signals given by the price system. This was not Fessenden's way: to him the technical challenge was enough in itself. If he could solve the technical problem people would buy the results—if they did not, it was because they were

[26] For my interpretation of Edison I have drawn heavily on the insights of Thomas P. Hughes. See his *Thomas Edison: Professional Inventor* (London, 1976) and "Edison's Method," in *Technology at the Turning Point*, ed. William B. Pickett (San Francisco, 1977), pp. 5-22.

stupid or malevolent or both. It never occurred to him that an invention might have technical merit, and clients might still not be interested in it—and close association with Edison, then at the height of his success and apparently incapable of inventing anything that would not sell, may have aggravated his uncritical self-confidence. Technical achievement, to Fessenden, was not only necessary for commercial success; it was a sufficient condition. How to market the products of his genius was a problem that he never solved.

Edison paid most of the expenses of the West Orange laboratory personally, from the earnings of his holdings in the various Edison companies. This income, amounting to some $125,000 in 1888, was sharply reduced in the year following, when these companies were merged into the new Edison General Electric Company, controlled by a syndicate of German electrical and banking interests organized by Henry Villard with the blessing of J. P. Morgan.[27] Edison had originally endorsed the reorganization, hoping that it would relieve his chronic cash flow problem. The result was exactly the opposite. Not only was he reduced to a minority stockholder in the new firm, with power to choose only one director; the income he drew from his holdings fell abruptly. His reaction, as he expressed it to Villard, was one of "absolute discouragement" and a determination to retire from the electric light business and devote himself to "things more pleasant."[28] One immediate consequence was a sharp cutback in operations and personnel at the West Orange laboratory. Fessenden was one of those let go.

In the circumstances we can only speculate as to the lines that Fessenden's career might have followed if he had been retained in Edison's service. He tells us in his unfinished autobiography that in 1890, before Edison departed on a grand tour of Europe, Fessenden asked whether he could take up work "on the lines of Hertz's experiments, which had recently been published." Edison said yes, but to wait until his return from Europe. By the time he did return the financial situation had changed and so had Edison's plans. It is curious to think of Edison taking any interest at all in alternating currents, and particularly in the very high frequency currents that would have been involved in an attempt to extend

---

[27] See Dietrich G. Buss, *Henry Villard: A Study of Transatlantic Investments and Interests, 1870-1895* (New York, 1978), pp. 188-220. Majority control of the Edison General Electric Company was in the hands of the Deutsche Bank, Allgemeine Elektrizitäts Gesellschaft, and Siemens & Halske. This syndicate controlled the company until its merger with the Thomson-Houston Company in 1892.

[28] Thomas Edison to Henry Villard, 8 February 1890, as cited in Buss, *Villard*, p. 210.

Hertz's experiments. But it is not impossible. If Fessenden in 1890 had begun work on Hertzian waves, he would have been one of the first in North America to do so. Hertz's key findings had been published in Wiedemann's *Annalen* only two years before. In the United States there seems to have been little interest in his work among physicists and electrical experimenters until Oliver Lodge delivered his famous lecture on "The Work of Hertz" at the Royal Institution in 1894.

For the time being experimental work on Hertzian waves had to be put aside. Fessenden needed a job—he had married in September 1890 and bachelor rooming houses would no longer do—and he found one as assistant electrician with the United States Electric Company, the eastern subsidiary of Westinghouse. The position was short-lived, but it was important in Fessenden's career for several reasons. It involved him for the first time in the design of alternating current machinery. It brought him to the attention of George Westinghouse. And it enabled him to extend his scientific reading, with the Newark public library substituting for the resources of Llewellyn Park. Alternating current theory and the papers of Hertz got most of his attention.[29] A developing interest in alternating current electricity—still a novelty for industrial use—was also responsible for his next move.[30] This was to the position of electrician—with a vague promise of an eventual partnership—with the Stanley Company in Pittsfield, Massachusetts. This firm operated a local power plant and street railway, manufactured transformers, and hoped to develop new inventions. Fessenden held the job for one year, working on insulating materials for transformers and on the design of alternating current motors. The most important benefit he derived from the experience may well have been a company-sponsored visit to England to inspect the new Ferranti power station outside London. Fessenden seized the opportunity to make two side trips: one to the Cavendish Laboratory in Cambridge to meet J. J. Thomson and discuss, among other scientific topics, the "electrostatic doublet theory of cohesion" which Fessenden had devised and in which he took great pride; the other to Newcastle to inspect a new Parsons steam turbine—a prime mover admirably suited for the driving of alternating current generators. The pieces were beginning to fall into place: Fessenden was clearly fascinated with Hertzian waves; he had acquired a comprehensive knowledge of alternating current theory;

[29] Fessenden, "Autobiography," *Radio News* 7 (October 1925), 557.

[30] As Passer reminds us, "Alternating-current power was not yet important in 1891, and the significance of the Westinghouse a.c. power patents was not generally understood." (Passer, *Electrical Manufacturers*, p. 327)

and he had gained useful experience in the design of alternating current motors and generators.

Returning to Pittsfield in 1892 he found the Stanley Company in financial difficulties, the promise of a partnership conveniently forgotten, and an offer awaiting him from the president of Purdue University which he promptly accepted. Purdue already had good programs in mechanical engineering and physics. Since 1887 it had been striving for equal success in the new field of electrical engineering, but it had been having trouble retaining qualified faculty. The first man appointed stayed only one year, his replacement only three; both had been university-trained physicists. Fessenden, the third to be appointed, lacked academic qualifications as a scientist but he had other qualities that commended him.

A university establishing a new program in electrical engineering in these years had two choices: one was to link it to an existing department of physics and staff it with men who were academically trained physicists; the other was to treat it as an extension of the engineering program and find an experienced practical electrical engineer to lead the course. Purdue had tried the first method without much success. The offer to Fessenden marked adoption of the second.[31] But Purdue may also have been trying to get the best of both worlds. Fessenden, it is true, could claim no formal training in pure science, but he had published in recognized scientific journals, and his electrostatic doublet theory, though it made little impression on the field, was at least an attempt to devise a new model of molecular bonding. To balance this, he had his three years in Edison's laboratory, a series of publications in engineering journals, and practical experience in business. He was known to be a rising star in the electrical world. And he had teaching experience—not, to be sure, at the university level but enough to ensure that he would not be a disaster in the classroom.[32]

Fessenden spent one academic year at Purdue, setting up the new laboratory and lecturing on electrical theory, with particular attention to alternating currents and high frequency oscillations. Toward the end

[31] Horton B. Knoll, *The Story of Purdue Engineering* (West Lafayette, Ind., 1963), chap. 9, esp. p. 269. See also Robert Rosenberg, "Physicists and Engineers at the Birth of a Discipline: Electrical Engineering Education" (Typescript).

[32] By the early 1890s almost all the larger and more substantial programs in electrical engineering had won their independence from physics (in an organizational sense). Very few programs were being taught by men who were physicists, the Massachusetts Institute of Technology being a conspicuous exception. I am indebted to Robert Rosenberg for guidance on these matters. Fessenden's publications are conveniently listed in Appendix III of H. M. Fessenden, *Fessenden*, pp. 353-62.

of the year he received a new offer, and this time there was no doubt as to who had suggested his name. The letter of invitation came from the chancellor of the Western University of Pennsylvania (later to be the University of Pittsburgh), but it was clear that the moving spirit was George Westinghouse. Fessenden had already been useful to Westinghouse. While employed by the United States Company, he had invented and patented silicon-iron and nickel-iron alloys for the lead-in wires in electric light bulbs and methods for sealing such wires in the glass envelope. Without these patents Westinghouse would have found it very difficult to fulfill his contract to light the Columbia Exposition, as platinum wires were controlled by patents he was not licensed to use.[33] Westinghouse, in short, already knew what Fessenden could do, and he wanted him in Pittsburgh. The chancellor explained, as Fessenden later recalled, that "Mr. Westinghouse had informed him that he had a particular regard for me and wished, if possible, that I should be offered the newly created chair of Electrical Engineering at the University." That was the kind of suggestion a prudent chancellor did not ignore. To underline the point, Fessenden received a letter from George Westinghouse personally, enclosing a check for $1,000 and stating that "he wished me to take up the gas secondary incandescent lamp if I should be offered and accept the chair at Pittsburgh." Suggestions as to the kinds of research likely to be acceptable to wealthy donors are, to be sure, not unusual in the annals of higher education, but few can have been as blunt and specific as this. Fessenden showed no signs of resentment; on the contrary, the prospect of being near to and connected with the Westinghouse works was a positive factor in his decision to accept.

Fessenden stayed at Pittsburgh for three years, with the Westinghouse Company providing half his salary.[34] It was in many ways an excellent arrangement from his point of view, providing institutional support with a minimum of institutional constraint. He had one foot in the world of industry, the other in the academy. Expectations from the institutions that employed him were manageable; he had considerable freedom to follow his own interests; and there were no financial worries. He lectured regularly on Hertzian waves, experimented with them in his laboratory,

[33] Fessenden, "Autobiography," *Radio News* 7 (November 1925), 715. The relevant patents are No. 452,494 (18 February 1891) and No. 453,742 (18 February 1891).

[34] H. M. Fessenden, *Fessenden*, p. 62. For information on the salary arrangements for Fessenden's appointment at the Western University, I am indebted to Professor Marlin H. Mickle of the Department of Electrical Engineering, University of Pittsburgh.

and apparently had a wireless communications system of some sort operating between Pittsburgh and the university campus in Allegheny City.[35] If so, this would be contemporaneous with Marconi's earliest experiments at Bologna. Fessenden's efforts were at first directed mainly toward developing a better detecting system. Lodge's lectures on Hertz were published in 1896 and these directed attention toward the limitations of the coherer. Two of Fessenden's students, Bennett and Bradshaw by name, collaborated with him on attempts to improve the device, and their work was published as a thesis in 1897.[36] Its principal conclusion was that the coherer could never be made reliable. Fessenden searched for an alternative—what he called a "continuously receptive receiver," in contrast to the on-off triggering action of the coherer. It was devices of this type that he described in his remarks to the American Institute of Electrical Engineers in 1899 (see above pp. 28-31), and it was for one of them that he received his first radio patent (No. 706,735, filed on 15 December 1899).[37]

But something more important was happening in these years than the development of a specific device. This was the crystallization of Fessenden's conception of Hertzian waves as high frequency alternating currents. It is impossible to date this kind of event; what is involved is a cumulative process that takes the individual from a vague hunch to a settled conviction. It is easy, in retrospect, to see how his previous experience must have inclined him toward such an approach. He came to wireless telegraphy, not by way of spark gaps as Hertz and Marconi did, and not from curiosity about lightning conductors and resonant circuits as Lodge did, but from working with electric motors, transformers, and dynamos—which is to say, from power engineering. And particularly relevant was his work with alternating currents, with the United States

[35] Fessenden Papers, 1140-27, memorandum entitled "More Important Fessenden Wireless Patents." Referring to himself in the third person, Fessenden wrote, "He worked between Pittsburgh and Allegheny, using his continuously receptive system, and was asked to allow it to be used to report the yacht races in the U.S. in 1899, but was too busy."

[36] Fessenden, "Wireless Telephony" (Paper presented at the 25th Annual Convention of the American Institute of Electrical Engineers, 29 June 1908), reprinted in Smithsonian Institution *Annual Report* for 1908 (Washington, D.C., 1909). Bennett later became a professor at the University of Wisconsin, while Bradshaw became manager of the Westinghouse Company's Newark works. See S. M. Kintner, "Pittsburgh's Contributions to Radio," *Proceedings* of the Institute of Radio Engineers, 20 (December 1932), 1849-59.

[37] Fessenden Papers, 1140-27; see also ibid., 1140-94, Fessenden to Herbert T. Wade, 18 May 1913.

Company and the Stanley Company and later in the laboratories at Purdue and Pittsburgh. By this time Fessenden was widely read in alternating current theory and had considerable practical experience with alternating current machines. He was at home in this medium and had the engineering achievements of Tesla and Westinghouse to inspire and challenge him. Hertzian waves required alternating currents of frequencies much higher than any used in power and lighting circuits. One did not have to work with such high frequency currents for long to realize that they raised problems for the engineer—the design of generators, of resonant circuits, of transformers—that were encountered only in milder and muted form at lower frequencies. But there was no essential difference. By 1898-1899 at the latest this had become the basic assumption that underlay Fessenden's conception of radio. It was to determine the whole course of his future work and the type of wireless communication system he was to build.

In his mind this conception set him distinctly apart from all those who followed in the footsteps of Marconi and Lodge and thought in terms of spark discharges and damped waves. What he was after was no incremental improvement on that system; it was (as he stated in 1908) ". . . an entirely new method . . . characterized by a return to first principles, the abandonment of the previously used methods and by the introduction of methods in almost every respect their exact antitheses."[38] And the heart of these new methods was to be the emission and reception of continuous waves.

* * *

Fessenden resigned from the Western University of Pennsylvania in 1900 to accept a contract from the U.S. Weather Bureau. He was later, in moments of depression, to express regret at having left the academic world, and there is no doubt that he could have continued his experimental work at Pittsburgh. But the prospects offered by the new contract were exciting. The bureau was interested in developing a network of wireless stations on the eastern seaboard for the exchange of meteorological data. The contract promised Fessenden greater research resources than he had at Pittsburgh, a better location for wireless experiments, and—he thought—freedom to develop his system as he wished and to control any patents that might result.

He took with him from Pittsburgh a commitment to the development of what we would now call a system of continuous wave radio and a

[38] Fessenden, "Wireless Telephony" (1908), p. 166.

determination to set wireless technology on a new track. His dissatisfaction with the Marconi system as it then existed arose in the first
instance from a conviction that it was "essentially and fundamentally
incapable of development into a practical system."[39] This was with reference to wireless telegraphy: the transmission of dots and dashes in the
Morse or some similar code. But by 1900 he had also become convinced
that such a system, with its damped wave transmitters and coherer-
equipped receivers, labored under insurmountable limitations with reference to wireless telephony: the transmission of the human voice. For
this continuous waves—undamped oscillations that could be modulated
by sound—were almost indispensable. As Fessenden's attention turned
increasingly to wireless telephony, therefore, his belief that damped wave
systems were leading wireless technology in a fundamentally wrong direction became intensified. No crisis had yet been reached; indeed, the
possibilities of spark telegraphy were only beginning to be exploited. But,
farther down the road, such systems were bound to reach a dead end.[40]

How could such continuous waves, suitable for wireless telephony, be
generated? In 1899-1900 three methods were thought to offer possibilities. The first called for a radical increase in the spark frequency. To
modulate a train of spark discharges by the sound frequencies of the
human voice was, in the normal case, impossible, for the voice would
be drowned out by the machine-gun-like hammering of the sparks themselves. But if the frequency of the spark discharges could be raised well
above the audible range—say, at the very least, above 5,000 sparks per
second—telephony might be possible.[41] The practical problem, if this
approach were followed, was to devise an "interrupter" that could generate a spark wave train at this rate or higher. Elihu Thomson in America
had done work along these lines, as had Nikola Tesla. In Europe Oliver
Lodge had tried to smooth out the spark wave train even further, by
adding inductance to the antenna circuit, and Ferdinand Braun had sought
the same objective by tuning the local oscillating circuit to a frequency
slightly different from the natural resonant frequency of the antenna.[42]

[39] Ibid., p. 174.

[40] This is to say that Fessenden had identified a "presumptive anomaly" in the
development of radio technology in the sense in which that concept is used by
Constant in his *Turbojet Revolution.*

[41] The range of frequencies audible to the human ear varies greatly among
individuals. The highest audible tone is about 30,000 Hz; the average for a male
voice is 500 Hz, but the overtones that make each voice distinctive go up to
20,000.

[42] Fessenden, "Wireless Telephony," *The Electrical Review* 60 (22 February
1907), 327. Compare Friedrich Kurylo and Charles Susskind, *Ferdinand Braun:*

The result in all these cases was not a true undamped wave but a close approximation thereto, depending on how high the spark rate could be carried.

The second method was the oscillating arc. Here again Elihu Thomson had shown the way with an 1892 patent (No. 500,630) describing a method of generating persistent oscillations from an electric arc. And in 1900 the English experimenter, William Duddell, demonstrated the so-called "singing arc," with which, by shunting a tuned resonant circuit across the arc, musical notes could be produced. But neither of these devices generated waves at frequencies high enough for wireless transmission. Thomson's circuit fed the oscillations from the arc to a spark gap, and it was the waves generated by the spark, rather than those created by the oscillating arc itself, that were radiated. And Duddell's arc oscillated only at audio frequencies. Ways were to be found, as we shall see, to make the oscillating arc a true generator of radiofrequency currents, but in 1900 these were still in the future.[43]

And thirdly there was the possibility of building a high frequency alternator—a piece of rotating machinery, identical in principle to those that generated alternating currents for lighting and power but operating at much higher speeds and generating currents of much higher frequencies. This was, as a matter of abstract theory, the straightforward approach: it involved taking a machine the theory of which had already been worked out for power frequencies and adapting it for use as a radio transmitter. The adaptations required, however, were far from simple or obvious, and in 1900 it was by no means certain that such a machine could be built at all. Thomson in 1889 had designed and built an alternator for use with arc lights, moving up in frequency out of the audible range in an attempt to eliminate an annoying hum. And Nikola Tesla in 1890 and later had constructed several high frequency alternators of novel design, including one with a disk-shaped armature that anticipated later General Electric designs.[44] This last machine had reached a frequency of 15,000 cycles per second, and this seems to have been the highest frequency attained by any alternator before 1900.

A Life of the Nobel Prizewinner and Inventor of the Cathode-Ray Oscilloscope (Cambridge, Mass, 1981), chap. 7, pp. 94-178. Kurylo and Susskind give no special emphasis to the feature of Braun's coupled circuits that Fessenden emphasized.

[43] See below, pp. 110-18, and C. F. Elwell, The Poulsen Arc Generator (New York, 1923), pp. 22-26. Thomson's oscillating arc is shown in Fessenden, "Wireless Telephony" (1908), 172; for Duddell's work, see W. Duddell, The Electrician 46 (14 and 21 December 1900) and Fleming, Principles, pp. 97-99.

[44] For illustrations, see Fleming, Principles, pp. 6-8.

In conception at least these three methods of moving toward true continuous wave transmission were clear to Fessenden when he began his work for the Weather Bureau in April 1900. The challenge was to move from idea to reality, to test all three methods and find out the limitations and possibilities of each. Equally important was the devising of better apparatus for receiving. Fessenden would have nothing to do with coherers or "imperfect contact" detectors.[45] He wanted a device that would be "continuously receptive," rather than requiring the resetting that a coherer did; it should not upset the tuning of resonant circuits; and it should give a response proportional to the energy received. Above all it should be more sensitive. The devices he had worked with at Pittsburgh met these criteria, but they were not suitable for field service.

Inventing better detecting apparatus was therefore high on Fessenden's agenda. Longer range and more reliable service depended on it. So did his hopes for continuous wave telegraphy and radiotelephony. More was involved than just higher sensitivity. For wireless telephony it was essential to have a detector that could follow the variations in amplitude of the transmitted wave—that could, as we would say today, demodulate the signal. No filings coherer could do that. Furthermore, if continuous waves were to be used for telegraphy, some means had to be found to convert the incoming dots and dashes into audible sounds. This was no problem with normal spark telegraphy, in which the sparks followed each other at an audible rate; that kind of signal carried its own modulation with it. But a true continuous wave signal had no modulation and would produce in an earpiece nothing but clicks and thumps. The closer Fessenden approached to true continuous wave transmission, the more inadequate conventional detectors would prove.[46]

[45] His colleague at Pittsburgh, S. M. Kintner, says that he refused to use coherers even for comparative tests and calls attention to the "real courage" that was required to develop other forms of detector: ". . . remember the coherer was generally supposed at that time, to have an order of sensitivity not even approached by any other known device. It was the very heart of the then young wireless system." (Kintner, "Pittsburgh's Contributions") Kintner in 1932, of course, was addressing a generation of electronic engineers that took continuous wave radio and the vacuum tube for granted.

[46] There is a problem with reference to crystal detectors. Ferdinand Braun in Germany had discovered the unidirectional conducting characteristics of certain metal sulfides in 1874: electric current would pass through very easily in one direction but only with great difficulty in the other. In 1883 he recognized that this effect could be used as a rectifier: that is, to convert alternating current into direct current. And in 1901, according to his later account, he recognized that this same effect could be used to detect Hertzian waves. See Kurylo and Susskind,

What Fessenden took with him when he left Pittsburgh was, in short, the concept of a new type of wireless system; and the systemic aspect is what calls for emphasis. New types of receiving apparatus were needed because new modes of transmission were to be used. There is a danger that, in describing the particular devices that Fessenden invented to give physical form to his system, our attention may be diverted from the overarching conception that linked them together and guided his research strategy. This was the conception of radio waves as high frequency alternating currents; implied by this was an abandonment of the ways of generating and detecting signals that were associated with spark telegraphy; and a corollary was the belief that radio telephony, as well as more efficient telegraphy, was possible.

The ten years that followed Fessenden's departure from Pittsburgh were the most productive period of his life. If his move from Bermuda to New York is when he starts to become an interesting personality, his decision to leave Pittsburgh marks the time when he becomes historically important. One has the impression, once again, of a change in trajectory, as if the very act of making the decision, and accepting the risks that came with it, had moved him onto a higher level of creativity.

It was with detectors that the first breakthrough came. Writing to his patent attorney from Roanoke Island, N.C., where he had set up his main experimental station, Fessenden made no attempt to conceal his delight. "What do you say," he asked, "to a receiver which gives telegrams at the rate of a thousand words per minute and is so sensitive that it gives that rate when the coherer will not even give a click. . . . Also that it is perfectly positive and gives these results in its very crudest form and on the very first trial. Well, that is what I have now. . . ."[47] Some of Fessenden's reaction can be ascribed to the euphoria of the successful inventor—"The new receiver is a *wonder*!!!" he wrote a few weeks later— but there was more sober evidence. Louis Dorman, official observer for the Weather Bureau, reported to his chief on 1 April 1902 that he had used the device with great success in working between Cape Hatteras and Roanoke Island: "The receiver is positive in its action, and entirely

---

Braun, pp. 28-29, 44-45, and 131. In the United States, however, knowledge of the detecting (as distinct from rectifying) function of crystal diodes awaited H. C. Dunwoody's discovery of the carborundum detector in 1906 and G. W. Pickard's discovery of the silicon detector in the same year. It seems clear, therefore, that the formal identity of a rectifier and a detector was not easily recognized—further evidence of the difficulty involved in recognizing Hertzian waves as alternating currents.

[47] Fessenden to Wolcott, 28 March 1902, as cited in H. M. Fessenden, *Fessenden*, p. 93, quoted by permission of The Putnam Publishing Group.

and absolutely reliable. It is entirely different in nature and action from the coherer, and gives no false signals like the latter does. I could hear every single dot and dash made at Hatteras with the utmost clearness. . . . It is possible for any expert telegrapher to receive by it as fast as the key can be handled."[48]

What evoked this enthusiasm was a device called the barretter—from an old French word meaning "exchanger," since it changed alternating into direct current. There were several variants. The original, the hot-wire barretter, Fessenden had brought with him from Pittsburgh; it was essentially a length of Wollaston wire (very fine platinum with a silver cladding) enclosed in a glass envelope, rather like a miniature incandescent lamp. A short section of the wire had its silver coating removed by immersion in nitric acid. When the antenna current from an incoming wave passed through the wire, its resistance changed; and if one connected a telephone earpiece between the barretter and ground, the change in resistance could be heard as a click. A succession of spark discharges, as from a normal spark transmitter, would be heard as the dots and dashes of the Morse code. The liquid barretter, which is what caused the excitement in March 1902, evolved from this device in a classic example of the carefully observed and analyzed laboratory accident. While etching the silver coating off a number of lengths of Wollaston wire, Fessenden noticed that one was responding particularly vigorously to a small test oscillator then running in the laboratory. Investigation showed that the acid, in this one case, had eaten right through the wire; what was left was a short stub, as it were, of thin platinum wire dipped in nitric acid.[49] This was the central principle of the liquid barretter, although it was later much refined and improved. Many people referred to it as a kind

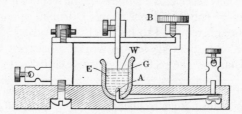

Fig. 2.1: Fessenden's liquid barretter.
Source: G. W. Pierce, *Principles of Wireless Telegraphy* (New York: McGraw-Hill Book Company, 1910; reproduced by permission), p. 202

[48] Fessenden Papers, 1040-2, Louis Dorman to Chief, U.S. Weather Bureau, 1 April 1902.

[49] Kintner, "Pittsburgh's Contributions," p. 1851.

of electrolytic detector, of which there were several competing varieties, but Fessenden rejected that description. He believed the action was thermal, and some later experts agree with him.[50] Whatever its principle of operation, it was a highly sensitive detector of Hertzian waves and it became the standard against which all others were compared. Unlike the coherer, it required no "tapping back."[51]

Devices that looked very similar were invented by others—Schloemilch in Germany, Ferrié in France, Vreeland and de Forest in the United States. The question of priority was hotly debated, and defense of Fessenden's patent required extensive litigation. Few devices, indeed, can have been so widely imitated and few patents so generally infringed (in Fessenden's case, with the aid and encouragement of the U.S. Navy).[52] As to its sensitivity, there was no question. Its stability, however, was another matter, and marine operators in particular, contending with the roll and pitch of the vessel, found it no easy task to keep the tiny platinum filament and cup of acid in correct adjustment.[53] But it did give Fessenden what he wanted—a continuously active detector—and it did provide radio technology with its most sensitive detecting device until the invention of the triode vacuum tube.

[50] Phillips, for example, discusses the liquid barretter under thermal detectors rather than under electrolytic detectors. The liquid barretter depicted in his text on p. 154 is a later and refined version. See V. J. Phillips, *Early Radio Wave Detectors* (London, 1980), pp. 150-71.

[51] For descriptions of the liquid barretter, see U.S. Patent No. 727,331; Phillips, *Detectors*, pp. 151-54; Fessenden, "Wireless Telephony" (1908), p. 169; Fleming, *Principles*, 2nd ed. (London, 1908), pp. 395-96; and W. Rupert Maclaurin, *Invention and Innovation in the Radio Industry* (New York, 1949), pp. 60-61. Pupin had patented a somewhat similar device in 1898 and stated in his patent that it would rectify at Hertzian frequencies.

[52] Susan Douglas, "Exploring Pathways in the Ether," (Ph.D. diss., Brown University, 1979), pp. 133-84; L. S. Howeth, *History of Communications-Electronics in the United States Navy* (Washington, D.C., 1963), pp. 99-101. The Navy purchased liquid barretters from the De Forest Company in 1905 despite the fact that Fessenden's patent had been upheld by the courts. In response to Fessenden's protests, the secretary of the navy stated that he was asking too high a price and therefore the department "feels that it is relieved of any moral obligation that might otherwise exist." Clearly it was the Navy's morality that had a price tag, not Fessenden's.

[53] The Marconi magnetic detector, invented in 1902, was also a continuously active detector in Fessenden's sense but showed just the reverse characteristics: great stability but poor sensitivity. Some marine operators alleged, indeed, that with a magnetic detector signals could be copied from another ship only when it was passing within sight. See Manuel Goulart, "More on Lightships and Magnetic Detectors," *Sparks Journal* (Society of Wireless Pioneers) 2 (Fall 1979), 5.

It did not, however, solve the problem of how to receive continuous wave telegraphy. The barretter could produce a sound in the earphone because it generated pulses of direct current at the spark frequency, which was normally within the audible range. But continuous wave telegraphy, unlike spark, carried no such modulation. Received through any kind of rectifying detector, it would be heard in earphones only as a succession of clicks. This was, in fact, exactly what Marconi said he had heard in Newfoundland in 1901, by connecting a coherer and telephone earpiece between his kite antenna and ground. This was not good enough for practical telegraphy.

Fessenden's awareness of this difficulty led him to an invention of much broader scope and significance than the barretter, one that went far beyond any particular hardware. This was the heterodyne principle, which has remained fundamental to radio technology ever since. The word, now part of every radio engineer's vocabulary, was Fessenden's coinage, reflecting his early training in Greek: to heterodyne meant to mix two different forces—in this case, two waves of different frequency. He seems to have been led to the discovery by reflecting on the fact that no detector, not even the liquid barretter, could approach the sensitivity of the ordinary telephone earpiece. But how to make Hertzian waves, oscillating at radio frequencies, move the diaphragm of an earpiece at audio frequencies? Fessenden's answer was to feed two currents into the earpiece: one the antenna current, a train of oscillations at the signal frequency, and the other a train of oscillations generated either locally at the receiver or at a second transmitter, with the two frequencies differing slightly from each other. The two frequencies would mix or "beat" against each other, and if the difference between them were correctly chosen, the result would be a wave train that the metal diaphragm of the earpiece could follow and that the human ear could hear. If, for example, the signal frequency were 1,600,000 cycles per second and the second oscillator set at 1,601,000 the "difference frequency" would be a somewhat shrill but clearly audible 1,000 cycles per second.

The principle should be familiar to anyone who has ever heard a piano tuner at work, and at audio frequencies it was well known and of ancient origin. Fessenden's contribution was to apply it at radio frequencies, and no one had done that before. He was perhaps fortunate to find judges who would uphold it as something that could be patented—who would acclaim it, indeed, as "invention of a high order . . . a new contribution to the knowledge of the time."[54] The difference between radio and audio

---

[54] Fessenden Papers 1140-4, verdict of District Judge Julius M. Mayer, southern district of New York, in *Kintner and Barrett v. Atlantic Communications Com-*

frequencies was, after all, merely one of degree, and the courts have not always been willing to recognize such mental steps as patentable. On the other hand, questions of property rights aside, that Fessenden did introduce into radio technology a concept and a method of the very first importance is not to be questioned.

The particular method Fessenden first used to implement heterodyning was, of course, primitive. He was using the telephone earpiece and the human ear as a "mixer."[55] Later, with vacuum tubes and crystal diodes, the mixing would be done electronically. Even in the original form, however, he considered it "the receiver *par excellence* . . . the most effective form of receiver in existence." He did not foresee the problems that the general use of heterodyning in radio design would later cause. When two signals were mixed, heterodyning generated not only the sum and difference signals but also harmonics of both signals being mixed plus the sum and difference signals generated by mixing the harmonics. The result was a proliferation of "spurious" signals which had to be suppressed by screening and filtering. But, at the time, these were trivial matters.

The heterodyne receiver was an invention that grew in importance as radio technology and the radio industry developed. At the time of discovery, despite Fessenden's enthusiasm, it had little impact. There were two reasons for this. First and obviously, there were no continuous wave transmitters on the air in 1900-1901 and few in the next decade and a half. The period from 1900 to 1915 was, indeed, the golden age of the spark transmitter, and to receive spark transmissions many types of detector were available—particularly, after 1906, the cheap, simple, and reliable crystal or carborundum detectors. As long as spark reigned supreme there was no need for Fessenden's heterodyne and therefore no market for it. In that sense the invention was about ten years ahead of its time. This was not true, of course, from Fessenden's point of view. His objective was a system of continuous wave radio and for that system to be complete he had to have an appropriate receiver.

The second reason calls for some explanation. The heterodyne receiver required a local oscillator to beat against, or mix with, the incoming signal.[56] This oscillator had to generate true continuous waves, at a

---

pany, in equity. The principle was described in U.S. Patent No. 706,740 and the method in Patent No. 1,050,728.

[55] In the original method, two bobbins of fine wire were wound on a single iron core in the earpiece, each carrying a slightly different frequency. See Fessenden, "Wireless Telephony," *The Electrical Review* 60 (1 March 1907), 369.

[56] This is true as the heterodyne principle is generally described. It was also possible, however, to transmit two continuous wave signals on frequencies spaced slightly apart, and receive them with a straightforward diode detector or even a

precisely regulated frequency; and it had to do this without introducing noise. Where was Fessenden to find such an oscillator in 1901? There were only three possibilities—the same three that he confronted in the design of continuous wave transmitters. High frequency sparks were too noisy for the purpose. And, if one were to fall back on spark reception, why not use a spark transmitter in the first place? Radiofrequency alternators were not yet available. This left the arc as the only candidate, and the first heterodyne receivers were in consequence equipped with small arcs as local oscillators. But arcs were inherently noisy (in the electronic sense); they were difficult to keep on frequency; and, although they did generate continuous waves, these were not undistorted sine waves. Seen from this perspective, the heterodyne receiver was a brilliant conceptual breakthrough, but one made before the hardware was at hand to implement it properly. The situation was to change drastically in 1912, with the invention of the vacuum tube oscillator; but in 1901 that was neither available nor contemplated.

In designing his receivers, therefore, Fessenden found himself facing essentially the same problem as confronted him in designing transmitters. If he was to break away from the spark tradition, he had to have a device that would generate continuous sine wave oscillations. If this was desirable for telegraphy, it was essential for telephony. Here, apart from the higher power level, the requirements were identical to those for the local oscillator in a heterodyne receiver: it had to be quiet, introducing no extraneous noise into the signal; and it had to be stable in frequency.

＊　＊　＊

If he had done nothing else, Fessenden's invention of the heterodyne receiver would be enough to earn him a permanent place in the annals of radio technology. But in some ways his development of the radio-frequency alternator was of more immediate importance. It was, perhaps, a less creative achievement, if such things can be measured. Fessenden was following a trail that had been blazed by others, Thomson and Tesla in particular. His contribution was to insist that the thing could be done and had to be done, at a time when others were convinced neither of

coherer. This was the principle described in Fessenden's original patent No. 7023740 (filed 28 September 1901, issued 12 August 1902). In the later version, more familiar to modern radio engineers, one of these "transmitters" was built into the receiver as a local oscillator. See Marius Latour, "The Heterodyne Method of Wireless Reception: Its Advantages and its Future," *Radio Review* 2 (January 1921), 15; Pratt Papers (Bancroft Library), Lloyd Espenschied to Haraden Pratt, 20 September 1954 and Espenschied to Benjamin F. Miessner, 23 March 1963.

feasibility nor of need. Without that stubborn insistence the history of the American radio industry would have followed a very different course.

The high frequency alternator was, however, only one of the techniques that Fessenden tried in his search for a continuous wave transmitter, and for several years it was anybody's guess whether an alternator could be built to meet his requirements. The problem was not in the conception but in the execution. It was all very well for him to assure his colleague Kintner that he knew how it could be done: "Take a high-frequency alternator of 100,000 cycles per second, connect one terminal to the antenna and the other one to ground, then tune to resonance." But Kintner was properly skeptical: "I didn't know of any 100,000-cycle machines—neither did he."[57] And to be in the market for one, even to have the funds to finance the building of one, did not necessarily mean that you would get what you wanted. Fessenden was pressing against the limits of the manufacturing capabilities of his day.

At Pittsburgh he had experimented with high-speed "interrupters"— devices that one could insert in the primary circuit of an induction coil, to produce from the secondary a train of sparks far more rapid than the eight sparks a second that Pupin had attributed to Marconi. The hope was that this would give an approximation to a continuous wave and, with the spark frequency far above the audible range, permit voice modulation. The Wehnelt interrupter was the latest such device in 1899 and for a while the electrical journals were full of it.[58] Fessenden worked with one for a while and got encouraging results; he also, by scribing fine grooves longitudinally on a phonograph cylinder, devised a mechanical interrupter of his own which was supposed to give him a spark frequency of 10,000 per second.[59] By placing a carbon microphone (or, as it was then called, a "transmitter") in the antenna circuit, he was able to modulate the wave train and with this equipment, in the fall of 1900, he transmitted speech between two stations one mile apart at Rock Point, Maryland. This was, he later claimed, the first time that intelligible speech had ever been transmitted by radio.[60]

Intelligible it may have been, but it was also very noisy. "The character of the speech was not good," Fessenden later reported, "and it was accompanied by an extremely loud and disagreeable noise, due to the

[57] Kintner, "Pittsburgh's Contributions," p. 1852.

[58] Fleming, *Principles*, pp. 51-52.

[59] Fessenden, "Wireless Telephony" (1908), 177-78. Fessenden believed that the spark frequency actually obtained was probably less than this.

[60] Fessenden "Wireless Telephony," *The Electrical Review* 60 (15 February 1907), 252.

irregularity of the spark." Two limitations, he thought, caused the trouble: the spark frequency was not nearly high enough—he now estimated that nothing less than 20,000 per second would do; and the discharges were still too sharply damped.

There were several ways of tackling the problem and Fessenden tried them all. First he developed a form of spark gap in which each spark, as soon as it was formed, was blown out by a blast of compressed nitrogen.[61] This was, in conception, very similar to the "quenched spark" that Telefunken was later to use effectively: the central idea was to reduce the damping coefficient by leaving the transmitter's antenna circuit free to resonate, with the spark providing only very short pulses of energy. Secondly, he began a series of experiments with Elihu Thomson's oscillating arc, working to raise and stabilize its frequency of operation. And thirdly, in early 1901, he took out a patent for a very high frequency alternator, to be used for wireless transmission, and he submitted an order for its manufacture to the General Electric Company.

In the short run the nitrogen spark gap gave him the best results. It reduced the noise and distortion of telephony somewhat, and it also proved well adapted for telegraphy (which was, after all, what the Weather Bureau was interested in). The oscillating arc he found difficult to work with at first, as in its original form it fluctuated unpredictably in frequency and intensity. By operating the arc under pressure he was able to stabilize its frequency, and the problem of how to key an arc for telegraphy without extinguishing it was solved by allowing it to run continuously and using the key to vary the electrical constants of the antenna circuit (frequency shift keying, as it was later to be called). By these and other means he was able to generate continuous waves that were, he believed, "absolutely constant in frequency and intensity" up to 3 million cycles per second (3MHz) with "an absence of harmonic frequencies."[62] Used for telephony, it gave a transmitted signal considerably quieter than the original 10,000-cycle spark, but still with a good deal of "foreign noise in the telephone."[63]

Fessenden was following the rules of the game, as learned from Edison. He was testing all possibilities before committing himself to one. But, if any doubt had existed before, it was quickly being dissipated: for te-

[61] Patent No. 706,741. See also Simon Papers (Bancroft Library), Fessenden to S. M. Kintner, 29 January 1932.

[62] Fessenden, "Wireless Telephony" (1908), 172. Fessenden's basic arc patent is No. 730,753 (9 April 1903).

[63] Fessenden, "Wireless Telephony," *The Electrical Review* 60 (15 February 1907), 252.

lephony to be possible, there was really no alternative to continuous undamped waves. And for quiet, distortion-free telephony to be possible, they had to be perfect sine waves. That, for Fessenden, ruled out the arc and the high-speed quenched spark, except as temporary expedients. There remained the alternator. The problem was to find someone able and willing to build the machine he wanted.

With his Pittsburgh background, his Westinghouse contacts, and the commitment of the Westinghouse Company to alternating current technology, nothing would have been more natural than for Westinghouse to tackle Fessenden's assignment. He asked, and he was rebuffed. In June 1900 the chief electrician told him that, because of the amount of work on hand, Westinghouse was in no position to undertake "special work of this kind."[64] Whether this was the true reason or an excuse for turning down an unpalatable job is not clear. Events were to show that this was a critical error for the corporation. Twenty years later Westinghouse was to find itself scrambling to establish some kind of patent position in radio, on pain of being left out of the field entirely. In 1900 it could have got in on the ground floor. It is far from clear why Fessenden's inquiry was so summarily rejected, nor why he took the refusal so calmly. Did he try to approach George Westinghouse personally? Or had relations with the industrialist so cooled that an appeal was out of the question? The record is silent on these matters.

In any event the decision took Westinghouse out of consideration. That left General Electric. Formed in 1893 from the merger of Edison General Electric and the Thomson-Houston Company, General Electric had impressive design and production facilities for both alternating and direct current equipment—indeed, acquisition of the Thomson-Houston alternating current patents had been one of the main reasons for the merger, and most of the top management of the new firm came from Thomson-Houston, not from the old Edison GE.[65] There might be smaller firms competent to tackle parts of the project—Fessenden was to use

[64] Fessenden Papers 1140-2, Charles Scott to Fessenden, 6 June 1900. It is intriguing to note that two years later B. G. Lamme of Westinghouse was to design and build a high-frequency alternator—and a 10,000 cycle machine at that—for the French physicist, Leblanc. This only adds to the puzzle of why Fessenden's request was turned down. (My thanks are due to James Brittain for reminding me of the Lamme alternator.)

[65] Passer, *Electrical Manufacturers*, pp. 321-29. Passer goes so far as to say that "General Electric, at the top-management level at least, was in reality a Thomson-Houston organization." The merger was arranged by the Morgan banking firm; Villard and his German associates lost control of the company at this time.

some of them later—but, with Westinghouse out of the picture, he really had no choice but to turn to GE if he wanted his alternator built at all.

And with GE, as a member of its newly created Research Laboratory, there was Charles Proteus Steinmetz. Born in Breslau, Germany, in 1865, educated there and at Zurich in engineering and mathematics, Steinmetz had come to the United States in 1890 and joined GE shortly after its formation.[66] By 1900 he had firmly established his reputation as a leader in alternating current theory and design. Fessenden wrote to him in June 1900, conveying general specifications for a high frequency alternator to be used for wireless transmission and expressing the hope that he would design such a machine as an experiment and persuade GE to quote a price for its manufacture. He thought that perhaps forty or more might eventually be needed.[67]

Steinmetz was intrigued. He agreed immediately that such an alternator would be a much more satisfactory generator of wireless waves than any induction coil. But Fessenden's request was no easy one to respond to. It posed a major challenge both to Steinmetz's skill as a designer and to the manufacturing abilities of the Schenectady shops, and one suspects that it was that challenge, rather than any prospect of developing a profitable new line of electrical equipment, that induced him to undertake the job. Orders for the design and manufacture of a test alternator were approved early in 1901, and by early summer it was ready for test. It did not reach Fessenden, however, until March 1903.[68]

If Fessenden had hoped to use this machine in the way he had airily described to Kintner—connect one terminal to the antenna and the other to the ground and tune to resonance—he must have been disappointed.

[66] J. W. Hammond, *Charles Proteus Steinmetz* (New York, 1924); James E. Brittain, "C. P. Steinmetz and E.F.W. Alexanderson: Creative Engineering in a Corporate Setting," IEEE *Proceedings* 64 (September 1976), 1413-17. Steinmetz was working for Eickemeyer & Osterheld, a small electrical manufacturing firm in Yonkers, N.Y., when it was acquired by the newly formed General Electric Company. His abilities had been called to the attention of E. W. Rice, chief engineer and vice-president of GE, by Ernst Danielson, chief engineer of the Swedish Electric Company; see E.F.W. Alexanderson, "Reminiscences" (Columbia University Oral History Collection), p. 11.

[67] In addition to the Clark Radio Collection at the Smithsonian Institution and the Alexanderson Papers at Union College, I have relied heavily, for information on the relations between Fessenden and the General Electric Company, on James E. Brittain's biography of Alexanderson, currently being prepared for publication. I am deeply grateful to Professor Brittain for permitting me to consult and use his unpublished manuscript.

[68] Brittain, *Alexanderson*, chap. 2, p. 10.

Steinmetz's alternator had a maximum frequency of 10,000 cycles per second (10 kHz), and that was a long way from any frequency that could be radiated directly. The limiting factor seems to have been the rotating armature, which was of conventional design with wire-wound coils; such an armature could not be driven over a certain number of revolutions per minute (3,759 in this case) before it would start to disintegrate under centrifugal force.[69] Fessenden made what use of it he could: he connected its output to a spark gap and used the spark to excite the antenna. This gave him a spark frequency of 20,000 per second, considerably higher than he had achieved earlier with his mechanical interrupter.[70] The waves were, of course, still damped; the signal was still noisy; and he still had not freed himself from spark. But this hybrid arrangement—high-speed alternator and quenched spark—was a substantial improvement on anything available before. By 1904 Fessenden's National Electric Signaling Company (hereafter referred to as NESCO) was advertising wireless telephone sets using this combination for commercial sale, with a guaranteed range of twenty-five miles.[71]

Fessenden later said that he always felt as if GE's engineers were doing him a favor when they designed and built his alternators. Perhaps in a sense they were. On the other hand, the company was never out of pocket: Fessenden was charged full development and manufacturing costs and he paid without a quibble. And there were incidental benefits to General Electric in the form of greater knowledge of alternating current theory and practice and the chance to test and develop the abilities of new recruits. This last may well have been a consideration when Fessenden submitted his second alternator order. This time he wanted an order-of-magnitude increase in frequency: at least 150,000 cycles per second. And he wanted more power—25 kilowatts—with the alternator to be driven by a Curtis steam turbine. What he was after was clearly an alternator that could function as a transmitter in its own right, not merely as an exciter to drive a spark gap.

Steinmetz did not tackle this job personally. His assistant, Ernst Berg,

[69] Data on the design and performance of this machine are scarce, but see ibid., chap. 2, pp. 10-11.

[70] Each complete sine wave from the alternator would give two sparks, at the peak positive and negative voltages.

[71] Fessenden, "Wireless Telephony," *The Electrical Review* 60 (22 February 1907), 328. Fessenden described the alternator used in this arrangement in 1903-1904 as giving "about 10 amperes at 100 volts"; I take it that this is the same machine as Brittain describes as giving 14 amperes at 80 volts. (Brittain, *Alexanderson*, chap. 2, p. 11) Fessenden ordered three or four additional 10 kHz alternators in February 1905, but these were of a different design.

handled the initial correspondence with Fessenden, and the design work was assigned to a young Swedish engineer, Ernst Alexanderson, who had recently joined GE's Engineering Department. Alexanderson had other work to do; his major assignment was designing motors for interurban electric railways. But he had already given indications of inventive ability, notably a self-excited alternator that made its trial run in December 1904. Fessenden's alternator would give him a further chance to show what he could do. In fact, he had already started work on a preliminary design before Fessenden's order arrived.[72]

According to Alexanderson, the alternator was "one of the inventions that I had to make in order to hold my job."[73] He was still very much on trial in the Engineering Department when given the assignment in 1904, and this despite the fact that he was in a sense Steinmetz's protégé. His academic credentials were good—three years at the Royal Institute of Technology in Stockholm and one at the Koenigliche Technische Hochschule in Berlin—but so were those of many of the young Swedish engineers who came to Schenectady to see "how things were done in a big way in America." Alexanderson had found about twenty-five of his classmates already working for GE when he first went there in 1902; most of them held jobs in the drafting department for a year or two and then returned to Sweden. Alexanderson started out in the drafting department too, but after a year and a half he moved into the testing department which was, as he later said, "the opening door to the Company." Steinmetz was responsible for that transfer. Once in the testing department Alexanderson started inventing—a new switching system for power circuits was his first patent—and he was given a desk in the Engineering Department "just to finish that invention." But then, "when the time came to move out, I had made another invention, and I kept on making inventions. Thus I was allowed to stay."

Alexanderson was twenty-six years old in 1904, while Fessenden was

---

[72] For this piece of information we are indebted to the careful research of James Brittain. Alexanderson, in his "Reminiscences," p. 16, intimates that he started work on the alternator *after* Fessenden's request came in. But Fessenden sent his specifications to GE on 8 December 1904; he was informed on 10 December that the Engineering Department had already begun the design of a 100 kHz machine and would prefer to complete that before tackling Fessenden's 150 kHz project. Alexanderson stated in 1915 that he had completed his preliminary design on 3 December 1904. The intriguing but unanswered question is who initiated design work for the 150 kHz alternator. See Alexanderson Papers, Drawer 1, No. 2, "History of the Development of High Frequency Alternators," report to A. G. Davis, 29 June 1915, and Brittain, *Alexanderson*, chap. 2, pp. 11-12.

[73] Alexanderson, "Reminiscences," p. 16.

thirty-eight. They were to make a highly productive partnership. Alexanderson at first knew little about radio, although he had attended Adolf Slaby's lectures on Hertzian waves while at Berlin. His orientation was toward power engineering and the core of his training had been in the design of rotating electrical machines. That was why he had come to work at General Electric—"the veritable citadel of electrical power engineering" at that time, as it has been accurately described.[74] He had learned about Steinmetz's pioneering work on alternating current analysis while still a student in Berlin, and it was quite explicitly with the hope of working with Steinmetz that he had come to the United States. This perspective—that of the electric power engineer—was to inspire all his work on radio. He thought of electromagnetic radiation in terms of high frequency alternating currents, just as Fessenden did. The partnership between Fessenden and Alexanderson, in short, did more than involve the largest electrical manufacturing company in the country in the manufacture of radio equipment. It also opened up channels of communication and creative interaction between two engineering subcultures: power engineering and electronics.[75] And it placed the research and manufacturing capabilities of a major corporation behind the efforts of an individual inventor.

Alexanderson's work on the alternator was at first a marginal increment to his major assignments in the Engineering Department: "almost a diversion," he called it; "a rather fantastic thing" in the eyes of his colleagues, but one he was "crazy enough to undertake." Steinmetz gave him a free hand, and his first design was a radical departure from the 10 kHz machine built in 1901. Gone now were the rotating armature windings. In their place was a stationary armature winding on an iron core, located between two rotating steel disks, with projecting poles or teeth cut in their circumferences. As the disks revolved, they induced alternating currents in the coils of the fixed armature, the frequency depending on the speed of rotation and the number of poles on the disks. This was an inductor-type alternator; the concept had been used before, both at normal power frequencies and, by Duddell and Tesla and B. G. Lamme of the Westinghouse Company, at high frequencies.[76] Its great

[74] James E. Brittain, "The Alexanderson Alternator: An Encounter Between Radio Physics and Electrical Power Engineering" (Paper presented at a joint meeting of the Society for the History of Technology and the History of Science Society, 31 October 1982), p. 5.

[75] For an extended discussion of this theme, see Brittain, "The Alexanderson Alternator.".

[76] Fleming, *Principles* pp. 6-14; Brittain, *Alexanderson* chap. 2, p. 13; B. G. Lamme, "Data and Tests on a 10,000 cycle-per-second Alternator," AIEE *Trans-*

advantage from Alexanderson's point of view was that it freed him from the restraint on rotor speed implicit in Steinmetz's design. To obtain the required speed of rotation Alexanderson proposed that the alternator be driven by a DeLaval steam turbine. And, to get the order-of-magnitude increase in frequency, the poles on the circumference of the rotating disks were to be cut very close together, only one-eighth of an inch apart. The cost was estimated at $1,200, exclusive of the turbine. Many of the central concepts of later Alexanderson alternators were implicit in this first design. The proposal was sent to Fessenden at the end of 1904, with a covering letter stating that GE would not guarantee the performance of such a machine but was willing to build it if Fessenden would pay all the costs. Fessenden approved the design, saying that he had suggested an inductor-type alternator earlier but Steinmetz had not cared for the idea. He made several minor suggestions for changes but dug his heels in on only one issue: he wanted a wooden armature instead of the laminated iron that Alexanderson had specified.

This difference was not a matter of mere personal whim. On most issues, when he disagreed with Fessenden's suggestions, Alexanderson was able to prevail by showing that they were either impractical or incorrect from a theoretical point of view. But on the question of wooden as against iron armatures at high frequencies there were no experimental data to cite, and precious little theory. Fessenden, used to working at radio frequencies, distrusted iron because he anticipated large losses from hysteresis and eddy currents. Alexanderson, with his background in power engineering, took the use of iron for granted. At the frequencies he was accustomed to, in the design of motors, dynamos, and transformers, he knew it worked, and Fessenden's objections, based largely on theory, did not impress him. In this case, however, Fessenden was paying the bills and he was entitled, within limits, to set the specifications. Alexanderson, without conceding the principle, revised his design so as to place the stationary armature windings on a ring made of wood, instead of laminated iron, and a machine built along these lines was completed in January 1906. Its nominal frequency was 100 kHz and, compared to the giant alternators to come later, it was a small machine, with a rotor diameter of twelve inches—not as small, however, as some of the designs Fessenden had been toying with, which had rotors only three inches in diameter.

Although there were to be many later changes in design, this machine was the prototype for the alternators to follow, and much was learned

actions 23 (May 1904), 417-28. For details of Alexanderson's design, see U.S. Patent No. 905,621 (issued 1 December 1908).

during its development and testing. There were, for example, serious problems of mechanical resonance and vibration as the rotor passed through certain critical speeds; these were solved by mounting the disk inductors on a thin, tapered shaft and adding auxiliary bearings near the disks—a solution that had the additional virtue of helping to correct the air gap between rotors and armature as the shaft heated up. Alexanderson also explored in some depth the question of disk design, emerging with the conviction that a tapered, prestressed nickel steel disk would give optimal results. And he tackled a problem that had perhaps been passed over too lightly in earlier planning: how the alternator would be driven. Fessenden now opposed the use of steam turbines, since they would call for a steam generating plant at each transmitter site. This meant, initially, driving the alternator by an electric motor through a system of belts and countershafts. And that in turn meant problems with belt slippage and heating. Alexanderson proposed to use in future a DeLaval turbine reduction gear in reverse—that is, not to decrease shaft speed but to increase it. Fessenden opposed this idea; he said he had seen others try to run a DeLaval gear in reverse and it would not work. In the meantime, however, using belt drive, Alexanderson was unable to raise the frequency higher than 50 kHz. This was only half the nominal design frequency, and the output power was also less than expected, for reasons not immediately apparent. On the other hand, it was an operating radiofrequency alternator. It had not disintegrated as it was brought up to operating speed, and no one had been killed during the tests—both eventualities that some of Alexanderson's assistants had anticipated.[77]

The machine was shipped to Fessenden's experimental radio station at Brant Rock, Massachusetts, in late August 1906 and, after minor delays due to a bent shaft and lack of a suitable drive motor, was connected to the antenna system and put on the air. Fessenden was enthusiastic about the results. True, even with new hard cotton belting, the frequency could be raised no higher than 76 kHz, and the output power at that frequency was less than 50 watts instead of the 250 that had been hoped for. But these were problems of development and refinement. As far as Fessenden was concerned the major victory had been won. He and the engineers of General Electric had done what Fleming and the other advocates of the spark system had called impossible: the problem of generating con-

[77] Brittain, *Alexanderson*, chap. 2, p. 19. Alexanderson, mindful of the risks involved in testing a high-speed machine of novel design, had required that the alternator be placed in a pit surrounded by sandbags. Nevertheless, the air gap between disks and armature had to be adjusted while the machine was running, and there was some trepidation lest the rotors be thrown out.

tinuous radio waves mechanically, at significant power levels, had been solved. There was reason for euphoria.

* * *

At this point in Fessenden's business career a victory was badly needed. His contract with the Weather Bureau had been terminated in August 1902, amid a flurry of accusations and counteraccusations. Fessenden charged that Willis L. Moore, chief of the bureau, had demanded a half-share in his patents without having contributed anything to them. The bureau, on the other hand, seems to have claimed that, as an organization, it had rights under those patents that Fessenden wished to deny. Whatever the truth of the matter, Moore notified Fessenden at the end of July 1902 that he had recommended his dismissal from government service for "disobedience to orders, insubordination and publishing contrary to my directions, wilful extravagance and untruthful statements [regarding] achievements of wireless system owned jointly by the weather bureau and yourself."[78] Fessenden resigned with effect from 1 September without waiting to be formally discharged. He does not seem to have been unduly distressed by the experience. "I find that it is fairly easy to get money, in fact very easy," he wrote to Elihu Thomson." . . . I am trying to get the cable companies interested. . . . It would be advantageous to have the General Electric in, on account of the manufacture of the apparatus."[79]

These and other plans less ambitious came to nothing, but in November 1902, with the help of his patent attorney, Fessenden secured the support of two well-to-do Pittsburgh businessmen, T. H. Given and Hay Walker, Jr. Neither of these individuals knew anything about wireless—they were chairman and president respectively of the Farmers Deposit Bank, while Walker also ran a brick company—but they were willing to underwrite Fessenden's experiments in return for an option to acquire majority ownership of the wireless patents he already possessed or might later secure. A corporation was formed—the National Electric Signaling Company—and terms of agreement were drawn up that, at the time, were acceptable to all parties, though they were to be the source of acute dissension in later years. There was no offering of stock to the general public—a striking contrast to most wireless promotions at that time.

The terms of agreement under which NESCO was formed strongly

[78] Fessenden Papers 1040-2, telegram, Willis L. Moore to Fessenden, 31 July 1902.

[79] Fessenden Papers 1040-2, Fessenden to Elihu Thomson, 11 May 1902.

suggest that Given and Walker believed they were investing in a radio system that was essentially complete. The funds they advanced were intended to construct, equip, and operate a chain of wireless stations by means of which the commercial use of Fessenden's patents could be demonstrated. Whether they were prepared, psychologically or financially, for the heavy developmental expenditures of the next ten years is very doubtful. From the beginning their conception seems to have been that they had a system to sell. The business problem was to find a suitable buyer and negotiate suitable terms. Revenues from operating the system or from the sale of equipment were of minor importance. This attitude lay behind their consistent reluctance to sell particular components of their system, even to the Navy. They wanted it sold entire, as a perfected, proven system.

Perfecting and proving the system, however, was more of a task than they anticipated. It does not appear that they ever tried to curtail Fessenden's expenditures, even when their investment had grown from the original $30,000 to some $2 million. On the other hand, neither did they give him the commercial guidance that he badly needed. This was evident in Fessenden's chronic difficulties with such potentially important customers as the Navy and the United Fruit Company. Whatever Fessenden's other abilities, he was no salesman. The record of his dealings with the Navy makes dismal reading and suggests strongly that he should not have been left responsible both for the technical development of his system and for its marketing.[80]

An uncertain sense of the market is also evident in the way Fessenden's efforts seemed to shift in orientation. The original chain of stations (at Washington, D.C., Collingswood, N.J. and Jersey City) was intended primarily as a demonstration project for relatively short-range wireless telegraphy. Whatever its technical success, there was no possibility that it would ever earn an operating profit, competing as it did with a well-established wired telegraph and telephone system on the eastern seaboard. But neither did it produce a buyer for the whole system, as Given and Walker had hoped. Fessenden then turned to something more dramatic: transatlantic wireless telegraphy. This was important to him personally because it gave him a chance to show that he could do anything Marconi could do, and better. It also provided an opportunity to test his new 100 kilowatt synchronous rotary spark transmitter—a compromise, in a sense,

[80] See, for example Douglas, "Pathways," pp. 133-84 and 216-80, and Howeth, *History*, pp. 85-106, 133-52, and 167-86. Since Douglas's manuscript is currently being rewritten for publication, I have felt at liberty to curtail my discussion of Fessenden's relations with the Navy.

with his long-term goal of a true continuous wave transmitter but nevertheless an impressive piece of equipment. And, one presumes, success in transatlantic operation was expected to generate the publicity that would attract a buyer for the system, perhaps one of the large cable companies. Again, the technical success was impressive: successful two-way communication between Brant Rock and its sister station at Machrihanish, on the Mull of Kintyre in the west of Scotland in the early weeks of 1906. But, if Western Union or Postal Telegraph took note of the accomplishment, they did nothing about it. Hopes for further success in that direction collapsed when the Machrihanish antenna blew down in a December gale. It was not rebuilt. Fessenden then pinned his hopes, at least for the immediate future, on wireless telephony. This was something Marconi had not attempted, and could not with the spark equipment to which he seemed committed. It was something that the new alternator just received from General Electric should be able to handle.

Constancy of purpose, in a commercial sense, was clearly not a characteristic of NESCO's operations in its first four years. This contrasts vividly with the history of the Marconi enterprises. Marconi was not interested in selling equipment to others. Far less was he interested in selling the Marconi system as a whole. The Marconi Wireless Telegraph Company existed to provide a communications service. But what business was NESCO in? Providing communications? Selling equipment? Or building up a potentially valuable piece of property that might eventually be sold for a capital gain? Given and Walker at least were clear that the third of these objectives dominated their expectations. But in the interim Fessenden was left to shift from one line of endeavor to another, without concentrating on any market long enough to develop it.

Fessenden's constancy was of a different nature: it was to a technological goal, not a commercial one. That long-run goal was continuous wave radio, and particularly radiotelephony. Seen from that point of view all his experiments with spark telegraphy were short-term expedients—even his big rotary spark, which he referred to almost deprecatingly as his "ultra commercial model." He was waiting for his true continuous wave generator; and by the fall of 1906 he had it.

There followed one of the more important technical successes in NESCO's history, and one of its most serious commercial disappointments. Technically the objective was to demonstrate the feasibility of radiotelephony; commercially it was to sell Fessenden's system to the Telephone Company. The first tests were conducted between the station at Brant Rock and a small schooner cruising offshore. These were highly successful. Next a new station was constructed at Plymouth, and during the winter of 1906 regular radiotelephone communications were maintained between the two stations, a distance of some eleven miles. The frequency used was between 50 and 60 kHz, this being limited by slippage in the belt drive, and the power output from the alternator never more than 500 watts and frequently much less. A problem was already beginning to show itself in the power-handling ability of the system. Fessenden's circuit placed the microphone directly in the antenna circuit, and there was a limit to how much current a carbon microphone could safely carry. Better methods for modulating the alternator's output would have to be found. But the test results satisfied Fessenden: even with the limited power available, speech could be transmitted clearly and distinctly over appreciable distances. In fact, in his judgment, the quality of reproduction was better than over wire lines.[81]

A public demonstration followed. Representatives of the news services and of the technical press were present, including, significantly, the editor of the *American Telephone Journal*, together with distinguished guests, among them Arthur Kennelly, Elihu Thomson, and, from the Boston laboratory of the Telephone Company, G. W. Pickard. To judge from Pickard's report to his superiors, the tests went well. Fessenden was able to demonstrate direct transmission of speech and music between Brant Rock and Plymouth and also an indirect circuit, using the regular telephone lines to carry the signal to the transmitter and to relay it automatically from the receiver. Variations in the audio volume of the received signal were large, apparently because of difficulties with the carbon-button microphone and "repeater." But Pickard considered the quality of reproduction very good. "Taken as a whole," he reported, "I should consider the speech transmission as distinctly commercial." With the apparatus he saw in use, he thought a range of a hundred miles could be attained, preserving commercial-quality speech; and with further development several hundred miles should be possible.[82]

Although Fessenden chose not to publicize it, an even more remarkable technical feat had been achieved several weeks before, while the Machrihanish antenna in Scotland was still standing. He had shut down the rotary spark transmitter for adjustment, but in the meantime used the Brant Rock antenna for radiotelephone tests with Plymouth. No changes

[81] Fessenden, "Recent Progress in Wireless Telephony," *Scientific American*, 19 January 1907, p. 68; Fessenden, "Long Distance Wireless Telephony," *The Electrician*, 4 October 1907, pp. 985-88.

[82] Selections from Pickard's report, dated 24 December 1906, are reprinted on pp. 205-14 of Ernst Ruhmer, *Wireless Telephony in Theory and Practice*, trans. James Erskine-Murray, with an Appendix by the translator (London and New York, 1908).

were made in the antenna circuit, so the frequency was the same as had been used for radiotelegraph tests with Machrihanish. As he later told the story, toward the end of September he received a registered letter from one of the operators at Machrihanish describing how, at a specified date and hour, he had heard one of the Brant Rock engineers, Stein by name, whose voice he was able to identify, giving instructions regarding the operation of a dynamo. The operator at Machrihanish conjectured that this had happened because Stein had been standing next to the rotary spark set. It turned out, however, that this was not the case: the rotary spark had been out of action at the time, but the alternator had been running. Stein had in fact been giving instructions by radiotelephone to the operator at Plymouth and, on checking his log, was able to confirm the date and time of transmission. What had actually happened, it appeared, was that the low-powered radiotelephone transmission between Brant Rock and Plymouth had been heard in Scotland, over three thousand miles away.[83]

Fessenden did not make this event public at the time. He had intended to make further tests, but the collapse of the Machrihanish antenna prevented them. He was well aware that only unusual propagation conditions could account for the anomaly. According to Pickard, the Brant Rock antenna radiated only 12 watts when the alternator was in use and, at 70 kHz, that was hardly enough to reach Scotland under normal conditions. And, of course, there were no neutral observers present. Fessenden had several times expressed his profound skepticism about Marconi's claim to have received signals from Britain in Newfoundland in 1901 and was not about to expose himself to the same kind of criticism. On the other hand, if he reported the events correctly, the circumstantial details were impressive: Stein at Brant Rock had not been told the date and time given by the Machrihanish operator when he checked his log. The evidence for Fessenden's achievement was at least as strong as that for Marconi's.

A second demonstration of technical capability was widely publicized. Indeed, it could hardly be otherwise. This was the now famous broadcast of speech and music from Brant Rock on Christmas Day and New Year's Eve, 1906. There may be room for argument as to whether this was the first true "broadcast" but the success of the experiment is beyond dispute. The intended audience, notified of the event by radiotelegraphy three days before, were shipboard operators off the Atlantic coast, particularly on Navy vessels and ships of the United Fruit Company—hardly a group

[83] Fessenden Papers 1140-94, Fessenden to Editor of *Scientific American*, 7 September 1918, p. 189; compare H. M. Fessenden, *Fessenden*, pp. 154-55.

easily deceived. Most of them by this time were using liquid barretter detectors, the patent on which had been widely infringed, so the ability to receive amplitude modulated continuous wave signals was not in question. The transmissions brought an enthusiastic response from the marine operators: few if any of them can have heard a human voice through their headphones before.

Clearly the alternator was a success. Its frequency was lower than desired. Its power output was much less than planned, and there were far too many losses in the antenna circuit. But it worked, wooden armature and all. And for a brief moment it looked as if the Telephone Company would buy the system. Pickard's report had had an effect and there were good reasons for AT&T to make the investment.

The Telephone Company's attitude to radio up to this time is best described as one of wary skepticism. There was no research program in that field; indeed, from the company's incorporation in 1880 until 1906 its commitment to original research had been minimal. Hammond V. Hayes, head of the company's Mechanical Department, after initiating a number of original investigations in the late 1880s, seems to have decided in 1892 that his department would serve the company better if it left original research to others. That, at least, was the policy he told his superiors he intended to follow—the type of assistants he hired and the type of work he had them do are not entirely consistent with his stated policy, and he may have been merely telling top management what they wanted to hear. Referring to research on new methods of transmitting speech, he reported to President Hudson that he had "determined for the future to abandon this portion of the work of the department, devoting all our attention to practical development of instruments and apparatus. I think the theoretical work can be accomplished quite as well and more economically by collaboration with the students of the [Massachusetts] Institute of Technology and probably of Harvard College."[84] And in 1906 he voiced the same philosophy in a report to President Fish, assuring him that "Every effort in the department is being exerted toward perfecting the engineering methods; no one is employed who, as an in-

[84] Federal Communications Commission, Proposed Report, Telephone Investigation Pursuant to Public Resolution No. 8, 74th Congress (Washington, D.C., 1938; referred to hereafter as Walker Report), p. 206. Compare Leonard S. Reich, "Industrial Research and the Pursuit of Corporate Security: The Early Years of Bell Labs," *Business History Review* 54 (Winter 1980), 507-508. Reich's version of Hammond's letter differs somewhat from that found in the Walker Report as published in 1938. See also Lillian Hoddeson, "The Emergence of Basic Research in the Bell Telephone System, 1875-1915," *Technology and Culture* 22 (July 1981), 512-44.

ventor, is capable of originating new apparatus of novel design."[85] Given this inhospitable climate, it was hardly to be expected that Telephone Company engineers would be found working on the frontiers of wireless technology. Some important work was, as a matter of fact, done by individuals associated with the company's Boston headquarters, though in some cases working on their own time: John S. Stone on tuned circuits, G. W. Pickard on wireless telephony and crystal detectors, G. A. Campbell on wave filters. But it fell far short of a sustained research effort.

By 1906, however, there were new factors to consider. In the first place, largely through Fessenden's work, telephony without wires had been shown to be feasible. Pickard's report removed any doubts on that score: he wrote not of laboratory experiments, but of commercial-quality speech transmission over several hundred miles. This posed a threat to a corporation with a multimillion dollar investment in a wired system. But secondly, the geographic extension of that system had by 1906 reached a point at which the conventional technology of wired telephone communications seemed to be reaching a definite limit. The problem was one of attenuation and distortion as distances increased. The invention of the loading coil in 1899-1900 had made it possible to telephone without serious distortion up to 1,700 miles—say from New York to Omaha or Kansas City.[86] But for distances greater than that—transcontinental telephony, for example—something more was needed. That could be some kind of line amplifier—what telephone engineers called a repeater. Or it might, just possibly, be an alternative technology that dispensed with wires. Either way, whether radio was a threat or a promise or something of both, in 1906 it was beginning to look as if telephone technology would have to reach out beyond the kind of engineering that had served it well in the past. That meant it would have to buy technology from persons outside the organization.

This was nothing new. The Telephone Company had a longstanding policy of paying for patent rights it needed; and its strong patent position usually meant that outside inventors had no one else to sell to. But a move into radio was a different proposition entirely. If it became a major corporate commitment, and not merely a kind of protective maneuver,

[85] Walker Report, p. 212.

[86] John V. Langdale, "The Growth of Long-distance Telephony in the Bell System: 1875-1907," *Journal of Historical Geography* 4 (1978), 145-59. On the invention of the loading coil, see James E. Brittain, "The Introduction of the Loading Coil: George A. Campbell and Michael I. Pupin," *Technology and Culture* 11 (January 1970), 36-57, and, in *Technology and Culture* 11 (October 1970), 596-603, the comments by Lloyd Espenschied, Joseph G. Jackson, and John G. Brainerd.

it could change the very conception of how a telephone system was built and operated.

Fessenden's liaison with General Electric had opened up channels of communication between two engineering subcultures: power engineering and wireless. His emerging liaison with the Telephone Company held out the prospect of a second linkage: between the technologies of wired and wireless systems. These three technological traditions had developed in relative isolation from each other. Their practitioners belonged to different professional societies; they used different vocabularies; and they had recognizably different styles of design and construction. Normally there was little interaction between them. The distinctive feature of the alternator as a technological artifact was that it tended to break down the isolation and increase the interaction. It was clearly, in its design and construction, a product of the power engineering tradition; yet it existed only because of Fessenden's vision of what radio could be; now in 1906 it seemed to hold the answer to the future of telephone communication. And in an organizational sense also, the alternator seemed to be bringing into existence a new alignment of corporations. General Electric had the manufacturing and design capability; the Telephone Company had its wealth of experience in the construction and operation of communication systems; NESCO had its know-how, backed up by patents, in continuous wave radio. It did not take much imagination to see in this alignment the makings of a new community of interest, perhaps even a new radio corporation—one that would be distinctively American in its orientation and affiliations. The element of national identity may not have been an important consideration in 1906; it was to become critical in the near future.

What was required was some person or some group who could bring the parties together and make them see the commonality of their interests. And for a short time it looked as if something of that nature might indeed come to pass. A follow-up report by E. H. Colpitts, one of Western Electric's engineers, was favorable and positive. And in April 1907 Hammond V. Hayes, by then chief engineer of AT&T, recommended to President Fish that an offer be made to Fessenden. "I feel that there is such a reasonable probability of wireless telegraphy and telephony being of commercial value to our company that I would advise taking steps to associate ourselves with Mr. Fessenden if some satisfactory arrangement can be made."[87] If Fish and Hayes had kept their positions, it is as certain

[87] Memorandum of 2 April 1907, as quoted in W. Rupert Maclaurin, *Invention and Innovation in the Radio Industry* (New York: Macmillan, 1949; copyright 1949 by Macmillan Publishing Co., renewed 1977 by Elfriede C. Maclaurin; quoted by permission). p. 65.

as such things can be that an offer would have been made; and what we know of Fessenden's financial backers suggests that, after appropriate bargaining, it would have been accepted. But this must remain hypothetical: the financial panic of 1907 caught the Telephone Company badly overextended. In the ensuing financial reorganization control shifted from Lee, Higginson & Company of Boston to George F. Baker and J. P. Morgan in New York, and with this shift in investment bankers came drastic changes in top management and policy. President Fish was replaced by Theodore N. Vail, and Hayes by John J. Carty, formerly chief engineer of the New York Telephone Company. Accompanying these changes was a sharp cutback in research expenses and personnel, the elimination of the Mechanical Department in Boston, and a centralization of control over technical development in the New York office under Carty. One of the first casualties was the proposed contract with Fessenden.[88]

Analysts of the history of AT&T, even those inclined to be critical of the company, usually portray the advent of the Vail-Carty regime as a progressive development. They do so largely because of Hammond Hayes's deliberate and perhaps prudent downplaying of innovative research before 1907, and because they are impressed by the aggressive program of technical development initiated by Carty after 1910. What gets lost to view is the damage done by the drastic retrenchment of 1907-1909. The closing down of the Boston Mechanical Department, where there had been active interest in radio, meant abandonment not only of negotiations with Fessenden but also of a proposed investigation of Lee de Forest's newly invented triode vacuum tube as a potential telephone amplifier.[89] What slipped through the Telephone Company's fingers, in short, was a unique opportunity to come to grips with electronic technology. The time lost had to be made up in a hurry after 1912. But the damage extended beyond the boundaries of that one corporation. Failure to reach agreement with the Telephone Company was a grievous blow to NESCO; indeed, from that setback the company never fully recovered. And it also eliminated for the next decade the possibility of organizing a consortium

[88] For analyses of the 1907 reorganization and its effects, see N. R. Danielian, *A.T.&T.: The Story of Industrial Conquest* (New York, 1939), pp. 50-77, 98-107; John Brooks, *Telephone: The First Hundred Years* (New York, 1975), pp. 122-32; Walker Report, pp. 96-103 and 207-16; Maclaurin, *Invention*, pp. 65-66; Lloyd Espenschied, "Reminiscences" (Columbia University Oral History Collection), pp. 16-18.

[89] Lloyd Espenschied to C. F. Elwell, 23 September 1952 (Espenschied Papers), quoted in Brittain, *Alexanderson*, chap. 3, pp. 6 and 22. Compare Espenschied, "Reminiscences," p. 17.

of American manufacturing and communications interests capable of challenging the supremacy of the British-based Marconi Company.

* * *

Hammond V. Hayes, without intending to do so, left one important legacy to NESCO. Before leaving office he asked General Electric whether Alexanderson would design an alternator for the Telephone Company. This was not for radio use but rather with the idea that it might meet the need for a line amplifier, using telephone currents to modulate the magnetic field. General Electric agreed and Alexanderson set to work, drawing on the experience he had gained from Fessenden's machine but free now to follow his own ideas. The result was an alternator different in important respects from its predecessor: there was now only a single disk rotor, instead of two, and the armature was of laminated iron instead of wood. Each of these changes meant higher power output. One reason for the falling-off in power in Fessenden's machine as the frequency increased had turned out to be a small but significant deformation of the disk rims as the rotors speeded up. Adoption of a single disk, rotating between a split armature, solved that problem. Choice of a laminated iron armature, of course, reflected Alexanderson's firm conviction that Fessenden was wrong in predicting serious losses from hysteresis and eddy currents. There was no way to prove this deductively or from experimental data; the new machine itself would provide the demonstration.

If it had not been for this contract with the Telephone Company, Alexanderson after 1906 might well have moved out of radio engineering completely. Even while working on Fessenden's machine much of his time had been spent on problems relating to railroad electrification. In April 1906, with work on the NESCO alternator completed, he was assigned full-time to the Railway Engineering Department.[90] The corporate assumption was that the experimental phase of work on the radiofrequency alternator was now over; future contracts with NESCO would be handled through regular channels. This is in fact what happened. Further orders for alternators received from Fessenden in late 1906 and early 1907 were forwarded to the Alternating Current Engineering Department, where they became the responsibility of another GE engineer named Conway Robinson. These were regarded as normal manufacturing orders, not assignments for research and development. For a time, therefore, General Electric had two high-speed alternator projects

[90] Brittain, *Alexanderson*, chap. 2, p. 25.

under way: an experimental project under Alexanderson for the Telephone Company, and a manufacturing project under Robinson for NESCO. And the engineers working on the one project knew nothing of the existence of the other.

This situation reflected little credit on GE's internal communications system and probably resulted in some duplication of effort. It could have had serious long-run consequences. First, it threatened to divert Alexanderson's creative energies into another field; and if that had happened what later came to be known as the Alexanderson radio system—the alternator, the magnetic modulator, the multiple-tuned antenna, the barrage receiver—would not have come into existence in that integrated form. It was around that system, developed and controlled by General Electric, that the Radio Corporation of America was later to be constructed. And second, when GE turned Fessenden's later orders over to the Alternating Current Engineering Department, it essentially froze the design as Alexanderson had left it in 1906. That machine, however, had serious limitations, particularly with respect to power output. Future designs for radio alternators, up to the giant 200 kilowatt machines that GE began building after 1917, traced their ancestry to the single-disk model with iron armature that Alexanderson designed for the Telephone Company. This was a model that could be "scaled up" in frequency and power as the 1906 model could not.

What saved the day was the fact that Alexanderson and Fessenden had developed a high regard for each other and kept up a personal correspondence. A letter to Fessenden in June 1907 described the new alternator Alexanderson was building. It caused some confusion, since Alexanderson did not think to mention that he was building it for the Telephone Company, not for Fessenden. By the time that misunderstanding was cleared up, the Telephone Company's contract had been abruptly terminated as part of Carty's retrenchment drive. Fessenden came to Schenectady to see Alexanderson's new alternator under test and, much impressed, promptly cancelled his contract for the machine GE was building for him—much to the dismay of the A.C. Engineering Department—and took over responsibility for financing Alexanderson's project. He still insisted, however, that both iron and wood armatures be tested, and development went ahead on that basis. What emerged by the closing months of 1908 was a 100 kHz alternator with a rated power output of 2 kilowatts. It had an iron armature, Alexanderson having finally ruled on the basis of comparative tests that wood was decidedly inferior, and a single-disk rotor, twelve inches in diameter, with slots cut in the circumference. Problems with air friction at very high rotor speeds were minimized by filling the slots with phosphor-bronze wire. Fessenden asked that the machine be sent to Brant Rock for on-the-air testing immediately, and authorized the construction of a second. Shortly thereafter he wrote asking for the development of a 35 kilowatt model.

This 2 kilowatt machine completed in 1909 was a landmark in the development of radio alternators. In service it performed very well: a "thoroughly practical piece of apparatus," Fessenden called it. Technically it was an achievement of which Alexanderson could legitimately be proud; his paper describing its characteristics presented at the annual meeting of the American Institute of Electrical Engineers in 1909 established his reputation as one of the leading radio designers of the day.[91] And commercially it opened up prospects that had not existed before. General Electric now had a design that could be standardized, a machine whose performance characteristics it could guarantee. For this particular power level the market would be limited: a 2 kilowatt 100 kilohertz alternator might be useful for short-range work, for tests and demonstrations and for scientific experiments, but it was not the kind of machine around which you could build a transoceanic radio system. But the great virtue of the 1909 model was that it could serve as the archetype for larger and more powerful alternators in the future. Alexanderson and

Pl. 1: Alexanderson's 2 kilowatt alternator.
Source: General Electric Company

[91] E.F.W. Alexanderson, "Alternator for One Hundred Thousand Cycles," AIEE *Transactions* 28 (June 1909), 399-412.

his co-workers—and no one was more ready than he to acknowledge the help he got from machinists and fitters on the shop floor at Schenectady—could now see where they were going. There would be no surprises: higher power was essentially a matter of increase in scale.

But, if higher powered alternators were to be built, who would buy them? None of the radio operating companies at the time showed interest: they were either satisfied with spark or lacked the resources to invest in new technology. Navy contracts offered a possibility: but when the Navy in 1909 invited bids for its new high-power station at Arlington, Virginia, it set performance standards that no alternator GE then had on the drawing boards could hope to meet. NESCO won that contract, but with a 100 kilowatt synchronous rotary spark set, not an alternator. As long as NESCO existed, and as long as Given and Walker supplied Fessenden with funds, GE had an outlet it could count on. But if that outlet disappeared, what would become of the alternator project? Best perhaps to abandon it. There was no obvious market, and in terms of GE's total business it was a drop in the bucket.

There was, of course, no thought at that time that General Electric might enter the radio communications business itself, or form a subsidiary to do so.[92] That was too radical a step to consider—as readily might GE build its own railroad or streetcar system. GE was in the electrical manufacturing business, not the communications business, and one did not compete against one's own customers. At least, not as long as one had customers.

Failure to come to terms with the Telephone Company, however, had left NESCO in a difficult position, not made easier by the growing dissension between Fessenden and his Pittsburgh financiers. Relations with the Navy improved substantially when NESCO secured the services of Col. John Firth, vice-president and sales manager of the Wireless Specialty Apparatus Company, which supplied United Fruit with much of its equipment. Firth was a master salesman, on excellent terms with the leading members of the Navy's Bureau of Equipment, and it was largely if not wholly through his efforts that NESCO won the important 1909 contract for the Arlington station. Indeed, there were many who said that the specifications had been drawn up so that only NESCO could win; such consideration would not have been shown if Fessenden had handled the negotiations. But Firth spent at least as much energy trying to mediate between Given and Walker, increasingly concerned about the security

[92] In 1904-1905 GE had considered building an interplant wireless telegraphy system; by 1916 there were regular radiotelephone transmissions from Schenectady to New York and Pittsfield.

and marketability of their investment, and Fessenden, increasingly suspicious that he was about to be forced out of the company entirely.

Given and Walker by this time held a 70 percent interest in NESCO; they had also advanced substantial sums as working capital, for which they held interest-bearing demand notes. There was no doubt who controlled the company. Fessenden, on the other hand, although he had been paid a salary and had been given a free hand with research and development expenditures, had never received any of the $330,000 which, under the original agreement, was to have been paid him for his patents. That sum was to have come from the company's first profits, and of course profits had never been realized. The situation was a tolerable one only granted the existence of considerable mutual trust and optimistic expectations about the future. By 1908 both trust and optimism had dissipated. Firth patched up a compromise, putting the company's debts to Fessenden on the same interest-bearing basis as its debts to Given and Walker, but the truce was a fragile one.

Fessenden, who did not easily admit defeat, continued his efforts to sell his system to the Telephone Company, but he made no progress. The attitude of the new regime at AT&T's headquarters was well summed up in a report of the company's patent counsel who, while admitting that the idea of replacing a wired telephone system with a wireless one was very attractive, expressed his conviction that it could not and would not reach practical realization "within the term of years yet remaining to Fessenden's fundamental patents."[93] In other words, by the time AT&T needed the new technology, it would be able to get it without paying for it. Frustrated in that endeavor, Fessenden turned once again to transatlantic telegraphy. This time, however, his plans were more ambitious, in an organizational if not a technical sense. He wanted to form a company in Canada, with a board of directors made up of distinguished Canadians. He wanted the British government to issue a license to that company, recognizing it as authorized to conduct radio communications between the mother country and Canada in competition with Marconi. And, needless to say, he wanted to use his radio patents to build and equip the necessary stations.

If Fessenden intended that Given and Walker should play any role in this new enterprise, he did not make that fact clear to them. Nor did he explain on what terms he intended that the Canadian company should acquire the use of patents that Given and Walker had assumed were the exclusive property of NESCO—and, indeed, that company's only sub-

[93] Thomas D. Lockwood to T. N. Vail, 8 July 1907, as quoted in Maclaurin, *Invention*, p. 66.

stantial assets. Their reaction was to dismiss him from the company.[94] He was offered a position as technical adviser, on condition that he divorce himself completely from the company's business affairs but, characteristically, he refused even to consider the possibility. After 8 January 1911 Fessenden and NESCO went their separate ways. He brought suit for breach of contract and, in the lower courts, was awarded damages of $400,000. NESCO went into voluntary receivership to conserve its assets pending appeal, and continued its technical development work on a reduced scale while in receivership. Except as owner of the Fessenden and certain allied patents, however, which it cross-licensed to the Marconi Company in 1914, it played no major role in American radio thereafter.[95]

To Fessenden there was no mystery about why Given and Walker had kicked him out of NESCO. His Canadian venture had nothing to do with it. The plain fact was that his radio system was now complete, its marketability had been proven—as for example by sales to the Navy and to United Fruit—and the Pittsburgh capitalists had got rid of him just as soon as they thought they could dispense with his services.[96] A more generous interpretation, and one that does not imply endorsement of the strong-arm tactics used in taking control of the Brant Rock station, is implicit in the verdict of Judge Julius M. Mayer of the Eighth Circuit Court of Appeals who, in a patent infringement case, congratulated the Pittsburgh businessmen for the "courageous investment of substantial sums of money" in the days when the radio art was in its infancy, for the carte blanche they had given Fessenden in conducting his experiments, and for their scrupulous avoidance of any kind of stock speculation, in contrast to other wireless companies of the day.[97] And there may have been something more than self-serving sentimentality in Given's comment

[94] The manner in which this was done seems to have been crude in the extreme, involving the threat or reality of physical violence to Mrs. Fessenden after her husband had been summoned to Pittsburgh for a conference. For an account redolent of nineteenth-century melodramas (of the second or third class), see H. M. Fessenden, *Fessenden*, pp. 182-87.

[95] NESCO's patents were transferred to the International Signaling Company in 1917 (see below, pp. 456-57) and finally were acquired by RCA. As a company, it continued to exist throughout World War I, building equipment and operating stations, with headquarters at Bush Terminal, Brooklyn. S. M. Kintner ran its business affairs and J.V.L. Hogan was responsible for technical matters.

[96] Compare Fessenden Papers 1140-4, Fessenden to J. J. Carty, 11 January 1911.

[97] Fessenden Papers 1140-4, verdict of Judge Julius M. Mayer, Southern District of New York, in *Kintner and Barrett v. Atlantic Communication Company*, in equity, 2 April 1917.

to Kintner that, if the radio business failed to turn out as expected, "I, at least, will not have on my conscience the thought that I've wasted the savings of poor scrub-women, widows with dependent children, or others who fall such easy prey to the high-powered stock salesmen."[98] If Given and Walker are to be faulted, it should perhaps be for their failure, since NESCO was first organized, to define Fessenden's authority and responsibility more precisely and to direct the company's business affairs more actively. They, after all, were the businessmen, not Fessenden.

Fessenden had little to do with radio thereafter, except as a litigant.[99] We do not need to follow him through his later inventions in other fields. As far as radio technology was concerned he had by 1911 accomplished much of what he had set out to do when he left Pittsburgh in 1900. Intellectually and psychologically, he had broken through the mind-set that identified radio with spark. Earlier than anyone else in the United States he had seen and identified the inherent limitations of spark and the cul-de-sac that spark technology would eventually have to face. More than that, he had demonstrated an alternative technology through which radio could escape those limitations. He had brought into existence a system of radio communications based on the continuous wave and by so doing he had set radio technology in motion along a new and radically different vector. By 1911 it had not travelled very far along that vector, and for that Fessenden's failure to show how the new technology could be integrated successfully into the marketplace was largely responsible. With larger resources, different management, and perhaps just better luck, the outcome might have been very different.

But the development of the radiofrequency alternator did not stop when Fessenden left NESCO. He had done more since 1900 than demonstrate in an abstract way that continuous wave radio was possible, that it could do everything spark could do and one thing in particular that spark could not: transmit the human voice. He had also, through his persistence and his personal relationship with Alexanderson, involved the largest manufacturer of electrical equipment in the country in his project. That involvement did not end when Fessenden left NESCO. The General Electric Company now became the carrier of the innovation.

[98] Kintner, "Pittsburgh's Contributions," p. 1853.

[99] Fessenden derived little if any immediate financial gain from NESCO. In later life, however, his radio patents brought him considerable wealth, notably in a $500,000 out-of-court settlement of a civil antitrust suit he had brought against RCA in 1926. Newspaper reports of that settlement set the figure at $2.5 million, but that appears inflated. See Maclaurin, *Invention*, p. 63, and compare *Boston Advertiser*, 23 September 1928.

Alexanderson and others at GE knew that they had in the alternator a machine that, once adopted, could make spark obsolete and revolutionize the art of long-distance radio. The problems ahead were no longer problems of design or fabrication; those had been solved during the Fessenden era. What remained to be faced were problems of marketing, of managing the economic deployment of the machine.

THREE

Elwell, Fuller, and the Arc

ON 13 February 1913 the United States Navy placed in commission its first high-powered radio station. Located at Arlington, Virginia, this installation was intended to be the first and central element in a network of powerful stations by which the Navy would be able to maintain communications with its remote bases and with units of the fleet wherever they might be. Congress had provided funds for the Arlington station in 1911. In the following year it appropriated $1 million for the next six stations in the system: in the Canal Zone, on the California coast, in the Hawaiian Islands, on Guam, in American Samoa, and in the Philippines. The amount was later increased to $1.5 million.[1]

Much depended on the success of the Navy's venture. It was, in a sense, the American analogue to the much-discussed British Imperial Chain, planned to link the colonies and dominions with the mother country.[2] Civil and military groups in Germany and France, at about the same time, were planning powerful radio stations to connect their home countries with North America and with their colonies in Africa and the Far East. And in Holland and the Dutch East Indies there were a few individuals already thinking in similar terms. All these plans, put forward and partly implemented between 1910 and the outbreak of the First World War, had a strong strategic cast to them. They were intended to extend the range of control of metropolitan governments; to minimize the strategic risks inherent in dependence on submarine cables that could

[1] Howeth, *History of Communications-Electronics in the United States Navy* (Washington, D.C., 1963), pp. 182-85.

[2] For short accounts of the British Imperial Chain, see W. J. Baker, *A History of the Marconi Company* (New York, 1971), pp. 137-38, 143-48, and 204-15; W. P. Jolly, *Marconi* (New York, 1972), pp. 190-220 and 247-62; and Frances Donaldson, *The Marconi Scandal* (London, 1962).

be cut in the event of war; and to provide communications where no cables existed. The American scheme was distinctive primarily for its emphasis on the deployment of naval power and for its orientation, originally, toward the Panama Canal and the Pacific.

In each instance the requirements for reliable communication over great distances, no matter what the season of the year or time of day, pushed radio technology to its limits. In the case of the Navy's system, for example, the hard truth was that, when the Arlington station was first planned, there was no radio transmitter in existence capable of providing the performance called for. As published in 1909, the contract specifications required that the station "be capable of transmitting messages at all times and at all seasons to a radius of 3,000 miles in any navigable direction from Washington, D.C."[3] This was demanding more than was possible, given the state of the art at that time.

The National Electric Signaling Company (NESCO) won the contract for the Arlington station. The transmitter to be installed was a Fessenden-designed 100 kilowatt synchronous rotary spark, which NESCO guaranteed would meet the Navy's specifications. Unfortunately, it failed to do so. The Navy accepted the machine knowing that, in 1909, it was the best that could be had. It would suffice for communication between shore stations, which had large directional antenna systems, but for long-distance communications with ships at sea, where antenna dimensions were limited, it was not adequate. Extended tests from NESCO's station at Brant Rock, while the antenna towers at Arlington were being erected, put the matter beyond dispute.

When, in 1913, the Arlington station was placed in service, it was the big rotary spark that dominated the transmitter room.[4] It was an impressive-looking machine, even when stationary; and when in operation, with its forty-eight radial copper electrodes spinning at 1,250 revolutions per minute, it must have been a remarkable sight, for its pyrotechnics alone. Certainly it was very noisy; all high-powered rotary spark machines were—the Marconi disk discharger being one of the worst offenders—and pains had to be taken to protect the eardrums of the operating personnel and to insulate the room where the receiving apparatus

[3] As quoted in Howeth, *History*, p. 139. The specifications also required that messages must not be interrupted by atmospheric disturbances or intentional or unintentional interference by neighboring stations, and that the station be capable of transmitting and receiving messages with entire secrecy.

[4] See the illustrations and descriptions in William H. G. Bullard, "Arlington Radio Station and Its Activities in the General Scheme of Naval Radio Communication," IRE *Proceedings* 4 (October 1916), 421-46. For the circuit diagram see Howeth, *History*, p. 140, fig. 11-1.

Pl. 2: Fessenden's rotary spark transmitter.
Source: Alan Douglas

was installed. But on the air its 500-cycle generator gave it a high-pitched, almost musical note, easily read through static, and up and down the eastern seaboard the distinctive note of the big rotary spark, first from Brant Rock and later from Arlington, became very familiar to naval and commercial operators.

Like Marconi's disk discharger and Telefunken's quenched spark, Fessenden's transmitter installed at Arlington carried high-powered spark technology to its practical limits. There is irony in the fact that Fessenden, committed as he was to the concept of the continuous wave, won his greatest commercial success with a machine that represented the ultimate development of the spark technology from which he was trying to escape. The rotary spark was not a true continuous wave generator, but it came as close to being one as any spark transmitter could. The fact that it failed to meet the Navy's requirements served to demonstrate that the inherent limits of the technique, which Fessenden had been among the first to identify, had now been reached. Spark had gone as far as it could go; future development would have to be along other lines.

What should these be? Fessenden had pinned his hopes to the alternator and, given another five years or so of development time, General Electric might well have produced an alternator able to perform as the Navy wanted its high-powered transmitters to perform. But in 1909 the only radiofrequency alternator that GE or NESCO had to offer was the 2 kilowatt model. That was still true in 1912. Even for a continuous wave machine, 2 kilowatts were only a fraction of the power necessary to meet the Navy's requirements.

NESCO's rotary spark was not, however, the only occupant of the transmitter room at Arlington. There was also a 5 kilowatt spark set from the Wireless Improvement Company, intended for short-range marine radio. More important, over to one side of the room and secured originally to a wooden framework that was clearly temporary, there was a third transmitter of completely different design. Apart from a small motor to rotate one of its electrodes, it had no moving parts at all. It was virtually noiseless in operation. And the signal it sent out over the Arlington antennas sounded quite different from the rotary spark's penetrating note. In fact, unless you were using one of Fessenden's heterodyne receivers or an equivalent device, you could hardly hear its signal at all—perhaps a kind of "shushing" sound as the transmitter was keyed, but no more than that. It cannot have seemed, to the uninformed observer, an impressive device. It was small in size. It had none of the sound and fury of the rotary spark in operation. And its power rating—only 30 kilowatts—hardly seemed to promise performance comparable to the 100 kilowatt spark machine. Yet this third transmitter at Arlington embodied the technology that was to make spark obsolete. It was a continuous wave transmitter—in this case, an oscillating arc.

This particular arc had been designed by Cyril F. Elwell, an Australian-born engineer working with the Federal Telegraph Company of California, and getting it installed at Arlington had not been easy. Elwell had come east in 1912, after successfully opening a long-distance radio circuit between San Francisco and Honolulu, in the hope of interesting the Navy in buying Federal equipment. Demonstration of a small 12 kilowatt arc transmitter had won him two supporters in the Radio Division of the Bureau of Steam Engineering: Lt. Comdrs. S. C. Hooper and A. J. Hepburn. But the chief of the bureau, Adm. H. I. Cone, was less easily convinced, and Dr. Louis Austin, head of the Naval Research Laboratory, objected strenuously even to the idea of comparative tests. Unlike Hooper and Hepburn, who were already converts in principle to continuous wave radio, Cone and Austin found it difficult to entertain the possibility that an arc transmitter, operating with much lower power, could outperform the rotary spark. Austin, in the course of his research on long-distance propagation, had worked closely with Fessenden and may well have felt that to admit this late entry, after NESCO had won the contract, was irresponsible. But for Admiral Cone it was a matter of common sense. In a confrontation that Elwell later delighted to recall, Cone inquired how many amperes of current the arc could feed into the antenna. Elwell replied, "About fifty." Well, said Cone, the Fessenden set was already supplying twice that. And when Elwell asserted that one ampere of continuous waves was worth two of damped waves, the admiral replied, "An ampere is an ampere and you cannot change it."[5] This, of course, was undeniable: what the admiral did not grasp was the way in which any spark transmitter, no matter how sophisticated, dissipated its power in unnecessary harmonic radiation. Elwell's arc transmitters were by no means innocent of harmonics either, but they came much closer to the theoretical ideal of single-frequency radiation than any spark set could.[6]

In the end Elwell's persistence paid off and he was given permission to install a 30 kilowatt transmitter, similar to those the Federal Company was using at San Francisco and Honolulu, in the Arlington station—with the admonition that he was "not to put any nail holes in the floors, walls or ceiling." This was, of course, the reason for the temporary wooden framework on which the arc and its generator were placed. That would have been a serious handicap to impose on the installation of an alternator or any other piece of high-speed rotating machinery, but it made little difference to Elwell. He had his equipment in place and ready to operate when the station was officially opened. The result was a remarkable demonstration of what continuous wave radio could accomplish, a turn-around in the fortunes of the Federal Company, and a policy decision of major importance on the design of the Navy's high-power chain.

The Navy had some prior experience with arc transmitters. Several low-power arc radiotelephone sets had been purchased from Lee de Forest in 1907 for shipboard use; their performance had been unsatisfactory, not entirely for technical reasons, and this may have had something to do with the skepticism Elwell encountered.[7] That skepticism, however, was soon dissipated. Even before the official tests began, the operators at Arlington were reporting successful two-way communication over remarkable distances. Their first attempt to call the Federal station in San Francisco was acknowledged immediately. After several messages had been exchanged, someone suggested calling Honolulu; but it turned out that the Honolulu operator was already listening to the conversation and

[5] Cyril F. Elwell, "Autobiography" (Stanford University Library), p. 76. This autobiography exists in several versions; unless otherwise indicated, future references will be to that contained in Box 1, folder 2, of the Cyril F. Elwell Papers (M49), Department of Special Collections, Stanford University Library.

[6] However, as will be explained later, early methods of keying arc transmitters involved the radiating of a compensating or "back" wave which carried no information, wasted power, and occupied spectrum space.

[7] For the de Forest radiotelephone sets, see Howeth, History, pp. 169-74.

promptly broke in with congratulations and a signal report.[8] For a station on the eastern seaboard to communicate with Honolulu, 4,500 nautical miles away, and with the latter station in daylight, was unheard of.

Convincing evidence was also obtained, under more carefully controlled conditions, when the U.S.S. *Salem* sailed from Philadelphia to Gibraltar in February 1913 with representatives of NESCO and Navy radio experts aboard. The purpose of the cruise was to conduct comparative tests of the two Arlington transmitters, and the results were impressive. The rotary spark performed well, being consistently heard at night during the voyage to Gibraltar and back. But the arc easily matched this performance, and on one occasion was heard during daylight hours while the *Salem* was at anchor off Gibraltar. Confirming evidence came when Louis Austin, still dubious, sent a trusted assistant to Key West and then to the Canal Zone to compare signal strengths under the difficult receiving conditions of the Caribbean area. Signals from both transmitters were of equal strength at Key West, although the arc was using only a quarter of the power; at Colon the arc's signals were readable while the rotary spark could not be heard at all.[9]

The impression these tests made on one well-informed participant can be gauged from the official report on the *Salem* cruise submitted by George H. Clark, the Navy's first civilian radio aide. Clark's particular concern was with receiving equipment, and in particular with the reception of continuous wave signals—a problem the Navy had not had to face up to this time. In addition to its regular receiving equipment (a crystal detector), the *Salem* carried two receivers for continuous wave signals: a Fessenden heterodyne detector, with a small arc serving as local oscillator; and one of the Federal Company's "tikkers," a mechanical device for interrupting the received signal at an audible rate so that it could be heard in headphones. (For details, see below, pp. 120-21.) Clark, using the regular receiver, found signals from the rotary spark much easier to copy. He admired its "clear musical note" and compared it with that of the arc, which gave a signal not at all clear but "more of a combined 'hiss' and 'whisper.'" To be sure, this made the signal from the arc very easy to identify amid interference from spark stations; but when signals had to be read through atmospheric interference the advantage lay entirely with the spark set.

But when the heterodyne receiver was used, Clark sang a different song. In fact his comments became almost lyrical. "The use of the heterodyne . . . transfers all of the advantage of the spark system to the arc.

[8] Elwell, "Autobiography," p. 77.
[9] Ibid., see also John Hogan in *Electrical World*, 21 June 1913, pp. 1361-66.

... The combination of the heterodyne receiver and the arc transmitter constitutes the most noteworthy advance in the development of practical radio-communication that has been made in the history of the art." And his final recommendation was, for an official report, remarkably lacking in ifs and buts: what Navy radio needed was greater distance and immunity from atmospheric disturbances; these could be gained "only by the combination of the heterodyne and the arc transmitter." And a chain of high-power intercommunicating stations embodying these features "would afford the first reliable radio service dependable, not by night and occasionally by day, but by night and day and every day."[10] It was apparent that continuous wave radio had won a convert.

And Clark was not the only one. From 1913 on there was no question in the mind of any responsible Navy official of reverting to spark equipment for high-power long-distance service. That continuous wave equipment of one kind or another would be installed in the stations of the high-powered chain was taken for granted. And there seems to have been a clear sense that an important watershed had been passed. Radio in the Navy would never be the same again. Lieutenant Commander Hepburn put the issue forcibly in his report to Admiral Cone, chief of the Bureau of Engineering, in April 1913. The Navy alone, he claimed, now possessed the sure knowledge to foretell a revolution in the art of radio. That revolution was bound to come no matter what the Navy did. But the Navy could get the credit that came from "clear-cut scientific investigation, confident judgment and decisive action" if it acted quickly enough. Alternatively it could be labelled as indecisive, timid, and inefficient.[11] As far as Hepburn was concerned, there might be a question as to whether the Navy could move fast enough, but there was no question as to the direction in which it ought to move.

Commitment to continuous wave radio, however, was not necessarily the same as commitment to the Federal Company's arcs. The Navy was well aware of General Electric's work on radiofrequency alternators. And it was known that the Telefunken organization in Germany was working on an alternator of its own. But Elwell was at hand and hungry for business; and it looked as if the Navy could get the equipment it wanted promptly, at low cost, and with minimal risk. Hepburn took the initiative in suggesting that Elwell be asked to build a 100 kilowatt arc. If he would

[10] Clark Radio Collection (Smithsonian Institution), Cl. 5 (1922), Box 67, copy of report from G. H. Clark to Chief, Bureau of Steam Engineering, 3 April 1913, on tests of the heterodyne method of receiving, carried out on the U.S.S. *Salem*.

[11] A. J. Hepburn to Chief, Bureau of Steam Engineering, 3 April 1913, as quoted in Howeth, *History*, p. 147.

agree to do this, was willing to guarantee performance, and would supply the equipment at cost or a little over, then the Navy could advertise for bids for a continuous wave transmitter, with the specifications drawn up so that only the Federal Company could meet them. He tried the idea out on Louis Austin of the Naval Research Laboratory and Austin "threw up his hands and would have none of it. He said all he had was his reputation and he couldn't think of lending approval to a proposition of that sort on the basis of such information as he then had." But Admiral Cone was willing to go along. He thought it "did look like a killing if it would work" and told Hepburn to go ahead, despite the latter's warning that "within forty eight hours he would have every responsible manufacturer of spark sets in the country descending on him."[12]

And so the contract was drawn up and put out for bids. Protests indeed there were, but the Navy stuck to its guns and the contract was awarded to the Federal Company on 30 June 1913. The transmitter was intended for the new station at Darien, in the Canal Zone; it went into service on 1 July 1915 with a rated power of 100 kilowatts.[13] Other contracts followed: a 500 kilowatt unit for Cavite in the Philippine Islands and a similar one for Pearl Harbor, Hawaii; 200 kilowatt sets for San Diego, California, and El Cayey, Puerto Rico; 30 kilowatt units for relay stations in Guam and Samoa; a 200 kilowatt set to replace the alternator at Sayville after the Navy took the station over from its German owners; a 500 kilowatt set for Annapolis, Maryland; and finally a 1,000 kilowatt transmitter for the Lafayette station, built by the Navy at Croix d'Hins in France[14] By 1918 the Navy was equipped with an impressive network

[12] This account is based on Hepburn to Hooper, as quoted in Howeth, *History*, pp. 183-84.

[13] R. S. Cranshaw, "The Darien Radio Station of the U. S. Navy (Panama Canal Zone), IRE *Proceedings* 4 (February 1916), 35-40. The total cost of the Darien station was about $400,000; the prospect of similar costs for later stations in the chain required the Navy to go back to Congress for an additional appropriation.

[14] Note that arc stations were normally rated by power *input* (i.e., the power supplied to the arc), while alternator stations were rated by power *output* (i.e., the power supplied to the antenna). Arc efficiencies were never higher than 50 percent. Contract dates for the transmitters of the Navy's high-power chain are listed in Clark Radio Collection, Cl. 100, v. 1, p. 215. The most powerful arc transmitter ever placed in service was built, not by the Federal Company, but by Cornelius de Groot in Malabar, Java. Often referred to as a "3,000 kilowatt arc," it usually operated at 1,600 kilowatts and occasionally at 2,400. See Kaye Weedon, "PKX-Bandung: The Story of de Groot's Mountain Gorge Antenna and Giant Arc Transmitter at Malabar, Java, 1917-1927" (Paper presented at

of high-powered continuous wave radio transmitters interconnecting all its major bases and the Navy Department in Washington, D.C., and capable of communicating with units of the fleet in any location on the Atlantic and Pacific Ocean or the Caribbean. This network far surpassed in coverage anything the British or any other government could claim and, because of the Navy's early and decisive commitment to continuous wave operation, functioned at a level of technical efficiency markedly superior to that of the Marconi or any other private system. Built in the short span of five years, the Navy's high-powered network used arc transmitters exclusively. And all of these transmitters had come from a single supplier: the Federal Telegraph Company.

Clearly Elwell's visit to Washington in 1912 had important consequences. Had it not been for the successful testing of the Federal arc, the Navy would probably have committed itself either to NESCO's rotary spark or to quenched spark transmitters built by Telefunken, five of which it had purchased in 1911.[15] Such a decision would have delayed the shift to continuous wave operation for at least five years, or until high-power radio alternators became available. And it would have jeopardized the feasibility of the high-power chain as the Navy conceived it. Between 1912 and 1917 it was the arc transmitter alone that made the Navy's long-distance radio system possible. No spark transmitter, however powerful, could have covered the distances and provided the reliability of service that the Navy demanded.

The officers of the Navy's Bureau of Steam Engineering who supported Elwell's proposals were undoubtedly risking something: their professional reputations, their career prospects, their credibility with their superiors. It took courage to endorse continuous wave radio when all existing equipment was built around spark. But Elwell was risking at least as much. He had complete confidence in the transmitters he had brought to Washington for demonstration purposes; and the 30 kilowatt arc installed at Arlington was also a known quantity, tried and proven

---

the annual meeting of the Antique Wireless Association, Canandaigua, N.Y. November 1981); "Radiotelegraphy in the Dutch East Indies," *Radio Review* 2 (November 1921), 574-82; "The High Power Station at Malabar, Java," IRE *Proceedings* 12 (December 1924), 693-722; Pratt Papers (Bancroft Library), Federal Telegraph Company file, Box 1, report by R. A. Lavender, 22 December 1923.

[15] Howeth, the official historian of Navy radio, states that, but for the development of the Federal arc, Telefunken would probably have become the Navy's sole source for radio transmitters after 1912. He does not speculate as to what dependence on a German supplier might have meant for American naval preparedness after the outbreak of World War I in Europe.

on the San Francisco–Honolulu circuit. But the Navy wanted for the Darien station an arc rated at 100 to 150 kilowatts. Federal Telegraph had never built a transmitter with that power rating.

This in itself might not have been cause for concern if Elwell had been able to assume that the design techniques he had used in the past would also serve in the future. Unfortunately, he had reason to know that this was not the case. Up to this point each increase in the size and power of Federal arc transmitters had been achieved by scaling up proportionately the dimensions of all the critical components, including the magnetic field in which the arc operated. In this way they had moved progressively from 5 kilowatts to 12 and then to 30. Shortly before Elwell came to Washington, however, he and his staff, in an attempt to improve the reliability of the Honolulu service, had tried their hands at building a 60 kilowatt arc transmitter, essentially by doubling the dimensions of the 30 kilowatt unit. It had proved a dismal failure. It was easy enough to pour more power into the larger unit; the problem was that no more usable power came out. The 60 kilowatt transmitter, in short, delivered no more radiofrequency energy to the antenna than did the 30 kilowatt one. The energy not fed to the antenna was dissipated as waste heat in the arc's cooling system. As one of Elwell's colleagues put it, the 60 kilowatt arc turned out to be "a good hot water heater" but a most inefficient transmitter.[16]

The Federal Company had, in fact, run into a major technical barrier in the design of high-powered arcs, and at a most inconvenient time. What an arc transmitter did, essentially, was convert direct current electricity into high frequency alternating current. In this conversion there were always losses. In moving from 5 kilowatts to 12 and then to 30 these losses had been kept within reasonable limits. In fact, by refinements in design, they had been reduced somewhat. But, beginning at about the 30 kilowatt level, it was a different story: losses began to increase rapidly and the efficiency of the arc fell off drastically. Why should this be so? Elwell in 1912 could not answer that question. And when, early in the following year, he recommended to the directors of the Federal Company that they submit a bid on the Darien station, he was no better off. He was, in effect, offering to build a machine he did not know how to build.

16 "Leonard Franklin Fuller: Research Engineer and Professor," an interview conducted by Arthur L. Norberg in 1973, 1974, and 1975 (Bancroft Library, University of California, Berkeley), p. 48. These and other passages from the extended interview with Leonard Fuller are quoted by permission of The Bancroft Library.

Nor did he inform the Navy that major problems in the design of large arc transmitters still remained unsolved.

Elwell, however, had never been one to follow the prudent course. The way to solve problems, to his way of thinking, was to commit yourself to solving them, to place yourself in a situation where a solution *had* to be found. To have backed out of the Navy contract because of technical uncertainties would have been completely out of character. Furthermore, although he had designed all the Federal-built arcs up to this point, he was not the only engineer on the company's payroll. If he personally lacked the skills in scientific analysis necessary to break through the 30 kilowatt barrier, there were others who possessed them. And on this issue at least he could count on the support of the company's board of directors. Building transmitters for the Navy's high-power chain could turn out to be a very profitable business, and one in which Federal Telegraph would have little or no competition. That could hardly be said of the business it had been in up to this point: handling press and commercial traffic in competition with Western Union, the Telephone Company, and the Pacific cable.

* * *

When, many years later, Cyril Elwell was living in semi-retirement in Palo Alto, California, scene of his earlier work for Federal Telegraph, he tried his hand at writing his autobiography. The manuscript that resulted never found a publisher, and it is not hard to understand why. For all its colorful detail, it lacks coherence: one can follow Elwell from place to place and from one line of work to another, but it is seldom clear why the shifts are happening. And the reader finds it hard to avoid the impression that, to Elwell himself, the dynamics of his life in retrospect were somewhat inexplicable. Time and again he falls back on the concept of fate. He tells us that "what may be considered as my life work was selected for me almost against my will." Discussing his search for a thesis topic at Stanford, he says "fate took a hand." He asks why he moved out of electro-metallurgy into radio and answers, "Perhaps it was a push of my fatemaster." Of his decision in 1913 to quit the Federal Company and move to Europe, he says, "Fate decided. . . ." In all this there is a hint of self-dramatization, and certainly no very profound philosophy; but there is also a suggestion that Elwell himself was more than a little puzzled by the track his life had followed.[17]

[17] Quotations are from the Elwell autobiography (see above, n. 5), and from the Elwell Papers (Stanford), folder 12.

He was born in Melbourne, Australia. His father, Henry Matthew Rogers, originally from Rochester, New York, had come to Australia and joined the South Australia Police Force in 1876.[18] His mother's name was Clotilde Gutman; she traced her family back to noble emigrés from France during the Revolution, thence to Cornwall, and thence to Nuremberg, Germany, where her father had been born.[19] She and Rogers met in Melbourne and, after the latter's discharge from the South Australia police, a child, Cyril Frederick, was born on 20 August 1884. There appears to be no official record of the marriage.[20]

Rogers either died or departed.[21] Clotilde then contracted a marriage to an Englishman, Thomas Dudley Elwell, and her boy took his new stepfather's name. Later in life he was to remember vividly having to change the initials on his school satchel. Thomas Elwell died in 1894, when Cyril was ten, and his mother married again, this time one Rodolf Tudor, owner of the Grosvenor Hotel in Sydney. Tudor, we are told, was not this individual's original name. C. F. Elwell described him as "a Hungarian nobleman whose father had killed himself over gambling losses."[22] Be that as it may, he seems to have known the hotel business.

Elwell, in his autobiography, never speculated as to what effect these kaleidoscopic changes, amounting almost to abrupt shifts in identity, might have had on his later personality and behavior. In his pragmatic way, he would probably have dismissed any such speculations as idle and unproductive; and possibly he would have been right. Certainly, in his reminiscences, he had nothing critical to say about his childhood days—and nothing in a positive vein either. At the very least one can say that he became accustomed to abrupt change, and perhaps learned not to commit himself too completely to any given state of affairs. He made friends where he could. Attendance at Melbourne and Sydney schools

[18] Elwell Papers (Foothill College), folder marked "Elwell, 1920-1950, founder and director of Mullard Radio Valve Company."

[19] Elwell Papers (Stanford), Box 1, folder 5 and folder 11. Folder 5 contains a typed transcript of Elwell's tape-recorded reminiscences; folder 11 contains notes in Elwell's handwriting on his family background and early career. Elwell stated in his autobiography (p. 12) that both his parents were Americans, but it is not clear when or how his mother acquired American citizenship.

[20] Elwell Papers (Stanford), folder 19: secretary, Victoria Police, Melbourne, to Miss Helena Miller, St. Christopher's Convent, Canberra, A.C.T., 3 January 1946.

[21] There appears to be no record of Rogers's having died in the State of Victoria between 1884 and 1939. See Elwell Papers (Stanford), secretary, Victoria Police to Miss Helena Miller, 3 January 1946.

[22] Elwell Papers (Stanford), Box 1, folder 5.

presumably provided some, but the one he remembered particularly in later life was a German electrician named Otto Bauer who ran the gas engine that generated power for the hotel. Bauer became Elwell's hero, and taught him much about electricity.[23] And that was the career he took up when his schooldays ended, going to work as an apprentice in the electrical branch of the New South Wales Railways.[24] There were no opportunities in Australia at that time for a young man to get formal education in electrical engineering; on-the-job training was the most that could be hoped for. It was not enough for Elwell. A chance encounter with the published Register of Stanford University (then Leland Stanford Junior University), sent to one of his work-mates by a sister who lived in California, gave him the opportunity he was looking for. It described a four-year course of study leading to the degree of A.B. in electrical engineering; and it stated that students could work their way through.

Elwell seems to have decided then and there, without any inner uncertainty, that come hell or high water he was going to Stanford to study electricity. He got no help, moral or monetary, from his mother, but by this time Elwell had accumulated some savings of his own from odd jobs of electrical wiring and there was no way she could stop him. Besides, the purser of one of the ships of the Spreckles Line that ran between San Francisco and Sydney was a frequent visitor to the Grosvenor Hotel, and he assured Elwell that he could work his way across. Letters of introduction were arranged. A certain Mr. Smith, an American who always seemed to have plenty of money to spend around the better-class hotels, gave Elwell a letter of introduction to Timothy Hopkins, adopted son of Mark Hopkins and a trustee of Stanford University; and he even prevailed upon Thomas W. Stanford, who lived in Melbourne, to give Elwell a letter of introduction to Governor Leland Stanford's wife. Whoever Mr. Smith really was—and Elwell never found out—he clearly was not unknown to San Francisco's business and political elite.

It cannot have been unusual in Australia in 1902 for a young man with a technical bent to want to study electrical engineering. What was distinctive about Elwell was his bullheaded determination that, with no financial support from his family, with the slimmest of cash resources of his own, on the basis of the barest information on the course of study ahead of him, and with no assurance whatever that his previous education would gain him admission, he was going to study electrical engineering in America, and not at just any university, but at Stanford in particular. And it must have been this determination, this confidence that the thing

[23] Elwell Papers (Stanford), folder 13.
[24] Elwell Papers (Foothill College), folder, "Elwell, 1920-1950."

could and would be done, that gave him the aid and encouragement of people on whom he had no claim except friendship. The same pattern was evident in later years, as for instance in the support given him by members of the Stanford faculty and administration. It came to be taken for granted that, when Cyril Elwell said something would be done, it got done.

He arrived in San Francisco in November 1902, with his savings much depleted. It was a rule of the steamship company, apparently, that anyone able to pay for a ticket would not be allowed to work his way across, so Elwell had to pay for his passage. But his letters of introduction were still intact. Timothy Hopkins—whose curiosity about Mr. Smith's current affairs Elwell could do little to satisfy—directed him to the dean of engineering at Stanford and that gentleman, who was on the point of leaving for Cornell, passed him in turn to C. D. Marx, head of the department of civil engineering. And from Marx, later to be a good friend, he at last began to get some solid information. His preparation, it turned out, was not good enough for admission to Stanford. But there was a private school in Palo Alto, Manzanita Hall, where he could make up the deficiencies, waiting on table and helping in the kitchen in exchange for a waiver of half the tuition. Odd jobs could pay for the rest. Hard work throughout that year and the following summer won him admission to the university in August 1903. Mathematics was, it appears, the subject in which he had to work hardest to pass the entrance examinations.

Elwell spent four years as an undergraduate at Stanford, the last two devoted to courses in electrical engineering. There is no evidence that he achieved any particular scholastic distinction, but he did become a well-known figure on campus, both to fellow students and to the faculty. This was partly due to athletic prowess but mostly, it seems, to the enterprise and energy he displayed after the San Francisco earthquake of 1906. Immediately following the disaster Elwell and a number of other Stanford students headed for the city to help with cleanup operations and fire-control. Elwell ended up holding two jobs, one as electrician to the street railway company, the other with the city water department. The Stanford campus also had suffered considerable damage and, although it was hoped that the buildings could be repaired in time for the start of classes in the fall, work was held up by a strike of stonemasons and electricians. Elwell was asked if he could round up a work force from the electrical engineering students who lived in the vicinity, and he promptly did so. With this addition to the labor supply, the strike was quickly broken, and the university opened in time for the fall semester with all the buildings rewired. Elwell was rewarded by appointment as night electrician to the university, with an office in the powerhouse, peace to do his studying from 5 p.m. to midnight, and a regular paycheck.[25]

There was no course in wireless engineering at Stanford at this time, and nobody on the faculty with special knowledge of the subject. Anyone interested in that branch of electrical engineering had to rely on the periodical literature (the London *Electrician* was a particularly useful source of information) and on the books in the library. Elwell, like most young electricians of this day, was intrigued by the work that was being done with Hertzian waves, and tried to keep up with the latest developments, but his autobiography does not mention any experimentation along those lines, not even the tinkering with spark gaps, coherers, and antennas that was a common feature of student life on many campuses at the time. His training was mostly oriented toward public utility electrical systems—street railways, urban lighting systems, power generating plants—rather than toward communications, and his aptitudes seem to have been toward the practical side of engineering design and construction rather than toward theory. The young man who graduated from Stanford with a bachelor's degree in May 1907 was a well-trained practical electrical engineer with a good local reputation for hard work and initiative, and for "following through" on anything he tackled. He was the kind of man you would be glad to have working for you: give him an assignment and you could be sure that it would be done—and probably a little sooner and a little more thoroughly than you had expected. Neither by training nor by temperament, however, was he a speculative thinker; there was little to suggest an aptitude for theoretical analysis; and if there was a fund of creativity in the man, it had not yet been tapped.

He did, however, want to learn more. The usual thing for electrical engineering graduates to do in those days, Elwell later recalled, was to take a job "on test" with one of the big electrical manufacturing companies—General Electric at Schenectady or Westinghouse at Pittsburgh. Elwell, in contrast, chose to stay at Stanford for an additional year and study for a graduate degree. This would require a thesis, describing some piece of original engineering design, and would entitle him to put the initials "E.E." after his name. He does not mention that, when he made the decision, he had any particular project or line of research in mind; he was just not yet ready to leave Stanford. And, with the assurance that he could keep his job as night electrician, he had a modicum of financial security he had not enjoyed before.

Uncertainty about a thesis topic did not last long. The summer of 1907 was spent visiting Australia, where his parents now owned two hotels in

[25] Elwell, "Autobiography," pp. 19-28.

Brisbane. On his return he was approached by two of the professors in Stanford's Department of Metallurgy, D. A. Lyon and G. H. Clevenger. These men were serving as consultants to the Noble Electric Steel Company, an experimental venture in the electrical reduction of California iron ores financed mostly by H. H. Noble, president of the Northern California Power Company.[26] A sizable deposit of ore had been discovered on the Pit River, in Shasta County; Noble's project involved using off-peak power from his hydroelectric generating system to reduce the ore to pig iron. The immediate problem was to draw up the specifications for a transformer capable of delivering the very large currents required for an experimental electric furnace that Lyon and Clevenger had designed, following the failure of earlier trials by the famous French industrial metallurgist, Paul T. Heroult. It was a limited assignment, requiring no great originality in design, such as one might readily turn over to a bright graduate student. Elwell gave them what they asked for: specifications for a transformer to deliver up to 8,000 amperes of current at voltages ranging from 20 to 80.

And there the matter might have rested, except for the fact that the General Electric Company, when invited to quote a price for building such a transformer, gave a figure far in excess of what had been budgeted and, even more important, specified a delivery time of ten months. Elwell called this ridiculous and, characteristically, offered to build the transformer himself in six weeks. And, with the advice and assistance of the chairman of Stanford's electrical engineering department, he did. The Noble Electric Steel Company got the transformer it wanted; pig iron was indeed produced from the new furnace, though a larger one would have to be built before commercial viability could be tested, and Elwell got first a $1,000 fee and then a salaried position as chief engineer— plus, by permission of his department chairman, a topic and data for his thesis.

This was not a bad beginning for Elwell's career as a professional engineer, and in some respects it foreshadowed the future. Close interaction between the academic, financial, and industrial communities was to become characteristic of the San Francisco Bay area, and in particular of the area around Stanford University and Palo Alto—the Silicon Valley of later years.[27] The establishment of the Federal Telegraph Company

[26] Elwell "Autobiography," pp. 30-32; Elwell Papers (Stanford), Box 1, folder 5.

[27] For elaboration of this point, see Jane Morgan, *Electronics in the West* (Palo Alto, Calif., 1967) and Arthur L. Norberg, "The Origins of Electronics Industry on the West Coast," IEEE *Proceedings* 64 (September 1976), 1314-22.

was to provide another example, with Stanford faculty members playing an even more active role. But, in terms of Elwell's personal history, his work for the Electric Steel Company had little in common with his later accomplishments, beyond the use of electricity and, perhaps, the brash self-confidence with which he tackled the assignment. There was, it is true, a large arc in the electric furnace he helped to build, but its function was to generate heat, not Hertzian waves. To draw any connection between that and Elwell's later work with the oscillating arc would be to strain credulity. At this stage in his career there is no evidence of any inclination to work with high frequency oscillations. In terms of the distinction commonly made in engineering circles at the time, Elwell was a "large current" man, not one of those who worked with the small currents of telegraphy and telephony.

This orientation, however, changed very quickly and drastically, in one of those sudden shifts that Elwell in later life found so hard to rationalize. He was doing very well in electro-metallurgy; his job for the Steel Company was bringing him ever-increasing responsibilities and compensation; his first professional paper, presented to the San Francisco section of the American Institute of Electrical Engineers, described his work on transformer and furnace design and gave him the beginnings of a reputation among fellow engineers. Yet in 1908 he moved out of metallurgy completely and never returned to it. His entire later career was devoted to radio communications and electronics. And this change in direction was made at a financial loss and without any reasonable assurance that it would be to his long-run professional or financial benefit.

The facts themselves are clear enough; it is Elwell's motivation that remains opaque. Several years before, in 1902, a young inventor named Francis Joseph McCarty had developed what he claimed to be a workable system of wireless telephony. His experiments had created quite a stir in the Bay area, and in 1905 he had succeeded in interesting certain local capitalists in financing his work. Among these were William and Tyler Henshaw, bankers in Oakland. Unfortunately, McCarty was killed in an automobile accident in 1906 before his system could be completed or thoroughly tested. This left the Henshaw brothers with a few pieces of equipment they did not understand, and little else to show for their investment.[28]

[28] This account is based on Elwell's autobiography, particularly the "first-person singular" version in Box 1, folder 1 of the Elwell Papers in Stanford University Library. (Elwell, however, attributes the invention to Ignatius McCarty, brother of Francis). See also C. F. Elwell to W. H. Hewlett, 24 December 1953 (Pratt Papers, Bancroft Library) and Foothill College, Federal Telegraph

They did the sensible thing: they got in touch with Professor Harris J. Ryan, head of the electrical engineering department at Stanford, and asked him to recommend someone who could look over the remains of the McCarty system, get it to function if at all possible, and make a report to them on its technical performance. If it looked as if the system could be made to work, they might invest more money in it. If not, they would write the investment off.

Ryan referred them to Elwell, who declined the assignment because (as he later put it) "the design of the new 15 ton electric furnace interested him more than a wireless telephone he did not believe in."[29] The Henshaws persisted, however—Ryan apparently declined to recommend anyone else—and Elwell finally agreed to accept the job, figuring that it would take several months to make and test the transformers for the new furnace and in the meantime he could carry out his tests, make his report, and pocket the fee the Henshaws were offering. A few pieces of test equipment were purchased on a trip east to Schenectady and in the summer of 1908 Elwell began his tests.

Elwell tells us that he was reasonably confident, before starting his experiments, that the McCarty system would not work. McCarty had used a more or less conventional spark gap to generate Hertzian waves, modulating the signal with a carbon microphone. Elwell had picked up enough theory to know that any such system radiated damped waves and that it was immensely difficult to modulate such waves with voice frequencies. His tests, first in an attic above the Stanford engineering laboratory and later at various sites in Palo Alto and Los Altos, confirmed this. It was indeed possible to transmit speech, at least over short distances and with lower power; but any attempt to get greater distance and higher power by lengthening the spark gap and strengthening the spark discharge resulted in intolerable noise and distortion.

This was predictable. Scores of experimenters had tried to do exactly what McCarty had attempted; and scores had emerged with the same conclusions as Elwell did. Wireless telephony was possible only given a source of true continuous wave radiation; it could not be grafted on to

Company file, clippings folder, unidentified clipping headlined "Boy's Dream of a Great Invention Come True," dated 5 June 1904. Jane Morgan (*Electronics in the West*, pp. 39-40) gives an account that differs in unimportant details. According to Morgan, McCarty's chief assistant was Charles Logwood, who later assisted de Forest with his audion experiments while de Forest was working for Federal Telegraph.

[29] Elwell, "Autobiography," p. 36. Elwell states that Ryan "knew of Elwell's long standing interest in wireless." There has been no mention, however, of any interest in wireless until this page of the autobiography.

the damped waves of spark technology. Elwell did not give up easily; following Fessenden's example—an indication that he was keeping an eye on the periodical literature—he tried for a spark frequency of 100,000 cycles, hoping that at a frequency above the audible range the spark noise would be less troublesome.[30] But the results were not much better. Elwell reported these findings to the Henshaw brothers and they, prudently, decided that their interest in wireless telephony was at an end.

Elwell's interest, however, was not, and this is, in retrospect, the most intriguing feature of the episode. He had accepted the assignment, originally with some reluctance, essentially as a means of occupying his time productively and profitably while the transformers for the new electric furnace were being built. Now the job was done, and nothing would have been easier and more reasonable than for him to have turned his back on wireless telephony and gone back to electric furnaces. In fact he did exactly the opposite. His report to the Henshaws was not just a statement of negative findings: it was also a suggestion that they finance him in the search for a really commercial system of wireless telephony, based on continuous waves. And when they vetoed that suggestion, they also offered to sell Elwell, for a nominal charge, all the equipment McCarty had used, together with the test gear that Elwell had acquired at their expense. He accepted the offer; for now, as he puts it in his autobiography, "Elwell was in wireless to stay."[31]

Why? If his findings on the McCarty system were negative enough to discourage the Henshaws from any attempt to salvage their investment, how could they also be positive enough to make Elwell discard his promising career in electro-metallurgy and commit himself to radio? An answer in terms of personal motivation can be no more than speculative. Elwell probably identified one key factor when, having made his usual reference to fate, he added, "More likely it was the feeling that I was to head the effort and be my own boss and make my own decisions."[32] But it was also true that he had found the technical challenge of wireless telephony irresistible. He knew the McCarty system could never be made to work, and that was the negative side of the matter. But he also knew precisely why it would not work, and that was the positive side. Like Fessenden, he had already solved the problem—in principle. All he needed was a

---

[30] Elwell stated explicitly that his goal was 100,000 cycles, "the same as Prof. Fessenden." Fessenden's articles on long-distance radiotelephony appeared in *Scientific American* on 19 January 1907, in the *Electrical Review* or. 15 February, and in the *Electrician* on 4 October of that year.

[31] Elwell, "Autobiography," p. 42.

[32] Elwell Papers (Stanford), folder 12.

generator of continuous high frequency oscillations that could deliver appreciable current to an antenna.

Several factors combined to lead Elwell to the certainty that the problem was soluble. First, his tests of the McCarty system had not been complete failures. The "boy inventor" really had transmitted speech, as he claimed, and so had Elwell. In fact, as long as the balls of the spark gap were very close together and the spark discharge practically continuous, so that it approximated to an arc, the quality of the transmitted speech was reasonably good. Trouble began only when, to get greater distance, the spark gap was made wider and the sparks became clearly intermittent.

Secondly, Elwell knew that the generation of true high frequency oscillations was not impossible. He knew about Fessenden's work from the engineering periodicals. But, more than that, he had seen a radiofrequency alternator in operation. General Electric had made quite a fuss over Elwell when he visited Schenectady in 1908, since they were interested in electric furnaces too and liked to sell big transformers, and among the other products they had shown him was an Alexanderson alternator—the 2 kilowatt model—running under test. Elwell's reaction was interesting: he concluded that the alternator system was "not practical" because the machine had to be run at such high speeds. But in addition (as he later expressed it), "he did not see any place in the Fessenden or General Electric Company organization."[33] Elwell, in short, wanted to run his own show. Tagging along behind Fessenden and General Electric was not what he had in mind; "I knew that this development held no opening for me."[34] But thirdly, he knew that the alternator was not his only possibility. There was another way of generating continuous wave high frequency current: Valdemar Poulsen's oscillating arc.[35] The device had been known for several years, and limited experimental use had been made of it, both in Europe and the United States. Fessenden had used arcs in his radiotelephony experiments from Roanoke Island; and Lee de Forest's ill-starred radiotelephone sets, installed in the ships of Roosevelt's "Great White Fleet" in 1907, had been arc transmitters. No one

[33] Elwell, "Autobiography," p. 45.

[34] C. F. Elwell to W. H. Hewlett, 24 December 1953 (Pratt Papers).

[35] Valdemar Poulsen, "System for Producing Continuous Electrical Oscillations," *Transactions* of the International Electrical Congress, St. Louis, 1904, 2: 963-71, reprinted in Absalon Larson, *Telegrafonen og Den Tradløse og opfinderparret Valdemar Poulsen og P. O. Pederson*, Ingeniørvidenskabelige Skrifter, No. 2 (Copenhagen, 1950), pp. 255-62; V. Poulsen, "A Method of Producing Undamped Electrical Oscillations and Its Employment in Wireless Telegraphy," *The Electrician* 58 (16 November 1906), 166-68.

yet, however, not even the Danish inventor and his backers, had tried to use the device for really high-powered long-distance radiotelephony or telegraphy. And no one had acquired the U.S. rights to Poulsen's patents. It was not quite true, as Elwell later claimed, that the invention had gone ignored and unused from 1902 to 1908; but its potentials had certainly not been fully exploited.

To Elwell, by coincidence, Valdemar Poulsen was more than a name. He had seen the man in the flesh, at the Paris Universal Exhibition of 1900, when Poulsen had won a grand prize for his invention of the Telegraphone, technological ancestor of the modern tape recorder. Elwell had been touring Europe with his parents at the time, and although he did not meet Poulsen, the Danish scientist apparently made an impression on him. Now, in 1908, he had no hesitation in cabling him directly: if Poulsen had not yet sold the U.S. rights to his continuous wave wireless patents, how much would he want for them? Poulsen in reply quoted a figure of $250,000. That was rather more than Elwell's personal savings could cover, but he went ahead anyway. He cabled Poulsen that he would arrive in Copenhagen within two weeks.[36]

As Elwell later told the story, events moved very quickly after his report on the McCarty system was completed, and the question can well be raised whether (as his autobiography suggests) he was acting entirely on his own at this point or whether perhaps there had not already come together an informal syndicate of Elwell and his Palo Alto friends and associates. Certainly the quarter-million dollars that Poulsen originally asked for his patent rights was more than Elwell had any prospect of raising from his own resources. The costs involved in travelling from Palo Alto to Copenhagen and back would in themselves have made a serious dent in his savings. The evidence is clear that, when Elwell returned from Copenhagen with a proposition that looked as if it might be viable, a small group of Stanford University faculty and alumni was assembled quickly and without serious difficulty to provide the initial capital. It is entirely possible, although Elwell provides no evidence on the point, that this group already existed before he went to negotiate

[36] Elwell, "Autobiography," p. 46. There is an element of doubt about this version of the story. Haraden Pratt, writing in 1965, reported that the initial approach to Poulsen was made, not by Elwell, but by John C. Coburn, a stock salesman, who later acted as agent for the Federal Telegraph Company. He cited as his authority a conversation he had held with Coburn, and stated that this account was partly verified by Peter Jensen, a Danish engineer who worked for Federal Telegraph in its early years. Elwell, however, does not mention Coburn's activities until after the Federal Company was established. See Haraden Pratt to Leonard Fuller, 31 May 1965 (Fuller Papers, Bancroft Library).

with Poulsen and that it was partly their encouragement and their resources that led him to move with such alacrity.

In Copenhagen, Poulsen and his colleague, P. O. Pedersen, had only a receiving station. Elwell was taken there first and heard radiotelephone transmissions from an arc transmitter in Lyngby, ten miles away. He was deeply impressed: in striking contrast to the McCarty system, the speech quality was excellent—better, in Elwell's judgment, than was usual in a wired telephone system. At Lyngby he inspected the arc transmitter and also heard high-speed radiotelegraphy signals from an arc station at Esbjerg, on the North Sea coast. These were transcribed with the aid of a photographic recorder that Poulsen had developed and, on checking, were found to be at the remarkable rate of 180 words per minute. There was no doubt, in short, that Poulsen had a workable system of radiotelephony and telegraphy, at least at low power and for relatively short distances.[37]

Poulsen and his financial backers in the meantime had been checking on Elwell. A request to the State Department, through the U.S. ambassador, for a report on his financial status produced a response in the unusually short period of twenty-four hours. Someone in the State Department had, it appears, noticed that David Starr Jordan, president of Stanford, was staying at the Willard Hotel in Washington, D.C., and, on telephoning him to ask about Elwell, was told that he could be trusted to carry out anything he undertook.[38] This was perhaps not quite the kind of information the Copenhagen group had been hoping for, but it sufficed.

What Elwell wanted was an option on the U.S. rights to Poulsen's arc transmitter patents. Poulsen, however, refused to sell these alone: if Elwell wanted the arc patents, he would also have to buy the patents on the high-speed photographic receiving system. And, for the package, the price was not the $250,000 originally specified, but $450,000. Elwell had little interest in the photographic recorder but, confident at this time of his ability to raise whatever funds were needed, he agreed to the deal. He returned to New York with the option in his pocket, together with a letter of introduction to a certain Mr. Lindley, who had raised the capital for the American Telegraphone Company.

He was back in Copenhagen again within six months, arguing for more realistic terms. The details are obscure, but it is clear that, in the spring of 1909, there was no money to be raised in New York for wireless

---

[37] Elwell, "Autobiography," pp. 46-48; compare Clark Radio Collection (Smithsonian Institution), Cl. 14, General History, 001-250.

[38] Elwell, "Autobiography," p. 47.

telephony—at least, not by Elwell. He asked Poulsen to modify, not the price to be paid, but the terms of payment. After some bargaining, this was accepted. Poulsen and his associates agreed to a schedule of payments of gradually increasing amounts, to begin after a successful daylight demonstration of wireless telephony for a distance of fifty miles over land.[39] In effect they agreed to help finance Elwell's venture. With this revision in the contract, signed on 17 August 1909, Elwell returned to the United States, but this time to Palo Alto, not New York.[40] Elwell also took with him a small 100 watt arc transmitter which he had purchased for $1,000 cash, and he placed an order for two larger models, one of 5 kilowatts, the other of 12, for a price of $6,000, half payable as soon as he returned to the United States, half when the transmitters were set up and ready to operate. And he made preliminary arrangements for two Danish engineers, trained by Poulsen, and a Danish mechanic to come to Palo Alto to set up the apparatus and help run the demonstrations.

Pl. 3: Elwell's 100 watt Poulsen arc. Dimensions: Base 6″ × 6⅜″, Overall height 8½″, Overall length 10½″.
Source: Foothill Electronics Museum

[39] Ibid., p. 50.
[40] Ibid.; Clark Radio Collection (Smithsonian Institution), Cl. 14, General History, 001-250.

Elwell's attempt to take the New York capital market by storm may have been somewhat ingenuous, but the rest of the transaction showed hardheaded good sense. It might have been possible to conduct the whole affair by letter and cable; but Elwell, with the McCarty system fresh in mind, wanted to see and hear the Poulsen transmitter in operation. And it might have been possible to import from Denmark only the specifications, blueprints, and circuit diagrams; many attempts to transfer technology from one country to another have been made in precisely that way. Elwell, however, intended to import his first two transmitters and the engineers to set them up and run them. Everything depended now on whether he could raise enough capital from his friends and associates to carry out the initial demonstrations. If they were successful, and made enough imprint on local opinion, he could raise additional funds by selling securities to the general public in the Bay area.

＊　　＊　　＊

Elwell once described the radiofrequency arc as an additive invention, meaning that it came into being by putting together elements previously known but not previously related. It might be argued that this is true of all inventions, but in the case of the Poulsen arc the process is particularly clear. The arc itself—or rather, the phenomenon of arcing—must have been known since batteries capable of delivering large currents first became available: when two conductors previously joined were separated, a flamelike arc would appear between them. The current, in other words, would continue to flow across the intervening space, provided it was not too large. And it would be recognized that the low-voltage, high-current discharge of the arc was a phenomenon different from the high-voltage, low-current disruptive discharge of the spark. As early as 1802 we find the arc being used as a means of illumination, with carbon electrodes substituted for metal conductors to improve the quality of the light.[41] The brilliance of the arc light was always impressive, though its harshness was unpleasant; but as long as batteries were the only source of electric current, its use was limited. The invention of the Gramme ring dynamo in 1871 provided an alternative source of electricity, capable of delivering large currents for extended periods and therefore admirably suited for the arc. It is no coincidence that the first successful commercial use of arc lighting was in Zenobe Gramme's factory in Paris in 1873. From

[41] C. M. Jarvis, "The Distribution and Utilization of Electricity," in *A History of Technology*, ed. Charles Singer, E. J. Holmyard, A. R. Hall and T. I. Williams, 5 vols. (London and New York, 1958), 5:208.

1875 on there were growing numbers of municipal and private installations in Europe. These were mostly for the interiors of large buildings
and for public streets and plazas—attempts to develop arcs of less than
1,500 candlepower for private residential lighting were not successful.
In the United States C. F. Brush, using dynamos and arc lamps of his
own design, completed the first commercial installation in Wanamaker's
department store in Philadelphia in 1878.

In all these cases the arc was used as a source of light. If it hissed, or
sputtered, or hummed, that was a defect to be eliminated by better design
of dynamos and purer carbon in the electrodes. But the arc was also a
matter of intense interest to experimenters and scientists. Experimenters
were curious as to whether the arc could be used for purposes other than
illumination. And scientists were intrigued and, to a degree, baffled by
the behavior of the arc in an electric circuit and by the physics of the arc
discharge.

It was discovered, for example, that if the arc were fed with alternating
current, or with imperfectly smoothed direct current, it radiated sound
waves through the air. This might be caused, for example, by the commutator in a direct current generator, the pitch of the sound being determined by the number of segments on the commutator. And what
seemed to be happening was that the arc flame itself fluctuated in cross-
section as the current through the arc fluctuated.[42] This was originally
thought of as a defect to be eliminated; but it was not long before
experimenters were linking a microphone to the arc circuit—either by
direct connection or by inductive coupling—and using the arc to transmit
and amplify the human voice. Here was an embryonic public address
system. And others tried to use the arc as a telephone repeater or line
amplifier, superimposing the small voice currents of the telephone line
on the direct current through the arc.[43] These and other adaptations were
handicapped by the fact that the arc itself was noisy, unless very carefully
adjusted, provided with very pure electrodes, and run at low power. But
they did call attention to the fact that the arc was not just a source of
light; the arc flame could be used for other purposes, particularly when
it carried alternating or fluctuating currents.

In the meantime, physicists and electrical engineers had been tackling

[42] William Duddell, "On Rapid Variations in the Current through the Direct-
Current Arc," IEE *Proceedings* (London) 30 (1901), 237.

[43] Duddell, "On Rapid Variations," pp. 239-43; George W. Pierce, *Principles
of Wireless Telegraphy* (New York, 1910), p. 254. The discovery of the sound-
amplifying and -transmitting abilities of the arc is attributed to H. T. Simon of
Göttingen University; see Wiedemann's *Annalen* 64 (1898), 233.

the problem of the anomalous behavior of the arc—or so at least it seemed—as part of an electrical circuit. The problem was that the arc apparently violated Ohm's Law. Simply stated, Ohm's Law could be made to seem almost a matter of common sense: the electrical current passing through an element in a direct current circuit was directly proportional to the voltage across it and inversely proportional to its resistance. But in fact the law was widely misunderstood and misinterpreted, if not by scientists of the first rank, then by many applied scientists and engineers. It was frequently cited, for example, to prove that the subdivision of the electric current for lighting purposes was an impracticable goal; this was the dogma that Thomas Edison had to combat in designing his system of incandescent lighting.[44]

Ohm's Law had originally been stated in 1826, but in the late 1870s it was still a novel scientific principle, only recently elevated to rank with the law of gravity as a fundamental law of nature in the strictest sense.[45] It was, therefore, disconcerting to discover that an arc lamp did not obey Ohm's Law. As the current passing through the arc increased, the difference in electrical potential between its two electrodes decreased. Current and voltage were inversely, not directly, proportional to each other. This could be handled at a verbal level by saying that an arc had "negative resistance," but this hardly added to anyone's understanding.

This was how matters stood when William Duddell, an English scientist, began his investigations of the arc in the late 1890s. His particular interest was in investigating how the difference in electrical potential across the arc varied as the current through it varied. In a solid conductor—a length of wire, for example—the current would be directly proportional to the difference in electrical potential at its two ends. Plotted on a graph with current on the vertical axis and voltage on the horizontal, the relation would be represented by a straight line rising from bottom left to top right. But if similar measurements were taken on an arc, the result was a curve that fell from top left of the graph to bottom right. This was referred to as the arc's characteristic curve; the fact that the arc had a "falling characteristic"—that it had negative resistance in that sense—was of great scientific interest and practical importance.

Duddell wanted to find out how the voltage varied with small changes

[44] Harold C. Passer, "Electrical Science and the Early Development of the Electrical Manufacturing Industry in the United States," *Annals of Science* 7 (December 1951), 382-92.

[45] George Chrystal in *Encyclopedia Britannica*, 9th ed. (New York, 1878), 13:12, cited in Passer, "Electrical Science," p. 387.

in the current; that is, he wanted to ascertain the slope of the characteristic curve and what factors determined the slope. The effect of slow changes—say at the rate of 250 cycles per second or less—had already been investigated by others; but their findings, in Duddell's opinion, were distorted by the fact that the shape of the electrodes changed while the fluctuations in current were taking place. What was needed was a measurement of the true resistance of the arc while all other conditions remained unchanged. This called for superimposing on the direct current flowing through the arc an alternating current of much higher frequency than had been used before—of the order of 5,000 cycles per second at least. How was this to be done? The simplest way was to connect an alternating current generator in parallel with the arc, adding appropriate inductances and capacitors to isolate the alternating and direct current circuits; and this Duddell did, generating by these means sounds all the way up to the limits of audibility. (See Fig. 3.1)

The question then arose whether the arc itself, without the alternator in parallel, could not be used to generate high frequency currents—whether, that is to say, advantage could not be taken of its falling characteristic curve to turn it into a true oscillator, so that one could, as it were, feed direct current into it and get alternating current out. This suggestion had been made to Duddell by George Francis Fitzgerald, the highly respected Irish physicist. What was needed was a method of making the arc intermittent, so that it would extinguish and then relight itself very rapidly; this could be done, as French physicists had shown, by using a blast of air to blow out the arc, or alternatively a transverse magnetic field could be used. Duddell tried this, using a magnet to make a direct current arc, with an inductance in series, intermittent. The results were disappointing; the rate of oscillation turned out to be quite irregular and rather low.

This was the point at which Duddell, in an attempt to smooth out the irregularity, made one small but vital change in the circuit. He added a

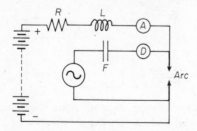

Fig. 3.1: Duddell's oscillating arc, with AC generator in parallel.
Source: Duddell, "On Rapid Variations," p. 238

capacitor across the terminals of the arc. This was not very remarkable; it was the kind of thing an experienced electrician would think of doing. But the results were remarkable. Duddell's account betrays his astonishment: "... to my surprise I found that the direct-current arc was intermittent even when *not* blown in any apparent way either by a stream of gas or by a magnetic field, and further that no self-induction in series with the arc was necessary."[46] And at first he found it inexplicable. "Here there was a puzzle—a direct-current solid arc burning under ordinary conditions with resistance in series, and supplied with current from accumulators, became intermittent and gave out a musical note on simply shunting the arc with a condenser."[47]

Matters, of course, were not quite that simple. Wire leads had been used to connect the capacitor to the arc, and these leads had some self-inductance. When Duddell twisted the leads to cancel out this self-inductance, the oscillations stopped. And when he added a coil in series, to increase the inductance, they became greatly intensified. In other words the true circuit called for shunting both a capacitor and an inductance in series across the arc. (See Fig. 3.2) These two elements, plus the arc itself, made up a resonant circuit that would oscillate at a certain definite frequency determined by the circuit constants. It was necessary that the resistance of this shunt circuit be low—lower than the resistance of the arc itself—but if that condition were satisfied, as long as the arc was supplied with direct current it would generate sustained oscillations.

Now, it had been known for some time that, if a condenser were discharged through a suitable low-resistance inductor, a train of oscillations of almost any desired frequency could be obtained. These oscillations, however, would be highly damped; their amplitude would decrease rapidly, the rate of decrease depending on the resistance of the circuit. But Duddell had developed a circuit in which the oscillations were

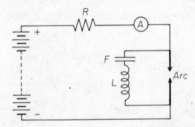

Fig. 3.2: Duddell's oscillating arc, with series-resonant circuit in parallel.
Source: Duddell, "On Rapid Variations," p. 248

[46] Duddell, "On Rapid Variations," p. 247.
[47] Ibid.

not damped but continuous, with constant amplitude, as long as the residual resistance of the circuit was cancelled out by the "negative resistance" of the arc. He had, that is, learned how to convert the arc's inherent instability into an asset; as he expressed it, "It must be remembered that although we have an alternating current through the condenser and self-induction, the source of supply is not an alternating one, and that *it is the arc itself which is acting as a converter and transforming a part of the direct current into alternating, the frequency of which can be varied between very wide limits by altering the self-induction and capacity* [emphasis in original]."[48]

This was Duddell's famous "musical arc." The name suggests an entertaining gadget, and it is true that it proved to be an excellent device for public lectures and demonstrations. But Duddell's purpose was more serious; he was not trying to develop an amusing or instructive gadget. He never personally succeeded in generating oscillations above about 10,000 cycles per second; and it was his belief (at least in 1903) that the arc could not be used to generate Hertzian waves, as the slope of the characteristic curve would become positive, he believed, at frequencies of about 100,000 cycles per second.[49] And the use of a magnetic field to blow out the arc he seems to have regarded as quite incidental. The essential feature, to him, was the discovery that, if you connected inductance and capacitance across an arc, you had a device that would convert direct into alternating current, with oscillations of constant amplitude. Such an arc was, for him, essentially a converter, and it was by that name that many preferred to call it, even when its primary use had become that of a radio transmitter.

Was it an original invention? One reputable experimenter did not think so. On 4 July 1893 Elihu Thomson, of the (American) Thomson-Houston Company, had been issued a patent on a method of producing undamped electric oscillations that looked very similar to Duddell's, and as soon as he heard of the Englishman's findings he wrote to *The Electrician* to claim prior discovery.[50] There is no evidence that Duddell was aware of

[48] Ibid., p. 248.

[49] W. Duddell, "The Musical Arc," *The Electrician* 51 (18 September 1903), 902. This conclusion held, he believed, for ordinary solid carbon electrodes and for all the conductors he had tried, including "gases, vapours, electrolytes, etc."

[50] See *The Electrician* 46 (18 January 1901), 477. Duddell's paper, "On Rapid Variations," was delivered on 13 December 1900; he inserted a footnote in the printed version to express his regret at being unaware of Thomson's experiments and omitting to give him credit for them. The Thomson U.S. patent is No. 500,630, applied for 18 July 1892. For Thomson's background and career, see Harold C. Passer, *The Electrical Manufacturers 1874-1900: A Study in Com-*

Thomson's work, and indeed it would be easy for anyone working on arcs to overlook the patent, for it appears, on casual inspection, to be a form of magnetic blowout for a spark gap. The illustrations certainly leave that impression. (See Fig. 3.3) The text, however, refers unambiguously to an "arc or spark"; it describes the purpose of the magnet as "to break any arc between the balls"; and it states that the magnet could be replaced by an air-jet. Above all, it clearly describes a "feeding circuit," containing inductance and capacitance, shunted around the spark or arc, the function of which is to maintain high frequency alternations. In all these respects Thomson certainly anticipated Duddell, and to the extent that priority is important, it should be awarded to him. Duddell, however, clearly arrived at his discovery by a different and independent route, and later developments, particularly in Europe, stemmed from his work more than they did from Thomson's.

Duddell's singing arc aroused considerable interest. That it worked was obvious, but why it did was not. A common analogy was to an organ pipe. There was a unidirectional blast of air coming from the bellows, corresponding to the direct current through the arc. When this jet of air met the lip of the organ pipe, it set up vibrations in the column of air inside the pipe, corresponding to the oscillations in the resonant circuit around the arc, and these controlled the motion of the airflow, causing it to play inside and outside the lip of the pipe. This action was

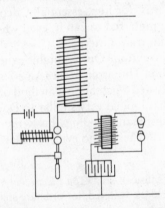

Fig. 3.3: Elihu Thomson's magnetic blowout.
Source: Thomson, *The Electrician*, 21 December 1906

petition, Entrepreneurship, Technical Change, and Economic Growth (Cambridge, Mass., 1953) pp. 21-57. The Thomson-Houston Company eventually became a major component of the General Electric Company.

self-sustaining as long as the air jet supplied energy.[51] This physical image was all very well, but it failed to explain why there appeared to be an upper limit to the frequencies attainable. It seemed to be the case that the arc's characteristic curve flattened out as the direct current through the arc increased. Consequently, if you wanted high frequency oscillations, you had to run the arc at very low power, operating where the characteristic curve had a steep slope. In that way you could use a small capacitor in the resonant circuit—necessary for high frequency oscillations—and still get large variations in voltage across the arc terminals. Larger arc currents would push you farther out on the characteristic curve, and in that relatively flat region you would have to use a larger capacitor to get the necessary voltage swing. This implied longer time constants in the oscillating circuit, since a larger capacitor took a longer time to charge and discharge than a smaller one, and that meant lower frequencies.

This was a serious stumbling block for anyone thinking in terms of Hertzian waves and, as we have seen, Duddell was pessimistic on that score. The problem was essentially how to get a steeply sloped characteristic curve at high currents. Valdemar Poulsen found the solution, or at least the essential elements of one. He took over from Duddell the resonant circuit in parallel with the arc and the idea of operating the arc in a strong magnetic field. But, instead of letting the arc burn in air, he operated it in an atmosphere of hydrocarbon vapor. Any kind of hydrocarbon vapor would do—ether, alcohol, coal gas, or pure hydrogen. Even ordinary steam could be used. In his first successful experiment—and the setup carries some of the indications of a lucky laboratory accident—he merely let the arc burn in the vapor of a spirit lamp. (See Fig. 3.4) The result was a substantial increase in power output. Adding a magnetic field stepped up the strength of the oscillations even further. And, most significantly, it proved possible to raise their frequency by reducing the size of the capacitor. Poulsen got to 150,000 cycles per second without any difficulty.[52]

Once again, making it happen was easier than explaining how it happened. Why did running the arc in a hydrogen atmosphere make so much difference? Was it solely due to the cooling effect of the gas or was

[51] J. A. Fleming, *The Principles of Electric Wave Telegraphy*, 1st ed. (London, 1906) p. 646; compare J. A. Fleming, "The Electric Arc as a Generator of Persistent Electric Oscillations," in *The Yearbook of Wireless Telegraphy and Telephony, 1917* (London, 1917), p. 666.

[52] V. Poulsen, "Continuous Electric Oscillations," in Larson, *Telegrafonen*, p. 261.

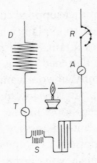

Fig. 3.4: Poulsen's arc in hydrogenous vapor.
Source: Poulsen, "Continuous Electrical Oscillations," p. 255

something more complex involved? Did the hydrogen tilt the character-istic curve, and if so how? And what exactly was the function of the magnetic flux? Did it serve merely to lengthen the arc discharge, by curving it to one side, or did it somehow "scavenge" the discharge, so that the oscillations in the resonant circuit could be stronger and more rapid? Poulsen's announcement of his discovery did not speculate on such matters. His paper at the St. Louis Electrical Congress in 1904 described his apparatus and reported a variety of measurements, but it did not theorize as to what was going on inside the arc chamber.[53] Nor did Ambrose Fleming do much better when, in 1906, he inserted in the final section of his *Principles of Wireless Telegraphy* a brief description of Poulsen's arc. It was clear, he wrote, that one element in the discovery was the effect of hydrogen or hydrocarbon vapor on the slope of the characteristic curve; but "the reason for this has not yet been fully ex-plained."[54] (See Fig. 3.5).

Clearly technology in this instance was outrunning science. In the short run this was of little consequence. The device was simple to build, and it worked. Poulsen, like Duddell before him, had proceeded in a highly empirical manner, guided (as far as we can tell) by no particular physical theory, and he had emerged with dramatic results. But the absence of theory, the lack of understanding of what was taking place within the arc plasma, promised trouble in the future, as experimenters moved to larger arcs and higher power. Upscaling the arc, increasing all its di-mensions proportionately, was an effective procedure for small incre-mental changes, but beyond a certain point it would no longer be safe

[53] Ibid., pp. 255-62.
[54] Fleming, *Principles*, p. 649. See also Fleming, "On the Poulsen Arc as a Means of Generating Undamped Oscillations," *The Electrician* 58 (16 November 1906), 166.

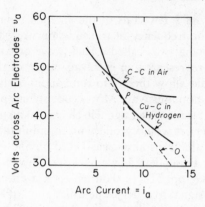

Fig. 3.5: Static characteristic curves of arcs in air and in hydrogen.
Source: Elwell, *Poulsen Arc Generator*, p. 28

to assume that an expansion of all dimensions in the same ratio would work. At that point there would be no substitute for a thorough understanding of the arc itself, so that the effect of changes could be calculated and designs modified as the scale increased.[55]

Poulsen, already well known in Europe for his invention of the telegraphone, seems to have realized quickly that in the "hydrogenic arc" he had a device with commercial potential. His laboratory arc was soon modified to become a radio transmitter capable of sustained power. This involved in particular dealing with problems of cooling. The positive electrode became a copper rod, internally cooled by water circulation, while the negative electrode was of carbon, rotated slowly to give even burning. And the entire arc chamber had to be cooled, either by circulating water or, in the case of smaller units intended for portable use, by large cooling fins. This attention to heat dissipation, of course, reflected the low efficiency of the arc converter.

Since the arc generated continuous waves, it could be used for wireless telephony, and this was one of its earliest claims to attention. Problems of modulating the signal were solved by inserting a carbon microphone—or rather, several such microphones in parallel—in series with the antenna inductance. Receiving signals of that sort presented no special difficulty, since any of a variety of detectors could be used in conjunction with a telephone earpiece. But when the arc was used for Morse code telegraphy, new problems arose. It was hardly possible to key the arc by extinguishing it and relighting it again. How then was it possible to send the dots and dashes of the Morse code? A simple solution was found by inserting the

[55] Compare Fuller (interview), p. 48.

telegraph key across a few coils of the antenna inductance. When the key was closed, the frequency of transmission was shifted a small percentage, creating what was called a back wave or compensating wave. This, to be sure, represented an extravagant use of the radiofrequency spectrum, but it did allow the arc to run continuously.

Reception of continuous wave Morse code signals presented the usual problem. Since the transmitted signal itself carried no modulation (unlike a spark transmitter), it would produce no sound in the earpiece, beyond a dull click when the transmitter was keyed on and off. It was to solve this problem that Fessenden, as we have seen, invented the heterodyne receiver (see above, pp. 58-60). Poulsen took a different approach: he fed the incoming signal to a device he called the "tikker" and used that to produce either an audible tone or a permanent record on photographic tape. This proved particularly well adapted for high-speed reception, and it was the system shown to Elwell in Copenhagen in 1909.

The Poulsen tikker—actually it was invented by his co-worker, Professor P. O. Pedersen of the Royal Technical College—was an ingenious device. Lee de Forest, who had wide experience in such matters, described it as "by far the most sensitive and efficient detector in existence" when used in conjunction with an ordinary telephone earpiece. In its simplest form it consisted of a small reed vibrated by magnetic action, as in an electric bell. To the end of this reed, and insulated from it, there was attached a gold wire, which vibrated against a second gold wire and acted as an interrupter. Connected in a telephone circuit, this would give a note that could be heard in the earphone. In the high-speed version (which Elwell and his associates paid heavily to acquire but which they seldom used in service) the gold wire was set in a magnetic field. Antenna currents passing through the wire caused it to bend slightly; these slight movements of the wire were magnified by lenses and projected onto photographic tape which, moving continuously, passed through a developing and fixing bath and became a permanent record, to be transcribed later when the film was scanned at slower speeds.[56]

Poulsen and Pedersen worked closely together, and what came to be called the Poulsen system should probably be regarded as a joint creation,

[56] There were many varieties of tikkers (also sometimes spelled tickers). See Vivian J. Phillips, *Early Radio Wave Detectors* (London, 1980), pp. 172-87; Lee de Forest, "Recent Developments in the Work of the Federal Telegraph Company," IRE *Proceedings* 1 (January 1913), 40; George C. Blake, *History of Radio Telegraphy and Telephony* (London, 1928), pp. 94-98; Elmer Bucher, *Practical Wireless Telegraphy* (New York, 1917), pp. 277-78; Fleming, *Principles*, pp. 702-704. The description of the high-speed version is based on an article in *The Wasp* (Foothill College, Poulsen Wireless Corporation file), 3 August 1912.

with Poulsen contributing most heavily to the transmitting elements and Pedersen to the receiving apparatus. Between them, they had by 1903 a workable continuous wave radio system, adaptable both for telephony and telegraphy. And they had patents to protect it (Danish patent No. 5,590 of 1902; British patent No. 15,599 of 1903). True, the oscillating arc itself could not be patented, in view of Duddell's work. And Elihu Thomson's patent of 1893 seemed to rule out any claim to originality in the use of the magnetic field.[57] But the use of the hydrocarbon atmosphere was unique, distinctive, and original. A syndicate was formed in Copenhagen to exploit the patents and develop the system commercially. The Lyngby station, ten miles from Copenhagen, was set up in 1905 and the Esbjerg station, 180 miles distant, in 1906. By this time they were operating the arc at about 2.8 kilowatts input power.

All capital up to this point had been furnished by the Copenhagen group. Outside capital made its appearance in 1906 when the Amalgamated Radio Telegraph Company, organized by Lord Armstrong and working closely with the American De Forest Company, acquired rights under the Poulsen patents. In 1906 a station was constructed at Cullercoats, near Newcastle, England, and successful tests conducted between there and Lyngby, a distance of some 530 miles. A beginning was also made with a powerful station at Knockroe, on the west coast of Ireland. This was equipped with a 30 kilowatt arc and was intended for transatlantic telegraphy. These plans were abandoned in 1907, when the Knockroe antenna blew down, but the Amalgamated Company's rights were acquired by the Lorenz Company of Berlin, which constructed an arc station at Weissensee, near Berlin. By the summer of 1908 this station was reported as being in constant communication with Lyngby, with both speech and music being transmitted with great clarity. And the editor of *Modern Electrics*, commenting on these achievements, reported that Poulsen—the "Danish Edison" who never gave a promise he could not keep—was very confident that soon he would be able to talk across the Atlantic Ocean.[58]

This was the situation when Elwell arrived in Copenhagen and it is not hard to see why he was given a warm welcome. Despite a series of

---

[57] Thomson vigorously reasserted his claim to priority in a letter to the editor of *The Electrician* dated 30 November 1906. See *The Electrician* 58 (21 December 1906), 378-80.

[58] "New Record in Wireless Telephony," *Modern Electrics* 1 (June 1908), 94; C. F. Elwell, "The Poulsen System of Radiotelegraphy," *The Electrician* 84 (28 May 1920), 596-99; C. F. Elwell, *The Poulsen Arc Generator* (New York, 1923), pp. 22-23; William Duddell, "The Arc and the Spark in Radiotelegraphy," *Nature*, 22 August 1907, 426-30; Fleming, *Principles*, pp. 647-52.

uniformly successful tests and demonstrations, the Poulsen system was not yet in commercial service anywhere, nor had Poulsen and his backers realized any return on their patents. Elwell offered them a foothold in the North American market, and at very little risk. There were no other bidders for the United States rights to the system. If Elwell could stage his demonstrations, raise his capital, and start making his scheduled payments—well and good. If not, the rights would revert to them. In the meantime, the young American engineer was clearly full of energy and initiative, he knew something about wireless telephony, and it seemed that he was vouched for by the American authorities. There could be no harm in letting him see what he could do.

* * *

When Elwell returned to Palo Alto in August 1909, his intention was to set up a low-power arc transmitter, give public demonstrations of wireless telephony, and raise enough money to start building a more extensive system. His longer-term plans were ambitious. Within the United States he envisaged "a continuous wave system of wireless telegraphy and telephony over thousands of miles in competition with the wire line and cable companies."[59] But this was not all. Then as later he never lost sight of the potentials that radio offered for communication across the Pacific, where submarine cable service was notoriously inadequate. In Copenhagen he had bargained hard over territorial rights. Poulsen had wished to retain the rights to his patents in American territories and possessions outside the continental United States, but Elwell had insisted that they be included in the deal. Already he had in mind communications with Hawaii and the Philippines—perhaps eventually to Japan and Australia.[60] In the short run, however, something more modest would have to suffice.

The first requirement was to raise enough money to pay for the two small transmitters ordered from Denmark and for the passage and wages of the two Danish engineers. At the suggestion of Professor Harris J. Ryan, then head of the electrical engineering department at Stanford, and with the promise of financial support from President Jordan and other members of the Stanford faculty, Elwell formed a company, the Poulsen Wireless Telephone and Telegraph Company, which came into existence in September 1909 with offices on Post Street in San Francisco. It was capitalized at $5 million, with shares of $1 par value. Elwell became

[59] Elwell, "Autobiography," p. 50.
[60] Ibid., p. 48.

president and chief engineer; directors were C. D. Marx, head of Stanford's civil engineering department and at that time mayor of Palo Alto, R. W. Barrett and F. A. Wise (both graduates of Stanford Law School) and J. P. Smith of Stockton. No list of the original stockholders has survived, but it can safely be assumed that it was made up mainly if not exclusively of Stanford University faculty and alumni.[61]

To raise more money than this small group could muster, Elwell got in touch with a stock salesman named John C. Coburn, and contracted with him to market the company's shares, with sixty cents of each dollar to go to the company and forty cents to Coburn and his staff. But Coburn had to have more than empty promises to sell, and this is where the little 100 watt transmitter Elwell had brought back from Copenhagen came into play. With that kind of power—enough to run a household light bulb—Elwell would have been imprudent to attempt anything very ambitious, but he could give prospective investors something to listen to. Roland Marx, Professor Marx's son, was an amateur radio operator and had built his own spark station in his father's barn. Elwell set up his arc there, connected a microphone in the antenna lead, and was ready to go. A receiving station with an antenna, a simple detector, and earphones was installed in a house about a mile away. Coburn brought each prospective buyer to the receiving station, Elwell talked into the microphone, and if all went well the Poulsen Wireless Company had another stockholder.[62]

As a technique for mobilizing small amounts of capital quickly, it worked remarkably well, though anyone familiar with Fessenden's work at Brant Rock should not have been greatly impressed. Enough money was raised to bring the two higher-powered arcs over from Copenhagen, at a total cost (including customs duty and freight) of $11,500—burdensome enough to convince Elwell and his associates that from then on they would build their transmitters themselves.[63] Elwell installed them at Stockton and Sacramento, just fifty miles apart, for reasons that were sufficiently obvious: each city had "many prosperous citizens." Tall antennas were erected—each 180 feet high, of the "umbrella" type that Poulsen had used in Denmark—and on 24 February 1910, with appropriate fanfare, two-way wireless communication between the two loca-

---

[61] Ibid., p. 51; Foothill College, file "Cyril F. Elwell"; C. F. Elwell to W. H. Hewlett, 24 December 1953 (Pratt Papers, Bancroft Library). Elwell also mentions as early contributors Professor L. M. Hoskins, Dr. T. Williams, and David Curry "of Yosemite Park fame."

[62] Elwell, "Autobiography," p. 52.

[63] Ibid., p. 54.

tions was begun. President David Starr Jordan officiated at the proceedings. The mayors of both cities were present and conversed with each other. "Wealthy Chinese" of Stockton talked with their compatriots in Sacramento. A representative of the U.S. Army was present and, of course, a flock of reporters. When it was all over (as Elwell later recalled), "Coburn and his henchmen descended on Stockton and Sacramento like a cloud. . . . Rich Chinese and a leading restaurant proprietor in Sacramento helped quite a lot. And a lot of priests had faith in something else besides the resurrection."[64]

There was, indeed, a good deal of the carnival atmosphere about the whole process. But, unlike other wireless promotion schemes of the day, there was no faking of results and no chicanery. Stockton and Sacramento were indeed linked by telephony without wires; the quality of speech was reportedly better than on the wire lines; and the Poulsen Wireless Company was open for business. This was the first time that a commercial radiotelephone service had been offered to the general public anywhere in North America, and it was a significant landmark.

Equally important, much had been learned by Elwell and his associates. In the first place, they had come to appreciate the mechanical simplicity of the arc. There was no reason at all why, given minimal machine shop facilities, they could not manufacture future transmitters themselves, and at a fraction of the cost of importing them from Denmark. This simplicity of manufacture, of course, masked the complexity of what was going on within the arc chamber; but that was a matter that could be left for the future. In this respect the arc stood in sharp contrast to the radiofrequency alternators that General Electric was building. The theory of the alternator was fully worked out; there were no mysteries there. It was the manufacturing that was tricky—the precise machining and delicate balancing of the high-speed rotor in particular. The theory of arc transmitters was imperfectly understood, and remained so for several years; but they were not difficult to construct.

One implication was that Elwell's company very quickly became a manufacturing and not merely an operating business. This was to have important consequences. In the short run it meant acquiring a factory. This was originally a very unassuming facility—a corrugated iron shed with only 900 square feet of floor space behind a small house at the corner of Emerson Street and Channing Avenue in Palo Alto.[65] This was

---

[64] Elwell Papers (Stanford), folder 14.

[65] The house was bought for $1,500 from D. M. Perham, a mechanic employed by the Poulsen Company. For an illustration and the dimensions, see F. J. Mann, "Federal Telephone and Radio Corporation: A Historical Review, 1909-1946," *Electrical Communication* 23 (December 1946), 383.

to remain the company's only laboratory and manufacturing plant until the Navy's order for a 100 kilowatt arc in 1913 made a move to larger quarters unavoidable.

From this modest facility there emerged a second generation of arc transmitters, designed and built by Elwell, which became the progenitors of a distinct evolutionary line, adapted to American operating requirements and the American market. The Danish personnel brought over to help set up the first few stations may have been useful for that purpose, but they played no important role in later developments. Two of them, indeed—Peter Jensen, one of the engineers, and F. Albertus, the mechanic—quit the Poulsen Company early in 1910 to start an enterprise of their own: the manufacture of the first loudspeaker, which they christened the Magnavox. This was the origin of the Magnavox Company, and later of the distinguished line of Jensen speakers.[66] From this point on, until Leonard Fuller joined the company in 1912, Elwell personally designed the company's transmitters and supervised their manufacture. Four 5 kilowatt arcs were the first to be built, intended for the Stockton and Sacramento stations. Two 12 kilowatt units followed.

The circuitry of these Elwell transmitters was very simple. (See Figs. 3.6 and 3.7) Direct current from a motor generator at about 500 volts was fed through large iron-cored chokes to the arc, burning in an atmosphere of coal gas or alcohol and in a strong magnetic flux. The anode or positive electrode, always of copper, was connected to the antenna through a large inductance. The capacitance, which in Duddell's original circuit and in Poulsen's had been connected in series with the inductance and which was essential if the arc were to oscillate, was now provided by the antenna itself. This implied, first, that every arc transmitter had to be designed for and matched to a particular antenna, since the capacitance between antenna and ground was an intrinsic part of the oscillating circuit. And it also meant that there was little filtering out of harmonics and other undesired frequencies, since the antenna circuit was very broadly resonant. Simplicity, in short, was bought at a cost. The same was true of the keying system. To short out a few coils of the antenna inductance every time the key was depressed was an easy so-

---

[66] Elwell, "Autobiography," p. 34; Thorn Mayes, *The Federal Telegraph Company, 1909-1920*, Antique Wireless Association Monographs, n.s. 3 (1979), 3; Mann, "Federal Telephone and Radio Corporation," p. 381; Robert Lozier, "Twenty Years of the Magnavox Story—1911-1931," *The Old Timer's Bulletin* 23 (June 1982), 6-9. For the role of Commander George Sweet, U.S.N., in the formation of the Magnavox Company, see Lillian C. White, *Pioneer and Patriot: George Cook Sweet, Commander, U.S.N., 1877-1953* (Delray Beach, Fl., 1963), pp. 73-77.

Pl. 4: Five kilowatt Federal arc.
Source: Foothill Electronics Museum

lution, but it did mean that every arc transmitter radiated on two wave-lengths, only one of which carried information. This wasted power and frequencies. These and other limitations were to be remedied later. But, in the period during which Elwell was building arcs in the little shed on Emerson Street, they were minor problems, easily dismissed. Later engineers would deal with them, when power levels were higher and harmonic radiation less readily tolerated.

From the technical point of view, then, the major lesson learned from the Stockton—Sacramento demonstrations was that the Poulsen system did work, even far distant from the careful ministrations of its inventor, and that in future transmitters could be and should be manufactured locally. There was some residual skepticism among potential investors as to whether communication would be possible when more than one station was transmitting, and the activities of a large spark station in Sacramento, which disrupted Elwell's voice transmissions, did nothing

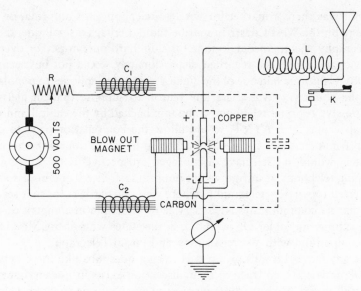

Fig. 3.6: Basic arc transmitter circuit.
Source: Bucher, "Wartime Wireless Instruction," p. 26

to remove the doubts. To prove that, with continuous wave transmission, close tuning was possible, a third station was constructed on a block of land near Ocean Point, San Francisco, and a pair of 12 kilowatt arcs, built in the Palo Alto plant, was installed there. Placed in service in July 1910, this came to be known as the Beach station and helped to prove to skeptics that, even when both Stockton and Sacramento were transmitting at the same time, the operator in San Francisco could, by a minor adjustment of the tuning dial, copy signals from one and exclude the other.

This was all very encouraging in a technical sense, but could the same be said of the company's commercial prospects? It was the hope of transmitting the human voice without interconnecting wires that had taken Elwell to Copenhagen, and wireless telephony continued to be the feature that intrigued engineers, impressed investors, and got publicity in the newspapers. Coburn would have found it much harder to sell shares if all he could let his prospects hear was the buzzing of a tikker in the earphones. And Elwell related, with some chagrin, how they had hoped to impress the newspaper reporters by demonstrating high-speed reception of Morse code—a feature that should have interested people in the news business. But all they would write about was telephony. That was news; telegraphy was not.

But was there a market for wireless telephony in California, or anywhere on the West Coast? In marine radio, perhaps a small one, though telegraphy was the normal mode for shipboard operators. For overland service the prospects, regarded dispassionately, could not but seem dubious. California had one of the highest ratios of telephones to population in the country.[67] Was there any reason to suppose that an alert and aggressive regional telephone company, backed by the capital and technical resources of AT&T, would allow the Poulsen Company to carve out a market share for itself? Lack of interconnection with the wired system would in itself have discouraged prospective users. But if, abandoning telephony as intriguing but unprofitable, Elwell and his associates reverted to wireless telegraphy, where were the markets for that service? In marine communications? In service to Honolulu in competition with the submarine cable? Or overland, in the states west of the Mississippi, in competition with Western Union and Postal Telegraph?

If anyone in Elwell's group was raising questions like these between 1908 and 1910, no trace of their discussion exists in the surviving historical record. Dreams there were in plenty, but one looks in vain for any dispassionate analysis of market prospects. For Elwell and his Stanford associates the technical challenge was sufficient motivation. To prospective investors, the prospect of speculative gains was what Coburn and his men emphasized, not dividends. And yet, in the end, the question could not be ignored. What market was the Poulsen Wireless Telephone and Telegraph Company going to serve? What business was it in?

In the short run it was not even clear that the company would survive long enough to find answers to these questions. Cash was always short: to purchase materials, to meet the payrolls in Palo Alto, Stockton, and Sacramento, to forward to Copenhagen the royalties to which Poulsen and his friends were now entitled. Ordinarily these expenses were met from the funds that Coburn raised from the sale of securities.[68] Subscriptions to the common stock, in other words, were used to meet current expenses. When this was not enough, Elwell borrowed: from his mother-in-law, who lived nearby, from Professor Marx, from anyone who would lend to him. By late 1910 he estimated that he personally had invested $89,000 in the company, in the form of securities purchased and short-

[67] John V. Langdale, "United States Telephone Industry: 1876-1930" (Mimeo); U.S. Bureau of the Census, *Telephones and Telegraphs, 1907* (Washington, D.C., 1910), p. 51.

[68] When the company was first formed, its shares sold for 10 cents each, but by the time the demonstrations were completed and the Beach Station built Coburn had got them up to $1.00. See Elwell Papers (Stanford), folder 14.

term loans. Some of this represented his own savings; the rest was money he had borrowed.

The dilemma was clear. If the company did not extend its system, it could not make money. There was not, and was not likely to be, enough traffic available between Stockton, Sacramento, and San Francisco—linked as they were by two telegraph companies and the telephone—to cover expenses, far less pay dividends. But expansion called for capital the company did not have—large amounts of capital, not the nickels and dimes that Coburn managed to scrape together. And for Elwell personally the imperative must have been obvious. If he was ever to get his money out of the company and pay his debts, someone would have to put a lot of money into it.

Elwell had received one lesson in corporate finance in New York; he was now to receive a second. In his autobiography he writes about the experience with rueful amusement, but it was clearly a painful experience at the time. Shortly after the Beach station was completed he was approached by a San Francisco financier named Beach Thompson, who had recently made a name for himself by running a new electric power line into the city. Thompson had seen the demonstrations and knew of the Poulsen Company's financial problems. He asked Elwell how much was needed immediately to keep the company afloat, and Elwell told him $100,000. Thompson then suggested that a new company be formed, with a capitalization of $10 million, to take over the existing shares of the Poulsen Company on a share-for-share basis. Since there were at the time about $3 million (par value) of these shares outstanding, that would leave $7 million, on paper, for future expansion. Elwell thought that this sounded good. Thompson asked for an option to buy the outstanding shares; with this in hand he would see what he could do to raise the needed cash.[69]

Elwell and his fellow directors agreed, on the understanding that the existing shareholders would retain a 30 percent interest in the new company, and Thompson got the option he asked for. Shortly afterwards, however, he appeared in Elwell's office and announced that he had registered a company under the laws of the State of Arizona with the title of the Poulsen Wireless Corporation. Its nominal capitalization was not $10 million but $25 million. Thompson proposed to exercise his option on behalf of this new company. Elwell protested, since this meant that the interest of the old stockholders would be reduced from 30 to 12 percent. They finally agreed on a compromise figure of 18 percent; this meant that the stockholders in the old company would receive $4.5

[69] Elwell, "Autobiography," p. 56.

million (par value) in shares of the new company in exchange for the shares they held in the old.

Thompson went ahead on this basis. The Poulsen Wireless Telephone and Telegraph Company was liquidated as of 25 January 1911 and all its assets were transferred to the Poulsen Wireless Corporation, of which Beach Thompson became president. Elwell and C. D. Marx joined the board of directors, representing the shareholders of the old company, but they were a minority interest, the other seats on the board being held by associates of Thompson.

The nominal capitalization of the new company was $25 million, represented by shares of $100 par value. At the first meeting of the board it was decided that the new shareholders would subscribe $150,000 in cash, in return for which they would receive 60,000 shares of stock with a nominal value of $6 million. In other words, they would receive their shares for two-and-a-half cents on the dollar. And it was also voted that all the shares issued in exchange for shares in the old company—including those issued to Elwell, Marx, and their associates—would be placed in escrow, so that they could not be sold on the open market. No such restriction was placed on the shares issued to the Thompson syndicate.

It was clear that, in return for the infusion of cash, Elwell and his friends had lost control of their company. Control now rested with Thompson and his associates in the San Francisco financial community. Elwell for a time would continue to play an important part as chief engineer, but the Stanford-affiliated group that had encouraged and supported him in the initial phases of the venture was now reduced to a passive role—a pattern of events that, since then, has become common among "high technology" firms created as offshoots of academic institutions. That Elwell felt he had been outmaneuvered is obvious, and perhaps he had. What particularly rankled was the fact that the new subscribers were receiving shares valued at $100 for a mere $2.50. The company's treasury, he felt, should have got more than that.[70]

On the other hand, seen from the viewpoint of Thompson and his syndicate, the fact that the shares they received had printed on them the statement that their nominal or par value was $100 was of very little significance. They were investing $150,000 in a company that was in deep financial trouble, that had never earned a profit, and that faced a highly uncertain future. What did they receive in return? A list of the property taken over from the old company shows wireless stations valued at $63,400; a factory valued at $5,000; apparatus and supplies worth

[70] Ibid., pp. 56-57.

$392; office equipment worth $50; and patents valued at $25,430,418.[71] But the stations at Palo Alto, Stockton, and Sacramento had been built for experimental purposes and for demonstrations; to carry them on the books at cost was quite unrealistic, and one of the first acts of the new board of directors was in fact to write their value down very substantially.[72] What it came down to was this: The only potentially valuable assets the company had were its rights under the Poulsen patents. And these were encumbered by the scheduled royalty payments to Poulsen.[73]

Investment in the Poulsen Wireless Corporation was, in short, a risky speculation; the heavy discount at which the shares were originally distributed reflected that fact. It is also true that, from this point on, the Poulsen Corporation's directors seldom if ever decided upon a policy without giving thought to the effect that policy would have on the market price of the company's shares. And why not? It was by disposing of those shares at the appropriate moment that they expected to reap the rewards of their enterprise.

✳   ✳   ✳

The Poulsen Wireless Corporation was the instrument by which Beach Thompson took over the original company, and it was Poulsen shares that were actively traded in San Francisco. Management of the wireless business was turned over to a wholly owned subsidiary, registered in California. This was originally entitled the Wireless Development Company, and had a modest capitalization of $100,000. Its name was changed shortly thereafter to the Federal Telegraph Company. If Poulsen Wireless was the important company to speculators and investors, Federal Telegraph was the one that counted to wireless operators, and the arc transmitters that it built and used were always known as Federal arcs.

Thompson's intervention not only provided a badly needed infusion of new cash (incidentally taking Elwell "off the hook," as he put it).[74] It also gave Federal Telegraph an explicit market orientation that it had previously lacked. Thompson's plans did not include wireless telephony,

[71] Foothill College, folder marked "History of Electronics–Perham," typescript "Federal Telegraph Company–Poulsen Wireless Corporation: History—1909-1912." The sum of these values equals the capitalization of Poulsen Wireless plus the $500,000 bond issue (see below, n. 73) minus $700 in "directors' shares."

[72] Ibid.

[73] These royalty payments ended when the Poulsen Wireless Corporation was formed. In their place Poulsen and his associates received $500,000 worth of bonds in the new company.

[74] Elwell Papers (Stanford), folder 14.

which had received most emphasis up to this point. Nor did they include marine radio, nor the manufacture of radio apparatus for sale to others. His design was essentially simple: he intended to construct a network of radio telegraph stations connecting the major cities of the Pacific coast and extending eastward into Texas and as far as Chicago. And he had in mind particularly the handling of press traffic, although the service would also be available to the public.

This was an ambitious project. To offer a press service—one that newspaper publishers could count on—called for a degree of reliability that wireless telegraphy had not shown up to this point. The distances involved were large; transmission would be over land, believed to be more difficult than over water; and the operators would have to copy traffic through static that was often heavy. The technical challenge was therefore a large one. But commercially too Thompson was setting the company a difficult task. The cities he intended to serve were already linked by wired telegraph systems. If Federal Telegraph was to attract traffic, it would either have to undercut Postal Telegraph and Western Union in rates, or interest a kind of customer that these companies were not currently serving. Prior history did not suggest that they would allow their share of the market to be eroded without retaliating.

The stations were built in rapid succession over the next two years. Los Angeles was the first to be completed and, working with the Beach station in San Francisco, served to test equipment and procedures. El Paso was next, closely followed by San Diego, Fort Worth, Kansas City, Chicago, and Seattle. A relay station was established at Phoenix to assist in moving traffic between Los Angeles and El Paso—a circuit much troubled with static in daytime; and for the same reason a station was built at Medford, Oregon, to assist the Portland and Seattle circuit. By early 1912 the Federal Company had a system of thirteen stations in operation (Stockton and Sacramento being retained as traffic feeders).[75] With the exception of the antennas and supporting towers, they were built to identical designs, each being equipped with a 12 kilowatt arc transmitter (except for San Diego, which had 5 kilowatts) and a "tikker" receiver; all equipment was made at the little factory in Palo Alto. They provided communications service over distances of 500 miles in the daytime and

[75] Elwell, "Autobiography," p. 57; Elwell Papers (Stanford), folder 13; Mann ("Federal Telephone and Radio Corporation," p. 388) give a total of fourteen stations by 1912, but he includes the new South San Francisco station, built in 1912, with Honolulu traffic in mind. This represented a shift to a different strategy. Mayes (Federal Telegraph Company, p. 5) omits the Seattle station.

1,000 miles at night—remarkable distances for a commercial radio service at that time.[76]

This system was quite explicitly in competition with the wired telegraph companies. Competition took the form both of lower rates per words and more varied and convenient forms of service. At San Francisco, Los Angeles, and Chicago, for example, offices were maintained in the downtown business district and linked to the transmitter site by direct wire connection (in Chicago, in fact, the station itself was located on the 26th floor of the Transportation Building in the central business district, and the antenna masts were erected on the roof). Eventually, in San Francisco and Los Angeles, the company offered service directly from the floor of the stock and produce exchanges; between these two cities agricultural market information made up a large proportion of the traffic. Special rates were also set to attract the general public. Federal charged the same price for fifteen words plus address as the wire companies did for ten words plus address; and a special "day letter rate" was introduced for fifty words plus address, for transmission at any time of the day at the company's convenience. Night letters of seventy-five words plus address could be sent for the same price as ten-word telegrams by Western Union.[77] For press traffic, Beach Thompson negotiated a contract for a special volume rate with the Publishers' Press Association.[78]

Much of this business called for high-speed operation—not so much because of the volume of traffic to be handled in each twenty-four hour period, but because of the need to pass as much traffic as possible during the periods when atmospheric static was low. Poulsen's high-speed photographic recorder did not prove usable for such service. In its place Elwell and his assistants developed a high-speed telegraphy system using perforated Wheatstone tape for transmission and Poulsen's telegraphone for reception. When the telegraphone tape was played back at lower speed for transcription, of course, the volume and pitch of the received signal dropped substantially; this was to have interesting consequences for the development of vacuum tube circuits. (See below, pp. 234-35.)

Technically, the construction and operation of this system was a notable achievement. Was it also profitable? The available evidence does

[76] Mann, "Federal Telephone and Radio Corporation," p. 388.

[77] Ibid.

[78] "Federal Telegraph Company–Poulsen Wireless Corporation: History—1909-1912." Federal paid for construction and operation of the stations while Publishers' Press undertook to furnish the commercial business to make them profitable. Perham comments: "This was a very complicated financial scheme and the expected benefits arising to the company were probably not all of the story."

not permit a certain answer. The Federal Telegraph Company was absorbed by the Mackay cable system in 1928 and subsequently became part of IT&T. In 1930 its manufacturing operations were transferred to the East Coast. At some stage most of its corporate records were lost or destroyed and the data now surviving do not permit a confident estimate of the company's earnings during the Beach Thompson regime.

An internal report prepared in 1914 by the company's secretary gives the following revenue and expenditure figures for the first three-and-a-half years of operation:

Year	Revenue	Expenses
1911	$ 3,641.68	$ 58,485.75
1912	$39,948.87	$179,093.87
1913	$79,922.56	$156,516.83
1914 (to June)	$62,453.96	$ 72,418.17

The dollar totals for "Expenses" in this report, unfortunately, include both construction and operating costs, and it is therefore not possible to determine the company's operating profits, nor what rate of return on capital it was earning.[79] The guess may be hazarded, however, that if the capital and operating budgets could be separated, they would show, between 1911 and 1914, a steeply rising trend in revenue from operations, together with substantial outflows of cash for construction. If Beach Thompson and his associates profited from their investment in this period, it was probably in the form of capital gains on the shares they held, not from dividends paid.

It would be interesting, too, to have a regional breakdown of the company's earnings. Thompson had originally intended to have a total of seventy cities in the network. In fact the maximum number in operation at any one time was fourteen, and that included two relay stations that did not generate traffic directly. The eastern section of the network, with the single exception of Phoenix, was abandoned in 1913, after high levels of atmospheric static during the summer of 1912 had shown that commercial service could not be guaranteed. And there are hints that this strategy of expansion eastward involved the company in serious losses. It appears, therefore, that the ambitious plan to link the West Coast with the plains states and Chicago proved uneconomic. And this is easy to believe. Survival in competition with the wired telegraph systems, not to mention the telephone, depended on obtaining volume business, and the eastern section of the network, from San Francisco and Los Angeles east

[79] Foothill College, folder "Report on Poulsen Wireless Corporation and Federal Telegraph Company."

to Chicago, never attained that, despite the contract with Publishers' Press. The Pacific Coast segment seems to have done better. Those stations continued to function until 1917, when, with American entry into World War I, all private radio stations were closed down by government order. Even then, Federal maintained its Pacific Coast service, using wire lines leased from the telephone company.

Elwell personally considered the attempt to build and operate a continental radiotelegraph network a major strategic error. "The route to bankruptcy," he called it.[80] And he associated it with what seemed to him irresponsible behavior on the part of Beach Thompson and his brother-in-law, Veeder, who was secretary of the company. Money raised by the sale of shares, he believed, was being used to line the pockets of the company's officers instead of to build up business.[81] But, quite apart from such evidences of personal alienation, Elwell had strong economic arguments on his side. Thompson's strategy, in his judgment, meant pitting the Federal Company against the toughest competition there was, in a market already well served by the wire networks, and in a context in which the technical advantages of radio could not be properly exploited. And in the meantime alternative strategies, which promised higher rates per word and freedom from land-line competition, were being ignored. As against Thompson's strategy of continental expansion, Elwell argued for a maritime strategy: marine radio, of course—a field in which Federal had so far done nothing—but above all transpacific radio. From San Francisco to Honolulu first, then to Midway, and then to Guam. Between the United States and the Orient there was only one submarine cable available for use and it was highly unlikely that another would be laid. Rates per word were high: 35 cents per word from San Francisco to Honolulu (16 cents for press service) and $1.08 per word from San Francisco to Manila.[82] Meanwhile, for the traffic it was sending from

[80] Pratt Papers (Bancroft Library), Box 4, folder "C. F. Elwell Autobiography, 'Radio Changed Horses.'"

[81] See Elwell Papers (Stanford) folder 13: "B. T. and the Publishers Press idea (lousy) held the field and lead [sic] to my discontent as much as the treasury squeeze—taking money for shares sold *to help develop* (not to buy a palatial home in Atherton for B. T. and maintain his brother-in-law Howard Veeder at the Bohemian Club)."

[82] There were two submarine cables across the Pacific: a British-owned cable from Vancouver to Fanning Island, Fiji, Norfolk Island, New Zealand, and Australia, laid in 1902; and an American-owned cable from San Francisco to Hawaii, Midway, Guam, and Manila, laid in 1902-1903, with an extension to Shanghai added in 1906. The U.S. cable also had a "spur" from Guam to Bonin, where it connected with a Japanese cable to Tokyo. See F. J. Brown, *The Cable and*

San Francisco to Chicago, Federal was receiving, on the average, only 3 cents per word. The cost of establishing at Honolulu a station capable of round-the-clock operation Elwell estimated at less than $40,000. The whole Pacific chain could be built for less than the $150,000 Thompson and his friends had originally subscribed.

If Elwell's plans were to be implemented, transmitters of higher power were needed. An operator with receiving equipment was sent to Hawaii soon after the Beach station, with its 12 kilowatt transmitter, opened for service. He reported that it could be heard in Hawaii at night, but not during daylight hours, and that reception was poor.[83] Late in 1911 a 30 kilowatt transmitter was imported from Denmark and installed in the San Francisco station, but it made little difference to reception in Hawaii. It did, however, give the Palo Alto factory a model to work from and it convinced Elwell that higher power in itself was not the answer. Using as his principal argument the need for greater reliability on the land circuits, he persuaded Federal's directors to authorize the construction of a new station in South San Francisco with a much larger antenna system—two wooden lattice masts 600 feet apart, each 440 feet high, with a huge triangular network of phosphor bronze wires strung between them.[84] This station was equipped with 30 kilowatt arcs built in the Palo Alto plant and its greater efficiency was immediately evident on the continental circuits. For the first time it became possible to send messages directly to Chicago without relays.

For Elwell the new South San Francisco station, with its more powerful arcs and larger antenna, had a special significance. It reopened the possibility of communicating with Hawaii. The problem was to convince Federal's directors to do something about it. Until the continental network was completed and its profitability either proved or disproved, they had little interest in expanding westward. Elwell used the threat of competition to move them. In May 1912 he received word through a business associate that the American Marconi Company, which already operated a powerful spark station at Bolinas, near San Francisco, intended to open a Pacific circuit by constructing a sister station in Hawaii. This, as a matter of fact, was correct information, although the threat was not as

---

Wireless Communications of the World (London, 1927), pp. 22-23. For the rates between San Francisco and Honolulu, see C. F. Elwell, "Autobiography," p. 58 and Mann, "Federal Telephone and Radio Corporation," p. 389.

[83] Mayes, *Federal Telegraph Company*, p. 4.

[84] This description of the antenna at South San Francisco is based on Lee de Forest, "Recent Developments in the Work of the Federal Telegraph Company," IRE *Proceedings* 1 (January 1913), 41. It does not wholly agree with that given in Elwell, "Autobiography," p. 58.

imminent as Elwell made it out to be in his report to Federal's directors. He asked for authority to construct immediately a station on Oahu that would be a duplicate of the new station at South San Francisco. The transmitter was already available; other materials could be obtained locally; and, he argued, whoever was first in the field would have the choice of sites—an important consideration, since Oahu was not a large island.

His urgency convinced them, and Elwell sailed for Honolulu immediately. He had the station built and ready to operate within forty-seven days—a remarkable achievement, and testimony both to Elwell's driving energy and to the ease of installing a Federal arc. One doubts whether a big rotary spark, or a Marconi disk discharger, or one of GE's alternators, could have been shipped to Hawaii, installed, and placed in service in such a short time. The Marconi high-powered station at Kahuku was not ready for testing until September 1914.

The Honolulu station—actually at Heeia, with an office in Honolulu—was completed in May 1912 and opened for press service on 12 August. Technically it was a qualified success. Communication with South San Francisco was good by night, but not reliable by day. This was enough, however, to offer a commercial service at rates substantially below those charged by the cable. Federal charged 25 cents per word for telegrams to San Francisco; the cable rate was 35 cents. For press traffic the cable rate was 16 cents a word; Elwell offered the Honolulu publishers a rate of 2 cents a word and guaranteed a daily press bulletin averaging 1,500 words.[85] And, after demonstrating that he could make good on his promises, he found ready customers. By October 1912 Federal was handling an average of 2,300 words of press traffic a day between San Francisco and Honolulu, and a steadily increasing volume of regular commercial traffic.

This achievement had a variety of important consequences. In the first place it put the Marconi Company on the defensive and called into question the Marconi commitment to spark technology. In 1912, to offer a commercial radio service over a distance of 2,100 nautical miles, and to maintain that service reliably, was unprecedented. The distance between Clifden, Ireland, and Glace Bay, Nova Scotia—the Marconi Company's transatlantic circuit at that time—was a little less: 1,939.5 nautical miles. More was at stake here than corporate prestige. The Marconi Company was at this time the prime contender for the contracts to build the stations of Britain's Imperial Chain. It intended to use for those stations its most advanced type of transmitter: the Marconi rotating disk

[85] Mann, "Federal Telephone and Radio Corporation," p. 389; Elwell, "Autobiography," p. 61.

discharger.[86] This represented spark technology at its most advanced. But now there had appeared on the scene an alternative technology—the Federal Company's version of the Poulsen arc—which seemed to promise performance at least as good as any Marconi transmitter, with considerably lower power requirements, and with a great gain in simplicity and economy.

Beach Thompson had been in London in 1911, arguing for the superiority of the arc, but he had nothing very remarkable to report in the way of performance at that time and few took him seriously. The opening of the Honolulu circuit changed all that. The Marconi Company publicly maintained its faith in spark, but a perceptible defensiveness crept into the tone of its pronouncements—as for example in the letters that Godfrey Isaacs, its managing director, wrote to the *Electrician* about the claims being made for the arc.[87] And for the first time the company felt it necessary to reassure its own stockholders. Marconi himself, they were told, had tested continuous wave systems many years before and had preferred to develop his own system, which was "a compromise between the continuous waves and spark systems, combining the best points of both."[88] And there was no truth, they were assured, to the recurrent assertions that their company had recently tried to buy the Poulsen patents. Indeed, Marconi himself had flatly denied it in testimony before a select committee of the House of Commons.[89] This invocation of Marconi's personal prestige may have dissipated some doubts but suppose that Marconi and his advisers were wrong? If indeed the future lay with the continuous wave, the position of the Marconi Company was not a comfortable one. Unlike the Germans, the French, and the Americans, it had no radio alternator under development. Poulsen's British patents had still five years to run in 1912 and, despite Marconi's denials, his company could still have acquired them. But it was beginning to look as if it was,

[86] For a description, see Baker, *History of the Marconi Company*, pp. 117-20. Baker presents the disk discharger as a continuous wave transmitter. This it was not, any more than were Fessenden's synchronous rotary spark or Max Wien's quenched spark. These were spark transmitters which, by generating damped spark wave trains that overlapped with each other and were in synchronization, produced an approximation to a continuous wave.

[87] See, for example, *The Electrician*, 6 September 1912 and 1 November 1912. Elwell replied on 3 October and 27 December 1912 and on 21 February 1913.

[88] Circular, Henry W. Allen, Marconi's Wireless Telegraph Company Ltd. to shareholders, 14 December 1912, as reproduced in Elwell, "Autobiography," pp. 64-65.

[89] As quoted in A. H. Morse, *Radio: Beam and Broadcast* (London, 1925), p. 77.

not the arc itself, but the way the Americans used the arc, the modifications they made in design as power levels increased, that made the difference. And to these American developments the Marconi Company did not have access. To an uncomfortable degree, its commercial future in long-distance radio, and its hope of winning the contracts for the Imperial Chain, depended on the disk discharger, the ultimate refinement of Hertz's spark gap.

But it was not just in Britain that the performance of Elwell's arcs was attracting official attention. British authorities worried about transmitters for the Imperial Chain. In the United States the analogous problem was the choice of transmitter for the Navy's new station at Arlington. Officially that decision had been made in 1909 when, as we have already seen, the contract was awarded to NESCO for a Fessenden-designed synchronous rotary spark set. Privately, however, some of the officers of the Navy's Radio Division were uneasy with that choice, just as some observers in Britain were beginning to look askance at the disk discharger. Fessenden had been preaching the virtues of the continuous wave for years, and he had made converts in principle. But there was no high frequency alternator available in 1909 or in 1912 that could generate the kind of power the Navy wanted.

Was there another possibility? There was no secret about the arc's ability to generate continuous waves. The technical literature put that beyond dispute. And the Navy itself had some experience with arcs, in the shape of the de Forest's ill-starred radiotelephone sets. But the requirement now was for high-powered, long-range radiotelegraphy. The Federal Company's overland circuits had made little impression on official opinion; but the San Francisco-Honolulu circuit was a different matter entirely. That was much closer to the kind of radio link that the Navy had in mind.

As far as we know, no official or even semi-official invitation was extended to Elwell or to the Federal Company to bid on Navy contracts. And, until the closing months of 1912, there is no indication that the directors of the Federal Company ever contemplated such action. Their production facilities up to that time had been fully employed manufacturing equipment for the company's own stations, and it was on the success of the overland intercity network that they had pinned their hopes for the company's future. Government contract work had been no part of their plans.

This does not mean, however, that their activities had escaped the Navy's attention. On the contrary, the San Francisco stations had been thoroughly examined by personnel from the Mare Island Navy Yard. Lt. E. H. Dodd, for example, was given a guided tour by Elwell and Lee de

Forest in April 1911. He was particularly struck by the simplicity and quiet operation of the transmitter—not an uncommon reaction from people accustomed to high-powered spark sets. "Everything about the station is very simple but well made," he wrote, "and the operating is practically noiseless."[90] His formal report, made on 23 May 1911, made the same point and others: "The installation is simple but well made and compact. The operating is practically noiseless. The apparatus can be operated swiftly and accurately, and can be selective when so desired. . . . The apparatus is not dangerous to handle . . . the messages are practically as secret as land telegraph communication." And he reported on performance: "Daylight communication, as well as night, is certain between San Francisco and Los Angeles, and probable between San Francisco and El Paso and San Francisco and Honolulu."[91]

This kind of information was not likely to be ignored by officers of the Navy's Radio Division in 1911. Elwell, in his autobiography, leaves the impression that it was he who opened negotiations for the sale of transmitters to the Navy; and that may well be true. But he had no difficulty, on his visit to Washington in September 1912, in gaining access to Hooper and Hepburn, nor in getting them to introduce him to Admiral Cone. And it may be that securing permission to install a 30 kilowatt Federal arc in the Arlington station did not require quite as much argumentation as Elwell, for dramatic purposes, suggested it did. The ground had been prepared; the reports from Mare Island had been noted; above all there was an undercurrent of dissatisfaction with the Fessenden rotary spark and the incontrovertible evidence of the successful functioning of the Honolulu circuit. The prospect of reliable radio communication between Mare Island and Pearl Harbor was not one that the Navy could ignore; any transmitter that could do that looked like a good prospect for Darien and possibly for the other stations of the high-powered chain.

There were, however, two problems. The first was that twenty-four-hour continuous service between San Francisco and Honolulu had not in fact been achieved. For reasons imperfectly understood but known to be typical of all very low frequency long-distance circuits, signals that were perfectly readable by night could not be heard by day. For commercial service this was inconvenient; it meant that all traffic had to be moved during the hours when the channel was open. Hence the emphasis

[90] Foothill College, folder "Reports of Lt. Dodd," letter to "My Dear Geo," 7 April 1911.
[91] Foothill College, folder "Reports of Lt. Dodd," letter, Lt. E. H. Dodd, U.S.N., to Commandant, Navy Yard, Mare Island, California, 23 May 1911.

on high-speed operation. But for naval purposes it was much more serious. Not all naval messages could safely be held for twelve hours.

As long as it was assumed that long-distance communication required very long waves, and as long as there was no known technique for amplifying signals at the receiver, there were only two ways of tackling this problem: more extensive antenna systems, and higher power. Elwell tried both, raising the antennas at San Francisco and Heeia to a height of 606 feet and stepping up the transmitter input power to 60 kilowatts. Neither had done the trick.[92] The 60 kilowatt arc was, in fact, a failure. Since the Navy's contract for the Darien station called for a 100 kilowatt unit, this was hardly encouraging.

The second problem was that Elwell's relations with the Poulsen Wireless Corporation and the Federal Telegraph Company were finally approaching the breaking point. Many factors contributed to this. Haraden Pratt, who knew Elwell personally, once said that he was "one of those individualists with whom it is hard to get along unless both parties see eye to eye."[93] Elwell and Beach Thompson never had seen eye to eye. Thompson had launched the company on a course that Elwell considered disastrous and he had neglected fields of action that in Elwell's judgment demanded decisive initiative. Beyond that there was a clash of temperament and personality, between Elwell the engineer and Thompson the financier and speculator. And beyond that again were Elwell's motives for getting into radio in the first place: he had wanted to be in charge, to make the decisions, to run his own show. Under Beach Thompson, that was exactly what he had not been able to do.

The stage was set, in short, for another of those lateral shifts that were characteristic of Elwell's career. And the issue that precipitated it could have been predicted. In the spring of 1913 Elwell went to Federal's board of directors and proposed that the company should extend its service westward from Honolulu to the Philippines, Guam, and eventually to Japan and China. To him the suggestion made perfect sense. The Honolulu circuit worked, at least at night. From Honolulu to Manila would be a longer jump, but it could be done. But the real market was the Orient. The directors would have none of it. Mindful of the failure of their eastern network, they were in no mood to authorize expansion

[92] Mann, "Federal Telephone and Radio Corporation," p. 389. Day-and-night service between San Francisco and Hawaii was finally achieved in January 1914, after the transmitter power had successfully been raised to 100 kilowatts. See Mayes, *Federal Telegraph Company*, p. 6.

[93] Haraden Pratt to Lloyd Espenschied, 12 July 1963 (Pratt Papers, Bancroft Library).

westward and they rejected his proposal. Elwell thereupon resigned and severed his connection with the company.

Elwell's split with Federal Telegraph invites comparison with Fessenden's difficulties with NESCO. There was, in Elwell's case, no dissension over patents, no prolonged litigation, none of the distressing and melodramatic events that marked Fessenden's departure. Elwell had simply had enough, so he left. He did not turn his back on further work on radio, as Fessenden did. Quite the contrary: he went to Europe and joined the Universal Wireless Syndicate, which had elaborate plans for a transatlantic radio service using arcs. And when that venture collapsed with the outbreak of World War I, Elwell became a kind of freelance engineer, building arc transmitters and antennas for anyone who would pay for them: the Royal Navy, the French and Italian governments, the British Post Office. Later he helped to found the Mullard Valve Company, one of the first commercial manufacturers of vacuum tubes in Britain. And still later he played a role in building the first antennas for Britain's longrange radar system. He ended his days back in Palo Alto, serving as engineering consultant for the Hewlett-Packard Company, manufacturers of high-quality electronic test equipment. His resignation from Federal Telegraph was merely an incident in a long and productive career; it had none of the traumatic quality that surrounded Fessenden's divorce from NESCO.

Nevertheless, there are parallels. In abstract terms, Fessenden and Elwell were both performing the same functions. They were technological translators, men working at the interface between the laboratory and the marketplace. Believing as they did that the continuous wave held the key to the future of radio, they took it upon themselves to perfect the necessary technology, to reduce it to practice, and to show that it could outperform available technical alternatives. To do these things required a shift from the world of purely technical criteria—designing an arc that would oscillate, showing that an alternator could generate Hertzian waves—to a world where market considerations played a major role. Where should stations be built? What service should they provide? Above all, who was to provide the necessary resources, and in return for what measure of control over decision-making? Fessenden and Elwell both successfully engineered that shift, but they paid a price. They had thereafter to work under institutional constraints. Specifically, they had to yield control over the innovation to individuals and organizations that reacted to different incentives and responded to different signals. Given and Walker were to Fessenden what Beach Thompson and his associates were to Elwell. NESCO was the institution created to manage Fessenden's innovations, as Federal Telegraph was to manage Elwell's. And in both cases, as control shifted from the individual innovator to the corporate institution, as technical development became increasingly a function of market performance, stresses appeared that in the end made joint action impossible.

This is not to say that no engineer or inventor could work with institutions like NESCO and Federal Telegraph: merely that Fessenden and Elwell could not. To complement Reginald Fessenden and carry on his work there was Ernst Alexanderson, who found General Electric an excellent environment for creative engineering. And to complement Cyril Elwell there was, as we shall see, Leonard Fuller, designer of the third and greatest generation of arc transmitters, a man no less talented than Elwell but with skills and a temperament that Elwell lacked.

*　　*　　*

Elwell left with the Federal Company a legacy of interrelated problems and opportunities. There was a problem of corporate strategy. The company had abandoned its eastern network, and it had refused to extend westward beyond Hawaii. Where did its commercial future lie? Building transmitters for the government was an attractive possibility but that brought up the second problem. How were the larger transmitters the Navy wanted to be built? So far Federal had not succeeded in breaking through the 30 kilowatt barrier. A solution to both these problems was offered by the positive elements in Elwell's legacy: the body of experience in the construction of arc transmitters that had been built up under his tutelage and the engineering talent that he had brought together in the Palo Alto plant.

There were two engineering groups working for Federal Telegraph in 1912, largely independent of each other. One was made up of Lee de Forest, Charles Logwood, and Herbert Van Etten. They were working on vacuum tube development, and their activities will be described in Chapter Four. The other group was headed by Leonard Fuller, then twenty-three years old. Elwell had hired Fuller in September 1912: his assignment was to work on improving arc transmitters, and specifically on the problem of the refractory 60 kilowatt arc that had been scaled up from the 30 kilowatt units used at South San Francisco and Honolulu.[94]

[94] Biographical information on Leonard Fuller is drawn largely from "Leonard Franklin Fuller: Research Engineer and Professor," an interview conducted by Arthur Lawrence Norberg in 1973, 1974, and 1975 (Bancroft Library, University of California, Berkeley: History of Science and Technology Project; 1976; reproduced by permission). Many of Fuller's papers are also on deposit in the

If Federal Telegraph was to become more than a regional telegraph service, its future depended on its manufacturing capabilities. These depended on its success in winning Navy contracts. And this in turn, after Elwell's departure, depended largely on Leonard Fuller. Fortunately, he came well prepared for the task.

Fuller had first come in contact with the Federal Telegraph Company in the summer of 1910. He was an undergraduate engineering student at Cornell University at the time and, while home in Portland, Oregon, on his summer vacation, he was asked by his family doctor if he would go down to San Francisco and look over the Federal Company's stations and equipment. The request was a sensible one: the physician and, one presumes, some of his friends and associates had been approached by representatives of the Poulsen Company with an invitation to purchase shares, and they wanted an independent report from someone they knew and trusted before parting with their money. Fuller was a good choice. He had been an active "ham" operator in Portland before going off to college and, as a Cornell engineering student, he could reasonably be expected to cast a knowledgeable eye on equipment that was a mystery to physicians and attorneys. Fuller went down by boat and Elwell showed him the stations at San Francisco, Stockton, and Sacramento—the 12 and 5 kilowatt transmitters, the crossed-gold-wire tikkers, the carbon microphones, and all the rest. For some reason he was not allowed to see the Palo Alto factory.

Fuller gave a favorable report when he got back to Portland. Like many others who were accustomed to spark sets and damped wave transmission, he was struck by the selectivity and sharpness of tuning that continuous wave operation made possible. And the Poulsen arc itself impressed him greatly: "its quietness in contrast with the noise of a spark; the simplicity of the stations, just a motor generator set to supply direct current, an arc, a helical loading coil and a telegraph key. That was it, as far as the transmitter was concerned."[95] Apart from the arc, the Federal stations were not, in fact, greatly different from amateur stations of the day.

Fuller had no thought at that time of going to work for Federal. It had been an interesting summer assignment, but he had his education at Cornell to finish. When he returned to Ithaca, however, he took the memory of the arc with him. Cornell had an amateur radio station at

Bancroft Library. See also Mayes, *Federal Telegraph Company*, pp. 6-13. Mr. Fuller is presently living in Palo Alto and, through correspondence, was of much assistance in the early stages of preparing this manuscript.

[95] Fuller interview, p. 41.

that time, located in Sibley College, used for instruction and demonstration. It was well equipped by the standards of the day—a fine flattop antenna, good receiving equipment, and a synchronous rotary spark transmitter that the students had built themselves. To someone who had seen and heard a Poulsen arc in operation, however, one can well imagine that the station may have seemed not quite as up-to-date as it could be. Fuller determined to build a Poulsen arc himself. With the help of a friendly faculty member he found a discarded direct-current generator that gave him the electromagnets he needed. They removed the armature and put an arc chamber in its place, machining the pole tips in the college's machine shop. And in short order they had a functioning Poulsen arc transmitter for the students to use and study—severely limited though it was by an inadequate direct current power supply.[96] It may well have been the first continuous-wave amateur station in the country.

Cornell required a senior thesis for the M.E. degree and, much as Elwell had used his work on electric furnaces to help meet academic requirements, so Fuller turned to good account his work on the Poulsen arc. His idea was that it might be possible to design an arc that oscillated at high frequencies like Poulsen's arc but did not require a hydrogenous atmosphere, a water-cooled anode, or an expensive magnetic field. One could substitute a rapidly rotating aluminum disk for the anode. That seemed to hold out several advantages: it would provide effective cooling; the arc would burn from a constantly changing spot on the rim of the disk; and it was believed to be the case that an arc with an aluminum anode and carbon cathode had a steeper characteristic curve than a copper-carbon arc in air. Fuller built such an arc in the Sibley College machine shop during the winter of 1911-1912. It functioned as expected, though the note sounded rough in a radio receiver (they had no oscillograph with which to check the waveform), but once again the limited power supply proved a problem. The arc could not be tested above 2 kilowatts input. Fuller had to discontinue working on it at that point.[97] He used his findings for the first section of a thesis and added second and third sections on other related topics. Technically, inability to complete the tests was a disappointment, but the thesis itself was commendable. Fuller was awarded his degree in June 1912.

It is possible that, when he left Cornell, Fuller already knew as much about the theory of the Poulsen arc as anyone in the country. Elwell and his staff in California knew how to build and operate the devices, but neither in Palo Alto nor anywhere else in the United States except at

[96] Ibid., p. 31.
[97] Ibid., pp. 31-32.

Cornell had the radiofrequency arc been subjected to systematic theoretical analysis and experimentation. The point is not that Fuller in 1912 had already worked out an adequate mathematical model of the arc's functioning, for he had not; that was not to come until completion of his doctoral dissertation for Stanford University seven years later. But he had begun to apply to the arc a type of disciplined analysis it had not received before. For Fuller it was not just an interesting and useful device; it was also an intellectual challenge.

Fuller had the offer of two jobs when he left Cornell. One was from the Great Western Power Company, then engaged in building a large hydroelectric generating station on the Feather River. The other was from the National Electric Signaling Company. It was not an easy choice. Fuller loved the West, and he liked electric power work. On the other hand he had a longstanding fascination with radio and NESCO offered him a chance to continue his research on arcs. That was enough to make even the prospect of working at the Bush Terminal in Brooklyn attractive, and Fuller accepted NESCO's offer. He felt, as he later recalled, that it would give him "an opportunity to participate in the research and development of new things, new challenges."[98] And if it turned out that he had made a mistake, there was no harm done; he could always get into the electric power business.

Fessenden, it will be recalled, had parted company with NESCO by this time, and the company's affairs were in some disarray. It was maintaining two operations: the experimental station at Brant Rock and, in Brooklyn, a manufacturing plant with machine shops, laboratories, and offices. Its bread-and-butter business was manufacturing 500-cycle synchronous rotary spark transmitters for marine use and for sale to the Navy, but it was also engaged in an intensive research effort to perfect the heterodyne or "beat frequency" receiver. This required, above all, a quiet and stable local oscillator. It was with this end in view that the company had hired Fuller. His assignment was to see whether a small Poulsen arc, without a magnetic field, could be developed for use in receivers. That would be a major breakthrough, if it could be done.

Fuller worked for NESCO for only a few months, testing various types of carbon and graphite electrodes. Whether the arc could ever have been made into a suitable local oscillator for receivers must remain an open question. Within a few years the "feedback" vacuum tube oscillator took over that role and no further work was done on arcs. NESCO, possibly because of Fessenden's successful litigation, found itself in financial dif-

[98] Ibid., p. 42.

ficulties in the late summer of 1912 and cut back its work force drastically. Fuller was among those laid off.

Here was his chance to get out of radio and into the more prosaic field of electrical power engineering if he chose. He had had his fling with wireless; now surely it was time to settle down to more conventional work. Instead, he sent a telegram to Elwell asking whether a position could be found for him with Federal Telegraph. And Elwell replied immediately, offering him a job. Fuller showed up in San Francisco in September 1912, and within a few days he was at work in the Palo Alto factory.

A question of motivation arises at this point, for Fuller's decision to stay in radio was to prove critical both to his career and to that of the Federal Company. It seems clear that what attracted him most was not the prospect of financial gain or of public reputation but rather the open-ended research opportunities, the unanswered questions, the intellectual puzzles. His work with arcs at Cornell and Brooklyn had merely served to whet his appetite. Now that NESCO had dropped its research program, Federal was the only organization in the country doing serious work with arcs. One has the impression once again that it was the arc itself, that remarkable generator of high-frequency oscillations, that was the major inducement.

As for Elwell, he undoubtedly realized that he needed more engineering talent on the premises if he was to build high-powered arcs for the Navy and extend the San Francisco–Honolulu circuit. And if, as is possible, he already felt in the closing months of 1912 that he and Federal Telegraph must soon part company, it became all the more important to have someone on the payroll who could take over when he left. To this it should be added that Elwell was a big enough person to know his own limitations. With the failure of the 60 kilowatt arc he was up against a major problem that he did not know how to solve. Controlled experiments and mathematical modelling were not his strong points.

Fuller's first assignment was to familiarize himself with the Marconi four-circuit tuning patent—the famous "four sevens" patent—and the high priority given to this matter is in itself interesting.[99] The most desirable method of coupling a transmitter to an antenna was to feed the transmitter's output into a resonant circuit and couple that to the antenna (itself a resonant circuit). Among other advantages, such an arrangement helped to filter out undesirable harmonics of the frequency on which the transmitter was supposed to be radiating. If a similar arrangement were used at the receiver, one had the "four-circuit" system covered by the

[99] Ibid., p. 47.

Marconi patent. The Federal Company did not use that method: their practice was to feed the arc's radiofrequency output through a tuning inductor and then into the antenna. This avoided the Marconi patent, but at a price—loss of much of the filtering action. Elwell's purpose in insisting that Fuller, before he did anything else, master the details of the Marconi patent was to make sure he understood why Federal built their antenna circuits as they did.

But it was not long before Fuller was working on arcs again—first a small laboratory-type arc such as he had used at Brooklyn, a convenient source of continuous-wave oscillations to have around the shop. And when Elwell left for England in early 1913, Fuller was appointed chief engineer in his place. He was young for the job—only twenty-three—a new recruit to the company, and barely a year out of Cornell. Other engineers in Palo Alto had worked for Federal Telegraph longer and knew more about how the company functioned, but there seems to have been a consensus that Fuller was the man for the job. Perhaps, in the circumstances, no one envied him.

There was already in existence a sizable literature on the theory of the oscillating arc. It referred primarily to low-powered units operating under laboratory conditions; it was mostly in German or Danish; and little of it was familiar to American radio engineers.[100] Fuller's task was to integrate what he knew of this literature with his own prior experience and the practical knowledge that the Federal Company had accumulated. And the immediate, urgent problem was, of course, the 60 kilowatt transmitter and the 100 kilowatt unit intended for the Navy's Canal Zone station.

The 60 kilowatt arc, installed at the South San Francisco station in the hope of making possible daylight communication with Honolulu, had been designed essentially by doubling the dimensions of the 30 kilowatt unit. In Fuller's words, "The Palo Alto drafting room had up-ratioed the 30-kilowatt arc as if under a magnifying glass, increased all its dimensions, made the electrodes larger, increased the water jackets of the chamber for cooling and so on, and thought they had a 60-kw design."[101] Yet the instruments at South San Francisco made it clear that it was feeding little if any more current into the antenna than its 30 kilowatt predecessor. Why was this? No answer was possible without better understanding of what was actually going on within the arc chamber and in the arc's oscillating circuit. "Understanding" in this context meant formulating a

[100] For details see P. O. Pedersen, "On the Poulsen Arc and Its Theory," IRE *Proceedings* 5 (August 1917), 256-316 and Larson, *Telegrafonen*, pp. 295-301.
[101] Fuller interview, p. 48.

mathematical model of the arc's behavior and then running tests to determine the parameters of that model.

This was the task that Fuller undertook, working in the Palo Alto laboratory with small-scale models and using the South San Francisco station for field tests. He enlarged the engineering staff, adding Harold Elliot, a Stanford graduate, as his mechanical engineer and chief draftsman, and established what was in effect an engineering research laboratory. His findings were presented to the engineering community in 1919 and presented as his doctoral dissertation at Stanford University in the same year.[102] To summarize them here would do violence to their technical and mathematical sophistication, but essentially they were a series of empirically derived formulas and curves plotted from test data that made it possible to predict how the efficiency of the arc would be affected by a given change in its design or operating conditions. They provided, in the aggregate, a classic example of engineering science; and they gave Fuller the theoretical basis for designing arc transmitters of any desired size—a basis that had not existed before. On this basis the Federal Company, within the next five years, was to provide the Navy with the transmitters for its high-powered chain.

Two matters turned out to be particularly critical: the strength of the arc's magnetic field, and the capacitance of the antenna into which the arc was feeding power. Little was known of either of these variables, or of how they were related to other operating parameters, such as frequency, direct-current power, antenna circuit resistance, or arc chamber atmosphere. It became clear that the essential function of the magnetic field was to blow out the arc flame once in each cycle. Regarded in this light, the strength of the magnetic field could be too great as well as too weak. And variations in the magnetic flux, in relation to other variables such as the length of the arc and the current flowing through it, made an important difference to the arc's mode of operation. Poulsen in Denmark and Elwell in California had tried to increase the arc's radiofrequency output by brute force: by crowding more and more direct-current voltage into the arc and stepping up, on a more or less hit-or-miss basis, the strength of the magnetic flux. Antenna current did increase somewhat as a result, but the curve eventually flattened out and finally refused to increase at all.

Fuller reasoned that the magnetic field strength, and the cross-sectional area or volume of this field, were the controlling variables, since it was

---

[102] Leonard Fuller, "The Design of Poulsen Arc Converters for Radio Telegraphy," IRE *Proceedings* 7 (October 1919), 449-65. Fuller's findings also provided the analytical content of C. F. Elwell's book, *The Poulsen Arc Generator*.

the function of the magnetic field to deionize the arc gap once every cycle. The trouble with previous attempts had been that, as more and more direct-current power was applied to the arc, more ions were produced in the arc gap than the magnetic field could remove in the time available. The route to higher power was correct flux density. It became evident that there was a best magnetic field strength for each set of conditions. Arcs had to be designed so that the flux density would either automatically be correct for a given station operating with a given antenna on a given frequency, or alternatively be easily adjustable for any change in those conditions. Fuller and his group came to call it "tuning the magnetic field." This was a new concept, but central to all later developments.[103]

Antenna circuits also proved tricky. There were two related problem-areas here. Recall, first, that at the heart of Duddell's original discovery had been the idea of connecting a series-resonant circuit (a coil and a capacitor in series) across the arc. There were two currents flowing through the arc: a direct current from the power supply, and an alternating current circulating through the resonant circuit (of which the arc itself was a part). One of Fuller's major discoveries was that an oscillating arc had three different modes of operation, depending on the relative strengths of the direct and alternating currents. Maximum current was delivered to the antenna when the two currents were equal (more precisely, when the effective value of the alternating current equalled the direct current). If the alternating current was larger than this, the current through the arc actually reversed during part of each cycle, producing strong harmonics and a wave train more like that of a spark transmitter than of a properly adjusted arc. If the alternating current was less than the direct current, so that the arc was never completely extinguished, efficiency fell off very rapidly. Properly adjusted, so that the effective value of the two currents was equal, the arc could attain a conversion efficiency of 50 percent, which was the theoretical maximum, corresponding to the Carnot cycle in thermodynamics.[104]

---

[103] For a technically correct account, see Fuller, "Poulsen Arc Converters." This attempt at a nontechnical version is based on the Fuller interview, pp. 49-50 and Clark Radio Collection (Smithsonian Institution), typescript "Radio in War and Peace," Cl. 100, v. 1, p. 283. See also Elmer Bucher, "Wartime Wireless Instruction," *Wireless Age* 6 (November 1918), 223-26. A certain mystique always clung to the theory of the arc; as late as 1923 Elwell thought it necessary to warn his readers that "The theory of the commercial Poulsen arc generator is quite complex and has not yet been thoroughly elucidated"—and this after both Fuller and Pedersen had presented their analyses. See Elwell, *Poulsen Arc Generator*, p. 34.

[104] Fuller, "Poulsen Arc Converters," pp. 461-62; Bucher, "Wartime Wireless Instruction," p. 25.

Now, the relative size of the two arc currents depended not only on the magnetic flux but also on the parameters of the resonant circuit. These included the capacitance and resistance of the antenna. In typical Federal circuits (see Fig. 3.7), there was no separate capacitor shunted across the arc, as there had been in Duddell's circuit. Instead, the antenna itself acted as one plate of a condenser, the other plate being the ground. This implied that the performance of an arc transmitter was very sensitive to anything that affected the electrical characteristics of the antenna it was feeding. Ideally, the two had to be matched to each other very closely. Further, the characteristics of the antenna were liable to change from day to day, as changes in the weather affected the antenna insulation and the conductivity of the ground.

It became apparent, in short, that the mechanical simplicity of the arc transmitter masked its electrical complexity. Its efficiency and the relative purity of the signal it radiated depended critically on informed design and informed adjustment of a number of interdependent operating parameters. Fuller and his team at Palo Alto were building the knowledge base that made this possible. And they succeeded brilliantly, as the successes of the next five years were to show. In 1912 Federal transmitters were stuck at the 30 kilowatt level, and had attained that figure only by close imitation of Danish models. By 1917 the Navy was running a 500 kilowatt Federal arc at Pearl Harbor. And by 1919 it had a 1,000 kilowatt transmitter ready for operation in France. Practical problems there were aplenty in securing these order-of-magnitude increases in power, but no further research effort of any consequence was called for.

*　　*　　*

The availability of higher power enabled Federal Telegraph to expand its activities as an operating company. In January 1914 two 100 kilowatt transmitters replaced the 30 kilowatt arcs at South San Francisco and Honolulu, making day and night service possible for the first time. In mid-1915 Federal entered the marine communications business, installing 2 kilowatt arc sets on the *Yale* and the *Harvard*, passenger steamers operating between San Francisco and Los Angeles. And later in the same year arc transmitters were placed on board the *Sierra*, the *Sonoma*, and the *Ventura* of the Pacific Mail Steamship Company, running between San Francisco and Australia, as well as on several Union Oil tankers.[105] These new services were in addition to Federal's overland circuits on the Pacific coast, which continued to handle large volumes of business, prin-

[105] Fuller interview, p. 55; Mayes, *Federal Telegraph Company*, p. 9; Mann, "Federal Telephone and Radio Corporation," p. 394.

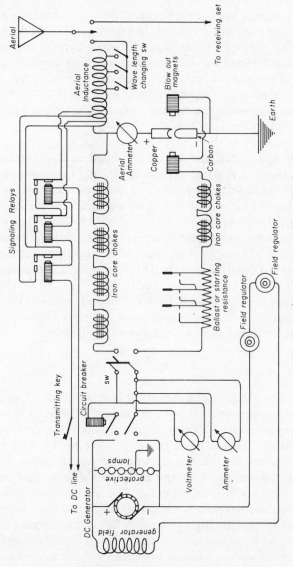

Fig. 3.7: Complete arc transmitter circuit.
Source: Bucher, "Wartime Wireless Instruction," p. 26

cipally from corporate accounts and the California Fruit Growers Association. The new element in the situation, however, was the expansion of the company's manufacturing business, building arc transmitters for the U.S. Navy.

L. S. Howeth, official historian of naval radio, depicts Federal Telegraph in this period as responding reluctantly and with hesitation to the Navy's call for high power. Federal accepted the contract for the Canal Zone transmitter in 1913, he writes, only at the Navy's risk, claiming that such a huge device would generate excessive heat and would never be satisfactory. And, after the Darien arc had proved a resounding success, he describes the people at Federal as "horrified" when asked to build a 200 kilowatt unit for San Diego and two 350 arcs for Pearl Harbor and Cavite.[106] Leonard Fuller's recollections do not confirm this. It is true that, after the Darien station was completed, Navy orders came thick and fast. Lt. Comdr. Stanford Hooper, then in charge of the Navy's Radio Division, at first wanted to explore other alternatives, inviting General Electric to install a 200 kilowatt alternator in the San Diego station and Telefunken to install one of their "frequency doubler" alternators at Cavite. This would have made possible a comparative test of the three leading continuous wave generators available at the time. General Electric, however, refused to guarantee the performance of a 200 kilowatt machine and, even when assured that the Navy would assume the risk of unsatisfactory performance, declined to bid for the contract on the grounds that they preferred not to do business that way. As we shall see, there may have been other reasons for GE's reluctance. Telefunken did submit a bid for the Cavite station but promptly withdrew it on the reasonable grounds that any alternator they shipped to the Philippines would probably be seized by the British blockade. This left the Federal Company with a clear field as far as the Navy's high-power chain was concerned.

What Howeth interpreted as hesitancy and trepidation was more probably the simple fact that in 1915 Federal Telegraph did not have the production facilities to manufacture the larger transmitters. The 100 kilowatt unit for Darien was the largest that could be built in the little factory on Channing Avenue, where there was not even an overhead crane nor convenient access to a railroad siding. The new arcs were large units. The Navy insisted that they had to meet AIEE specifications for general electrical machinery. These called for a temperature rise of no more than 40° C. when operated continuously, and the ability to operate with a 25 percent overload for two hours with a temperature rise of no

[106] Howeth, *History*, p. 222.

more than 50° C.[107] This implied a substantial increase in size and weight and more elaborate cooling facilities; and Fuller's closed magnetic circuits, with their large cast steel yokes, worked in the same direction. The Cavite and Pearl Harbor arcs each weighed approximately 60 tons, and the unit built for the Bordeaux station 85 tons. The new generation of arcs, even to the unsophisticated eye, looked quite different from their Elwell-designed predecessors—more like the vertical-shaft turbogenerators found in hydroelectric installations than any conventional radio transmitter.

This was not the kind of device that you could build in what was essentially an enlarged garage. One of the first consequences of the new Navy contracts was therefore a move to larger quarters. This was a new and much larger plant between El Camino Real and the Southern Pacific

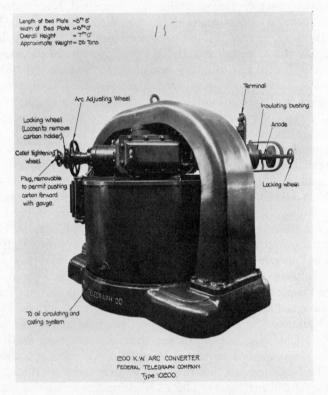

Pl. 5: Two hundred kilowatt Federal arc.
Source: Foothill Electronics Museum

[107] Mayes, *Federal Telegraph Company*, pp. 9-10.

Railroad in Palo Alto, with approximately 25,000 square feet of floor space on the ground floor and 3,500 square feet of office space above that. This was first occupied in 1916. It provided much better facilities and made it possible to tackle the new transmitters. The work force, which had numbered less than 20 at the Channing Avenue plant, grew to 300 and remained at that level during World War I.

Until this plant was ready there was good reason for Federal to be less than enthusiastic about building larger transmitters. And, even though the key design problems had been solved, there was still much learning to be done. The very high radiofrequency voltages generated by the new transmitters caused many headaches. G. H. Clark, the civilian radio aide whose memoirs provided Howeth with much of his information, later recalled his experiences with Federal arcs at Tuckerton in the fall of 1914. The insulation of the copper electrode, as he remembered it, was made from California redwood, and after each few hours of operation "the standard procedure consisted of knocking off the heads of the bolts with a sledge, tearing off the smoking and in some cases flaming wooden bushings, and putting some new ones 'on the fire' for the next schedule."[108] Navy operators were notorious for rough treatment of equipment, and Clark loved a good story; but on the other hand it is undeniable that insulation that was satisfactory for high-voltage power lines or for spark transmitters proved quite inadequate for the sustained high currents and high voltages generated by a large arc transmitter. Federal set up a high-frequency high-voltage testing program in Palo Alto, working in conjunction with the Ohio Brass Company and Professor Harris J. Ryan of Stanford University, and some of the pioneering research studies of radiofrequency high-voltage discharges emerged as a result.[109]

There were also problems with keying the transmitters and with adapting to varying antenna conditions. The San Diego installation provided good examples. There was considerable bureaucratic hoopla attending the commissioning of this station, and Howeth writes about a "faulty keying circuit" that led the naval authorities to send their official opening messages by land telegraph ahead of time, in case the transmitter should

---

[108] Clark Radio Collection, Cl. 100, v. 2, "Radio in War and Peace," p. 280. Elwell's comments on this passage read: "My God! This is news to me. At least I never perpetrated anything like this on the French, British, or Italians. I used silica bushings on which one could pour water while red hot!"

[109] Harris J. Ryan and Roland G. Marx, "Sustained Radio Frequency High-Voltage Discharges," IRE *Proceedings* 3 (December 1915), 349-70. Marx was one of Federal's engineers on loan to Stanford; the 12 kilowatt arc used in the experiments was also provided by Federal.

let them down.[110] The uncertainty arose in fact not from a faulty keying circuit but from the fact that the antenna and ground systems at the station were provided by the Navy and were not completed when the 200 kilowatt arc was shipped from Palo Alto. There was, in consequence, no way in which the Federal Company could tell beforehand precisely what the antenna circuit resistance would be. Fuller in Palo Alto had keyed the arc with a magnetic amplifier circuit, which was attractive because it generated no "back wave" and required no keying relays. But such a circuit was sensitive to changes in antenna circuit resistance and at San Diego that resistance varied widely when ground moisture changed. Rather than run the risk of lowering the transmitter output by magnetic amplifier keying, Fuller changed to the more conventional frequency shift keying, which had greater reliability and could withstand high overloads.[111] This did entail some last-minute work, but in fact the opening of the station and the transmission and receipt of the first messages went off without a hitch. The episode demonstrated, not

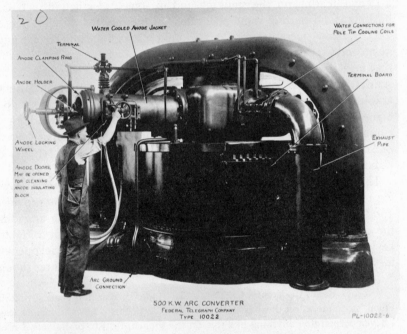

Pl. 6: Five hundred kilowatt Federal arc.
Source: Foothill Electronics Museum

110 Howeth, *History*, p. 224.
111 Fuller interview, pp. 61-62.

improvisation and uncertainty, but rather a reassuring technical flexibility.

None of this should have been surprising to any experienced engineer. What is remarkable, indeed, is the smoothness and speed with which the new transmitters were built, installed, and placed in service. The Arlington installation (30 kilowatts) was completed in December 1912; Darien, Canal Zone (100 kilowatts) in July 1915; San Diego (200 kilowatts) in January 1917; Pearl Harbor and Cavite (500 kilowatts each) in September and December of that year. The distances covered were unprecedented for the time—5,300 miles from Cavite to Pearl Harbor, 7,800 from Cavite to San Diego. And the confidence the Navy placed in the arcs is well attested by the decision to equip Guam and Samoa with only 30 kilowatt units, as higher power had proved to be unnecessary. By the end of 1917 the Caribbean and transpacific segments of the Navy's network were complete. Coverage of the North Atlantic lagged somewhat but the prospect of involvement in the European conflict stepped up the pace. The new station at Annapolis, Maryland, got a 500 kilowatt arc in 1917; the foreign-owned stations at Tuckerton, N.J., and Sayville, N.Y., were taken

Pl. 7: At the controls: Operating a 500 kilowatt Federal arc.
Source: Science Museum, South Kensington

over and equipped with arcs in the same year; and in January 1918 contracts were signed for the new "superpower" 1,000 kilowatt station at Croix d'Hins near Bordeaux, intended to provide secure communications with the Allied governments and the American Expeditionary Force.[112] All this was in addition to volume production of smaller arcs for shipboard use: some 300 for the U.S. Shipping Board and, for the Navy, enough to equip every battleship in the fleet.

*　　*　　*

Much had indeed been accomplished in a short time. Elwell went to Copenhagen in May 1909 and brought back a tiny arc converter rated at 100 watts input. Ten years later the Federal Company was building arcs rated at 500 kilowatts and stood ready to build arcs of twice that power if they were needed. These were the most powerful radio transmitters in the world at that time. The basic technology was, as we have seen, of European origin (with due recognition to Thomson's magnetic blowout of 1892). What the American engineers did was transfer that technology to the United States, develop and improve it, and use those improvements to create devices that, in terms of power, efficiency, and reliability, bore only a faint family resemblance to their European ancestors.

This had been done originally in the hope of establishing a private commercial radiotelegraph and telephone service. The big increases in power, however, came in response to government orders, and specifically in response to the ambitious performance requirements set for the Navy's telecommunications network. The availability of the Federal arc transmitters, particularly those designed by Fuller, made it possible for the United States Navy to step into the world of continuous wave radio almost ten years before any other governmental or commercial entity.[113]

If Reginald Fessenden had stayed with NESCO, if NESCO had maintained its interest in radiofrequency alternators, and if General Electric had permitted Alexanderson to push ahead rapidly with the development of 50 kilowatt and then 200 kilowatt machines, the Federal arcs would have faced formidable competition sooner than they did. General Electric, however, did not complete a 50 kilowatt alternator until early 1917, and

112 For contract dates, see Clark Radio Collection, Cl. 100, v. 1, p. 215; Mayes, *Federal Telegraph Company*, p. 11, gives dates of completion.

113 That is, if we take 1912 as the date for the Navy's acceptance of the Federal arc at Arlington and 1921 as the date when RCA opened its long-distance circuits, using the alternator, closely followed by the Marconi Company.

a 200 kilowatt machine was not ready for service until July 1918. For all their virtues of frequency stability and purity of emissions, GE's alternators were expensive and complicated machines to build. Federal arcs had a significantly shorter development time.

German and French engineers were not far behind the United States in alternator development. The French Goldschmidt alternator, installed at Tuckerton in 1914, and Telefunken's Von Arco machine, installed at Sayville in the same year, differed in concept from Alexanderson's alternator but were no less capable of operating at high power on the low frequencies. In high-powered arc transmitters, strangely, the Europeans showed little interest. The German Lorenz Company manufactured several models, mostly for marine use, but their most powerful machine was rated at only 50 kilowatts. There was nothing in Europe, before or after World War I, to compare with Federal arc transmitters, except those designed by Elwell.

The organization that might have been expected to take the lead in the development of continuous wave technology in fact did nothing about it. This was Marconi's Wireless Telegraph Company Limited, the company that at one time had seemed on the verge of attaining an effective monopoly of world radio communications. Some possible reasons for this have been suggested earlier: misleading scientific and technical advice; complacency as to the virtues of advanced spark technology, as represented by the disk discharger; and a corporate conservatism that may have stemmed from earlier successes with spark and overcommitment to spark equipment. Whatever the reasons, the facts are plain: the Marconi organization had no radio alternator planned or under development in 1914; and it was only in that year that, despite earlier protestations, it finally acquired the British rights to the Poulsen arc patents. That, in effect, put it in the same relative position as the Federal Company had occupied in 1909. It was not licensed under any of the patents that the Federal Company controlled and, in consequence, it did not have access to the technology that had enabled the Federal Company to convert the Poulsen arc into an efficient high-powered transmitter. Its American subsidiary was in even worse shape, since Federal controlled rights to the Poulsen patents in the United States.

It was to be expected, therefore, that when the directors and management of the Marconi Company turned their thoughts to the competitive environment that would face them when hostilities finally drew to a close, they would find it impossible to avoid the conclusion that access to continuous wave technology was indispensable. Having no resources in that technology themselves, they would have to acquire it from others. If this was expedient for their own corporate survival, it was essential if

Pl. 8: Fuller and his associates
(Leonard Fuller at left; original Poulsen arcs imported from Denmark in
foreground; 500 kilowatt Federal arc in rear).
Source: Foothill Electronics Museum

they expected to play any role in construction of the long-postponed
British Imperial Chain. Two possibilities seemed especially attractive:
General Electric's alternator, which had been installed in American Marconi's New Brunswick station in the latter phases of the war and had
performed magnificently; and the Federal Company's arc, thoroughly
tested and proven by the United States Navy over distances and under
conditions very similar to those the Marconi Company itself would face.
To acquire rights to either or both of these devices cannot have seemed,
initially, very difficult; it was surely only a question of money.

From the point of view of the United States Navy, however, and from
that of a few highly placed advisors in the United States government, it
was very far from being just a question of money. In continuous wave
radio technology American scientists and engineers had established an
important technological differential. It might not survive for long, for
the rate of change in such matters was rapid. But while it existed it could
be exploited in the American national interest. The Marconi Company
had aroused considerable resentment in the prewar years by its aggressive

attempts to eliminate competition, by its reluctance to accept national regulation, and by its refusal, until compelled to do so, to allow inter-communication with other radio systems. Above all, it was resented in the United States because it appeared to represent British imperialism and what was interpreted as British domination of the world's long-distance communications. The slight lead that American engineers had won in continuous wave radio technology might, if properly exploited, enable the United States to achieve in international communications a role more commensurate with its new industrial, financial, and military might. It was highly probable, therefore, that any attempt by the British-based Marconi Company to acquire American continuous-wave technology would evoke a vigorous response.

FOUR

De Forest
and the Audion

W LEE DE FOREST's doctoral dissertation, submitted to the Sheffield Scientific School at Yale in 1899, is sometimes called the first American dissertation to deal with "wireless." Yet the document itself makes no reference to wireless communication. It is a report of research on the "Reflection of Electric Waves of Very High Frequencies at the Ends of Parallel Wires." The behavior of high-frequency electric oscillations on wires, not in free space, was the scientific problem that de Forest was investigating.[1]

Nevertheless, there is some truth in the conventional description. De Forest was indeed working with Hertzian waves, generated by a spark oscillator. He was using parallel wires only as a transmission line or wave-guide. The waves generated by the oscillator travelled down the wires to the far end and were reflected back, as light would be by a mirror. The interaction of the outgoing and returning waves created what were called standing waves. (See Fig. 4.1) By measuring the distance

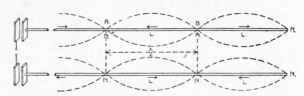

Fig. 4.1: Standing waves on wires.
Source: H. Poincaré and F. K. Vreeland, *Maxwell's Theory and Wireless Telegraphy* (New York: McGraw Publishing Company, 1904), p. 61

[1] The dissertation was published in somewhat abbreviated form in the *American Journal of Science*, 4th ser., 8 (1899), 58-71. The original handwritten version is in the De Forest Collection at the Foothill College Electronics Museum.

between successive peaks or troughs in the standing waves, one could arrive at a measurement of the wavelength.

The technique was not new. Heinrich Hertz, in the most famous of his experiments at Karlsruhe in 1887, had not used wire wave-guides. He had measured the distance between successive peaks as the waves travelled through empty space. That was one reason why his work had such dramatic impact. But in other experiments he had used a single wire. (See Fig. 4.2) Oliver Lodge in England, at roughly the same time, was measuring the "recoil kick" that took place when Leyden jar condensers were discharged down long parallel wires. And on the continent the Swiss physicists, Edouard Sarasin and Auguste de la Rive, following up Hertz's work, used a pair of parallel wires to make their measurements. This was also the technique used by the Austrian scientist, Lecher, with whose name the technique came to be commonly associated. Today we still speak of "Lecher wires." They provide a convenient and accurate method for measuring wavelengths, particularly at ultrahigh frequencies.[2]

The general topic for his doctoral research had been suggested to de Forest by Harry Bumstead, a young instructor at the Sheffield School who had befriended him. But in the dissertation itself the individuals to whom he acknowledges indebtedness are Professor A. W. Wright, "for

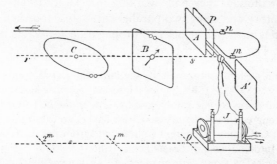

Fig. 4.2: Hertz's apparatus for locating standing waves.
Source: Heinrich Hertz, *Electric Waves: Being Researches on the Propagation of Electric Action with Finite Velocity through Space*, Authorized English translation by D. E. Jones (London and New York: Macmillan and Company, 1893), p. 108

[2] For Hertz's experiments, see Heinrich Hertz, *Electric Waves, being Research on the Propagation of Electric Action with Finite Velocity through Space*, trans. D. E. Jones (London, 1893), and (in summary) Hugh G. J. Aitken, *Syntony and Spark: The Origins of Radio* (1976; Princeton, 1985), pp. 48-75. J. A. Fleming, *The Principles of Electric Wave Telegraphy*, 3rd ed. (London, 1916), pp. 362-68, summarizes the early research. Lecher's work was reported in "Eine Studie über elektrische Resonanserscheinungen," Wiedemann's *Annalen* 41 (1888), 850.

his kindly interest in the work and many valuable suggestions," Professor Willard Gibbs, for "general aid in the theoretical study of the subject," and—surprisingly—Dr. Edwin Barton, of University College, Nottingham, England, for "advice and helpful suggestions." Barton had studied under Hertz at Bonn in the summer of 1893 and may well have guided de Forest to the relevant literature, which by 1898-1899 was quite extensive, and to some of the problems still unsolved. Certainly de Forest had received little direct help from the faculty of the Sheffield School. For their curriculum in electrical science he had little but contempt. And in the work for his doctoral dissertation he had been very much a loner, setting up his own apparatus, taking his own measurements, and interpreting the results by himself. This was not the kind of doctoral dissertation that fed into the research program of a senior professor.[3]

Hertz had left two puzzles. First, when using longer wavelengths, he had found that the velocity of propagation along wires was less than it was through air. And second, when measuring wavelengths along the wire, he had observed that the final quarter-wave, before the incident wave was reflected back, seemed to be shorter than the others. It was as if the wave was reflected, not from the surface of the reflector but from some distance behind it. Sarasin and de la Rive had cleared up the first anomaly, or thought they had: working in a larger room than Hertz's small laboratory and using two parallel wires close together, they found approximately the same velocity of propagation for waves on the wires and in air.[4] The second problem—that of the so-called "end effect"—

[3] For Barton's relationship to Hertz, see *Philosophical Magazine* 5th ser., 44 (July-December 1897), 151. Biographical information on de Forest is drawn from his diaries and drafts for his autobiography, now in the de Forest Papers at the Library of Congress; from the Lee de Forest Papers at Yale University Library; the de Forest Collection at Foothill College; and from Lee de Forest, *Father of Radio: The Autobiography of Lee de Forest* (Chicago, 1950). See also, however, Georgette Carneal, *A Conqueror of Space: An Authorized Biography of the Life and Work of Lee de Forest* (New York, 1930); Samuel Lubell, "Magnificent Failure" *Saturday Evening Post*, 17 January, 24 January, and 31 January, 1942, *passim*; and James A. Hijiya, "The De Forests: Three American Lives" (Ph.D. diss., Cornell University, 1977). Lee de Forest's father and grandfather both preferred to capitalize the "D" in DeForest, but both Lee and his younger brother changed to a lower-case "d" and separated the "de" from "Forest" while students at Yale.

[4] For a useful discussion, see George W. Pierce, *Principles of Wireless Telegraphy* (New York, 1910), p. 68. In fact, however, the velocity of propagation along a conductor may differ significantly from the velocity in free space. In the case of coaxial cable with polyethylene insulation, for example, the velocity is only some 65 percent of the velocity in air.

proved harder to resolve, and it provided one of the principal foci for de Forest's dissertation. There was an "apparent capacity" across the ends of the wires. How did this vary with frequency? What effect did it have on the length of the standing waves?

De Forest set up his spark oscillator and Lecher wires in the basement of the Sloane Physics Laboratory—not the best of environments, but he was lucky to be there at all. He had been kicked out of the laboratory of the Sheffield School in Winchester Hall—"scourged from the tabernacle," as he put it in his diary—by one of the senior faculty members for an offense perhaps not as trivial as appears at first sight. De Forest had hammered nails into one of the oak laboratory tables (to hold the ends of his Lecher wires, almost certainly), and this cavalier attitude toward the institution's property had been too much for Professor Charles Hastings, who promptly banished him from the laboratory and, but for the intervention of Bumstead and Wright, would have expelled him from Yale. There had been a prior history of disagreements with Hastings, and the incident of the laboratory table was merely the last straw. But neither then nor later was de Forest overly scrupulous about making use of what he needed when he needed it. The impression he made at this stage in his life is probably accurately described by his biographer: "awkward . . . graceless in his contacts with people, always saying the wrong thing at the wrong time and withdrawing to stand and wonder just what he had done this time."[5] Even within the small world of the Sheffield School, not to mention the broader community of Yale, Lee de Forest had always been something of an outsider, plunging with exaggerated zeal into those student rituals to which he was admitted, excessively grateful to those faculty members who treated him with respect.

But in the "cold, dark basement of Sloane," where de Forest worked six days a week during the winter and spring of 1898-1899, these personal problems could be forgotten. There were more immediate difficulties. De Forest, like Hertz before him, was working in a very confined space. This meant that, if he were to get more than one or two nodes in the standing waves on his Lecher wires, he had to generate Hertzian waves of short wavelength and very high frequency. To generate such waves he had only one device available to him: a spark oscillator. But a spark oscillator generated, not a single wave of a specific frequency, but many waves covering a wide band of frequencies. How was de Forest to ensure that, as far as possible, waves of only a single frequency travelled along his Lecher wires to produce the single set of standing waves he was looking for?

[5] Carneal, *De Forest*, p. 53.

The answer was tuning. The Lecher wires themselves made up a relatively sharply tuned circuit. They would resonate only at certain frequencies: a fundamental frequency, at which they were a quarter-wavelength long; the third harmonic; the fifth harmonic; and so on. They acted like an organ pipe that would "sound" only a specific note and its harmonics. The spark discharge itself, like the wind blowing into the organ pipe, had no single frequency of oscillation of its own. It was the resonant circuit that determined the frequency.[6]

To obtain even sharper tuning—that is, to approximate even more closely to a single wave—de Forest adopted what was known as the Blondlot oscillator, perhaps the most notable improvement that had been made in the Lecher system since its introduction. Lecher had coupled his spark gap to the parallel wires through a pair of capacitors. (See Fig. 4.3) This gave a system in which it was difficult to say exactly what frequencies were being fed into the Lecher wires. In a Blondlot oscillator the oscillations were generated in a circular wire circuit containing a spark gap and a capacitor; this circuit then was coupled inductively to the Lecher wires. (See Fig. 4.4) The essential feature was that the inductance and capacitance of the circular resonant circuit could be calculated beforehand; with these values known, the experimenter could know the length of the waves that were being fed into the Lecher wires—or, at least, the length of the most powerful waves. The circular wire and the encircling loop acted as the primary and secondary of a transformer with some degree of selectivity.

Choosing an oscillator and deciding how to couple it to the Lecher

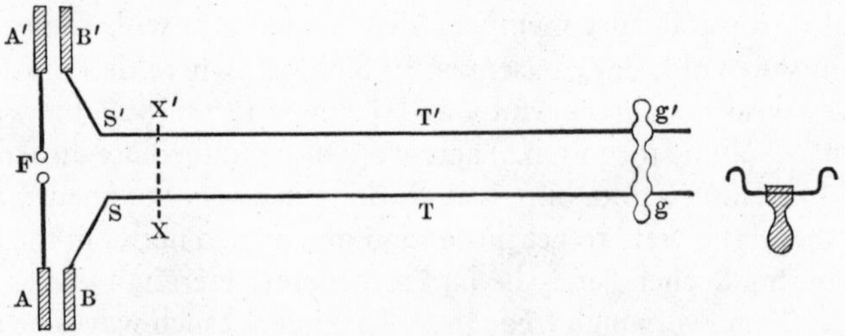

Fig. 4.3: Lecher's apparatus: Capacitive coupling.
Note: gg' represents a Geissler tube detector
Source: G. W. Pierce, *Principles of Wireless Telegraphy* (New York: McGraw-Hill Book Company, 1910; reproduced by permission), p. 70

[6] See H. Poincaré and F. K. Vreeland, *Maxwell's Theory and Wireless Telegraphy* (New York, 1904), pp. 61-65; Fleming, *Principles*, pp. 407-408; and Aitken, *Syntony*, pp. 71-73.

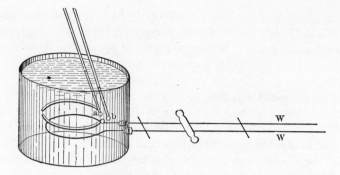

Fig. 4.4: Blondlot's apparatus: Inductive coupling.
Note: In this design the spark oscillator and pick-up coil are depicted
immersed in oil.
Source: G. W. Pierce, *Principles of Wireless Telegraphy* (New York: McGraw-Hill Book
Company, 1910; reproduced by permission), p. 72

wires were the first problems de Forest encountered. But a major problem remained: what detector should he use? Hertz, looking for standing wave peaks in free space, had used a ring resonator with a tiny spark gap. Voltage peaks were indicated when a spark jumped across the gap. That would not do for de Forest's work with wires. Nor would the device that had become by 1898 the standard method for detecting Hertzian waves: the coherer. Coherers either responded to impulses or they did not. They could never give an analog reading and hence could never be used to trace the gradual rise and fall of voltage levels along parallel wires. De Forest followed Lecher's example in using what he called a glow tube or vacuum tube, commonly known then as a Geissler tube. This was a partially evacuated glass tube containing two metal electrodes and some residual gas. Placed across the ends of the Lecher wires, it glowed when the oscillator was operating, as the gas inside the tube was ionized by the difference in electrical potential between the two wires. This made possible close identification of voltage peaks: you moved a conducting "bridge" up and down the Lecher wires, and when the bridge was near a voltage node, the vacuum tube would start to glow. With perseverance and patience it was possible to identify the whole system of standing waves.

Perseverance and patience were certainly the attributes required. De Forest's experimental apparatus was decidedly tricky and the measurements did not fall readily into his hands. His memoirs describe the "dreary hours in the cold darkness alone in that cellar, peeping at a glow tube that would not glow aright, running back and again to correct some unknown fault in a contrary interrupter."[7] But eventually it was done. April 30, 1899, saw the document completed. "It is quite a presentable and ponderous thesis," de Forest noted in his diary. "It shows well something of the enormous amount of conscientious work I have done for it."[8]

That indeed it did. But how good a dissertation was it? Good enough, obviously, to win a Ph.D from Yale in 1899. And although de Forest himself had no respect for Yale's curriculum in electrical science—"What a pathetic excuse for a course in Electrical Engineering was that at Sheff in those days"—we may presume that any doctoral research endorsed by Willard Gibbs would meet more than minimal standards of quality and scientific rigor. De Forest got the results he was looking for. He was able to measure the "end effect," describe how it varied with frequency, and relate it to the "virtual capacity" that existed at the end of Lecher wires. He had shown considerable experimental ingenuity in designing and constructing appropriate apparatus. And he had demonstrated a thorough knowledge of the relevant literature in the field. More than that is hard to say. His results were not strikingly novel. They provided more precise measurements of a phenomenon known to exist. But they called for no recasting of theory, no change in the way physicists thought about the propagation of Hertzian waves along wires.

A different question arises when we ask what the dissertation meant to Lee de Forest. He had come to Yale with a single objective: to train himself as an inventor. Much of his dissatisfaction with the educational offerings of the Sheffield School stemmed from his perception that what he was learning had no direct practical application. The curriculum in electrical engineering was faulted because, unlike the course in civil engineering, it had no "adequate plant." His training in mathematics struck him as irrelevant: ". . . how little of real practical knowledge as to the application of mathematics to engineering problems did I acquire at Yale. A thorough-going training in manipulation of mathematical equations I obtained, but absolutely no knowledge, or instruction, as to how to apply these keen tools to actual problems . . . most of the time I spent in higher mathematics in my last two years at Yale was totally wasted." Even the revered Willard Gibbs got his share of criticism: ". . . because he dealt solely with his own system of mathematical symbols and analysis the actual, practical value which I obtained from my years under him was indisputably far less than had he dealt with conventional mathematics.

---

[7] De Forest Papers (Yale), "The Diaries of Doctor de Forest: Part I, My Early Life."

[8] Ibid.

Gibbs did not undertake to instruct his pupils in practical methods for applying his system to such problems as they would encounter in commercial research laboratories."[9]

Such complaints are the common currency of engineering schools, particularly those located in institutions where most of the faculty are still committed to the liberal arts or the ethos of "pure" science. But in de Forest's case they went beyond normal student griping. He had had to battle his father to get where he was—not for permission to come to Yale, but for permission to enter the program of Yale's scientific school rather than prepare himself for the Congregational ministry. And, despite a fellowship, considerable financial sacrifice by the family had been required to send him there. De Forest wanted a return on the investment. He wanted Yale to give him knowledge that he could put to use as an inventor. And when it was not forthcoming he made his discontent known.

From that point of view, the apparatus he built for his doctoral research and the problems he encountered take on added significance. Consider the essential components. He had a generator of high-frequency waves: the Blondlot oscillator. He had a tuning and coupling system: the oscillator's resonant loop and encircling wire. He had a transmission line: the Lecher wires themselves. And he had a detecting device: the glow lamp or "vacuum tube." Consider, too, the problems he encountered. First, the difficulty of getting a single frequency of oscillation from a generator of damped waves. De Forest had no alternative but to use a spark oscillator. But that meant cluttering up his Lecher wires with multiple harmonic oscillations that must at times have driven him to distraction. How much simpler his task would have been if there had been available a device that generated continuous waves at a single frequency! And there was the problem of the detector. Since a coherer was useless in the circumstances, de Forest turned to a partially evacuated glass tube and relied on gas ionization to indicate resonance. Here was a detector quite different in concept from a coherer. In the future, glow lamps and gas ionization would play a central role in de Forest's thinking.

Whether or not, in 1899, de Forest thought of his research apparatus as a wireless transmitter—that is, as potentially part of a communications system—we cannot tell. His diaries hint at it, and we know that, in seeking employment after graduation, he contacted first Tesla and then Marconi. But in the dissertation itself he is silent on the subject. From his equipment in the Sloane basement to a functioning wireless transmitter was, however, but a short step. All he needed was a means of coupling the transmission line to an antenna, and a detector that was less sluggish than his

[9] Ibid.

glow tube. For a man like de Forest, impatient with knowledge that had no application to practice, not to take these further steps was unthinkable.

And within a couple of years he had taken them. There is, among the surviving de Forest papers, the rough sketch of an electrical circuit drawn on stationery with "De Forest Wireless Telegraph Company" at the letterhead and dated 1902. If you come fresh from reading the dissertation you meet this circuit with a jolt of recognition. There are the spark oscillator and the Lecher wires, just as in the Yale experiments. But there is one change: the wires are now coupled to an antenna and ground system. And under the sketch is written "Induction scheme, for exciting the antenna from the Lecher resonant circuit, not *directly*, but inductively."[10]

* * *

When de Forest was ending his undergraduate years at Yale, he completed a questionnaire for his class yearbook. One of the questions was, "Why did you come to college?" and to that de Forest responded, "To direct and temper my genius." Another was, "Next to yourself, whom would you prefer to be?" And de Forest replied, "Nikola Tesla."[11]

Wherever you look in the early history of radio technology, you run into the name of Nikola Tesla. Tuning circuits, high-frequency alternators, rotary spark transmitters—name almost any device that became important in the later history of radio, and you can find an anticipation by Tesla. Here was a man always trying great things, capable of profound insights and startling leaps of the creative imagination, and yet somehow limited in his ability to integrate his inventions into commercially viable systems. We remember Tesla today mainly for the one case in which he unquestionably surmounted that limitation—his invention of the polyphase system of alternating current—and for the familiar device—the Tesla coil—that replicates the spectacular demonstrations of high-voltage discharges that Tesla loved to engineer. His other achievements are little known except by specialists.[12]

[10] De Forest Papers (Foothill College), Box marked "1900."

[11] De Forest Papers (Library of Congress), Box 2, "Yale '96S: Statistical Questions."

[12] There is no adequate biography of Tesla. Of the two available, the earlier—John J. O'Neill, *Prodigal Genius: The Life of Nikola Tesla* (New York, 1944)—is preferable, but devoted Tesla fans, of whom there are many, will also wish to be familiar with Margaret Cheney, *Tesla: Man out of Time* (Englewood Cliffs, N.J., 1981). For Tesla's anticipation of Marconi in the field of radio invention, see Leland Anderson, "Priority in the Invention of Radio: Tesla vs. Marconi," *Antique Wireless Association Monographs* n.s. 4 (n.d.).

It is not surprising that de Forest should have wished to model himself on Tesla and later work for him. In 1896 alternating currents were the latest thing and Tesla's monetary and technical successes were enough to stir the imagination of any young electrical engineer. And there was a question of personal style. Tesla and de Forest were both inveterate romantics, always ready to identify what was with what should be. Both, as soon as they could afford it, revelled in the pleasures of an affluent and elegant life-style. And both loved publicity.

They had one more characteristic in common. Both were convinced they were geniuses. When de Forest wrote that he had come to Yale to "temper his genius," he meant to refer not to a specific talent that he possessed but to the special kind of person that he believed himself to be. Tesla and de Forest both believed that they were not as other men were. Such a conviction no doubt brings its moments of euphoria. But it also demands, if inner doubts are to be quieted, periodic confirmation from the world at large. When that is not forthcoming, the result can be spells of black despair. De Forest, on the evidence of his diaries, had his full measure of these. His sense of being different, which at some times was equated to a conviction of his own genius, at other times translated into a feeling of isolation, of being alienated from the common run of humanity.

He was born in Council Bluffs, Iowa, in 1873, first son of a Congregational minister. When he was three years old the family moved to Waterloo and two years after that to Muscatine, both in Iowa. The major change in location and in his way of life came, however, at the age of six, when his father accepted the presidency of Talladega College, in Alabama. This institution had been established by the American Missionary Association only ten years earlier, to provide schooling for the children of black freedmen and to make sure that they were raised in the Christian faith.[13] Lee de Forest's father was the first full-time administrator of the institution. During his seventeen years as president it grew from a "normal school" emphasizing the training of schoolteachers and preachers into a full-fledged college, on the verge of true academic distinction. The teaching staff was, of course, all white and all from north of the Mason-Dixon line.

Much of the time and energy of the Reverend DeForest (as he preferred to spell his name) were spent in raising money for his institution, and this entailed long absences from home. His three children—Lee had one older sister and a younger brother—were enrolled in the Cassedy School,

[13] For the foundation and early years of Talladega College, see Addie L. J. Butler, *The Distinctive Black College: Talladega, Tuskegee, and Morehouse* (Metuchen, N.J., 1977), chaps. 2 and 3.

the grammar school associated with the college, and received instruction side by side with the black pupils. The charter of the college required it to be "open to all of either sex, without regard to sect, race, or color." In fact there were few if any white pupils besides the DeForest children. Lee de Forest makes it clear in his autobiography that his family was ostracized by the white community of Talladega, particularly in the early years of his father's presidency. And it appears that at first they lived on the margin of poverty, at least by the standards of friends and relatives back in Iowa. For the first two years the DeForests lived in two small rooms on the second floor of the girls' dormitory, totally cut off from any interaction with local "society," except such as could be provided by the other teachers—all "educated, refined men and women from the best schools and influences of the North." And even when they finally got a house of their own, built from bricks made by the students themselves, the DeForests lived a life narrowly bounded by the college, its teachers, and its pupils. Lee and his brother and sister invented their own amusements. They found their friends and companions among the black pupils of the college, whom they did not greatly respect but with whom they felt safe. The local white youngsters called them "damned blue-bellied Yankees" and were not to be trusted. Lee and his young brother were "well frightened of the white boys of the town, knowing how they regarded us, and always felt safer in the company of two or three of the larger Negro boys of the school."[14]

From his companions and fellow pupils Lee de Forest picked up the accent, cadences, and colloquialisms of southern black speech—a characteristic of which he was slightly ashamed in later life. At the same time, however, his black associates provided a much-needed relief from the strict parental discipline he experienced at home. He describes his father as "stern and upright" and as having "erred on the side of severity." Corporal punishment was common, sometimes for Lee's own transgressions, more often (if his recollections on the point are to be trusted) for the offenses of his younger brother. When misery grew too strong he would retreat to "the dark place"—part of the cellar that had no win-

[14] De Forest, *Autobiography*, p. 22. Ostracism and occasional violence were the normal experience of northern teachers in southern black schools after the Civil War. One of the first teachers at Talladega, William C. Lake, had been killed by a mob. See Henry Lee Swint, *The Northern Teacher in the South, 1862-1870* (New York, 1967), Ronald E. Butchart, *Northern Schools, Southern Blacks, and Reconstruction: Freedmen's Education 1862-1875* (Westport, Conn., 1980), and Robert G. Sherer, *Subordination or Liberation? The Development and Conflicting Theories of Black Education in Nineteenth Century Alabama* (Tuscaloosa, Ala., 1977), especially chap. 12.

dows—there "with a gulping throat to sob away the ache resulting from the sense of injustice which tortured my boyish soul."[15]

There were, however, means of escape—windows onto a world ruled by a different ethic from that which his father espoused. Two of these he recalled with particular clarity in later life. One was the daily arrival of the express train on the East Tennessee, Virginia and Georgia Railroad.[16] The steam locomotive that hauled this train came to have a special fascination, and Lee de Forest studied it until he knew every linkage, lever, and valve. The second occurred when a group of northern engineers, financed by British capital, began the construction of a blast furnace for the smelting of iron ore at a site about a mile from the DeForest home. That venture did not survive for long, but while it functioned it gave young Lee de Forest a general education in nineteenth-century metallurgy. The furnace itself, the machine shop, the foundry, above all the narrow-gauge railway that moved ore to the furnace site—these were marvels to him and in later years he could recall every detail. And the same was true of other mechanisms he saw every day: the college's printing press, the two-wheeled plows that were beginning to appear in the fields, a lawn mower, even a grandfather clock. De Forest in later life remembered these machines with great clarity. They embodied a logic different from that which governed life in rural Alabama, and different too from that which inspired his father's authoritarian morality.

It was his father's firm intention that both his sons should attend Yale College, from which he himself had been graduated in 1857; and he fervently hoped that both would become ordained ministers in the Congregational Church, or if not that, at least teachers or professors. Lee de Forest opposed this design. By the age of thirteen he had decided that he would be an inventor. He had no objection to going to Yale; there was a scholarship available there, left by a distant relative, that would help considerably with finances. But Yale's Sheffield Scientific School was where he intended to be, not Yale College. His father opposed the idea vehemently—a "half-baked education" was all he saw his son getting out of a scientific school. Lee finally invoked his mother's intervention, and the Reverend DeForest gave in with as good grace as he could muster. Even a Sheffield School education, he reluctantly concluded, was better than no college education at all. But, he warned his son, the delights that a true Yale man could hope to enjoy would never be his. The cultural refinement that came from the study of the classics, the choice companionships and inspiring friendships that came from belonging to a real

[15] De Forest, *Autobiography*, p. 29.
[16] Carneal, *De Forest*, p. 16.

Yale class and being a member of Yale College—all these were joys that "Sheff men can never share."[17]

Paternal premonitions and the prospect of being at Yale but not really part of it did not deter Lee de Forest. There remained, however, the matter of winning admission to the institution, and the educational opportunities available in Talladega, even supplemented by personal tuition by his father and assiduous reading of the *Patent Office Gazette*, would not suffice for that. It was decided that he should go to Mr. Dwight L. Moody's School for Boys, at Mt. Hermon, Massachusetts, not far from Amherst College where his maternal grandfather had been a student. His sister was already attending Northfield Seminary (as it was then called), just across the Connecticut River from Mt. Hermon, and it may be that the expense of paying her board and tuition had something to do with the delay in sending Lee north to continue his schooling. He was nineteen when he went to Mt. Hermon in the fall of 1891; the average age for entering pupils was seventeen.

Mt. Hermon represented escape from Talladega in a geographical sense, but in most respects it was an all-too-familiar environment. As de Forest put it in his autobiography, "Mt. Hermon was a school founded on basic rock-ribbed Fundamentalism and rock-studded dirt farming. At Talladega I had had surfeit of one and plenty of the other."[18] And, just as in Talladega he had been, as a northerner, unacceptable to whites of his own generation, so at Mt. Hermon, with his southern accent and attitudes, he found it difficult to make friends. Some measure of acceptability came from athletic prowess and some from the fact that, with a sister at Northfield, he had visiting privileges beyond the Mt. Hermon boundaries that were denied to most of his classmates. And he was fortunate, too, in finding a teacher, Charles Dickerson, who was young, enthusiastic, and interested in his students. From Dickerson, de Forest for the first

---

[17] De Forest, *Autobiography*, p. 50. Russell H. Chittenden, *History of the Sheffield Scientific School of Yale University, 1846-1922*, 2 vols. (New Haven, Conn., 1928), and George W. Pierson, *Yale College, An Educational History 1871-1921* (New Haven, Conn., 1952). Pierson (p. 67) describes the Officers of Yale College in the last quarter of the nineteenth century as intending to meet the need for newer subjects and methods by creating a federated university in which "each school could stand for a different thing." This meant that at the undergraduate level the Sheffield School became a kind of safety valve. In some ways, however, this strategy worked too well, and Pierson adds, "Misusing Sheff as a trash basket . . . had serious University disadvantages. Intellectually it tainted whatever Sheff began; and it advertised to the outside world the inferior status at Yale of its Scientific School."

[18] De Forest, *Autobiography*, p. 56.

time began to receive systematic instruction in physics and chemistry, welcome relief from the farm work and Bible study that dominated the life of the school. And the completion of a new science building during his senior year provided a place where he could feel at home. In the laboratory he could experience, as he expressed it in his diary, "the sense of being in the right place doing the right thing."[19]

He completed his work at Mt. Hermon in two years instead of the normal three and passed the entrance examinations for the Sheffield School without difficulty in June 1893. As a student in the engineeering curriculum, he did not live in one of Yale's undergraduate dormitories but found accommodation in a private rooming house and ate his meals in local restaurants—when he could afford them. The DeForest fellowship paid for his tuition, but other spending money had to come from Talladega or from summer earnings. He threw himself with energy into his studies: analytical geometry, mechanical drafting, German, English, physics, and chemistry in his freshman year. In what spare time he had, he read philosophy and listened to public lectures, and for the first time he began to question seriously the religious dogmatism in which he had been reared. A lecture on the evolution of the brain set him to reading and thinking about Darwinism, and gradually he abandoned "the cherished tenets of my faith, the religious doctrines which had been ingrained within me."[20] Yale furnished nothing very profound to take their place: a vague agnosticism and a sense of "the significance of the scientific approach."

His principal interest at this time seems to have been in mechanical engineering. Electricity at the undergraduate level was taught as part of the physics course, and instruction was severely limited by inadequate equipment for demonstrations and experiments. Sheffield had a small dynamo, a storage battery, several galvanometers, a few standards of inductance and capacitance, and that was all.[21] Not until his senior year as an undergraduate did he begin to read Clerk Maxwell and to think about high-frequency currents as a field for invention. A public lecture by Harry Bumstead in which Hertz's key experiments were replicated fired de Forest's imagination. But by then his undergraduate years at Yale were almost over.

Throughout his years at Yale, as earlier in Talladega, he was trying to prove himself an inventor. At Talladega it had been a cotton-picking machine and a mechanical gate that the driver of a cart could open

[19] Quoted in Carneal, *De Forest*, p. 51. I have not personally found this phrase in de Forest's diaries.

[20] De Forest, *Autobiography*, pp. 68-69.

[21] Carneal, *De Forest*, p. 60.

without descending to the ground.[22] At Yale it was an improved compass-joint, a new type-bar movement for the typewriter, a parlor-game modelled on the Chicago World's Fair, but above all an underground trolley system submitted in competition for a $50,000 prize offered by the New York Metropolitan Railway Company. His motivation was partly to make money but partly also to validate his conception of himself. None of these ideas came to anything and these disappointments, coming on top of his growing doubts about revealed religion, knocked away two of the props on which he had built his life up to this point. By the end of his second year at Yale self-doubt had begun to replace his earlier self-confidence. It was at this juncture, too, that he began reading Maxwell and Hertz and his interests turned toward high-frequency currents. And this is when Tesla's name first appears in his diary. "His works," wrote de Forest, "are the greatest incitors [sic] to zealous work & study. How I *pray* that I may equal & excel him, that all the settled and forgranted [sic] beliefs in my genius & destiny are not idle visions of conceit. It would break my spirit to learn of it. I want *millions* of dollars."[23]

Accumulating pressures brought a serious breakdown in health in his senior year, requiring three weeks in the college infirmary with what his biographer identifies as typhoid fever.[24] He was released with a warning that he could not expect to function effectively on a diet of five-cent hash, toast, and warm milk. By this time, too, it had become evident that his father's anxieties had not been without foundation. Sheffield students were not ostracized at Yale as white children from the North had been in Talladega but neither were they fully accepted into undergraduate life. Lee de Forest did make some friends at Yale, a few of whom stuck by him during his financial and legal troubles in later life. But he did not find it easy. Looking back on his undergraduate years and the "camaraderie" of student life, he regretted that "I have had so pitifully little of it, a Sheff man with but three years in my class, out of a Society, rooming at a distance from my companions, and by poverty debarred from so many of their outings and gatherings, games, theaters."[25] Where he did feel at home and at ease was in the laboratory, preferably alone, working on his precious inventions. But, when he completed his undergraduate years at Yale in 1896, there was no hard evidence to show that he had any particular talent as a creator of inventions. It was still a matter of

[22] Ibid., pp. 34-37.
[23] De Forest Papers (Library of Congress), journal entry for 25 March 1895.
[24] Carneal, *De Forest*, p. 67.
[25] De Forest, *Autobiography*, pp. 84-85, quoting from his diary.

clever ideas, in a hodgepodge of unrelated fields, imperfectly worked out and, when put to the test, unmarketable.

Halfway through his senior year Lee de Forest learned that his father had died, after a fall from which he never recovered consciousness. They had met for the last time the previous summer, while Lee was working as a waiter in a Rhode Island summer hotel. Whether on that occasion they discussed his future career is not known, but it would be strange if they had not. Now, however, he was head of the family, responsible for his mother and elder sister and for his younger brother, still a pupil at Mt. Hermon. He was also, in a sense that had not been true before, in charge of his own future.

The Reverend DeForest left his family better off than they had expected. There was an inheritance of some $6,000, and with these funds de Forest purchased a rooming house on "Freshman Row" (Temple Street) in New Haven and established his mother and sister there, to earn a respectable income from student rentals—and, incidentally, provide him with decent meals. It was a sensible solution in the circumstances and, despite the bereavement, a timely one for Lee de Forest, for it enabled him to stay at Yale and work for his doctorate. Whether his father, if he had lived, would have continued to support him, with a second son about to enter Yale, we cannot be sure; but he could hardly have been faulted if he had told his eldest son that, after three years at the Sheffield School, he was now on his own.

In his work for the doctoral degree de Forest concentrated on electricity. There was a certain amount of practical work with the limited equipment available, but the emphasis in his courses was overwhelmingly on mathematics and theory. In contrast to his later recollections, his diary suggests that at the time he found his mathematical training of considerable value. After two courses under Willard Gibbs in his first postgraduate year and one in the second, he could write, "My mathematical training this year I find already of the greatest practical value. Without such, and every bit of it, I could not read these books leading up to Maxwell. I want another year, still higher. Then I can expect to deal intelligently with light and wave phenomena. . . ."[26] Instruction in electrical theory and practice, however, was under the supervision of Professor Charles Hastings, and between Hastings and de Forest there was a personal incompatibility that found expression in a variety of trivial and not-so-trivial ways. Work on high-frequency phenomena was almost entirely a matter of de Forest's personal reading—Maxwell, Hertz, Lodge, Tesla—with encouragement and some guidance from Harry Bumstead.

[26] Ibid., p. 87, quoting from his diary.

He began working on Lecher wire measurements in the Winchester Hall laboratory and, when ejected from that location by an irate Hastings, was given space in the basement of the Sloane Physics Laboratory. When he selected the Lecher wire project for his dissertation topic—a problem suggested originally by Bumstead—A. W. Wright agreed to serve as official supervisor, with Willard Gibbs following progress "by remote control."[27] But in carrying out the research, de Forest was on his own.

What led de Forest to concentrate his energy and attention on high-frequency alternating currents? This was not a field for which the Sheffield School was well equipped nor one in which the faculty had special expertise. Bumstead's interest in and repetition of Hertz's experiments certainly had an impact. And Gibbs's lectures on the electromagnetic theory of light provided the theoretical foundation. De Forest's diaries and autobiography, however, both highlight the example of Tesla's career and the influence of Tesla's writings. Tesla had achieved, and looked as if he might continue to achieve, the spectacular success as inventor that de Forest longed for. He had done so in a dramatic and startling way, in a field that was not cluttered up by the myriad prior inventions of others. And he had done it on his own, in his own way, so that there could be no doubt, when the thing was done, who had done it.

Lee de Forest, when he passed his doctoral examinations, had done well as a student of electrical science. But had he up to this point invented anything at all remarkable, anything that would justify those long battles with his father, anything that could validate those grandiloquent boasts (to himself and to others) that his true genius was to be an inventor? He had not. Even his dissertation showed few signs of originality: a harsh critic might have described it as pedestrian. High-frequency alternating currents, however, still offered a field in which a young electrical engineer with ambition and imagination might make his mark, however ordinary his talents in more conventional fields. But in choosing that field de Forest was raising the stakes. As he wrote, "Should I prove wrong I would be away behind my classmates and it would go hard with me for not knowing my engineering better; but I risked all on the cast of that die."[28]

* * *

Lee de Forest received his doctoral degree in 1899. What opportunities were available at that time for a young electrical engineer who hoped to

[27] Ibid., pp. 88, 96.
[28] Ibid., p. 89.

make his reputation in "wireless communication"? What was the state of the art, and what contributions could someone like de Forest make?

Five years earlier, in 1894, Oliver Lodge had demonstrated wireless telegraphy as a scientific curiosity at the Oxford meetings of the British Association for the Advancement of Science, using an induction coil and spark gap for a transmitter and a coherer and galvanometer for a receiver. That was the first public demonstration anywhere of the use of Hertzian waves for communication.[29] In the same year Guglielmo Marconi began his experiments in the attic of his father's villa in Bologna. By 1896 he had a workable communications system to demonstrate to the British Post Office. In 1899 he succeeded in telegraphing across the English Channel, and later in that year he used his apparatus to report the America's Cup yacht races for the New York *Herald*. Reginald Fessenden in 1899 delivered his address on "The Possibilities of Wireless Telegraphy" to the American Institute of Electrical Engineers (see above, pp. 28-29) and received his first patent for a new detector. Twelve months later he was developing a wireless telegraphy system for the U.S. Weather Bureau. Nikola Tesla in 1898 was demonstrating wireless control of model ships in Madison Square Garden and predicting the imminent completion of a system that could transmit both power and intelligence over long distances without wires. In Germany, Adolf Slaby and Ferdinand Braun were both working on wireless communication systems that might rival or surpass Marconi's. And in a score of other locations, in Europe and America, there were individual engineers and experimenters tinkering with induction coils, spark gaps, antennas, and coherers.

What differentiated de Forest from the many other individuals who, at the turn of the century, were fascinated by the new technology of wireless communication? Not very much. His scientific background was certainly no better than that of Lodge, Fessenden, Slaby, or Braun. His identification of the markets to be served was vague. In his ability to raise capital he lacked the family connections that were so critical to Marconi's early success. He had no access at first to government contracts or patronage as did Slaby. He was without the academic connections and prior record of successful invention that helped Fessenden.

These limitations might have been overcome if de Forest in 1899 had had a clear vision of what it was he wanted to accomplish technologically—if his dreams had been disciplined by a clear conception of what wireless technology should be like. This was what was needed if a bridge was to be built between the scientific training that was his principal asset, and the commercial and monetary success to which he aspired. But,

[29] Aitken, *Syntony*, pp. 115-24.

except in one respect, there was no such integrating idea. With Fessenden, as we have seen, it was his unshakable conviction of the superior efficiency of the continuous wave. With Elwell, a decade later, it was commitment to a particular technique for generating continuous waves: the oscillating arc. One looks in vain for something analogous in de Forest. This is why, for more than ten years after he left Yale, he gives the impression of a man scrambling for a foothold in a rapidly shifting technology, trying now one device and now another, inventing what he could, borrowing (to use a neutral term) what he could not. The irony lies in the fact that this was the man who was to invent the three-element vacuum tube, the device that was to dominate radio technology for the next half-century.

There was one specific exception to the general diffuseness of de Forest's interest in wireless communication in 1899. This was his rejection of the coherer. He was not alone in this. Fessenden also considered the coherer unsatisfactory, particularly for any system that used undamped waves. And the many attempts to find alternatives to the coherer suggest that dissatisfaction was general. De Forest had done some work with coherers in connection with his dissertation but had relied mostly on his "glow tube." Even for the Lecher wire experiments this was not entirely satisfactory: the tube tended to glow dimly all the time (an indication of the many wavelengths present on the wires) and it had called for some patience and practice on de Forest's part before he learned to distinguish the significant peaks from those that were irrelevant.[30] But it was certainly better than any coherer. For use in a communications system, however, the glow tube would not do. De Forest needed a device that could follow the dots and dashes of the Morse code and the glow tube was too sluggish in action for that.

When he left Yale he did not have such a device. Nor did he have any of the other elements that were necessary if he were to assemble a wireless "system" that could function at least as well as Marconi's without infringing Marconi patents. The hard truth was that in 1899 de Forest did not find himself readily employable in any capacity where he could put his training to use, far less one where he could experiment with wireless. An approach to Tesla had proved fruitless.[31] Equally futile was a letter to Marconi which did not even elicit an acknowledgment.[32] De Forest ended up working in the dynamo department of the Western Electric

[30] De Forest, "Reflection of Hertzian Waves," *American Journal of Science*, p. 60.

[31] De Forest, *Autobiography*, p. 90; Carneal, *De Forest*, pp. 80-81.

[32] A lengthy excerpt from de Forest's letter to Marconi may be found in W. P. Jolly, *Marconi* (New York, 1972), p. 74.

Company in Chicago for eight dollars a week—"much chasing of parts and mopping of grease," he called it, and not at all what a Doctor of Philosophy might be worth in the proper place.[33]

But it was a start. From the dynamo department he moved up to the testing laboratory and there, with the connivance of his supervisor, he was allowed to work in his spare time—which somehow expanded into full time—on his wireless equipment. Evenings were spent in the John Crerar Library, reading the European periodicals—*Science Abstracts*, Wiedemann's *Annalen*, *Comptes Rendus*. De Forest knew what he was looking for: a detector that would be faster and simpler than a coherer; one that would be self-restoring and not require the mechanical "tapping back" that most coherers needed; and one that would let the operator hear the received signal in earphones, dispensing with the Morse "inker" and paper tape that were standard Marconi equipment.

He found what he wanted, or thought he did, in the April 1899 issue of Wiedemann's *Annalen*. A German physicist named Aschkinass there described a detector of Hertzian waves consisting essentially of a thin piece of tinfoil laid on a glass plate and cut in two with a razor. A drop of water or alcohol was placed on the slit, and a battery connected to the two ends of the foil. When a spark coil was excited in the vicinity and an earphone placed in the circuit a faint ripping sound could be heard. This was, technically, an electrolytic anti-coherer, and it was not a new principle.[34] But, if it could be made less erratic, it seemed to meet de Forest's requirements for a detector that would be self-restoring and produce a sound in earphones. He took up the idea with enthusiasm and worked on it in a corner of the Western Electric laboratory, trying other metals and fluids in the attempt to overcome a persistent tendency for the device to "clog" after a short period of operation. After about a year's work he thought he had the problem solved. The device by this time consisted of a sandwich of two flat plates separated by a thin layer of liquid or of some porous material soaked in liquid. He called it a "responder" and, in association with a fellow worker, Edwin H. Smythe, who had aided him both technically and financially, he applied for a patent.

This device was never of any importance as part of a functioning communications system.[35] That does not mean, however, that it was

[33] De Forest, *Autobiography*, p. 102.

[34] The device was, in fact, often referred to as a "Schafer's plate," after the German scientist who first devised it. See V. J. Phillips, *Early Radio Wave Detectors* (London, 1980), pp. 65-67, and Fleming, *Principles*, pp. 491-92.

[35] Compare Robert A. Chipman, "De Forest and the Triode Detector," *Scientific American* 212 (March 1965), 94.

insignificant, for the episode tells us something about de Forest's emerging style as inventor and innovator. Note, for example, the systematic search of the relevant literature, and particularly of the European journals. De Forest could read the languages, and he knew where to look. If he had derived no more than this from his Yale training, his time had not been wasted. Then and later critics might question de Forest's talents as a scientist, but they never charged him with failing to keep up with the literature. Note, too, the highly empirical manner in which he went about improving the device. Neither de Forest nor anyone else at the time really understood how electrolytic detectors functioned.[36] Improving performance was essentially a matter of trying one thing and then trying something else. And in this respect the development of the responder was typical of almost all experimentation in wireless telegraphy going on at that time. The scientific community might generate the ideas, but the design of apparatus was a question of "cut and try."

A second "responder" followed within a year. This was also an electrolytic anti-coherer. Two metal electrodes were inserted in an insulated tube, like plugs, with a gap of about 1/200 of an inch between them. (See Fig. 4.5) The intervening space was filled with any one of several mixtures that de Forest elegantly referred to as "goo," the most common being a paste of glycerine or vaseline mixed with water or alcohol, with

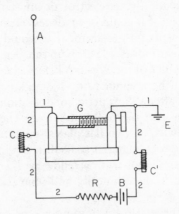

Fig. 4.5: De Forest's responder.

[36] De Forest's description of the action of the responder, seen under a microscope, with its "tiny ferry-boats" and "little pontoon ferry-men," is a curious gem of scientific imagery but hardly suggests that he understood the physics of the process. See Lee de Forest, *The Electrician* 54 (1904), 94, and Phillips, *Detectors*, pp. 66-67. Michael Pupin is a possible exception to the generalization in the text.

a small quantity of lead oxide or metallic filings added.[37] As in the earlier model, a battery and telephone earpiece in series were connected across the responder, as were the antenna and a connection to ground. When an electromagnetic impulse was received by the antenna, it broke the minute chains of metallic particles that had formed between the electrodes and this was heard as a click in the earpiece. A chain of impulses (such as a "dash" in the Morse code) would be heard as a continuous sound.

For this responder also de Forest and Smythe applied for a patent. It represented some advance in practicality and convenience over the earlier type, since the electrolyte was now totally enclosed and therefore less likely to dry up. But reliability was still a problem. Christmas 1901 found de Forest lamenting in his diary "If only my Responder would not clog." This was at a time when the newspapers were full of Marconi's reports of his reception of transatlantic signals in Newfoundland and when it was becoming imperative that de Forest be able to convince potential backers that he had the elements of a workable and patentable system. ". . . Time is short," he noted in his diary, "and Marconi sails fine and weatherworthy boats, and these boats are already headed toward America."[38]

By this time (December 1901) de Forest was committed to work in wireless telegraphy. He had quit his job with Western Electric in the spring of 1901 to become chief engineer with the newly formed American Wireless Telegraph Company in Milwaukee. His tenure of that position was brief, as he refused to make his new detector available for the company's use and was promptly fired. Back in Chicago he worked for a while as assistant editor of the *Western Electrician* and taught part-time at the Lewis Institute, earning five dollars a week for his teaching and accepting the same amount from his friend and associate, Ed Smythe.

He also became acquainted with Clarence Freeman, a professor at the Armour Institute, who arranged for de Forest to have the use of the institute's electrical laboratory in return for helping the students with their assignments. The first tests of de Forest's "system" were run between an antenna on the institute's roof and a receiving station at a hotel about half a mile away. De Forest's transmitter was completely conventional— a Ruhmkorff induction coil, a Wehnelt high-speed interrupter, and a spark gap—and the distance covered was not impressive. Marconi had, after all, demonstrated two-way communication over more than thirty-

[37] For descriptions see Fleming, *Principles* (1916 edition), p. 508 and Phillips, *Detectors*, pp. 68-69. The U.S. patents are Nos. 716,000 and 716,334, applied for on 5 July 1901.

[38] De Forest, *Autobiography*, pp. 117-18, quoting from his diary.

five miles to the U.S. Navy two years before.[39] But it proved to de Forest that his detector worked outside the laboratory, and further tests from a yacht on Lake Michigan to a receiving station on the shore confirmed his optimism. These activities brought de Forest his first newspaper publicity; he found that he liked it.

Freeman was himself interested in wireless telegraphy and had invented a transmitter of novel design. He was also in a position to finance further tests and demonstrations, which neither Smythe nor de Forest was. On condition that his transmitter would be used, he agreed to a shift of operations from Chicago to New York, in the hope that de Forest, through his Mt. Hermon and Yale connections, would be able to find someone to finance their system and that they would get a contract to report the international yacht races, as Marconi had done in 1899. The evidence indicates that both Smythe and Freeman would have preferred to stay in the Chicago area; de Forest insisted on the move because (as he later recalled the decision) he thought the prospects for commercial wireless were better in New York and he felt that his "independence of management" required it.[40]

Independence of management meant for de Forest escaping from a collaboration with Smythe and Freeman that he found more and more distasteful. "Not for Smythe did I toil six years at Yale," he wrote in his diary. But it was Smythe who paid for the patent applications and provided the five dollars a week that enabled de Forest to keep going. In Freeman's transmitter he had no confidence at all. It was new and untried; above all, it was not de Forest's. Unless he used it, however, Freeman would withdraw his financial support and without that support there would be no move to New York and no more demonstrations. The fact of the matter was that de Forest did not want collaborators. It was to be the de Forest system, not a hyphenated one. That implied, however, access to outside capital on a much larger scale, and from sources that would not insist on sharing the glory of future success.

The move to New York in August 1901 did not solve these problems. Through the good offices of certain of de Forest's Yale classmates, a small syndicate was organized to finance the building of Freeman's transmitter.[41] With this, the responder, and a tugboat provided by the Publishers'

[39] L. S. Howeth, *History of Communications-Electronics in the United States Navy* (Washington, D.C., 1963), pp. 28-32.

[40] De Forest, *Autobiography*, p. 123; Carneal, *De Forest*, pp. 120-22.

[41] Chipman ("De Forest and the Triode Detector") writes as if Freeman's transmitter had been used for the tests in Chicago. De Forest's autobiography, however, makes it clear that the first model was built only after the move to New York.

Press Association, de Forest undertook to report the international yacht races in competition with Marconi. The result was a debacle, as neither Marconi's equipment nor de Forest's contained provision for selective tuning; interference between the two transmitters, and from a third one whose ownership was unknown at the time, meant that no messages at all could be copied. Freeman's transmitter proved unworkable and de Forest junked it after the second race, substituting a conventional spark coil and interrupter. These events did not make it easier for de Forest to retain the confidence of the small group that was providing him with funds, now organized as the Wireless Telegraph Company of America. This organization was capitalized at only $3,000, with the stock distributed among the original subscribers, so there was no easy way to raise additional risk capital, even if prospects had been brighter.[42] To remedy this difficulty, a new company, the De Forest Wireless Telegraph Company, was incorporated in February 1902, under the laws of the State of Maine, with a capitalization of $3 million. This generated a small inflow of new funds from the sale of shares. To increase this inflow of capital de Forest joined forces with Abraham White.[43]

White was an uncommonly intelligent, enterprising, and unscrupulous financier who, a few years earlier, had acquired a reputation and the start of a personal fortune by successful speculation in federal government bonds—not what one would normally consider a speculative security. He knew nothing about wireless telegraphy. He was, however, a master at the uses of publicity, and he knew how to sell corporate securities to a gullible public. In that sense he was precisely what was needed at this juncture. De Forest had used his Yale and Mt. Hermon friendships to the full: there was no more money to be had from that quarter. "Respectable" financial intermediaries, whether commercial banks or investment houses, would have nothing to do with him. He had no business connections. And there were no family resources on which he could draw. White held out the prospect of raising large sums from the general public, not a source likely to inhibit de Forest by overclose monitoring of what he was up to technically. And he intended to do so by the shrewd use of publicity, by exploiting the public's growing fascination with wireless

[42] Thorn Mayes, "DeForest Radio Telephone Companies, 1907-1920" (Mimeo), p. 1.

[43] To distinguish the several de Forest companies and follow their fortunes calls for unusual patience and pertinacity. The most reliable guide is Mayes's unpublished manuscript (see n. 42), which is based on state incorporation papers and court records. De Forest, in his *Autobiography*, does not always distinguish between companies with similar names, and Carneal's biography is quite unreliable in this connection.

and by building up de Forest's image as a worker of wonders. Five years later de Forest was to unleash some of his most bitter invective against Abraham White; in 1902, however, they were made for each other.

The De Forest Wireless Telegraph Company, even with its $3 million authorized capital stock, hardly gave White enough elbow room, and late in 1902 the name was changed to the American De Forest Wireless Telegraph Company and the capitalization was increased to $5 million. In 1904 it was again increased, this time to $15 million. White, of course, earned his commission on each share sold. De Forest relied on the sale of shares to pay his own salary as chief engineer and to finance the company's operations. He also held much of the company's stock: 20 percent in 1906.[44] There was little thought of earning a profit from operations, in the sense of commercial message-handling or the sale of equipment. The "normal" revenue of the company was the money raised by the sale of stock: that was the business it was in. It was de Forest's responsibility to generate the publicity that made the sale of stock possible.

White did not interfere with the technical side of the business, nor did de Forest with the financial practices that eventually landed White in the Atlanta federal penitentiary. This is not to say that de Forest was ignorant of what was going on; merely that he saw nothing to object to. "Soon, we believe, the suckers will begin to bite," he noted in his diary in early 1902. " 'Wireless' is the bait to use at present. May we stock our string before the wind veers and the sucker shoals are swept out to sea."[45] If raising capital to build radio stations required the tactics of a snake-oil salesman, it made sense to have as your partner the best one around. And White was very good. No opportunity to tout the virtues of the de Forest system was lost, and if that called for misrepresentation or the diffusion of false information, White was not one to quibble over details. Apart from the dubious legality of his procedures—there was no Securities and Exchange Commission in those days, but using the mails to defraud was a federal offense then as now—the only problem, if it was a problem, was that White regarded the company as an instrument for raising money, not as an organization for rendering a communications service or advancing the art of wireless telegraphy. Demonstrations were held where they would get favorable publicity. Stations were built where stock could be sold. The Atlanta station, to cite one famous example, cost $3,000 to build and yielded $50,000 in stock sales. That meant, to White, that Atlanta was the right place to have a station, despite the fact that it never

[44] De Forest, *Autobiography*, p. 218.
[45] De Forest Papers (Library of Congress), diary entry for 9 February 1902.

generated any traffic.[46] Within his own frame of reference, that was completely rational economic behavior. The trick to raising money was to get favorable publicity, and White never let an opportunity pass to do that. Not until 1905, after a summer during which high levels of static had played havoc with the overland circuits, do we get the first hints of a disagreement over policy, with de Forest arguing for more emphasis in future on marine radio. Shipowners, however, were not great buyers of speculative wireless stocks, and White opposed the move.

Between 1902 and 1906, indeed, it looked as if the American De Forest Company was in a fair way to outdistance all its competitors. It had won the Gold Medal and the Grand Prize for the best wireless system at the St. Louis World's Fair. It had been chosen by the London *Times* to transmit its correspondent's dispatches during the Russo-Japanese war. It had survived lawsuits for infringement of Marconi patents. It had proved that, under favorable conditions, long-distance wireless communication could be maintained over land. It had won important Navy contracts for the construction of spark stations in Florida and the Caribbean. It had become a significant supplier of radio equipment to the Navy, underbidding both Fessenden and Telefunken. And it had placed in operation its own system of radio stations on the Great Lakes and down the Atlantic seaboard. In the process it had trained a new generation of commercial radio operators, not former Western Union or Postal Telegraph personnel but men who had learned their trade in "wireless" from the ground up.

* * *

How had de Forest been able to accomplish all this, after the catastrophic performance of his equipment during the yacht races of 1900? Not by any single invention or technological breakthrough. In 1902 he had for a detector nothing better than the second version of his responder. He had no circuits for tuning—certainly none that did not infringe the Marconi and Lodge patents. He had no particular theory of antenna construction or design. For a transmitter he had the ruins of Freeman's invention, and beyond that nothing but the classic induction coil and interrupter. Technically, these were not impressive foundations for a wireless system.

By 1906, however, de Forest had put together what he referred to as the "American system of wireless." He believed that it stood in sharp

[46] Susan Douglas, "Exploring Pathways in the Ether: The Formative Years of Radio in America, 1896-1912" (Ph.D. diss., Brown University, 1979), p. 112.

contrast to standard European practice, that it permitted significantly higher sending speeds, and that it enabled radio traffic to be copied through levels of static and interference that other systems could not handle. It was essentially simple, easily maintained, and rugged—those were, indeed, among its major virtues. And it was made up of elements borrowed either from allied technologies or from other inventors.

As regards antenna design de Forest had little to contribute. Get as much wire as you can into the air, either vertically or horizontally, and then tune for the hottest spark or the highest antenna current—that was about as sophisticated as his thinking about antennas ever became. But it worked well enough, whether in demonstrations from the Eiffel Tower or across Florida swampland, and sometimes useful directional effects could be obtained. Similarly, as regards tuning circuits, de Forest had nothing to add to existing knowledge. The Marconi "four sevens" patent and the Lodge syntony patents governed that field, and the principles and methods they embodied could not be evaded. De Forest felt personally confident, however, that they had been anticipated by Tesla and Pupin, and that was good enough for him. Then and later de Forest's attitude to patents depended on whose patents were at stake. His own were to be defended at all costs, but others could be disregarded with impunity if he personally thought them invalid. For a time the Marconi and Lodge patents were simply ignored, and de Forest's slide tuner and "pancake" tuning coils became standard features of "the American system."[47]

For transmitting equipment de Forest relied at first on the standard Ruhmkorff induction coil, with a Morse key in the primary circuit and a spark gap in the secondary. There was nothing original there. He soon grew impatient, however, with the conventional "hammer" interrupter, partly because of its tendency to stick and jam, partly because of the low spark frequency that it generated. Low spark frequencies produced a low-pitched sound in the receiving operator's headphones, and this was often hard to distinguish from the grumbles and crashes of atmospheric disturbances. To secure a higher spark frequency de Forest often used a Wehnelt or electrolytic interrupter, which could generate sparks at a rate

[47] The U.S. Supreme Court finally struck down the fundamental Marconi tuning patent in 1943, essentially on the ground that it had been anticipated by earlier patents granted to Tesla, John S. Stone, and Oliver Lodge. For an analysis of this decision, and of the difficulty Marconi experienced in getting his patent issued because of the examiner's knowledge of these prior disclosures, see Anderson, "Priority in the Invention of Radio."

of several hundred per second.[48] At the receiving station this was heard as a high-pitched, piercing note, quite different from atmospheric noise and easily distinguished from the signals of conventional spark stations on nearby frequencies.[49]

A more radical departure from conventional apparatus came when de Forest abandoned the induction coil completely and substituted an alternating-current generator and step-up transformer. According to his own account, this was the type of transmitter he wanted to use when first he moved from Chicago to New York but the commitment to Freeman's device prevented him. By 1902 at the latest he had such a transmitter in operation and after that date it became standard equipment at de Forest stations. The type of transmitter that resulted must, of course, be sharply distinguished from the radiofrequency alternators that Fessenden and Alexanderson were later to build. De Forest's alternating-current generators operated at ordinary commercial frequencies—60 cycles per second, usually. The voltage they generated was stepped up by a transformer and used to drive a spark gap, and it was the spark gap that generated the radiated wave, not the alternator directly. The distinction was an important one: de Forest could use commercial alternators, much cheaper and more readily available than the custom-built radiofrequency machines that General Electric made for Fessenden. The apparatus that resulted was rugged and effective, easily installed and maintained even on board ship.

These transmitters, whether driven by an interrupter or by an alternating-current generator, were firmly within the spark tradition. At this stage in his career de Forest was content to work within the confines of spark technology. Even his dreams of wireless telephony, at this time, rested on the assumption that intelligible speech could be transmitted if only the spark frequency were raised high enough. There is none of the striving to escape from spark that we find in Fessenden. Not until 1907, after quitting the Telegraph Company, did de Forest experiment with arc telephone transmitters. And not until 1911-1912, after conversations with Alexanderson, do we find him investigating radiofrequency alternators. De Forest was slow to abandon spark; his efforts were directed toward improving and refining spark technology, not escaping from it.

In striving for higher spark frequencies and for a signal that could be read by an operator using earphones, de Forest was already departing from European practice. Standard Marconi and German practice at this

[48] The Wehnelt interrupter had been first introduced in 1899 and excited considerable interest. See Fleming, *Principles* (1916 edition), pp. 63-65.

[49] De Forest, *Autobiography*, p. 149.

time called for the received signal to be printed on paper tape by a Morse "inker" so that it could be filed and checked later. And, until 1905-1906, this was also standard procedure in the United States Navy. The practice of copying Morse code by ear and then transcribing it by hand (either by typewriter or by handwriting) later became so accepted that we forget what an innovation it was. And it seems to have been of American origin, although further research on the subject is indicated. It implied, of course, trusting a human operator, and where rank was important that could be ticklish. There was no printed record to refer to afterwards if the accuracy of transcription should be challenged. Once accepted, however, the practice had important consequences for the design of radio apparatus and for the efficiency of radio circuits. Sending speeds, when a Morse inker was used at the receiving station, could not exceed a maximum of about twelve words per minute because of the mechanical sluggishness of the relay and the inker itself. A good American operator using earphones could copy at thirty-five words per minute. Further, a receiving circuit using a coherer or magnetic detector and an inker could not discriminate between the signal to be copied and atmospherics—the "strays" and "x's" that were the bane of every operator. Any received impulse that passed the tuned circuits and the detector left a mark on the tape. An operator using headphones—and that most remarkable of filters, the human ear— could discriminate not only between signal and noise but also between interfering signals, if they were of different audio frequencies.[50]

The fact that reliance was placed on a human operator using earphones rather than on a mechanical inker using paper tape had implications for the design of transmitters—high spark frequencies, for example—but even more specifically for the design of detectors. In a sense the coherer and the Morse inker were complementary devices. The coherer, with its on–off or "triggering" action, was well suited to drive an inker, which either made a mark on the tape or did not. The American hostility to coherers was all of a piece with the American preference for earphones. Earphones were more sensitive than any mechanical inker could be, but they were incompatible with coherers. A coherer could produce in earphones nothing but clicks. It was no accident that Marconi in Newfoundland in 1901, straining to hear signals from his transmitter in Ireland, used a coherer

[50] The story is told, in fact, of one American operator who, when asked which of several interfering signals he was copying, replied succinctly, "All of them." The ability of a skilled radiotelegraph operator to "copy in his head," even while sharpening a pencil or chatting with passers-by, and then, several minutes later, produce an error-free transcription, is a phenomenon probably appreciated to the full only by those who have attempted the feat and failed.

and a telephone earpiece; he needed the most sensitive device available. But it was also no accident that the signal he had ordered to be transmitted was the letter s—three dots in Morse code. What Marconi heard, or thought he heard, was three clicks; he was in no position to discriminate between dots and dashes. American experimenters, taking it for granted that received signals would be decoded by the human ear, wanted what Fessenden called a "continuously responsive" detector—one that would yield a sound in the earpiece as long as the transmitting operator held his key down. No coherer could do that.[51] And they wanted one that would enable the operator to discriminate between the signal he wanted to hear and the noise he did not. No coherer could do that either.

De Forest's concentration on electrolytic detectors makes sense in this context. Unfortunately, neither of the "responders" he had developed up to 1902 was satisfactory. Even the second version, with its enclosed "goo" electrolyte, tended to clog at unpredictable moments. If this was inconvenient for commercial message-handling, it was even more awkward if it happened during an important demonstration to potential investors. It was, therefore, a matter of extreme urgency in 1902 that the De Forest Wireless Telegraph Company locate and acquire a reliable detector.

De Forest was always perfectly candid about how he solved the problem. Early in the spring of 1903 he visited Reginald Fessenden, then working at Fortress Monroe, Virginia, and inspected his equipment, including particularly the "liquid barretter" (see above, pp. 55-57). He also talked with one of Fessenden's assistants, Frederick Vreeland, who informed him that he, and not Fessenden, was the true inventor of the device. Thereupon (as de Forest puts it in his *Autobiography*) "we ourselves resolved to use a Wollaston-wire rectifier detector or its equivalent."[52] A little research in back files of the electrical journals turned up the fact that Pupin in 1899 had disclosed an electrolytic rectifier using Wollaston wire dipped in acid. That was good enough for de Forest. Fessenden might have a patent on the liquid barretter, but it was not really his invention and he had been anticipated by Pupin anyway. De Forest set one of his assistants to work devising an electrolytic detector that would work on the same principle as Fessenden's without being too obviously an imitation. The result was the so-called "spade" detector. A fine platinum wire was flattened out and sealed in a glass tube; then the end of the tube was broken off and ground down until only a minute edge of platinum was exposed. This was then mounted over a lead cup

---

[51] A coherer could have been used, of course, to switch on and off a local oscillator, if one had been available.

[52] De Forest, *Autobiography*, p. 161.

containing acid, and the end of the glass tube was dipped in the acid.[53] It made a sensitive and reliable self-restoring detector and promptly replaced earlier "responders" in all de Forest stations. Large numbers were also purchased from de Forest by the Navy, despite vigorous protests from an aggrieved Fessenden. De Forest could afford to sell them very cheaply; his company made its money by selling shares, not detectors.

These, then, were the major elements in de Forest's "American system of wireless": the alternating-current generator, step-up transformer, and spark gap at the transmitter; spade detector, slide tuner, and headphones at the receiver. Within the limits of spark technology, it was an efficient and economical system, recognizably different in appearance and concept from standard Marconi and German equipment. The electrolytic detector replaced the coherer; the alternating-current generator and transformer took the place of the induction coil and interrupter; earphones and the operator's skill replaced the relay and Morse inker. De Forest personally had no doubt of its superiority; sending and receiving speeds were higher, and signals could be copied through heavy interference. It provided one of the two foundations on which the fortunes of the American De Forest Wireless Telegraph Company rested between 1902 and 1906. The other was the stock-selling ability of Abe White and his henchmen.

These foundations were not as secure as they seemed. Fessenden brought suit for infringement of his liquid barretter patent in 1903. The case dragged on for three years and was finally decided in his favor in 1906. A court injunction denied the American De Forest Company further use of the spade detector but prompt action in acquiring rights to the recently invented carborundum detector enabled its stations to continue functioning.[54] Meanwhile, however, Abraham White and certain of the other directors had formed a new company, the United Wireless Telegraph

[53] For more complete descriptions, see de Forest, *Autobiography*, p. 162; Carneal, *De Forest*, p. 150. Phillips (*Detectors*, pp. 70-82) cites a number of experimenters who developed very similar electrolytic detectors about the same time.

[54] Gen. H. C. Dunwoody of the U.S. Army applied for a patent on the carborundum detector on 23 March 1906; the patent was issued on 4 December of that year. Dunwoody became a vice-president of the American De Forest Company. For G. W. Pickard's role in showing the company's personnel how to use the carborundum detector effectively, see Alan Douglas, "The Crystal Detector," IEEE *Spectrum* (April 1981), 64-67, and (from the Patent Office files by courtesy of Alan Douglas) *Pickard vs. Ashton & Curtis*. Interference No. 31,649 (November 1911). Note that de Forest's two-electrode audion detector, although invented in 1906, was not ready for commercial use at that time. Attorneys for American De Forest in fact attached no value to it when Lee de Forest left the company.

Company, and to it had transferred all the assets of American De Forest (but none of its obligations), almost certainly with a view to shielding them from attachment for damages awarded to NESCO and Fessenden. De Forest, absent in England, had not participated in this decision and, on his return, found that the company, and his equity in it, had been in effect liquidated. He also found, much to his surprise, that White and his associates held him personally responsible for having deceived them about the spade detector. He resigned or was fired as vice-president and director on 28 November 1906. The stock he owned in the company was now worthless and he turned it in to the company's treasury. This meant that he had severed all formal connection with the company and gave him protection against its creditors, who were apparently numerous. He retained, however, his rights to the recently developed audion detector and accepted, in settlement of all claims, $1,000 in cash, half of which went to his attorney.[55] This left his personal finances in lamentable condition, and he was reduced to seeking a job from, of all people, Reginald Fessenden.

This was, in effect, the end of the American De Forest Wireless Telegraph Company. It was not, however, the end of United Wireless, to which ownership of the De Forest Company's tangible and intangible assets had been transferred, and it is necessary, if later events are to be understood, to follow the history of this organization a little farther. Despite, or because of, its dubious past, United Wireless was at the end of 1906 the major American operating company. It continued its policies of aggressive stock promotion and the expansion of its system in the years that followed, building on the technology as de Forest had left it but using the carborundum detector. By 1911 it had no less than 70 shore stations and some 400 ship installations. This compared with 176 ship installations by the American Marconi Company, 6 by Fessenden, and 5 by de Forest.[56] United Wireless in fact held a near-monopoly of commercial radio communications on the Atlantic seaboard and on the Great Lakes. Competition had been largely eliminated by price-cutting. White and his successor as president, Christopher Columbus Wilson, a former colonel in the Confederate Army whose stock-marketing tactics seem to have been even more flamboyant than those of his predecessor,

[55] De Forest, *Autobiography*, pp. 216-20. De Forest depicts the return of his shares to the company as a quixotic gesture, alleging that White had sold shares that did not exist and that de Forest's shares were needed to make up the discrepancy.

[56] S. Douglas, "Pathways," p. 270, citing *Annual Report of the Commissioner of Navigation* (Washington, D.C., 1911), Appendix M, p. 202.

would supply apparatus to shipowners without cost, pay the salaries of the operators, and rarely charge more than a nominal rental fee, relying on income from the sale of shares to keep the corporate treasury full. In the process considerable damage was done to the public's confidence in wireless as a means of communication and in wireless stocks as a form of investment.

By 1910, however, this patchwork garment was beginning to come apart at the seams. In June of that year the Department of Justice filed suit against the individuals who had been most conspicuously involved in the financing of the American De Forest Company, charging them with use of the mails to defraud the public. White and Wilson were convicted and sentenced to terms in the penitentiary. In 1911 the Marconi Wireless Telegraph Company, under the vigorous leadership of its new general manager, Godfrey Isaacs, belatedly moved to enforce its wireless patents against American competitors and sued United for infringement of the "four sevens" tuning patent. Judgment in the suit went to the Marconi Company. United was unable to pay the damages awarded, and its assets were taken over by the Marconi interests and transferred to the Marconi Wireless Telegraph Company of America.[57]

Up to this date the Marconi Company had been a minor factor in American radio. Its business had been almost wholly confined to trans-atlantic communications and, to a small extent, marine radio. This was true no longer. By the absorption of United Wireless, American Marconi, controlled by its British parent, became unquestionably the dominant firm in the American radio industry. Even those who had most strongly condemned the "circus psychology and cavalier financing" of United Wireless found reason to question whether this new alignment was necessarily in the national interest.

＊　　＊　　＊

Lee de Forest's search for an efficient detector of electromagnetic waves provides one thread of continuity running through his life between the time he left Yale and his resignation from the American De Forest Company in 1906. His failure to find such a detector without infringing the property rights of others was the major factor responsible for the demise of that company and the rise of United Wireless. It is ironic, therefore,

[57] A. H. Morse, *Radio: Beam and Broadcast* (London, 1925), p. 97. The Marconi Company paid $700,000 for the assets of United Wireless (presumably as an offset to damages due Marconi) and sold them to its American subsidiary for $1,488,800 in common stock.

that in 1906—only three days, in fact, before he resigned from the company—he had conceived the first triode vacuum tube and given orders for its manufacture.

This device was the ancestor of all later vacuum tube detectors, amplifiers, and oscillators. Its invention is one of the "great divides" in the history of radio technology; the whole basis of radio communication begins to shift with the introduction and diffusion of this device. Inevitably, therefore, the analysis of how this invention came about has attracted considerable attention.

An important issue of historical interpretation is involved. Lee de Forest himself provided a highly plausible and internally consistent account of how the invention was made. In fact, every time he told the story it became more "rational," in the sense that each step in the process of invention was presented as following reasonably and naturally from what had gone before. This account by no means ignores the influence of previous workers, nor does it exclude the role of chance; but it depicts de Forest as following a largely autonomous and self-directed course of discovery. And this is where the problem lies. There is evidence, ignored or denied by de Forest, that he was decisively influenced at a critical point in his work by John Ambrose Fleming, whose development in 1904 of the diode vacuum tube or "Fleming valve" had been made known through the scientific periodicals and by his application for patents. The suggestion has been made that de Forest's own version of the process of discovery was essentially an elaborate rationalization, its function being to deny the influence of the British scientist and enhance de Forest's patent rights and his image as an independent inventor.

Dispute over which interpretation is to be accepted began almost as soon as the development of the three-element audion was announced and has continued ever since. It is highly unlikely that, at this date, any new evidence will be uncovered to settle the issue. The problem is one of exegesis: we have to interpret evidence already known. Interpretation depends on the questions that are asked. In the present context our primary interest is not in deciding priority nor in allocating relative credit for creative ability but in analyzing the invention as a case study in the process of technological discovery. From that point of view the evidence provided by de Forest's version and the evidence unearthed by others are alike highly relevant.[58]

[58] Among recent contributions to the understanding of this issue, particularly noteworthy is Gerald F. J. Tyne's masterly *Saga of the Vacuum Tube* (Indianapolis, 1977), on which I have relied heavily for evidence and insights. Tyne not only had several interviews with de Forest but also was able to use the records

De Forest's interest in partially evacuated glass tubes as devices for the detection of Hertzian waves began in the basement of the Sloane Physics Laboratory at Yale. The "glow lamp" or vacuum tube—he uses the very term in his dissertation—was placed at the end of the Lecher wires and enclosed in a wooden box with a window in it, so that de Forest could gauge, by the strength of the glow, when the Lecher wires were resonant. It was not a very good detector for that purpose, since it did not give unambiguous readings, but it was the best de Forest could do. To see in this device an ancestor of the later audion calls for some imagination. It had no heated filament. It did not operate by electron emission (as we would now express the matter). It functioned by the ionization of the residual gas. Nevertheless it seems to have left an indelible impression on de Forest's mind: wireless waves could be detected by their effect on ionized gas.[59]

Gases can be ionized by heat, and in that condition they become conductors of electricity, a fact well known to de Forest in 1900. In the latter half of that year, it will be recalled, he was living in Chicago, working on wireless detectors at the Armour Institute and, in the evenings, in his own bedroom. He had in his room a small induction coil and spark gap, used for test purposes, and he noticed that, when this miniature transmitter was keyed, the gas light in the room seemed to flicker. The observation was accidental; he was working on the electrolytic responder at the time, not on gas flames. But it did arouse de Forest's curiosity.

---

of H. W. McCandless & Company, manufacturers of the early audions. See also, however, John W. Stokes, *70 Years of Radio Tubes and Valves* (Vestal, N.Y., 1982), especially pp. 1-9; Phillips, *Detectors*, pp. 205-15; Chipman, "De Forest and the Triode Detector," especially pp. 96-100; and George Schiers, "The First Electron Tube," *Scientific American* 220 (March 1969), 104-12. An authoritative survey of the scientific and technical background may be found in Kaye Weedon, "Av elektronrørenes historie: Røralderen begynte for alvor med overgangen til hoeyvakuum" ["Breakthrough in Electron Tube Devices: The Introduction of High Vacuum"], *Volund 1980* (Norwegian Technical Museum, 1980). The "canonical" version of the invention of the audion may be found in de Forest, *Autobiography*, pp. 210-15 and a more romanticized version in Carneal, *De Forest*, pp. 182-92.

[59] Ionization is a process by which atoms in a gas gain or lose electrons, usually through the agency of an electrical discharge, or high temperature, or the passage of radiation. If the gas is enclosed in a tube, and the pressure reduced, ionization is shown by the appearance of luminosity. See *Van Nostrand's Scientific Encyclopedia*, 4th ed. (Princeton, N.J., 1968), p. 944. Readers may wish to duplicate de Forest's use of a glow tube by moving a small fluorescent tube, such as is used for illumination, along the antenna of a VHF transmitter—not a bad way, in fact, of deciding how to trim the antenna to resonance.

The entry in his notebook under the date of 10 September 1900 testifies to his interest and excitement: "A welsbach burner so adjusted that there was a *surplus of gas* to oxygen, was noticed [to] become momentarily *dim*, then brighter than normal when the coil in the room was sparking. . . . It seems to be a detector to Hertz waves. The change in light was very noticible [sic] being several candles when lamp was properly adjusted. The first response was quite rapid, but the return to normal brilliancy rather sluggish. This can probably be bettered by small mantle etc & finer regulation of gas and air supply."[60] He speculated about what was happening. It had to be an effect on the mantle and the burning gas, not on the gas in the pipes since these were perfectly shielded. It might be due to "ozonizing" the air around the lamp ("No it is not this," he added later), or to some quality of the rare oxides in the mantle itself, or possibly the electrification of the heated gas atoms permitted their "more violent motion & more perfect combination with oxygen." Or it could be something entirely different. But already that evening he was visualizing how to use the effect for wireless reception. "A relay arrangement ought to be easily practicable, and a reciever [sic] for wireless telegraphy possible. Say run the receiving aerial from the top ring supporting the mantel [sic], and earth to the *burner* itself. Wrap the hot wire of relay in spiral within or without the mantel."[61]

The experiment was repeated on 19 September, this time with Ed Smythe present and with some pains taken to eliminate spurious effects. Entries in the laboratory notebook are in the handwriting of both men. Smythe noted that "by accustoming the eyes to the light of the mantle the flame could be seen playing on the inside of the mantle. The moment the induction coil circuit was closed the flame would appear to fan out, reach down, and lick the portions of the mantle which had been only at red heat and cause them to incandesce. The increase in brilliancy of the light seemed to amount to several c.p. [candlepower]."[62] To eliminate any possible ultraviolet rays, a tin box was held between the burner and the spark, but that made no noticeable difference. When the induction coil was placed in a closet, however, the effect was noticeably decreased— a finding that might have warned them of the disappointment ahead. But de Forest was "profoundly elated." The explanation, he now thought, lay in "the expansion of the cylindrical body of heated and highly sensitive

[60] De Forest Papers (Library of Congress), microfilm reel 2, laboratory notebook entry for 10 September 1900. The original documents are in the custody of the Foothill College Electronics Museum.

[61] Ibid.

[62] Ibid., laboratory notebook entry for 19 September 1910.

gas within and about the mantel—This latter serving merely as a *holder* to keep the gases spread out in their most sensitive condition. The electrification of those gases by passage of a Hertz train may cause the expansion and force the heated gases down upon the cooler and dark portions of the mantel as we noticed done." And, his imagination soaring into the empyrean, he wondered, "Is there not here at least an *analogy* between this effect of elect.-mag. waves on heated gases and the intimate connection between *sun spots* and the magnetic storms that accompany them?"

The prospects were, one must admit, intoxicating. What detector could be more sensitive than a gas flame? What could be more perfectly self-restoring or more continuously responsive? Since hot gases were known to conduct electricity, was it implausible to think of their conductivity as changing under the influence of Hertzian waves? If the effect could be harnessed, the result might be a device that surpassed in sensitivity any other form of wireless detector then available. It is hard not to sympathize with the excitement that Smythe and de Forest felt.

Their hopes, however, had no basis in fact, and the whole episode has become a classic example of imperfectly controlled experimentation and misinterpreted data. It was sound, not Hertzian waves, that caused the light to flicker. November 5 saw notebook entries in a more sober vein: "*Sound*, rather than *Hertz* waves, cause it because: shielded by brick and plaster partition 2 ft thick, but with door between the 2 rooms *open*, found no diminution, so long as the door was open a few inches. But when door was within 2-3 in. of being closed, or completely shut—to a point where the *sound* of the spark becomes muffled to one out in the room holding the lamp—effect falls off very decidedly and completely disappears." Clapping the hands together, in fact, produced exactly the same effect.

The entire episode would have been forgotten but for one thing. De Forest always thereafter insisted that, although the episode of the Welsbach mantle was a technical debacle, it nevertheless left him convinced that in some way or other heated gases could be used to detect Hertzian waves. The subject, he tells us, was always in the back of his mind; there remained with him "the firm conviction that in the heated gases surrounding incandescent electrodes there must nevertheless exist a response, in some electrical form, to high-frequency electrical oscillations."[63] The effect he thought he had found in his Chicago bedroom was an illusion,

---

[63] De Forest Papers (Library of Congress), microfilm reel 3, speech on accepting the Franklin Institute Medal.

but "the illusion had served its purpose."[64] That purpose was to leave with him the inner certainty that in some way he did not yet understand there did reside in the gases enveloping an incandescent electrode "latent forces, or unrealized phenomena" that could be used to create a detector of electromagnetic waves "far more delicate and sensitive than any known form of detecting device." And from that to the audion, which as conceived by de Forest *did* work by gas ionization, was but a short step.

Not all critics have been willing to take him at his word. It is pointed out, for example, that he made no mention of any such conviction when reporting his gas mantle experiment in the *Electrical World* in 1902.[65] And since such inner beliefs are in the nature of the case not public evidence, there is no way to prove or disprove de Forest's assertion. The suggestion is, of course, that this "prehistory" of the audion was created in order to show that de Forest was following an unbroken line of inquiry from 1900 on and was not turned on to a new track by learning of the Fleming valve. What happened in the Chicago bedroom is not in question; that is authenticated by laboratory notebooks. It is the alleged effect on de Forest's thinking that is in dispute.

Fortunately, there is other evidence. De Forest did not abandon his experiments with gas flames after this one disappointing experience. By 1903, according to his own account, he had developed a "flame detector" that really worked. (See Fig. 4.6) In this arrangement two platinum electrodes were held close together in the flame of a Bunsen burner; the antenna was connected to one and a ground connection to the other; and across the electrodes there were connected a battery and an earpiece. Application for a patent on a modified version of this device was filed

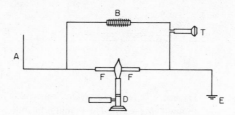

Fig. 4.6: De Forest's two-electrode flame detector.
Source: De Forest, "The Audion," *Journal* of The Franklin Institute, 1920; reproduced by permission of The Franklin Institute

[64] Lee de Forest, "The Audion—Its Action and Some Recent Applications," *Journal of the Franklin Institute* 190 (July 1920), 2. Quotations from this article are reproduced by kind permission of the editor of the *Journal*.

[65] Chipman, "De Forest and the Triode Detector," p. 97.

on 3 February 1905, and a patent (No. 979,275) was issued on 20 December 1910. In this later version (see Fig. 4.7) a significant change was made: the high-frequency circuit, from antenna to ground, was separated from the telephone circuit. When the antenna picked up a burst of Hertzian waves it caused a fluctuation in the flame of the Bunsen burner; this caused a change in the resistance between the electrodes in the secondary circuit that could be heard in the earpiece.

This device, although it may have worked after a fashion, was never used outside the laboratory. Its importance lies in the fact that de Forest later described it as his "basic American patent on the audion" and as the "earliest patent of the audion group."[66] And it was later to figure prominently as de Forest's "parent patent" in litigation between de Forest and the Marconi Company.[67] This is at first sight remarkable. We think of de Forest's audion and Fleming's valve and all later vacuum tubes as devices in which the electrodes are enclosed within an evacuated glass (or perhaps metal) envelope, and in which an incandescent filament supplies electrons to a positively charged plate. De Forest's flame detector of 1905 is not at all like that. Instead of an incandescent filament we have a Bunsen burner; instead of an evacuated envelope we have electrodes in a flame. And there is no hint of a control grid, the crucial element in de Forest's triode detector.

Yet de Forest meant to convey something when he called this the basic American patent on the audion. And analyzing what he meant throws much light on how he conceptualized the audion and on the processes

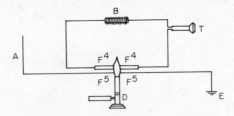

Fig. 4.7: De Forest's four-electrode flame detector.
Source: De Forest, "The Audion," *Journal* of The Franklin Institute, 1920; reproduced by permission of The Franklin Institute

[66] Lee de Forest, "Correspondence: The Audio-Detector and Amplifier," *The Electrician* 66 (23 January 1914), 659-60; de Forest, "The Audion—Its Action and Some Recent Applications," p. 6.

[67] See Federal Reporter, Vol. 236, District Court, Southern District of New York, *Marconi Wireless Telegraph Company of America v. De Forest Radio Telephone and Telegraph Company et al.*, decision of District Judge Mayer, 20 September 1916.

that led him to its discovery. What was "basic" about the flame detector patent was, first, that it seemed to offer a practical detector of Hertzian waves using ionized gases; and secondly that in its improved form the antenna circuit, with its radiofrequency currents, was separated from the earphone circuit, with its audiofrequency currents.

The first of these points is of critical importance: de Forest was led to the invention of the audion by his work with hot gases, and the audion, as he conceived it when he invented it, functioned only because it contained ionized gas. The incandescent filament was necessary, not as itself a source of charged particles, but only as a means of heating the gases. De Forest was quite explicit on the point: he writes of his conviction "that the same detector action which had been found in the neighborhood of an incandescent platinum wire . . . in a gas flame existed also in the more attenuated gas surrounding the filament of an incandescent lamp. In one case the burning gases heated the electrodes; in the other the electrodes heated the remanent [sic] gases."[68]

To a later generation, accustomed after 1920 to "hard" vacuum tubes, this was strange doctrine. Apart from certain special-purpose types, radio tubes were supposed to contain essentially no residual gas at all and great pains were taken during manufacture to eliminate any traces of gas that might linger, even in the metal or glass elements of the tube. If a tube became soft or "gassy" it was replaced immediately. De Forest's early audions, however, were not that kind of tube. That is not how he thought of them, nor how he had them manufactured. As he saw it, residual gas was necessary if the tube were to conduct; a perfect vacuum would have been a perfect insulator and no current could have passed through a perfect insulator. And in believing this he would have received some support, in 1905-1907, from professional physicists; to believe that a current could pass through a vacuum required familiarity with and acceptance of the then novel theory of the electron. Later generations of vacuum tubes traced their ancestry to the "hard" vacuum tubes developed by the laboratories of General Electric and the Telephone Company. But these were already an important evolutionary step beyond de Forest's audion.

The second respect in which de Forest considered the flame detector of 1905 to be "basic" was that it separated the radiofrequency from the audiofrequency circuits. This was related to his insistence that the audion "was never, strictly speaking, a *rectifying* device."[69] Here his wish to

[68] De Forest, "The Audion—Its Action and Some Recent Applications," pp. 2-3.

[69] Ibid., p. 3.

distinguish his invention from the Fleming valve, which was a diode rectifier, is very clear. The differentiating characteristic, as de Forest saw it, was that in the flame detector and later in the three-element audion the radiofrequency currents were not rectified (that is, converted into direct current) but were used to control the currents in the earphone circuit. He thought in terms of a relay or "triggering" effect, in which weak variations in one circuit could create stronger variations in the other. "From the beginning I was obsessed with the idea of finding a *relay* detector, in which local electric energy should be controlled by the incoming waves—and not a mere manifestation of the electrical energy of the waves themselves."[70] This is why he thought the local battery was important: it would provide a source of energy larger than that picked up by the antenna.[71]

If we are interested in the intellectual process by which de Forest eventually arrived at the triode vacuum tube, then the flame detector of 1905 merits the importance he attached to it, and the "distance" between that device and the audion is by no means as great as it initially appears to be. The reason why he laid such emphasis on what was in reality an unusable gadget lies in the timing. The application for a patent on the flame detector was filed on 3 February 1905 but signed and witnessed on 4 November 1904. This was before the filing date for Fleming's patent, which was 16 November 1904.[72] If the connections between the flame detector and the audion were as close as he said they were, then de Forest had proved his point: the essential ideas that went into the three-element audion were all present in the device that he had developed before learning of Fleming's valve. And this was true, he could argue, even though in physical form the flame detector and the audion were in no way similar. From that perspective, Fleming's aggrieved challenge to de Forest to prove from any prior published scientific book, paper, or patent specification that anyone before November 1904 had used or suggested "an electric incandescent lamp, having a metal plate or plates sealed into the bulb, as *a means of rectifying electric oscillations or as a receiver for wireless telegraphy* [emphasis in original]," was beside the point.

[70] Ibid., p. 3.

[71] De Forest has been seriously faulted for claiming that his *diode* audion of 1906 could provide gain. Chipman ("De Forest and the Triode Detector," p. 98) is correct in pointing out that the battery in that case merely provided forward bias, shifting the operating point on the diode's characteristic curve. The point to be emphasized, however, is surely that de Forest's search for a relay action did lead eventually to the triode audion, which could amplify.

[72] J. A. Fleming, "The Audion-Detector and Amplifier," *The Electrician* 65 (5 December 1913), 377-78.

Stated in those terms, that was a challenge de Forest could never have met, nor would he have tried to. Fleming wanted to argue about a specific device, de Forest about the ideas that underlay that device and from which it had been created. The two men were operating in different realms of discourse. The question may still be raised, however, whether even on his own terms de Forest did not owe more to Fleming than he would publicly admit.

If we grant de Forest his "anticipations" of the audion in the flame detector of 1905, we are granting him essentially all he asked for. The progression from there to the two-electrode audion, so similar to the Fleming valve, and from there to the triode with its crucial control grid, can easily be made to seem natural and almost inevitable. And that is the way de Forest wanted the story to be told and remembered. For consider: a gas flame burning in the open air was obviously unstable and likely to flicker. What could be more natural than to shield it by a glass chimney? But, when it came right down to it, why use a flame at all? All that was really needed was a means of heating rarefied gases.[73] An incandescent filament could do that. So eliminate the gas flame completely, use two filaments instead of the two platinum electrodes, and enclose them in a glass envelope that has been slightly evacuated. That is what de Forest did—quite independently of Fleming, he would insist. At this stage he had a device resembling an incandescent light bulb, but with two filaments instead of one. (See Fig. 4.8) But it was clearly unnecessary to have both of these filaments incandescent, since one was enough to heat the gas. Convert one of them to a flat plate, or better yet two interconnected flat plates, one on each side of the filament. (See Fig. 4.9)

De Forest christened it the audion—this model had only two electrodes but the three-electrode audion with its vital control grid was to follow

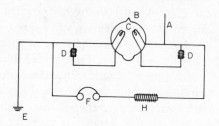

Fig. 4.8: De Forest's two-filament audion.
Source: De Forest, "The Audion," *Journal* of The Franklin Institute, 1920; reproduced by permission of The Franklin Institute

[73] Lee de Forest, "Milestones in Radio History," Clark Radio Collection (Smithsonian Institution), C1. 14.

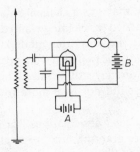

Fig. 4.9: De Forest's two-electrode audion.
Source: De Forest, "The Audion," *Transactions* of the AIEE (now IEEE) 25, copyright
1906; reproduced by permission

very soon. He described the device in a presentation to the American
Institute of Electrical Engineers on 26 October 1906. He was careful to
acknowledge his debts to certain previous workers, particularly the German physicists, Julius Elster and Hans Geitel, who in 1882 had begun a
careful series of experiments with ionized gases in evacuated glass con-
tainers. Nowhere, however, did he mention the name of Ambrose Fleming.

This is strange, since it is certain that he knew of the Fleming valve—
he had in fact referred to it in one of his earlier flame detector patents.
And we know that, beginning in the latter part of 1905, he had been
experimenting with Fleming valves and having them manufactured to his
order.[74] Later, after the controversy was well under way, he was to go
to some pains to differentiate the two-element audion from the Fleming
valve, even though physically they were practically indistinguishable. He
pointed out, for example, that his circuits called for a positive voltage
to be applied to the plate while Fleming had applied a small negative
voltage; and he claimed that the Fleming valve was a rectifier while his
device detected by virtue of "the asymmetry of its characteristic curve."
These points were hardly well taken. Functionally Fleming's valve and
de Forest's two-element audion were identical. De Forest's device would
actually have functioned better without the positive bias on the plate,
and it did detect by rectification, even though this rectification took place
in a slightly different manner.[75] From this point of view it was very

[74] Tyne, *Saga*, pp. 53-54; Simon Papers (Bancroft Library), H. W. McCandless
to G.H.C. [George H. Clark], 9 December 1949. McCandless manufactured
several Fleming valves for C. D. Babcock, de Forest's assistant, in the latter part
of 1905, duplicating a sample furnished by him.

[75] Lee de Forest, "The Audion: A New Receiver for Wireless Telegraphy,"

difficult to make any distinction between the device de Forest described in 1906 and the device on which Fleming had filed for a patent in 1904.

De Forest was asking too much. If he had been content to say that all the ideas necessary for a thermionic diode had been present in his flame detectors, that the only thing lacking was the notion of enclosing the rarefied gas in a glass envelope and heating it by means of an incandescent filament, and that word of Fleming's valve had led him to take that further step—then perhaps few would have objected. The route that de Forest had followed was, after all, quite different from the route that had led Fleming to his valve. His conception of how the device worked was different from Fleming's. Physically the Fleming valve and the two-electrode audion were virtually identical; electronically they functioned in almost exactly the same way. As physical artifacts the two devices were practically indistinguishable; but they had totally different intellectual histories.

❋   ❋   ❋

Fleming's valve was a linear descendant of a device that had no connection with wireless telegraphy at all. This was the famous "Edison effect." Working with his carbon filament lamps in the early 1880s, Edison noticed that after a period of use the inner surface of the glass bulb became progressively darkened, and that, when this happened, there was always a thin and lighter streak on one side of the bulb in line with the plane of the filament. It was as if particles of some kind were being emitted from one leg of the filament—particles that darkened the glass except where they were intercepted by the other leg. Now Edison was working with direct, not alternating, currents and knew that one leg of the filament would always have a slight positive electrical potential relative to the other. Curious about the darkening of the bulbs and always on the lookout for a useful effect, he inserted a second electrode into the glass envelope as a kind of probe, and connected it, first to the positive leg of the filament and then to the negative one. (See Fig. 4.10) He found that current flowed only when the second, or cold, electrode was positive with respect to the incandescent filament.

Edison assumed that the dark deposit resulted from carbon particles ejected by the filament but he had no theory to explain the phenomenon nor could he easily think up a good use for it. Clearly a force of some

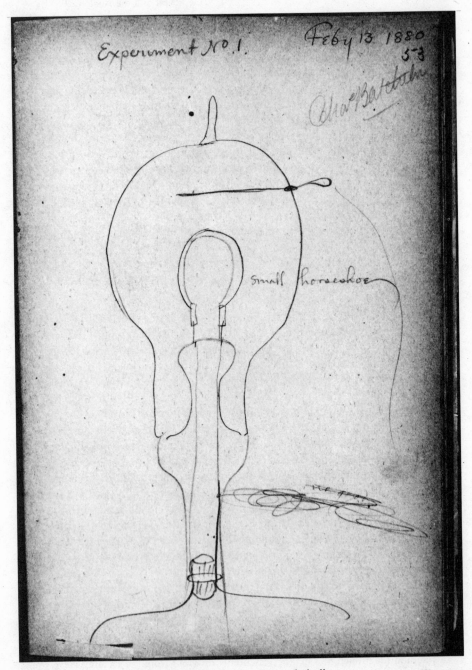

Fig 4.10: Edison's two-electrode bulb.
Source: National Park Service—Edison National Historic Site; reproduced by permission

kind was causing the particles to migrate from the heated filament to the cold electrode, and since these particles were attracted only when the latter was positive, they had to be carrying a negative charge. But what exactly they were and why they moved in this way he did not know. As was his habit, he took out a patent anyway (No. 307,031, issued on 21 October 1884), showing how the effect could be used to regulate and measure the flow of electrical current.

It is easy to read more into this patent than Edison intended to be there. The device (see Fig. 4.11) looks like a diode rectifier, such as would convert alternating current into direct current that could deflect the meter. But Edison was working exclusively with direct current. What function then does the diode serve? The patent does not even disclose the one-way conductivity that would entitle us to call it a rectifier. In fact, the effect does not depend on rectification. Variations in line voltage affected the temperature of the filament and the voltage on the plate and therefore the current flowing through the meter, but that is a long way from rectification. Edison invented the device and patented it (in the United States and in Britain) as an industrially applicable voltage-sensing device; the rectifying function was purely incidental. The heart of the device was the rapid variation in anode current as a function of filament voltage.[76]

Edison's effect attracted attention among scientists partly because it linked the new technology of incandescent lighting with a continuing line of scientific research into the conductivity of rarefied gases. This had origins that long preceded Edison. Edmond Becquerel had begun his research on conduction by ionization in 1853, noting paradoxical results that defied explanation at the time. The German physicist, Wilhelm Hittorf, had predicted as early as 1874 that an electrical current would pass through a vacuum if the incandescent element were made hot enough. Eugen Goldstein in 1882 had deposited a sealed letter with the Vienna Academy that, when opened in 1884, described his experiments showing electrical conduction through vacua.[77] The question was one of great scientific interest. If a unidirectional current could pass through a vacuum, what kind of particles served as carriers of the current? The straightforward answer, accepted by many physicists, was that a truly perfect vacuum was necessarily a perfect insulator. But, of course, no vacuum was

[76] Compare Stokes, *Tubes and Valves*, pp. 1-2, and Pratt Papers (Bancroft Library), Lloyd Espenschied to Haraden Pratt, 5 May 1963.

[77] Kaye Weedon, "The Prediction and Triple Discovery of the Edison Effect" (Typescript); Tyne, *Saga*, pp. 27-28. For a comprehensive survey of research on conduction through ionized gases, see J. J. Thomson, *Conduction of Electricity through Gases*, (Cambridge, 1903), particularly chap. 8, pp. 155-92.

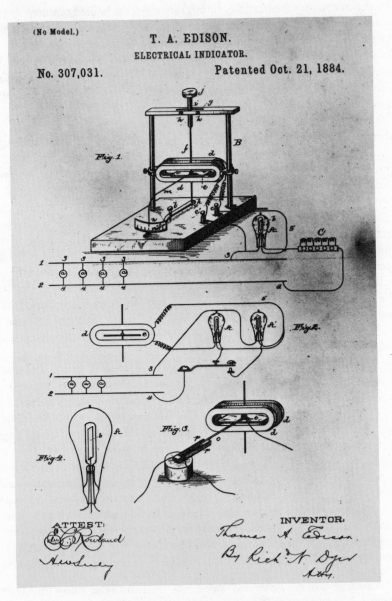

Fig. 4.11: Edison's electrical indicator.
Source: National Park Service—Edison National Historic Site; reproduced by permission

ever perfect. Any current that passed through a partial vacuum did so only because there still remained in the glass envelope traces of gas that the vacuum pump had been unable to remove completely, or perhaps minute particles of dust emitted by the incandescent filament. Make the vacuum truly perfect, it was argued, and no current could pass to the positively charged anode, no matter how hot the filament.

Fleming first became acquainted with the Edison effect around 1885. In 1882 he had accepted appointment as consultant to the new Edison Electric Light Company of London and in that capacity had observed the same darkening effect in bulbs as had troubled Edison and the same "molecular shadow" cast by one leg of the filament on the side of the bulb. He discussed the matter in papers presented to the Physical Society of London in 1883 and again in 1885.[78] His friend William Preece, chief engineer of the British General Post Office, visited the United States in 1884, saw Edison's indicator lamp on display at the International Electric Exhibition in Philadelphia, and persuaded the inventor to give him a few specimens to take back to England for further study. He reported the results of his own experiments with the lamps to the Royal Society in March 1885 (incidentally coining the term "Edison effect" by analogy to the well-known Crookes effect) and it is probable, though not certain, that Fleming received his first samples of the modified bulbs from Preece.[79]

Fleming reported his own findings in a lecture to the Physical Society and a long article in the *Philosophical Magazine* in 1896.[80] In general he confirmed what Edison and Preece had found, presenting measurements of current flow with a variety of different circuit arrangements. In each of his bulbs the vacuum used was "as perfect as in good ordinary commercial lamps" and he interpreted his results as confirming J. J. Thomson's theory that certain gases in rarefied condition were very good conductors of electricity, that the greatest part of the resistance to conduction was at the electrodes, and that this could be largely overcome

[78] Tyne, *Saga*, p. 33. Fleming visited Edison in the United States in 1884 but it is not known whether he saw any of Edison's two-electrode bulbs on that occasion or discussed the Edison effect. See Stokes, *Tubes and Valves*, p. 2.

[79] William H. Preece, "On a Peculiar Behaviour of Glow-Lamps When Raised to High Incandescence," *Proceedings* of the Royal Society of London, 38 (11 December 1884-18 June 1885), 219-30. Tyne states (*Saga*, p. 34) that it is uncertain how Fleming first learned of the Edison effect, but Stokes (*Tubes and Valves*, p. 2) is positive that Preece passed the bulbs to Fleming after carrying out his own experiments on them.

[80] J. A. Fleming, "A Further Examination of the Edison Effect in Glow Lamps," *The London, Edinburgh, and Dublin Philosophical Magazine and Journal of Science*, 5th ser., 42 (July-December 1896), 52-102.

by heating the electrode to incandescence. This was all quite conventional. The discussion was entirely in terms of conduction through rarefied gases, and Fleming did not even raise the question of what would happen if the vacuum were closer to being perfect. There was, however, one significant advance, namely that Fleming explicitly described the two-electrode vacuum tube as providing unilateral conductivity. "We have then a unilateral conductivity exhibited by this highly vacuous space bounded by two electrodes one of which is incandescent and the other of which is cold. Negative electricity is discharged at once out of the hot surface but not out of the cold, and a negative discharge can take place from hot to cold but not *vice versa*."[81] This clearly implied recognition of the device as a rectifier of alternating current, even though Fleming nowhere in his presentation described its use for that purpose. Indeed, he suggested no practical applications whatever.[82]

In 1899 Fleming accepted appointment as scientific adviser to the Marconi Wireless Telegraph Company. This placed him in a unique position, in command of three bodies of knowledge that might otherwise have remained unrelated. These were: first, the knowledge he had gained as adviser to the Edison Electric Light Company and in particular his familiarity with the Edison effect and the two-electrode bulbs developed to investigate it. Second, there was his knowledge of the technical requirements of the Marconi Company, specifically for a detector more sensitive than Marconi's coherer or magnetic detector. And thirdly there was his knowledge of the current state of research in physics. Particularly important in this last respect was his thorough familiarity with research on the conduction of electricity through ionized gases and, as an offshoot of that, the speculative thinking that, in the 1890s, was leading up to the theory of the electron—the negatively charged elementary particle emitted by heated cathodes in vacuum tubes.[83]

[81] Ibid., pp. 90-91.

[82] Stokes (*Tubes and Valves*, p. 3) states that in 1901 Fleming's experimental bulbs "had previously been found capable of rectifying local oscillations" and cites Fleming's *The Thermionic Valve in Radiotelegraphy and Telephony* as his only authority. Stokes appears to accept the usual view that Fleming did no further experimental work on the bulbs between 1896 and 1901, and certainly in the paper of 1896 there is no reference to their use for this purpose. Clearly Fleming had an interest in pushing the date for the first use of the bulbs as detectors of Hertzian waves as far back as possible.

[83] The word was introduced in 1891 by G. J. Stoney to denote the "natural unit of electricity" (either negative or positive). J. J. Thomson preferred the word "corpuscle" to refer to the carriers of electricity ejected by incandescent cathodes but showed by his research that each such corpuscle had a negative charge equal

Fleming's first work for Marconi involved designing the power plant for the new station under construction at Poldhu—a natural assignment in view of his earlier work for the Edison Company. But he soon turned his attention to detectors, the weakest link in the Marconi system. He was looking for the same kind of device as Fessenden was at about the same time: a detector that would give a quantitative indication of the strength of the signal being received, something no coherer could do. And Fleming's reasons were partly personal, in that he was growing hard of hearing at this time and preferred an indicator he could watch rather than one he had to listen to. This suggested a meter of some kind. No meter that he knew of, however, could respond to the high frequency waves of wireless transmissions.

It was at this point—by a "sudden very happy thought," as he put it—that he called to mind his earlier work with Edison glow lamps.[84] The lamps used in his 1895-1896 experiments were still there, stored in a laboratory cupboard. And he knew that, in an appropriate circuit, they could give "unilateral conductivity," or in other words that they would rectify. He did not know, however, that they would rectify alternating currents at the frequencies Marconi was using. Nor was it certain that they would make good practical detectors, able to respond to the rapid signalling speeds of the Morse code. Nevertheless, it was worth trying. In October 1904 he took one of the bulbs he had used nine years before— this particular one had a flat metal plate supported between the legs of a single loop carbon filament—and connected it into a simple resonant receiving circuit containing a sensitive mirror galvanometer. When a small test oscillator and spark gap were switched on, the meter responded immediately.

He called it an "oscillation valve"—not a happy term, since it did not normally oscillate—and had several new models made at the Ediswan Lamp Works, with a metal cylinder surrounding the filament—a much more effective geometry. (See Fig. 4.12) And he recognized the importance of what he had done. "I have found a method," he wrote to Marconi in November, "of rectifying electrical oscillations . . . so that I can detect them with an ordinary mirror galvanometer. . . . This opens up a wide

---

to one electron. The two words soon came to be used interchangeably. Later research disclosed the existence of positive electrons, commonly known as positrons. See H. J. van der Bijl, *The Thermionic Vacuum Tube and Its Applications* (New York, 1920), chap. 1; A. Schuster, "Experiments on the Discharge of Electricity through Gases: Sketch for a Theory," *Proceedings* of the Royal Society, 37 (1884), 317-39; J. J. Thomson, "On the Velocity of the Cathode Rays," *Philosophical Magazine* 38 (1884), 358-65.

[84] J. A. Fleming, *Memories of a Scientific Life* (London, 1934), p. 141.

Pl. 9: The Fleming valve.
Source: Institution of Electrical Engineers

field for work as I can now measure exactly the effect of the transmitter. I have not mentioned this to anyone yet as it may become very useful."[85] It was closer to what we would now call a field strength meter than to a detector suitable for use in a commercial receiving station. In other words, it was in the first instance a measuring device. In the 1906 edition of his manual, *The Principles of Electric Wave Telegraphy*, the oscillation valve was described as a "cymoscope"—literally, a device for looking at electric waves.

How did this device compare with de Forest's two-electrode audion? There is no question about priority, if we are thinking of the physical artifact. Fleming filed a provisional application for a patent in Great Britain on 16 November 1904 and it was granted on 21 September 1905 (British patent No. 24,850). De Forest at that time was still playing

[85] The letter is reproduced in Shiers, "The First Electron Tube," 110.

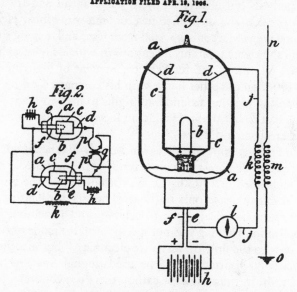

PATENTED NOV. 7, 1905.

J. A. FLEMING.
INSTRUMENT FOR CONVERTING ALTERNATING ELECTRIC CURRENTS
INTO CONTINUOUS CURRENTS.
APPLICATION FILED APR. 19, 1905.

Fig.1.

Fig.2.

Fig. 4.12: The Fleming valve

around with open-flame gas detectors. Secondly, they arrived at the conception in different ways. If both narratives can be taken at face value, de Forest was working out of the technology of gas burners, Fleming from the technology of incandescent lamps. Thirdly, they visualized their inventions differently. To Fleming what he had discovered was essentially a valve. It permitted electricity to flow in one direction but not in the other, and by virtue of that fact it was a rectifier and a detector of high-frequency currents. To de Forest it was essentially a relay—a means whereby a small current in the antenna circuit could control a larger reaction in the audio circuit. And finally, Fleming was led to his invention by his search for a measuring instrument.[86] Not so with de Forest: he wanted a sensitive and continuously responsive detector.

And there was one further difference. Fleming thought of his rectifier as a finished product, whereas to de Forest it was merely a temporary

---

[86] J. A. Fleming, "On the Conversion of Electric Oscillations into Continuous Currents by Means of a Vacuum Valve," *Proceedings* of the Royal Society, 74 (1905), 487.

and provisional stopping-place. Why this should have been so is hard to define. There is the difference in personalities: Fleming the careful, methodical physicist; de Forest the egotistical romantic, rather sloppy in his research methods but always ready to move ahead and try something new. And there is the difference in economic situations. Fleming had already done what Marconi expected from him; and, having signed over to Marconi the rights to all his inventions in wireless telegraphy, he had no financial interest in the further development of his rectifier. To de Forest in 1906, on the other hand, with his company (and his first marriage) collapsing in ruins around him, rights to his two-electrode audion were all he had to his credit. He had an incentive, indeed a necessity, to push on that Fleming did not.

How did Fleming think his valves functioned? Was he thinking in terms of ionized gas, as de Forest was, or in terms of the new theory of electron emission, published in 1902 by O. W. Richardson?[87] The bulbs he first used had been exhausted only to the level normal for lighting purposes; no special effort had been made to achieve a high vacuum. And in his patent application Fleming was not specific as to the mode of conduction, referring only to the flow of "negative electricity." But in addressing the Royal Society in February 1905 he used (though sparingly) the term "electron": "In the incandescent carbon there is a continual production of electrons or negative ions by atomic dissociation."[88] And in an important paper in 1906 he distinguished sharply between the mode of current flow when the vacuum was low (say one-thousandth of an atmosphere) and when it was high (one hundred-millionth of an atmosphere). When the vacuum was relatively low one had the complex but thoroughly investigated phenomena of conduction through an ionized gas. But with a high vacuum, if the cathode were heated to incandescence and a positive potential applied to the anode, the current flow was by the emission of negatively charged electrons. He stipulated that the bulbs he was describing had to be "exhausted as completely as possible" so that "the highest attainable vacuum is made." Under those conditions, he wrote, the current flow was explicable "only upon the electronic hypothesis of matter and electricity."[89]

The first Fleming valves, then, exhausted only to the level of ordinary

[87] O. W. Richardson in *Proceedings* of the Cambridge Philosophical Society, 11 (1902), 296; Richardson, "The Electrical Conductivity Imparted to a Vacuum by Hot Conductors," *Philosophical Transactions*, 201A (1903), 497-549.

[88] Fleming, "Conversion of Electric Oscillations," p. 487.

[89] J. A. Fleming, "On the Electric Conductivity of a Vacuum," *Scientific American Supplement*, no. 1568 (20 January 1906), 25130.

light bulbs, functioned by virtue of the conductivity of the residual gas they contained. But when Fleming came to analyze the phenomenon, he realized that the gas was by no means necessary. When the device was used to rectify high-frequency oscillations it would function with the greatest stability and predictability when it functioned purely by virtue of the emission of electrons from an incandescent filament. The filament was brought to incandescence not to heat the gas but to serve as a source of free electrons.

How good a detector was it? Good enough to be used in service, once the operator had learned to read Morse by watching the swings of the galvanometer.[90] Several were manufactured and shipped to Marconi at Poldhu.[91] They were sensitive devices—more so than a magnetic detector—and less easily injured or put out of adjustment by "atmospherics" than a coherer. The Marconi Company designed and manufactured special receiving sets in which they could be used, but since the company refused to license the use of its patents by others and was somewhat committed to Marconi's magnetic detector, there was little possibility of the Fleming valve coming into general use. Then, too, the timing was unfortunate. Fleming's detecting valve came on the scene at almost exactly the same time as the electrolytic and crystal detectors, which were at least as sensitive and much more readily available.

It is strange that Fleming never thought of inserting a third electrode in his valve. He came very close. In his experiments on the Edison effect he had found that the stream of electric particles coming from the hot filament could be deflected and controlled by a magnet held outside the glass envelope. The idea of a control electrode was implicit there. And in one of his bulbs he had placed, instead of a flat plate, a zigzag screen of platinum wire to see whether that was as effective in collecting the electrons. One could hardly come closer to the concept of a "grid." But he never placed a third electrode inside the valve. As he wrote later, ". . . sad to say, it never occurred to me."[92] Perhaps one reason why it did not was simply that, in his visual images and in his language, it was a valve, and there was no place for a third or control element in a valve.

---

[90] I have been unable to determine whether a Fleming valve was ever, in Marconi practice, used with earphones. Fleming in 1905 wrote exclusively in terms of "a sensitive dead beat galvanometer of the type called by cable engineers a 'Speaking Galvanometer,'" adding that: "Anyone who can 'read mirror' can read off the signals as quickly as they can be sent on an ordinary short submarine cable with this arrangement." See Fleming, "Conversion of Electric Oscillations," p. 480.

[91] Tyne, *Saga*, pp. 45-46.

[92] Fleming, *Memories*, pp. 143-44.

The long-term consequences were profound. The fact that the Marconi Company owned the patent on the Fleming diode was to prove of central importance in the later history of the radio industry. Anyone who wanted to manufacture radio tubes had to reckon with the legally validated property rights of the Marconi Company. If the Marconi Company had also owned the patents on the triode tube, that company would literally have been in a position to control how and when and for whose benefit the second major era of radio technology would proceed. The Marconi Company would have dominated the vacuum tube era just as once it had dominated the era of spark. Events did not work out that way: the reason they did not is that Ambrose Fleming, having discovered the thermionic diode, did not take the next step and invent the triode. The history of invention, of course, is full of similar stories. What is obvious in retrospect is not obvious at the time. What seems a small and natural step to later analysts can be an impassable chasm to the man of the spot. The chasm is impassable not because it is wide but because it is not seen.

De Forest, on the other hand, was thinking in terms of a relay. In itself that was just as deceptive and imperfect a physical image, especially for a two-electrode device, as was Fleming's image of a valve. But it had one great virtue. In a relay there are always at least two circuits: one that controls the relay, and one that the relay controls. Built into de Forest's thinking was the idea of two circuits: one, the weak current in the antenna circuit, that would do the controlling; and the other, the stronger current in the earphone circuit, that would be controlled.

A third or control electrode, in the form of a coil or flat plate, had already been shown in certain of the types of audion described by de Forest in his presentation to the AIEE on 26 October 1906. But it was outside the glass envelope and functioned by electrostatic or electromagnetic action, not by intercepting the current passing between filament and plate. He applied for a patent on a three-electrode audion containing a control grid placed inside the tube and between the filament and plate on 29 January 1907. It is scarcely to be believed, de Forest being the kind of man he was, that he would have refrained from mentioning the device in the course of his AIEE lecture if he had had it at that time.[93] October 26, 1906 and 29 January 1907 therefore bracket the period during which this critical step in the process of invention was taken. In fact, from the records of his manufacturer, McCandless, we can come closer than that: the first three-electrode audions with a control grid

[93] Compare the judgment of Gerald Tyne in Tyne to Lloyd Espenschied, 18 July 1965 (Pratt Papers, Bancroft Library).

between filament and anode were ordered on 25 November 1906. The patent application was sworn to on 21 December.[94]

Strictly speaking, these were not the first three-electrode audions, even if we confine that term to tubes with all the electrodes inside the glass envelope. On 26 October 1906 de Forest had filed for a patent on what he called, significantly, a "device for amplifying feeble electrical currents." (See Fig. 4.13) This is the first hint that he was thinking of the audion as an amplifier and not just as a detector. It is very doubtful whether the circuit as shown would provide any gain, but the use of the term was to prove important in later litigation. What the drawing in the patent application shows is an audion with one filament and two plates—in effect, two audions sharing one envelope and one filament. The incoming signal is coupled to one plate and the output is taken from the other.

The device itself is of no importance except as indicating that in 1906 de Forest was already thinking of amplification, not merely detection, and as suggesting the intervening steps he took in making the all-important move from the two-electrode to the three-electrode audion—the move that Fleming never made and that brought into existence what may be called, without hyperbole, one of the pivotal inventions of the twentieth century. It must be remembered that the idea of a control electrode was not new to the art; it must have been familiar to anyone who had examined the cathode ray tubes (such as those made by Thomson, Braun, and Zenneck in the 1890s) that were common features of any well-equipped electrical laboratory, and perhaps particularly the Zehnder "trigger tube" with which we know de Forest's assistant, Clifford Babcock, was familiar. There is a significant entry in de Forest's laboratory notebook under the date of 23 September 1906 that suggests that the construction of cathode ray tubes was much in his mind. "As a further analogy between the conduction in the Audion and that in the cathode ray tube," he noted, "I should expect with a perforated screen, grid, or the like placed between filament and anode, to find (as in a cathode tube having constrictions or diaphragms in its length) a reproduction on the side of the grid nearest the anode of the cathode drop, or variation layer."[95] Thinking along

[94] The patent was issued as No. 879,532 on 18 February 1908. The delay between the signing of the patent application on 21 December 1906 and its filing on 29 January 1907 is easily explained: de Forest had difficulty getting together the $15 filing fee. See Gerald Tyne to Emil Simon, 12 December 1954 (Simon Papers, Bancroft Library).

[95] De Forest Papers (Library of Congress), microfilm reel 3, laboratory notebook entry for 23 September 1906. On the possible influence of the Zehnder "trigger tube," see George Applegate, "An Adventure in Book Collecting," *The*

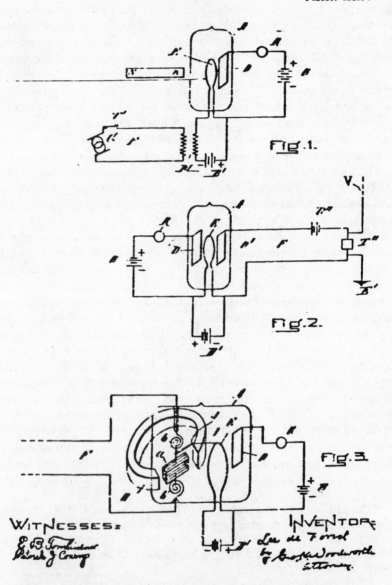

No. 841,387.   PATENTED JAN. 15, 1907.
L. DE FOREST.
DEVICE FOR AMPLIFYING FEEBLE ELECTRICAL CURRENTS.
APPLICATION FILED OCT. 25, 1906.

3 SHEETS—SHEET 1

Fig.1.

Fig.2.

Fig.3.

WITNESSES: INVENTOR:

Fig. 4.13: De Forest's "device for amplifying feeble electrical currents"

these lines, and having in front of him the two-plate audion, de Forest converted one plate into a zigzag grid of wires and placed it between the filament and the other plate. (See Fig. 4.14) The rest of the circuit remained essentially the same: the antenna circuit was fed to the new grid while the earphone circuit was connected to the plate. The analogy with a relay was there, in the two circuit connections; but so was the analogy with the older cathode ray tubes, in the positioning of the control grid.[96]

Did he understand how the device functioned? Some say he did not. Sir Edward Appleton, for example, tells us that "the inventor of the three-electrode valve did not know how his invention worked, at first. It needed some years before its mode of action and its astonishing potentialities were properly recognized; and all this was done chiefly by other people, not by the inventor."[97] This view rests upon a misinterpretation. It is true that, in his patent application, de Forest simply declined to discuss the principles by which the triode operated, stating merely that he had determined by experiment that the presence of the control grid increased sensitivity, that the phenomenon was exceedingly complex, and that he did not deem it necessary to enter into a discussion of the probable explanation. And it is true that his laboratory notebooks for 1906 show him still toying with the idea that dust particles carried the current: "The disappearance of the acoustic effect . . . lends weight to the carrier or *metallic dust* explanation, as against the ionic theory." But in the discussion that followed his October 1906 lecture to the AIEE he was asked specifically whether he thought the action depended on the ionization of the residual gases or whether the vacuum was so perfect that the ions or electrons came from the electrodes themselves. And his answer was unambiguous: "I think it is due to the ionization of the residual gases. . . .

Old Timer's Bulletin (Antique Wireless Association) (December 1962) and Lloyd Espenschied, "How did de Forest ever Invent the Grid Audion . . . ," ibid. 4 (June 1963).

[96] De Forest later gave a quite different explanation of how he hit upon the idea of the control grid, saying that he was concerned about the loss of signal energy in the two-electrode audion, since it was partially shunted through the battery and earphone. So he experimented with a control grid to prevent that loss. The insertion of simple chokes, however, would have taken care of this problem, and we know that de Forest was familiar with their use. See Lee de Forest, "Milestones in Radio History," Clark Radio Collection, Cl. 14, and Chipman, "De Forest and the Triode Detector," p. 100.

[97] Edward Appleton, "Thermionic Devices from the Development of the Triode up to 1939," in *Thermionic Valves 1904-1954*, ed. Institute of Electrical Engineers (London, 1955), pp. 17-25, quoted in Weedon, "High Vacuum," p. 33.

L. DE FOREST.
SPACE TELEGRAPHY.
APPLICATION FILED JAN. 29, 1907.

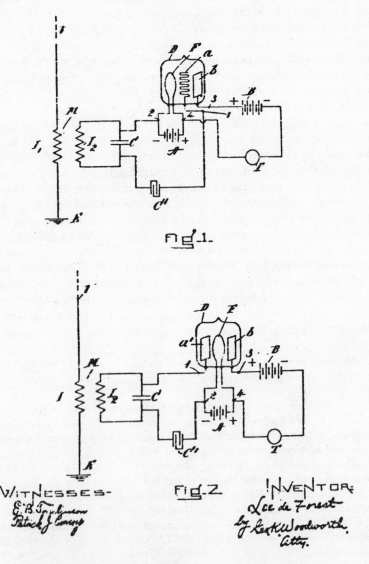

Fig. 4.14: De Forest's three-electrode audion circuits

If the exhausting process is carried too far, the audion loses its sensitiveness. The gas particles, rather than the particles of the metal dust, are the carriers."[98]

The misinterpretation, and it is a common one, lies in thinking that de Forest was wrong on this point. And that belief in turn rests on two implicit assumptions: first, that he was trying to produce a "hard" vacuum tube, such as those later produced by Langmuir and Arnold, but failed to do so; and second, that the action of the three-electrode audion was "really" by pure electron emission but de Forest mistakenly thought of it as by gas ionization. Neither of these assumptions is correct. De Forest was trying to devise a detector that functioned by ionized gases, and he did so; and his identification of the current carriers was not an error. An audion was not a high-vacuum tube that had become "gassy" through carelessness or poor manufacturing; it was an audion, and an

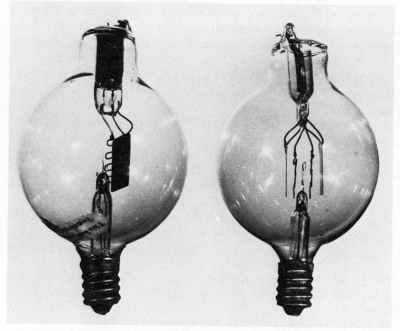

Pl. 10: De Forest audions.
Left: Double-wing audion, about 1909;
Right: Single-wing audion, about 1908.
Source: Smithsonian Institution

[98] De Forest, "The Audion," pp. 769 and 778.

audion, as conceived by its inventor, *was* a low-vacuum tube that contained gas.[99]

And de Forest thought it was wonderful. His patent attorney, G. K. Woodworth, interviewed many years later, recalled how he and de Forest met over breakfast, and de Forest "scratched out a rough diagram of the grid audion on a menu card. ... De Forest was not usually given to displays of enthusiasm, but on the occasion of his disclosure of his grid audion to me he was wildly exuberant and knew at that time . . . he had solved the problem that had been holding wireless telegraphy back."[100] Few others shared his enthusiasm. To many it was just a Fleming valve with an extra electrode—a description that infuriated de Forest. "Had Fleming thought of the grid, had he inserted it in the Edison Valve, he would have had exactly what he did have, a rectifier with a grid-shaped anode—nothing more. Had I come to this stage by the route Fleming followed, I should have done exactly as Fleming did—missed it exactly as he missed the Audion."[101] The triode, he insisted, acted like a relay; the "B battery," placing a positive voltage on the anode, was essential; that and the control grid made all the difference in the world.

Perhaps so, but the world was not impressed. The first public display of the triode audion was at a lecture before the Brooklyn Institute of Arts and Sciences on 14 March 1907 and the reaction of a young "ham" radio operator present on that occasion was probably typical. "The audion itself at the lecture was just one of those curious things, like all detectors, and was soon forgotten."[102] Some amateurs bought them—those who could scrape together the $5.00 purchase price, about half the weekly wage of a skilled laborer in those days. And the Navy bought some, with unimpressive results. Navy operators had a bad habit of turning up the filament voltage in an attempt to increase sensitivity, and once the filament was burnt out the audion was useless. It was true that an audion, properly adjusted so that it was operating on the "knee" of its characteristic curve, gave a degree of sensitivity that was remarkable; but each one was different, their characteristics changed over time, and

[99] Compare Irving Langmuir, "The Pure Electron Discharge and Its Applications in Radio Telegraphy and Telephony," IRE *Proceedings* 3 (September 1915), 261-86.

[100] Tyne to Simon, 12 December 1954 (Simon Papers, Bancroft Library). Tyne was quoting from a set of notes made by C. S. Thomson as transcribed by G. H. Clark.

[101] Lee de Forest, draft of article for *Radio Broadcast* (1922), in Simon Papers (Bancroft Library).

[102] Lloyd Espenschied to Emil Simon, 8 March 1954 (Simon Papers, Box 4, Bancroft Library).

they conspicuously lacked those qualities of uniformity and interchange-ability that the Navy prized. Even McCandless, sole manufacturer of these early audions, did not much like them: too tricky to make and too many customer complaints. Christmas tree bulbs were a lot less bother. For radio purposes the new crystal detectors, even the electrolytics, were simpler and more reliable, and they did not require batteries.

And there were other problems, although no one faced them at the time. Fleming had patented his valve in Britain, in Europe, and in the United States. Was it possible to manufacture and sell de Forest's audions without infringing Fleming's patents? Fleming and the Marconi Company thought not, and American courts were soon to uphold that view. On the other hand, de Forest had a valid patent on the triode in the United States—he allowed his British patent to expire through inability to pay the renewal fee. Possession of the Fleming patent did not give Marconi the right to use triodes (although in fact the American Marconi Company, facing difficult reception conditions in the Caribbean, did so). But did de Forest, or those he licensed, have the right to use triodes without permission from Fleming? How strong was Fleming's patent anyway? The effect on which it was based had been disclosed long before; its use as a rectifier had been disclosed by Fleming himself eight years before he applied for his patent; the most he could reasonably claim was its use as a detector of radio waves. The claims made in Fleming's patent went far beyond that, and seemed altogether too broad for what was in fact disclosed. And American courts were eventually to agree with that po-sition also.

These were, in 1906, only clouds on the horizon but they might have served as warnings of the conflicts that lay ahead, when both de Forest's patents and Fleming's had become the property of large corporations with millions of dollars at stake. Fleming's patent rights in the United States became the property of the American Marconi Company and then, in 1919, of the Radio Corporation of America. As long as the Fleming patent was upheld by the courts, no one could manufacture diode or triode vacuum tubes for sale to the public without a license from one of those corporations. De Forest's triode patents became in 1912 the prop-erty of the American Telephone and Telegraph Company, essential to its long-distance network and in 1919 to its plans for radio. No one could manufacture triodes for sale to the general public without a license from AT&T. What in 1907 seemed to be merely a squabble over priority between Fleming and de Forest developed by 1919 into an impasse that required for its resolution a major realignment of corporate property rights.

In 1907, however, the triode audion was only a detector, one of many available to the wireless operator, and the world could have got along without it. Its true potentials were not discovered until 1911-1912. For several years after the invention of the three-electrode audion, in fact, de Forest did little to develop the device further. And no one else did either. For a device that was to have such remarkable long-run consequences, the three-element audion entered radio technology very quietly.

De Forest's personal and business affairs during this period were in extreme disarray and it is easy to see why he personally had neither the time nor the funds to carry vacuum tube development further. To rebuild his fortunes after the dissolution of the American De Forest Wireless Telegraph Company, he formed the De Forest Radio Telephone Company (a New York corporation) in February 1907, with a modest capitalization of $200,000, and then in May of that year the Radio Telephone Company (incorporated in New Jersey) with the more grandiose authorized capital of $2 million. Half of the stock was issued to de Forest personally for his audion patents and for the tuned circuit patents of his friend, John S. Stone. To help manage and finance the new organization de Forest teamed up with one James Dunlop Smith, formerly a stock salesman for the Telegraph Company, who seems to have learned the tricks of the trade all too well from his former mentor, Abe White.

It is not necessary for us to follow the history of this company and its successor, North American Wireless, in detail. De Forest's primary goal was to develop wireless telephone sets, using small Poulsen arcs. (In his usual fashion, he cheerfully ignored the fact that the exclusive U.S. rights to the Poulsen patents were held by the Federal Company.) Several of these sets were manufactured, the largest single order being in 1907 from the U.S. Navy, for twenty-six complete radiotelephone sets to be installed on the ships of the Navy's "Great White Fleet," then about to embark on its goodwill tour around the world. Very little time was allowed, however, for installation of the sets or for training the crew in their use, and all but one were dismantled by order of the fleet commander as soon as the ships left port. The equipment was never used for its intended purpose (control of tactical maneuvering) and certainly deserved a fairer trial. De Forest also worked on equipment for wireless telegraphy and induced Georg Seibt to come over from Germany to introduce the quenched spark system. For both telegraphy and telephony, the audion was used for reception.

These activities were financed mainly by the sale of stock in the Radio Telephone Company and its operating subsidiaries. They received a setback in the spring of 1909 when Smith, then president of the company, announced that it was insolvent and promptly resigned. (The last 20,000 shares sold had, it appears, been sold from his holdings and to his account, rather than for the benefit of the company.) Formal bankruptcy was declared in March 1911 and a grand jury investigation began shortly thereafter. De Forest at the time was on the Pacific coast, supervising the installation of quenched spark transmitters on two Army transports. He decided that he should stay there; he liked the climate, but that was probably not the only reason.

De Forest's difficulties with the Radio Telephone Company and his decision to remain on the West Coast marked a kind of watershed in his life. Up to this point he had been very much a "lone wolf," teaming up with one or a few other individuals who could raise the necessary money while he did the inventing. He was certainly no stranger to corporations, but the corporations he knew and used were instruments for generating securities to be sold to the public, not structured organizations with specialized staffs and hierarchical levels of management. After 1909, however, we find him dealing more and more with organizations that he had no hand in creating. His inventions, and particularly the audion patents, become properties to be sold to others; his own role in developing them becomes a matter of exploiting whatever personal residual rights he can retain.

This shift in career track invites comparison with Fessenden's breakup with NESCO in 1911 and Elwell's departure from Federal Telegraph in 1913. The particular circumstances differ, of course, but there are common features. In each case the divorce was precipitated by difficulties over finance, but these in fact masked more fundamental problems of marketing. Where could an appropriate economic niche for the innovation be found? How was its deployment to be managed? In Fessenden's case these problems, unsolved by NESCO, were inherited by General Electric. In Elwell's case they were resolved by converting Federal Telegraph into a contractor for the Navy. In de Forest's, sale of the audion patents to AT&T at least converted them into matters for a large communications company, not an individual inventor, to handle. In all three cases, management of the innovation became the concern of large organizations—General Electric, the Navy, the Telephone Company—instead of the individual entrepreneurs and inventors of the pioneer phase.

As far as we can tell, no important development work was done on the audion between 1906 and 1909. The significance of the period lies rather in the fact that during these years de Forest was able to bring together in the New York laboratory of the Radio Telephone Company a small group of engineers who became familiar with the audion and the circuits in which it could be used. When the company went bankrupt and paychecks stopped arriving, this group dispersed. But its members took with them the knowledge of the audion they had acquired and, in some cases, a hunch that it could be used in ways other than just as a detector.

Two individuals in this small diaspora were to play critical roles in the next couple of years. The first was de Forest himself, on the West Coast and without a job but with a reputation of sorts and with certain claims on the audion patents (claims that were not unclouded, however, since he had signed over his rights to the Radio Telephone Company). The second was an engineer named Fritz Lowenstein. When the Radio Telephone Company failed Lowenstein started his own laboratory on Nassau Street in New York City, set himself up in business as a consultant on radio matters, and induced the Swiss-born engineer, Frederick Kolster, who also had worked with de Forest, to join him. Lowenstein had an interesting background. He received his engineering training in Europe, emigrated to America, and became assistant to Nikola Tesla in his work on wireless power transmission in Colorado. Thoroughly familiar with the audion, he had also kept in touch with European research on vacuum tubes. It is possible that he knew of the important research on vacuum tube amplifiers being done by the brilliant young Austrian physicist, Robert von Lieben. In the summer of 1911 Lowenstein had agreed to work as consultant for John Hays Hammond, Jr., who was experimenting with the radio guidance of boats and torpedoes at his summer home in Gloucester, Massachusetts.[103]

Hammond had made no great progress with his experiments up to this time. He had ample funds available, provided by his father, a mining engineer who had been with Cecil Rhodes in South Africa and had made a very large fortune in diamonds. He had a talented assistant, a former Navy operator named B. F. Miessner, and he himself had graduated from Yale's Sheffield School in 1910 and was no neophyte in technical matters. Hammond had a theory of radio guidance to work from and he knew the components needed to make a feasible system, but he lacked the knowledge and skill to design them. He does not seem to have been a highly creative individual in the conventional sense, but he did have a remarkable ability to gather round him people who were themselves creative and to provide them with the funds, encouragement, direction,

[103] For Lowenstein's background, see Benjamin F. Miessner, *On the Early History of Radio Guidance* (San Francisco, 1964), p. 6. For brief sketches of Kolster and Hammond, see Orrin E. Dunlap, *Radio's 100 Men of Science* (New York, 1944), pp. 212-15 and 229-31.

and organizational support that they needed. For a brief period the Hammond Laboratory outside Gloucester was one of the only two places in the United States where important development work on the audion was being carried on—the other being the Federal Company's laboratory in Palo Alto—and Hammond personally played a unique role, both as catalyst of the research effort and as the information channel between the two locations.

When Lowenstein first became involved in the radio guidance project, Hammond was using a spark set for his transmitter and a Marconi filings coherer as his detector. Results were poor, Hammond was despondent, and his father was beginning to balk at pouring more money into useless experiments.[104] The key elements needed if a workable system was to be developed were a stable continuous-wave transmitter, a sensitive detector, a highly selective tuning system, and an amplifier or relay. Lowenstein worked mainly on the tuning system, the detector, and the amplifier. Quenched spark and rotary spark transmitters were used at first, but by late 1912 Hammond and his team were planning to use Alexanderson alternators and by 1913 arc transmitters made by the Federal Company.

Lowenstein had been experimenting with audions while a member of the de Forest group, and his initial contract with Hammond provided that Hammond would help to finance that work in return for an option to acquire a 50 percent interest if it should be successful. Lowenstein's term for the three-element audion was the "ion controller," and it is easy to see why the development of these devices quickly assumed an importance second only to that of highly selective tuned circuits in the system Hammond hoped to build. Successful guidance of ships or torpedoes by radio waves called for the transmission of information on several frequencies without mutual interference, for immunity from jamming by hostile forces, for the selective reception of weak signals over considerable distances, and for the amplification of those signals so that they could do useful work—in this case actuate mechanical devices such as steering motors. It presented, in short, all the technical desiderata of a radio communications system in extreme form. The audion or ion controller

---

[104] Our knowledge of equipment and events rests largely on the memoirs of Lowenstein's assistant, B. F. Miessner, and the documentation provided by him. See Miessner, *Radio Guidance, passim*. Much additional information may be found in J. H. Hammond, Jr. and E. S. Purington, "A History of Some Foundations of Modern Radio-Electronic Technology," IRE *Proceedings* 45 (September 1957), 1191-1208, the discussion of that paper by Lloyd Espenschied in IRE *Proceedings* 47, (July 1959) and the rebuttal by Hammond and Purington in the same issue. The Hammond-Purington article is copyright 1957 IRE (now IEEE) and all excerpts from it are reproduced by permission.

could serve as a detector, if it could be made rugged and stable enough. Could it also serve as an amplifier, perhaps even as an oscillator?

De Forest had thought of his audion as potentially an amplifier and held a defensible patent on it in that use. Indeed, anyone who had the image of a relay in mind could hardly do otherwise, for amplification is precisely what a relay does. Lowenstein had already tried to use the audion as an amplifier while working for the Radio Telephone Company between 1909 and 1910 and had achieved some success. These efforts were continued and intensified after May 1911 when he became consultant to Hammond. Initial difficulties in obtaining a high enough vacuum were overcome and on 13 November 1911 he reported success. "At last a test over actual long distances. When I heard your voice I fairly jumped in delight; it came in so clear with every shade of its personal characteristics. . . . Your low voice spoken one foot from the transmitter came in as loud as conversation carried on between two extension phones on the same switchboard."[105]

As this letter suggests, Lowenstein and Hammond were using the telephone lines to test, not a radio circuit, but the gain of their amplifier and this confirms what we know of the circuits Lowenstein had used while working with the Radio Telephone Company, which clearly show a grid audion provided with input and output transformers connected as an amplifier in the receiving leg of a standard telephone instrument.[106] The key to success lay partly in the use of a higher vacuum but mostly in the fact that, in place of the grid condenser of the usual audion detector circuit, Lowenstein substituted a "C battery," as it came to be called, to place a negative bias on the grid. By shifting the operating point on the audion's characteristic curve (showing the relation between grid voltage and plate current) this change converted the audion detector into a true "Class A amplifier," to use the later typology, in which variations in the input signal were repeated identically in the output circuit but with higher amplitude. This became the heart of Lowenstein's famous "grid bias" patent (No. 1,231,764, applied for on 24 April 1912) which he eventually sold to AT&T for $150,000.

Lowenstein had taken considerable pains to stabilize his amplifier— that is, to minimize feedback between adjacent stages that would tend to make it break into self-oscillation. Laboratory notebooks make it clear, however, that during this same period Lowenstein also developed circuits

[105] Lowenstein to Hammond, 13 November 1911, quoted in Hammond and Purington, "Foundations," p. 1198. See also Espenschied Papers (Smithsonian Institution), Box LE-5.

[106] Espenschied, "Discussion," p. 1254.

in which his ion controller functioned as an oscillator at both audio and radio frequencies. To someone who really understood how the circuits worked, this possibly did not seem very remarkable. All one had to do was permit a fraction of the output signal to leak back into the input in proper phase and then control the frequency of oscillation by tuned circuits. Lowenstein apparently had a vacuum tube transmitter operating early in 1912, but he made no attempt to patent the invention. Miessner's comment is appropriate: "Perhaps he felt, in all modesty, that there *was* no invention, once the amplifying action of the tube was realized, especially since audions oscillated naturally." Not all inventors are so unassuming.[107]

These were remarkable achievements, anticipating anything de Forest and his group were doing on the West Coast. With due allowance for the talent and imagination of the individuals involved, major credit must also be given to the specificity of the technical task confronting them and its very demanding nature. Hammond was not the first to think of the possibility of guidance by radio. Two British inventors, Wilson and Evans, had controlled boats by wireless waves on the River Thames as early as 1897. Tesla applied for a patent on radio control (issued as No. 613,809) in 1898, and Lt. (later Adm.) Bradley Fiske of the U.S. Navy filed for a similar patent a few months after Tesla and was able to prove that he had anticipated him.[108] Hammond, however, took up the problem at a particularly critical moment, both from the point of view of demand—the Army and the Navy were both interested in his work—and from the point of view of technical feasibility. Radio guidance with damped-wave spark transmitters, broadly tuned circuits, and coherer detectors could never be feasible outside the laboratory. Continuous wave transmitters, highly selective circuits, sensitive detectors, and above all vacuum tube amplifiers brought it within the realm of operational feasibility for the first time.

Hammond went to London in June 1912 to attend the International Wireless Telegraph Conference and, after the conference was over, visited Berlin and was shown through the Telefunken laboratories. There he saw samples of the vacuum tubes that Robert von Lieben and his co-workers, Reisz and Strauss, had been working on. Von Lieben had applied for a German patent in March 1906 on what he called a "cathode-ray relay." This was a high-vacuum tube with magnetic control of the cathode

[107] Miessner, *Radio Guidance*, p. 23. Audions oscillated "naturally" because of their high inter-electrode capacity.

[108] Hammond and Purington, "Foundations," p. 1192 and Espenschied, "Discussion," p. 1253.

rays. By 1910, however, he and his associates had abandoned this approach and were working with a tube that functioned by ionization of rarefied gas, like de Forest's audions. It was intended for use as a telephone amplifier, and it worked. In 1911 von Bronk of the Telefunken organization used it for radiofrequency amplification of signals before detection, an advance of major importance; and early in 1913 Alexander Meissner, also of Telefunken, succeeded in making it oscillate.[109]

The progress that had been made in Germany surprised Hammond. He wrote to Lowenstein on 9 August, immediately after his return, warning him that ". . . the Telefunken company in Germany have exactly the same thing you have. In order to make any money out of this scheme we had better do business with it before we enter in patent complications. I am in no position to fight the Telefunken company over patent rights."[110] Doing business with Lowenstein's ion controller, however, did not prove easy. The amplifier had already been shown to F. B. Jewett and Otto Blackwell, AT&T's transmission engineers, on 27 January 1912, before Hammond went to Europe, but nothing had come of it, and this despite the fact that the Telephone Company, having committed itself to providing transcontinental telephone service in time for the Panama Pacific Exhibition in 1915, was in urgent need of a line amplifier (or "repeater," in the terminology of telephone engineers). Why the Telephone Company did not jump at the chance to acquire Lowenstein's amplifier is not clear. There was no doubt that it worked (though probably not up to Telephone Company standards). On the other hand, Lowenstein at first would not disclose how it worked.[111] The apparatus was demonstrated in a lead-lined padlocked wooden box—lead-lined to prevent covert X-raying—and Lowenstein was close-mouthed as to its contents. And it may be that he himself was not clear what he had that was patentable and could be sold. Whatever the reason, nothing was done, although further tests were held in June, at which time Lowenstein did disclose his circuit. When Hammond returned from Germany with news of the Leiben-Reisz amplifier, negotiations with AT&T seemed stalled on dead center.

The Telephone Company, however, was not the only potential user of Lowenstein's ion controller. In the spring of 1913 Hammond was visited in Gloucester by Ernst Alexanderson of the General Electric Company.

[109] Tyne, *Saga*, pp. 75-79 and Weedon, "High Vacuum," p. 76.

[110] Hammond to Lowenstein, 9 August 1912, as quoted in Lloyd Espenschied to Haraden Pratt, 2 June 1964 (Pratt Papers, Bancroft Library).

[111] Miessner, *Radio Guidance*, pp. 19-20; Espenschied Papers, Box 7, folder 3, copy of memorandum by Frank Jewett entitled "Hammond and Lowenstein Telephone Repeater," 3 February 1912.

Alexanderson had come to discuss Hammond's plans to use two of GE's alternators in his radio guidance work. The two men got on well together—they shared a common interest in yachting—and the conversation turned to Lowenstein's experiments with the audion. Alexanderson was immediately interested, particularly in the possibility of developing a highly selective and sensitive radio receiver by using several ion controllers in a "cascade" circuit, each tube being provided with a resonant circuit tuned to a single frequency. But he had heard that de Forest's audions were too "sluggish" for such use and he was concerned that, used in Lowenstein's circuit, the same limitations would show up. When he returned to Schenectady he did not forget the episode: General Electric, with the alternator, was building up a strong position in radio transmission but it had nothing of its own in receiving circuits or apparatus. Early in 1913 Alexanderson wrote to Hammond asking him to send one of the vacuum tubes Lowenstein had been using. Hammond complied, and Alexanderson sent the device at once to the General Electric Research Laboratory with some suggestions for modification, saying that he was eager to try it as a high-frequency relay.

The request was timely. Irving Langmuir had been working on the improvement of incandescent lamps (with particular reference to the Edison effect) and on high-voltage X-ray tubes. He had grown increasingly convinced of the validity of Richardson's theory of electron emission and suspected immediately that most of the problems with the de Forest audion stemmed from the fact that it was being operated at too low a vacuum—and this despite the fact that McCandless was warning his customers that a high vacuum decreased rather than increased sensitivity. Early tests supported Langmuir's hypothesis: the first tube made at GE proved much more sensitive than a de Forest audion, it would amplify high frequencies without distortion, and it showed none of the sluggishness that Alexanderson had feared. By April 1913 the GE Research Laboratory had a highly successful vacuum tube amplifier to demonstrate and Alexanderson was already beginning to speculate about its use in transmitters, and not merely as a receiving detector and amplifier.

This was the start of what quickly grew into a highly productive research effort on vacuum tubes at the GE Research Laboratory, a marked shift from its earlier concentration on incandescent bulbs and X-ray tubes. There is no doubt that the timing was determined by Alexanderson's visit to Gloucester. Lowenstein, Hammond, and Alexanderson served as the agents through whom information about the audion was transferred from one social locale to another: from de Forest's Radio Telephone Company through the Hammond Laboratory to General Electric; from a bankrupt and moribund operating company to the nation's largest manufacturer

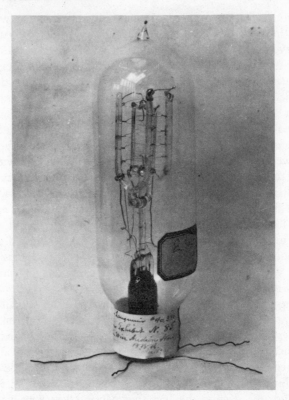

Pl. 11: General Electric's first successful high-vacuum triode.
Source: General Electric Company

of electrical equipment; from a single inventor, working with a handful
of assistants, to one of the first corporate research laboratories. Each of
the individuals involved had other objectives primarily in view: Lowenstein to develop his amplifier for the Telephone Company, Hammond
to perfect the art of radio guidance, Alexanderson to find a receiver to
accompany his alternator. These purposes meshed to create an effective
channel for the transfer of the new technology.

In terms of property rights, however, General Electric's position was
weak. It held rights neither under the Fleming valve patent nor under
the de Forest audion patents. However dramatic might be the progress
of its research staff in developing new and more efficient types of high-
vacuum radio tubes, the corporation was vulnerable unless and until it
secured rights under these basic patents. Langmuir's patent on the use
of high vacuum (eventually disallowed by the courts), Alexanderson's

patent on the multiple-tuned receiver circuit, and a number of other "development" patents represented a belated attempt to shore up a weak trading position in patent rights.

* * *

In California events had been following a different course. De Forest, after completing the installation of his quenched spark transmitters on the two Army transports, tried to set up a radiotelegraph service between San Francisco and Los Angeles in competition with the Federal Company but was unsuccessful. His affairs were, in fact, in desperate shape; he hardly dared return to the East Coast with a grand jury investigation in progress and an indictment in the offing; the equipment he used for his San Francisco–Los Angeles circuit was seized by the sheriff for unpaid debts and put up for auction. To help tide himself over he was reduced to selling his personal test equipment—an inventor's last recourse. In June 1911 he appeared in the office of Cyril Elwell, chief engineer of Federal, and offered to sell him his Seibt wavemeter. Not having such an instrument, Elwell bought it for cash. On the first of the next month de Forest showed up again, but this time he wanted a job.[112]

Elwell and Beach Thompson, president of Federal, put de Forest on the payroll but at first without any clearly defined responsibilities. His first job was helping to relocate the station at Medford, Oregon, where the company was having difficulty finding a good site. De Forest had a reputation as an inventor, and it would have been reasonable if, as a condition of accepting employment with Federal Telegraph, he had been required to agree that he would turn over to the company the rights to any invention he might make while in its employ. This was not done, although Elwell later claimed that he tried to persuade Beach Thompson that it should be. At first, indeed, there seems to have been no formal contract of employment at all. The first of which we have record was signed on 16 September 1912. A "memorandum of agreement" signed on the day following provided that de Forest should give the Federal Company exclusive rights to his inventions if he retained title to them; in return he was to receive, at his option, either 1,000 shares of Poulsen Wireless stock or $100,000.[113] De Forest in fact never exercised this

[112] C. F. Elwell, "Autobiography" (Stanford University Library), p. 70.

[113] Lloyd Espenschied to Lee de Forest, 8 September 1953, and Espenschied to F. E. Terman, 16 April 1946 (Pratt Papers); Espenschied Papers, Box 8, folder 1, copy of agreement between Lee de Forest and Federal Telegraph Company, 17 September 1912.

option, and the Federal Company never received exclusive rights to any of the inventions made while de Forest was in its service. It did, however, acquire "shop rights" to his inventions—that is, the right to use them without paying license fees or royalties—and these were to prove of considerable value later.[114]

De Forest's early work with Federal had nothing to do with vacuum tubes. The company in fact had little interest in audions. It had its own continuous wave transmitter, the Poulsen arc; its own receiver, the improved "tikker" invented by Charles Logwood; and its own system of high-speed reception, based on the Poulsen telegraphone, to handle the heavy volume of night press traffic and night commercial letters on the San Francisco–Los Angeles circuit. Its major problem was how to increase the power of its arc transmitters, and that problem was in the hands of the company's regular engineering staff, headed by Elwell and later Fuller. There was, however, one difficulty connected with the high-speed receiving system. Poulsen's "tikker" had never proved satisfactory; and his system for recording high-speed Morse signals on photographic film was too expensive for low-priced traffic. Logwood's tikker, which replaced Poulsen's, was simplicity itself: a short piece of bent steel wire (originally from a mandolin string, it is said) riding on a serrated wheel on the axle of a small electric motor. This produced a rough but readable note. For high-speed traffic it was Federal's practice to record these sounds on the magnetized wire of a Poulsen telegraphone, which could then be played back at slower speed for the operator to transcribe the message.

It was an effective system but it had one drawback. When the signal was played back at slow speed, it fell drastically in both pitch and volume. What the operator heard in the earphones was a low, rough, "grunting" note which was hard to copy in the presence of static. The remedy, in principle, was simple. To raise the volume of the signal one needed an amplifier. To raise its pitch one needed to take advantage of Fessenden's heterodyne principle and "beat" the incoming signal against a signal generated locally. A small arc, such as Fuller had worked with at NESCO, might in a pinch have been used as a local oscillator. But Federal had no amplifier.

Amplification was seen as the major problem, and in the early months of 1912 the company's management decided to do something about it.

---

[114] For de Forest's failure to exercise his option, see C. F. Elwell to W. H. Hewlett, 24 December 1953 (Pratt Papers). Federal's "shop rights" under the de Forest patents were rediscovered by Ellery Stone, president of the company, after 1924. See H. Pratt to F. E. Terman, 3 November 1964, and Pratt to Col. F. Babin, 19 March 1963 (Pratt Papers).

De Forest later stated explicitly that the assignment was given to Logwood first, and Logwood brought it to him.[115] De Forest told him that he had taken out a patent on an audion amplifier in 1907 but had never actually used the device. They decided to work on it together, and the company added a third member to the group: Herbert van Etten, a trained and experienced telephone engineer. With de Forest's knowledge of the audion and Logwood's experimental ingenuity, they made a good team.

It was probably not happenstance that led Federal's management to tackle the development of a vacuum tube amplifier at this time. John Hays Hammond, back in Gloucester, had written to Thompson in January 1912, asking for information on the Federal Company's activities and urging him to look into the equipment that Hammond's group had developed. He emphasized their possession of a superior radiofrequency oscillator and did not refer specifically to Lowenstein's amplifier. "Our method is far more reliable and simpler than the Poulsen arc method, or the high frequency alternator system as used by Fessenden and others."[116] Details were not provided, but one at least of Thompson's acquaintances had visited the Gloucester laboratory and knew of Lowenstein's ion controller.[117] De Forest later believed that this letter was very probably what induced Thompson to initiate the development of an audion amplifier and said that he personally had not realized the full potential of the audion as an amplifier until given this assignment.[118] Hammond's letter may therefore have had important consequences. He received no reply; with the inventor of the audion on the payroll, Thompson may well have felt that he needed no outside assistance.

De Forest, Logwood, and van Etten worked apart from the other engineers at Federal and their efforts seem to have been directed more toward developing a telephone amplifier and thereby winning the $1 million prize that AT&T was rumored to be offering than to solving any of the Federal Company's problems. Their approach was highly empirical and not guided by any particular theoretical insight. Audions were pro-

[115] Statement, "Charles Vern Logwood," by Lee de Forest, 17 February 1942, enclosed in de Forest to A. N. Goldsmith, 21 November 1945 (Pratt Papers).

[116] J. H. Hammond, Jr., to Beach Thompson, 25 January 1912, reprinted in Espenschied, "Discussion," p. 1255. The date on this letter is actually 1911, but this is clearly a typist's error, as Hammond's letter to Major Burnham (see n. 117), which is dated 25 January 1912, speaks of his intention to write to Thompson "herewith."

[117] J. H. Hammond, Jr. to Major F. R. Burnham, 25 January 1912, reprinted in Espenschied, "Discussion," p. 1255.

[118] Lee de Forest to Lloyd Espenschied, 15 September 1953 (Espenschied Papers, Box LE-8, folder 1), cited in Espenschied, "Discussion," p. 1255.

cured from McCandless in New York, and de Forest and his group experimented with them in a wide variety of circuits, trying to get them to "boost." In an attempt to improve performance, de Forest took some of the audions to a glass blower in San Francisco experienced in evacuating X-ray tubes and had them reexhausted. But there was no explicit recognition of the desirability of moving from conduction by gas ionization to conduction by electron emission, and no significant changes were made in the internal components of the tubes.

The situation was, however, significantly different from what it had been in the closing months of 1906. De Forest now had good working conditions, adequate financial security, a clear assignment (but no one giving him orders), very competent co-workers, and audions with a harder vacuum. He does not, however, appear to have approached the task of designing a vacuum tube amplifier with any greater understanding of the triode audion than he had when he first invented it. When success came, in July and August 1912, it came not from any new theoretical insight but because persistent tinkering with circuits finally gave a stronger signal in the audion's plate circuit than was fed to its grid. But there was no "eureka moment"; the process was a gradual one and it was only after three or four weeks of experimenting that they were sure they had a circuit that was stable and provided gain. (See Fig. 4.15) The reevacuated audions helped; so did the audiofrequency transformers that van Etten designed, drawing on his telephone experience; and so did the practice of feeding the output of one audion into the input circuit of a succeeding one, so that two- or three-stage amplifiers could be built.

But it was still a strange device. De Forest tested it by placing his watch—the laboratory notebooks show a "trusty Ingersoll" on the circuit diagrams—in front of the microphone and gauging the loudness of the sound in a telephone receiver. And dramatic demonstrations could be given of the sound of a handkerchief being dropped. This was all very well, but what such tests showed was the ability of the audion to handle very small signal levels, not its ability to handle the voltages and currents encountered in a normal telephone circuit (in contrast with the test procedures used by Lowenstein). Any attempt to apply larger voltages or higher signal levels brought the telltale "blue glow" that meant gas ionization, with resultant distortion, reduction in gain, and increase in noise. And the circuit still contained a capacitor in series in the grid circuit—a hangover from its use as a radiofrequency detector but detrimental to its use at audio frequencies. De Forest had produced an amplifier but, as he left it, it was not an amplifier the Telephone Company could use.

Van Etten wrote to J. J. Carty, chief engineer of AT&T, on 2 August 1912, telling him of the audion's potential as a telephone repeater and

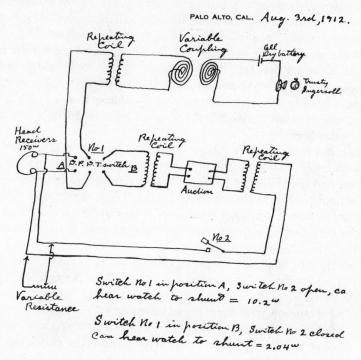

FEDERAL TELEGRAPH COMPANY

FACTORY
913 EMERSON ST.

PALO ALTO. CAL. Aug. 3rd, 1912.

Switch No 1 in position A, Switch No 2 open, ca
hear watch to shunt = 10.2ʷ

Switch No 1 in position B, Switch No 2 closed
Can hear watch to shunt = 2.04ʷ

Fig. 4.15: De Forest's amplifier circuit

of its ability to amplify the sound of a ticking watch by a factor of five, but the letter was never sent.[119] Instead, de Forest decided to go to New York himself, taking the audion with him.[120] He did not, however, approach the Telephone Company directly but rather contacted first his friend John S. Stone, whose wireless patents the Radio Telephone Com-

[119] Espenschied Papers, Box LE-7, folder 1, H. van Etten to Espenschied, 7 August 1956.

[120] In his *Autobiography* de Forest describes how he explained the situation to Beach Thompson, showing him the earlier patents on the grid audion and the "device for amplifying feeble electric currents" that had been assigned to the Radio Telephone Company, so that Thompson "clearly realized" that the Federal Company had no claim to them. One may speculate as to why this information had not been given to Thompson sooner and whether the interview was quite as amicable as de Forest makes it out to be. See de Forest, *Autobiography*, pp. 295-96.

pany had acquired some time before. This was a shrewd move. Stone had been associated with AT&T in one capacity of another for many years, was widely respected, and had the entree to a level of management to which de Forest would have had difficulty gaining access.[121] De Forest set up his amplifier in a room at the Fine Arts Club in New York City, where Stone was living, demonstrated it to him, and discussed the best way to approach the Telephone Company. Stone took charge of the arrangements; he would contact his friends in AT&T's management; in the meantime, he was scheduled to give an address to the Franklin Institute, and he would make a point of referring to de Forest's new telephone relay there.[122]

Stone did de Forest one further important service on this occasion. Throughout his work on the audion amplifier de Forest and his co-workers had been troubled by its tendency to break into self-oscillation. This should not have surprised them: the tube's input and output circuits were not isolated from each other, and it was to be expected that, as soon as they devised a circuit that had some net gain, it should start to "howl." The phenomenon was familiar to anyone who had tried the juvenile trick of holding a telephone earpiece up against its mouthpiece (as one could with the telephones of those days). There is ample testimony that de Forest at the time thought of this as a problem to be eliminated, not as an important new discovery to be exploited.[123] And, to someone whose single-minded purpose was to build an amplifier, this was natural. Nevertheless, what de Forest had discovered in Palo Alto was something just as important as his original discovery of the triode audion or his discovery that the audion could amplify: he had discovered that, in an

[121] George H. Clark, *The Life of John Stone Stone* (San Diego, 1946); Dunlap, *Men of Science*, pp. 149-53.

[122] Espenschied Papers, Box LE-8, folder 1, copy of J. J. Carty to C. E. Scribner, 30 October 1912. John S. Stone, "The Practical Aspects of the Propagation of High-Frequency Electric Waves along Wires," *Journal of the Franklin Institute* 174 (October 1912), 353-84. In a footnote on p. 375 of this paper Stone called attention to a "new telephone relay amplifier . . . entirely electrical in its action . . . [and] productive of great amplification." There is evidence in Carty's letter (see above) that Stone, either at this time or earlier, had acquired an option on de Forest's fundamental amplifier patent (No. 841, 387). Lubell also describes Stone as having been given a sixty-day option. If so, he was the key man in any deal de Forest might make with AT&T.

[123] For one among many examples, see Leonard Fuller to Edwin Armstrong, 20 November 1934 (Fuller Papers, Bancroft Library). De Forest did not disclose his discovery of the oscillating audion to Elwell or Fuller or Beach Thompson but only to three junior engineers of the Federal Company, who heard the whistling signals from a "black box" without seeing the circuit.

appropriate circuit, an audion could oscillate—that is, it could serve as a generator of continuous waves, and not merely as a detector.

Lowenstein made the same discovery; and, at almost the same time, Edwin Armstrong, Irving Langmuir, and (in Germany) Alexander Meissner made it also. In Worcester, Massachusetts, the young Robert Goddard filed for a patent on a vacuum tube oscillator on 1 August 1912—five days before the crucial notebook entry by Van Etten on which de Forest's claim was based.[124] Priority in the discovery was to become the most bitterly contested single issue in the entire history of radio technology. And this was not surprising. What bothered de Forest in his Palo Alto laboratory were howls, whistles, and squeals. But a circuit that oscillated at audio frequencies could also, with certain modifications, oscillate at radio frequencies: that is, it could become a radio transmitter (or the essential element in a heterodyne receiver). Property rights in that discovery came to have very great value, for it made the audion into a continuous wave generator—one that was to long outlast its larger competitors, the alternator and the arc.

From laboratory notebooks it can be established that on 6 August 1912, by coupling the input and output circuits of an audion amplifier, de Forest succeeded in generating a clear audio note in his earphones, that he knew what he had done and how he had done it.[125] The fact that he did not promptly file for a patent on the circuit—in fact he did not file until 20 March 1914 (on the oscillator) and 23 September 1915 (on the feedback circuit)—was to prove of major importance in later litigation but need not concern us here. The legal issues have been thoroughly explored elsewhere.[126] What is of present interest is whether de Forest realized at the time that, in addition to an audio oscillator that might help solve Federal's problems with high-speed reception, he also had a

[124] For the little-known Goddard oscillator, see Esther C. Goddard and G. E. Pendray, *The Papers of Robert Goddard*, 3 vols. (New York, 1970), 1:14-15 and 2:989; and Arvid E. Anderson, "Robert H. Goddard: Original Inventor-Patentee of the High Frequency Vacuum Tube Oscillator" (unpublished manuscript, copyright held by the Merle Collins Foundation).

[125] See the sketches and notes from van Etten's notebook as reproduced in James R. Gaffey, "Certain Aspects of the Armstrong Regeneration, Superregeneration, and Superheterodyne Controversies," *Patent, Trademark, and Copyright Journal of Research and Education* (Summer 1960), 179-81.

[126] See, for example, the opinion of the Supreme Court (delivered by Chief Justice Cardozo) in U.S. Reports, Vol. 293, Cases Adjudged in the Supreme Court at October Term, 1934, *Radio Corporation of America et al. v. Radio Engineering Laboratories, Inc.*; and Alfred McCormack, "The Regenerative Circuit Litigation," *Air Law Review* 5 (July 1934), 282-95.

radiofrequency oscillator. On this point the evidence is uncertain and the role of John S. Stone becomes critical.

When de Forest sketched for Stone the circuit of his audion amplifier, one of the first questions Stone asked was whether he had had any trouble with self-oscillation.[127] This showed insight: Stone was familiar with other kinds of "repeater" that the Telephone Company had tested and he knew that it was not enough for a repeater merely to amplify. It also had to be stable, and this despite the fact that it was required to pass signals in both directions. De Forest told him that indeed the amplifier had tended to "howl" and drew the relevant "feedback" circuit; but he said that he had managed to control the tendency. (See Fig. 4.16) Stone

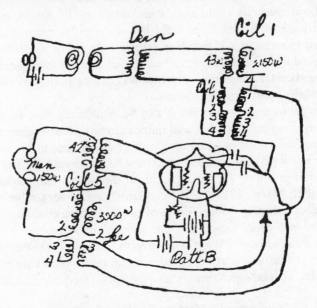

Fig. 4.16: De Forest's feedback circuit

[127] Compare the testimony of John Stone Stone in *RCA et al. v. University Wireless Communication Co., Inc.*, Equity 767, on de Forest patent 1507016/ 17, U.S. District Court, District of Delaware, testimony begun 13 March 1930: "I was not surprised with [*sc.* when] de Forest told me that his audion would generate sustained oscillations, because when he had told me that the audion would act as a repeater I foresaw that this meant it would generate oscillations, and I saw this as a limitation to its use as an amplifier for telephony. It was obvious to me as an experienced telephone engineer that any relay which amplified would produce sustained oscillations. . . . I told him he might get into trouble when trying to use the audion amplifier for two way amplification on telephone circuits."

then, according to his own later court testimony, asked whether de Forest knew if the oscillations extended into frequencies above the range of human hearing—whether, in fact, they could occur at radio frequencies. And de Forest replied that they did, that he had been aware of the fact, and that he had in mind the use of the audion as a generator of radio frequencies.[128]

The only evidence to support de Forest's claim that he knew in October 1912 that he had invented a radiofrequency, and not merely an audio-frequency, oscillator is Stone's account of that conversation in New York. And there were those who, in later years, were not slow to point out that, by the time the testimony was given in court, the relevant patents had become the property of the Telephone Company, which was de-fending them against the conflicting claims of Armstrong and the West-inghouse Company, and that Stone and the Telephone Company had always been closely associated.[129]

The implication is that Stone, consciously or unconsciously, read more into his conversation with de Forest than had originally been there. De Forest, after all, waited for almost eight months before filing for a patent on the oscillator, and longer than that before filing on the feedback circuit. This was not characteristic of him, if indeed he understood the full sig-nificance of his invention when he made it, and the explanations offered by his lawyers—lack of funds, preoccupation with the amplifier, a belief that he was a pioneer in the art with no rival in the offing—are not wholly convincing. Meanwhile Edwin Armstrong, somewhat delayed by his fa-ther's refusal to provide money for a patent application, managed to have his drawing of a feedback circuit notarized on 31 January 1913. (See Fig. 4.17) In legal terms this represented "reduction to practice" for Armstrong.[130] De Forest did not achieve controlled heterodyne reception until 17 April 1913.

[128] A copy of Stone's deposition in Suit in Equity No. 767 (U.S. District Court of Delaware) may be found enclosed in Espenschied to Pratt, 20 May 1953 (Pratt Papers).

[129] See, for example, Haraden Pratt to W. R. Hewlett, 4 August 1954 (Pratt Papers): "Some of the witnesses in that case, now deceased, were practically ordered to testify affirmatively by their employer so de Forest would win and or [sc. of] course the Company would benefit. One of them consulted me as to what his course of action should be." Compare Espenschied to Pratt, 20 May 1953 (Pratt Papers). It was through Pratt's intervention that the key Palo Alto labo-ratory notebooks establishing the date of de Forest's discovery were recovered. See Pratt to Espenschied, 14 May 1953 (Pratt Papers).

[130] Compare the opinion of Judge Manton in *Armstrong et al. v. De Forest Radio Telephone and Telegraph Co.*, Circuit Court of Appeals, Second Circuit, 13 March 1922, p. 590.

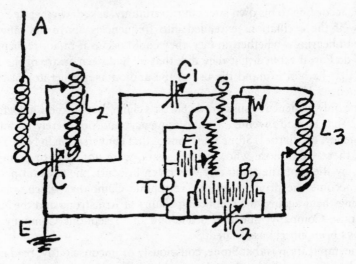

Fig. 4.17: Armstrong's feedback circuit.
Source: Armstrong's notebook, as reproduced in Gaffey, "Armstrong
Controversies," p. 179

Once the Palo Alto laboratory notebooks were located and introduced in evidence, no one doubted that on 6 August 1912 de Forest had invented an oscillator. What many radio engineers questioned was whether, in August 1912, he knew what he was doing, whether he understood that his audiofrequency oscillator could also be a generator of radio waves, and whether, as the courts finally held, the gap between that oscillator and a device actually useful in practice could be bridged by knowledge possessed by electrical experimenters and engineers in 1912. It hardly seems possible to resolve the issue at this late date. A reasonable interpretation might be, however, that de Forest knew he had a circuit that would oscillate but attached little or no importance to it until Stone's questions made him think for the first time about the broader implications of what he had done, and in particular about the possibility of harnessing the oscillator for useful purposes, at both audio and radio frequencies. This would also help to explain why, according to Stone, neither of them at the time perceived the commercial value of the discovery.[131]

In the short run, to both of them, the amplifier was the main thing and the problem was how to make sure that the Telephone Company engineers examined it without prejudice. Stone sent J. J. Carty a copy of

[131] Testimony of J. S. Stone, as cited in Gaffey, "Armstrong Controversies," p. 183.

his Franklin Institute paper on 21 October 1912, and Carty reacted immediately, setting up a meeting between Stone, de Forest, and Frank Jewett, transmission engineer in AT&T's Engineering Department, for 30 October. The Telephone Company engineers actually present on that occasion included neither Carty nor Jewett, and they seemed only mildly impressed. They did, however, ask de Forest if he would leave the apparatus with them overnight and return the next day in case they wished to ask further questions. De Forest raised no objection and, when he returned on 31 October, found the atmosphere markedly different. "Where before was indifference and an attitude of boredom, now an atmosphere of keenest interest pervaded the room."[132] He performed the usual demonstrations, dropping a handkerchief and speaking into the telephone receiver in a low voice. He was again asked if he would leave the amplifier with them for a while for further tests, and again he did not object, confident (as he later put it) that the circuits he had shown them were in the main covered by his earlier patents.

The impression made by the amplifier on the telephone engineers present was brought out in later court testimony. Frank Jewett, for example, on the basis of the demonstrations and the tests, was confident that "while the apparatus did amplify very weak sounds it would not in its then existing form handle the amount of energy which any telephone repeater must handle if it was to be a successful repeater."[133] He pointed out that in a telephone line there is an energy level below which one cannot go, a level determined by interference from high-tension power lines. When they attempted to run de Forest's amplifier at that level, it failed to function: the quality of speech became badly distorted and the tubes began to show a "blue haze" not present at other times. E. H. Colpitts, head of the Engineering Department's Research Branch, recalled that "with low input volumes of speech, the device operated, one might almost say, beautifully; it operated very, very well."[134] But to make it a device useful in the telephone business its energy-handling capacity would have to be increased manyfold. And he added, significantly, that de Forest knew all about the limitations of the amplifier; in fact, he brought out himself the fact that, when the plate voltage was increased beyond a certain point, the blue haze indicating ionization became evident.

The most interesting reactions, however, were those of Harold D.

[132] De Forest, *Autobiography*, p. 297.

[133] Testimony of Frank B. Jewett in *De Forest Radio Company v. General Electric Company*, Supreme Court of the United States, October Term, 1930 (in Pratt Papers, Box 4).

[134] Ibid., testimony of Edwin H. Colpitts.

Arnold, a young physicist who had joined the Research Branch in January 1911. Arnold had been trained at the University of Chicago under Robert A. Millikan and came to the Telephone Company in response to a request from Jewett that Millikan should send him a man trained in "the new physics" to work on the repeater problem—specifically "by utilizing somehow those electron streams which you have been playing with here in your research work in physics for the past ten years."[135] Since joining the research group Arnold had been trying to develop a suitable amplifier out of mercury-vapor discharge tubes. He had had some success, but the devices had a number of problems and were never used commercially.

Asked in court what impression the de Forest amplifier made on him when he first saw it, Arnold replied frankly, "I was amazed. I had made a study of repeaters and I thought that I had pretty well sized up all the repeater possibilities in the world at that time . . . and when I went into the room and saw this thing and saw how it worked I was much astonished and somewhat chagrined because I had overlooked the wonderful possibilities of that third electrode operation, the grid operation of the audion."[136] He was familiar with the Fleming valve; he had even followed the acrimonious dispute between Fleming and de Forest in the pages of the *Electrician*; but he had never thought about what a grid audion might do in telephone service. It was a device for wireless equipment, not for telephone lines.

Most important, Arnold knew exactly what to do about the audion's limitations. "I suggested that we make the thing larger, increase the size of the plate with the corresponding increases in the size of the grid but particularly at that time I suggested that we were not getting enough electrons from the filament." What he wanted to do, in fact, was convert the de Forest audion into a different kind of device. He wanted a much higher vacuum in the tube, with residual gas eliminated to the greatest possible extent; and he knew the newly invented Gaede molecular vacuum pump made that possible. He wanted more electron emission from the filament without an increase in filament voltage; and he knew Wehnelt's new oxide-coated filaments would do that.[137] Arnold, in short, looked at the audion and saw in it (as did Langmuir) something that its inventor

[135] Quoted in M. D. Fagen, ed., *A History of Engineering and Science in the Bell System: The Early Years (1875-1925)* (Bell Telephone Laboratories, 1975), p. 258.

[136] Testimony of Harold D. Arnold in *De Forest Radio Company v. General Electric Company*, Supreme Court, October Term, 1930.

[137] Thomas H. Briggs, "Arthur Wehnelt and His Wonderful Cathode," *The Old Timers' Bulletin* 23 (December 1982), 19-20.

did not see: the possibility of making it into a high-vacuum device, operating by pure electron emission. He was able to do this, not because he was more perspicacious than de Forest, but because he had access to different information. He was trained in the new electron physics and de Forest was not.

Arnold said that he was amazed when he first saw de Forest's amplifier. His candor did him credit, but one wonders where he had been, in the course of his worldwide study of repeaters, the previous January, when Lowenstein first demonstrated his amplifier, or in June, when Lowenstein opened up his "black box" and let the Telephone Company's engineers look at his circuit. True, Lowenstein's amplifier was erratic, but so was de Forest's. There was no substantial difference between them, except de Forest's idea of running two or three audions in "cascade." Knowledge of what Lowenstein was trying to do with the triode audion, irrespective of the degree of success he attained, should have been enough to alert Arnold to its possibilities. But this was not the first time the Telephone Company's engineers had missed an opportunity. Back in November 1906 H. J. Round of the New York Bell Telephone Company had heard de Forest's talk on the two-element audion to the AIEE and reported to his boss, C. H. Arnold, that it ought to be tested as a telephone repeater, using a magnetic field to modulate the cathode discharge.[138] And in 1907 G. A. Campbell of AT&T's Boston laboratory appears to have recognized the need to look into electronic devices to solve the amplification problem.[139] Nothing was done about either of these initiatives, and the closing of the Boston laboratory gave the quietus to research along those lines. AT&T's engineers may well have felt proud of themselves in 1912 for figuring out so quickly why the audion did not operate as they thought it should, but they might have spared a moment to consider how their organization had fumbled the ball five years before.

Why did the Telephone Company react positively to de Forest's amplifier but not to Lowenstein's? It does not appear that they differed greatly in performance; both required further work to make them usable for telephone service. Part of the answer may lie in the personalities of the two men. Lowenstein was described as "diffident," and no one ever called de Forest that. Lowenstein was reluctant to open up his box and let the Telephone Company see what he had to offer (partly because he had no patent protection even on the grid bias circuit at that time), while

[138] Memorandum, H. J. Round to C. H. Arnold, 15 November 1906, in Espenschied Papers, Box LE-8, folder 2; and in Simon Papers, Box 4, Espenschied to Emil Simon, 8 March 1956.

[139] Fagen, *History*, p. 256, n. 35.

de Forest showed no such hesitation. But partly too it was a matter of sponsorship. Lee de Forest came as the friend of the urbane and cosmopolitan John S. Stone, a man whom every telephone engineer admired and respected. Lowenstein was backed only by a young amateur who seemed to have more money than he knew what to do with and whose best-known achievement up to that time was the invention of an electric dog that could follow a moving light.[140] None of these things should have made any difference, but they did.

Even so, the Telephone Company was in no hurry to make a deal with de Forest and for almost a year he was left in uncertainty as to whether AT&T wished to acquire rights under his patents or not. The delay was partly understandable. There were always more tests to be conducted. More important, perhaps, title to de Forest's audion and amplifier patents had been conveyed to the Radio Telephone Company and then transferred to the North American Wireless Corporation, and de Forest was only one among several claimants to the assets of those organizations. Title to the patents had to be cleared up before AT&T would buy them, and that took time. AT&T turned the job over to an outside attorney, not identified as representing the Telephone Company, and he by 26 July 1913 managed to induce all parties to sign over to him exclusive rights to the use of the audion in all fields except wireless telegraphy and telephony. The price was $50,000. Stone and de Forest had asked for $100,000, and there is evidence that AT&T would have paid that price if necessary.[141] But there was no one else bidding for the rights, de Forest's companies were in desperate financial condition and, with a trial for fraud hanging over him, he needed the money.

By this purchase AT&T acquired the rights that it needed in the short run—that is, those necessary to carry out its commitment to provide transcontinental telephone service by the end of 1914. But, once Arnold and his associates got to work on the audion, its versatility quickly became apparent and so did the desirability of acquiring broader rights. On 7 August 1914 the company purchased for $90,000 a nonexclusive license in the field of wireless telephony.[142] And in March 1917 it purchased for

[140] Hammond and Purington, "Foundations," pp. 1193-94. Hammond had in fact successfully demonstrated radio control of seagoing vessels by the late fall of 1912.

[141] Samuel Lubell, "Magnificent Failure," *Saturday Evening Post*, 24 January 1942, p. 43. Lubell quotes Charles Neave, one of AT&T's outside attorneys, as saying that the rights were acquired "for about half of what we were prepared to pay."

[142] Tyne (*Saga*, p. 115) refers to this as a license in the field of wireless *telegraphy*, but this is a slip. AT&T had no interest in providing a telegraph service.

$250,000 an exclusive license to all the remaining rights, plus rights to any inventions de Forest might make in the next seven years, leaving to de Forest only personal, nontransferable, and nonassignable rights in certain specified fields.[143]

The Telephone Company's moves, from 1913 on, to acquire exclusive rights under the de Forest patents were part of a strategy aimed at controlling the technology of the vacuum tube. Originally this strategy had a limited objective: to extend the spatial coverage of the wired telephone service. This was successfully achieved by the end of 1914 with the opening of telephone service across the continent. But the range of desired objectives expanded as improved types of vacuum tube became available, as new circuits were devised, and as new uses for the device became apparent. Multiplex or carrier current telephony, over the wired system, with the audion serving as oscillator and modulator, was successfully tested before the end of 1914. But the most dramatic demonstration of the new technology came in the latter half of 1915 when Telephone Company engineers, using the Navy's antenna system at Arlington, Virginia, demonstrated long-distance radiotelephony by successfully transmitting speech and music to San Francisco, Honolulu, Darien, and Paris. This was far from a commercial service, and the technique adopted—feeding a modulated signal to a final bank of some 500 triode amplifiers—was not the method used in future transmitters. But, with the tubes available at the time, it was a remarkable achievement: part public relations stunt, part display of technical virtuosity, part a strong assertion of the role the Telephone Company intended to play in the future of radio communications.

To assemble the system of property rights that underlay this technological system, the Telephone Company had moved progressively to achieve control of what were believed to be the key vacuum tube patents. And, just as its technological position seemed impressive, so did its legal position. There were, however, at least three areas of vulnerability, and in

---

[143] De Forest's personal rights were to cause great trouble in the years ahead. He retained, under all his patents, a personal, nontransferable, and nonassignable right to make and sell all types of radio equipment to users for their use only; to make and sell to the U.S. government for its use; to grant a license for use to the Marconi Company; to make and sell radio apparatus for the distribution of news and music; and to make and sell apparatus for the reception of news and music. The clauses referring to "news and music" must have seemed of trivial importance at the time—certainly not fields the Telephone Company would ever wish to enter. De Forest proved highly ingenious at stretching his personal rights to fit changing opportunities.

each of them there were signs that, eventually, the Telephone Company would have to come to terms with outside interests.

First, there was the Fleming diode patent. In 1915 the Marconi Company had brought suit against de Forest, claiming that the triode audion infringed the Fleming patent. De Forest filed a countersuit, alleging that the Marconi Company had infringed his patents by using triodes without a license. The Marconi Company admitted infringement and was enjoined from using triodes without permission in future. But the court also held that the triode vacuum tube did infringe the patent on the Fleming diode, and de Forest (and therefore AT&T) was prohibited from using triodes without a license from Marconi. The result was an impasse: clearly some form of cross-licensing had to be arranged if any of the interested parties were to manufacture or use triode vacuum tubes, the essential element in the new technology.

Second, there were the patents on the vacuum tube oscillator and the feedback or regenerative circuits. Claims made by Langmuir and Meissner were dismissed by the courts as the dates claimed for invention were too late. This left the question of priority, and therefore the validity of the de Forest patents, in dispute between Armstrong and de Forest. The complexity of the issue and the bitter personal animosity between the two men guaranteed that there would be no resolution of that problem in the near future. And when in 1920 Armstrong sold his patents to Westinghouse, the stage was set for a second confrontation between two major corporations.

And thirdly there were the patents on the high-vacuum tube itself. Langmuir at General Electric and Arnold in the Telephone Company's laboratory had both concluded that further development of the audion required much higher vacua than de Forest had used. Langmuir, perhaps encouraged by a Patent Department that was well aware of GE's weak bargaining position in tube patents, had applied for and received a patent on the high-vacuum tube. Arnold, believing that moving from a lower to a higher vacuum was only a question of degree and an obvious mental step, at first made no application but, when Langmuir filed, belatedly submitted an application himself, claiming prior discovery. And that was only one, though perhaps the most important, of a score of patent interferences that quickly developed between GE and AT&T, each company conducting an active program of vacuum tube research, each trying to consolidate its position through patents.

After 6 April 1917, when the United States entered World War I, these conflicts became for a time irrelevant. Patent rights were temporarily ignored as all parties strove to cope with the explosive wartime growth in the demand for tubes. The return of peace brought them to the fore once again. The development of radio technology had created property rights of immense value; but the way those rights had come to be distributed threatened to impede further development and commercial exploitation. A realignment of corporate interests was clearly indicated.

FIVE

Radio, Cables, and the National Interest

BETWEEN 1900 and 1914 radio technology went through a radical transformation: the shift from spark to continuous wave transmission. This transition was by no means complete in 1914; spark sets far outnumbered continuous wave transmitters in that year. It was clear, however, to all informed observers that the normal technology of radio transmission would in future require continuous waves. Spark transmitters, whether the old-fashioned open spark sets still found on shipboard or the more sophisticated rotary sparks, quenched sparks, and disk dischargers, were obsolescent, if not already obsolete.

There were in 1914 three known methods of generating continuous high-frequency radio waves: the high-frequency alternator, the Poulsen arc, and the vacuum tube oscillator. Of these the third was so new that it was barely out of the laboratory. Vacuum tube transmitters, to be practically useful, required the development of tubes that could yield substantial power output, and these were not yet available. Development of the second method—the high-frequency alternator—was, in the United States, in the hands of the General Electric Company. The most powerful model actually built up to 1914 generated only 2 kilowatts, but a 50 kilowatt machine was under construction in 1914 in the Schenectady shops and there appeared to be no important difficulties in the way of building even higher power models when and if a demand for them appeared. The same could be said of Poulsen arcs. The most powerful model available in 1914 was the 100 kilowatt unit designed by Fuller for the Navy's Darien station. But the breakthrough in design had been made. The Federal Company could build much more powerful arc transmitters if called upon to do so.

With both the alternator and the arc the United States had achieved by 1914 a technological lead over other countries. For alternators, the lead was not a large one, if the comparison was with Germany and France. The Telefunken organization had developed two distinct types of radio-frequency alternator by 1914, based on design philosophies quite different from that followed by Alexanderson at GE. And in France the Latour and Bethenod alternators looked promising.[1] The Marconi organization in Britain had no high-frequency alternator at its disposal and no plans to build one. Poulsen arcs were well known in Europe but only low-power arc transmitters had been built. The ideas and techniques required to design and build transmitters of 60 kilowatts and up were, in 1914, exclusively an American possession. The Marconi Company had acquired rights to Poulsen's patent in 1914 but did nothing with them. Spokesmen for the company denied any intention of shifting to arc transmission.

It had at one time appeared probable that the Marconi organization would dominate the business of radio. Its position in marine radio was a powerful one. Its plans for an intercontinental system were impressive. In the United States, the Marconi Wireless Telegraph Company of America had become, since the absorption of United Wireless, by far the largest operating company. There were, however, two impediments—one technical, one ideological—to the dominant position that the Marconi companies hoped to achieve.

The first was the Marconi commitment to spark technology, symbolized by the fact that it neither possessed a high-power continuous wave transmitter nor had one under development. The Marconi organization had achieved something close to global supremacy in wireless by its technological leadership in the age of spark. Now that era was drawing to a close. It was becoming a matter of extreme urgency that the Marconi companies either develop a high-power continuous wave transmitter themselves, or come to terms with some other organization that had already done so.

The second impediment was nationalism, and there is some irony in this. The parent firm was a British company; most of the common stock was held in Great Britain and Ireland; and it was in London that the strategic decisions on company policy were taken. But the Marconi or-

[1] For the European alternators, see M. Latour, "High Frequency Alternators" (Paper read before the Societé Internationale des Electriciens) *Wireless World* (July 1919), 187-90; Emil E. Mayer, "The Goldschmidt System of Radio Telegraphy" IRE *Proceedings* 2 (March 1914), 69-108; E. E. Bucher, "Radio Frequency Changers," *Wireless Age* 6 (November 1918), 10; Bernard Leggett, *Wireless Telegraphy* (London, 1921), pp. 386-99; W. Dornig, "Der Hochfrequenz-Maschinen-Sender (400MK) Nauen," *Telefunken-Zeitung*, No. 17 (n.d.), 64-74.

ganization had never regarded itself as a creature of the British government, except in the most formal and juridical sense, nor as an instrument of British policy. Its organizational goals were global in scope. It thought of itself as architect and operator of a communications system that was potentially worldwide, and for that reason it had interests that transcended those of any single national government. This was not, however, how the Marconi Company was perceived by the outside world. In the United States it was taken for granted that Marconi policies were designed to advance British interests. In Germany the Telefunken organization had been deliberately created, by initiative from the very highest levels of government, to offset and neutralize Marconi, and therefore British, influence.[2]

Marconi corporate ambitions were suspect because they threatened to extend and perpetuate, in the new age of long-distance radio, the hegemony that Britain had achieved in the age of the submarine cables. By 1914 most of the world's submarine cables were British-owned. This state of affairs rested on several hard facts. In the first place, Britain had a near-monopoly of the world supply of gutta-percha, the durable natural plastic that was then the indispensable material for cable insulation. Most of this supply came from Singapore, the Malay Peninsula, and Borneo. These sources were controlled by British capital. Secondly, the world's largest and most experienced cable manufacturing company, the Telegraph Construction and Maintenance Company, was located in Britain and owned by British capitalists, many of them also associated with the sources of gutta-percha. Cable manufacturing facilities of very limited scale were to be found in Germany, France, and Italy; but the British firm commanded such a wealth of technical expertise as to give it, in combination with preferred access to gutta-percha supplies, something approaching a monopoly. Without its cooperation the laying of important new cables or the extension of old ones was difficult if not impossible. As for the cables themselves, most of the world's privately owned network—there were also several government-owned cables—was in the hands of the so-called Electra House group of British cable companies under the leadership of Sir John Denison-Pender. Financial connections among the cable companies, the Construction and Maintenance Company, and the sources of gutta-percha were close—so close that, to outside critics, it seemed like a single integrated system of ownership and control. British interests dominated the manufacture, laying, and financing of

[2] Amalgamation of the Slaby-Arco and Braun-Siemens-Halske companies was ordered by the Kaiser in 1903. The new company, Geselleschaft für Drahtlose Telegraphie, was commonly known as Telefunken.

cables. And in most cases they could count on the support of their government, sensitive as it was to the interdependence of communications, commerce, sea power, and imperial defense.[3]

The Electra House group had won a dominant position in long-distance submarine cables. Many thought that the rise of the Marconi Company would create a similar situation in long-distance radio. This was an unwelcome prospect to those who had looked to the new technology of wireless to create an alternative system to the British-controlled cables— a system that could be under national control, that would not be cut in the event of war by the exercise of British naval power, that could offer sustainable communications with distant colonies or neutral countries no matter who controlled the cable relay points. Marconi dominance of long-distance radio would negate these possibilities. Such considerations underlay the German hostility to the Marconi Company and German governmental support for Telefunken. They found their physical embodiment in the tall antennas at Nauen, outside Berlin, and in the high-powered stations that German engineers were building at Tuckerton in New Jersey and Sayville on Long Island.

In the United States, hostility to the Marconi Company was particularly strong in the Navy Department, though not confined to that organization. The Navy's involvement in radio stemmed from the fact that the major use of the new technology in its early years was for communication with and between ships at sea. The implications for the control of fleet movements and tactical maneuvering were obvious, though often resisted by line officers, accustomed to substantial freedom of decision-making once out of sight of land. Institutional recognition of the Navy's role came in 1904 when President Theodore Roosevelt, in an attempt to end bureaucratic infighting, appointed a board to study the radio activities of the various departments of government and make recommendations for their development. The board reported that the Navy's interest in radio was paramount and, while recognizing the Army's right to operate stations in military areas, recommended that other government departments transfer their radio installations to the Navy. This report, though never endorsed by Congress, was taken by the Navy as a mandate to exercise a general supervision over radio communications in the United States, and this conception of its role was little changed by passage of the Radio Act of 1912, which vested the general regulation of nongovernmental radio activities in the Department of Commerce. To the extent that the United

[3] On the British role in the submarine cable system, see especially Vary T. Coates and Bernard Finn, *A Retrospective Technology Assessment: Submarine Telegraphy* (San Francisco, 1979).

States government can be said to have had a radio policy in these years, it was a policy enunciated by the Navy Department.[4]

That policy, as stated by Woodrow Wilson's secretary of the navy, Josephus Daniels, was simple and unambiguous: it called for government ownership. And since the Navy was the government department with the greatest experience in radio matters and operated its own extensive system of coastal and long-distance stations, government ownership meant in effect ownership by the Navy. Daniels's advisers furnished him with a variety of arguments to support this position: the necessity for a "single hand" to allocate frequencies and minimize interference; the weakness of divided control when dealing with the telegraph administrations of foreign countries; and the probability that a naval monopoly of radio would prove to be "a good business proposition."[5] But Daniels's conception of what was at stake did not depend on fine-spun arguments. His ideological background—that of a southern Populist, an agrarian radical, a Bryan Democrat—predisposed him in favor of government ownership. As he saw it, the government, representing all the people, should have a monopoly of radio communication just as it had a monopoly of the mails.[6] The most succinct statement of his position was probably that entered in his diary after a meeting of the Council of National Defense in October 1917: "Logic is: Communication is a governmental function and government must own, control radio, waterways, telegraph and telephone."[7] Repeated attempts to persuade Congress of this logic and pass the necessary legislation were unavailing, but Daniels himself never retreated from that position.

Between the Marconi Company and the Navy Department there had been a long history of distrust. There were particular reasons for this. Until 1906, for example, the Marconi Company was willing only to lease its equipment, not sell it outright, and this was incompatible with Navy procurement policy. When Marconi would sell apparatus, it preferred to sell its system as a whole, not particular items that the Navy might then use, in conjunction with equipment from other suppliers, in its own

[4] The best recent analysis of the Navy's role in wireless in these years is Susan Douglas, "Exploring Pathways in the Ether" (Ph.D. diss., Brown University, 1979).

[5] "Government Control of Radio Communication," *Hearings* before The Committee on the Merchant Marine and Fisheries on H.R. 13159, A Bill to Further Regulate Radio Communication, 12, 13, 17, 18, and 19 December 1918 (Washington, D.C., 1919), pp. 5-27.

[6] *Hearings* on H.R. 13159, p. 26.

[7] Josephus Daniels, *The Wilson Era; Years of War and After* (Chapel Hill, N.C., 1946), p. 106.

composite system. And it refused to permit intercommunication between its stations and those of other systems until compelled to do so by the International Wireless Conference of 1906. Beyond these particular sources of friction, however, there was a profound suspicion of what the Navy perceived as the Marconi drive for monopoly. If there had to be a single entity controlling the use of the radio spectrum, as seemed probably necessary if interference were to be avoided, that entity should be the Navy and not a private company—particularly not one controlled by foreigners.

Opposition to the Marconi Company was to that extent part and parcel of the Navy's drive to control radio. But there was also a good measure of Anglophobia. Dislike of depending on foreigners did not prevent the Navy from purchasing large quantities of German-built equipment. Nor did it lead the Navy to adopt a policy of building up American equipment manufacturers or American operating companies. Greater consistency in purchasing and testing, and more respect for patent rights, might well have saved NESCO from disaster and helped it to develop into a strong American-based and American-owned radio company. Even de Forest's Wireless Telegraph Company, despite its shady financial practices, might have acquired respectability and economic viability with Navy support. But, just as the Navy's chain of coastal radio stations (which made no charge for their services) diverted marine radio traffic from private operating companies, so did the Navy's practice of spreading its purchases of radio equipment over a variety of small-scale suppliers prevent any one of them from achieving financial strength and stability. The major exception was the Federal Company, but Navy purchases from Federal did not begin until after 1912.

In a different political and social context the Navy might have nurtured the development in the United States of a private company that could have challenged the Marconi organization on its own ground, in marine and transoceanic radio, as the German government did with Telefunken. But this did not happen. Instead the Navy allowed itself to become mired in repetitive struggles to induce Congress to endorse government ownership—a program which, in the face of vigorous opposition, led nowhere. This did not make it easier to limit the growing role that the Marconi Company was coming to play in American communications. On the one hand government ownership, which could have been used to freeze out the Marconi Company, was politically unacceptable. On the other, there existed no private communications company in the United States competent to offer the Marconi Company effective competition.

This was true in 1914, when Europe went to war; it was still true in 1918, when the killing in Europe stopped. The tensions were unresolved.

The Marconi Company in 1918 desperately needed access to a high-powered continuous wave transmitter, both to modernize its own long-distance system and to place it in a position to bid for the contracts of the "Imperial Chain," Britain's long-delayed scheme to link the colonies and dominions with the mother country by radio. In the United States there was a new determination, by no means confined to the Navy, that the country had to take control of its external communications into its own hands. Experience with British censorship and interruption of submarine cable traffic during the war had not been pleasant. Radio, in contrast, had given American authorities direct communications with their armed forces in Europe and, in the last stages of the war, with civilian populations and government authorities in the enemy nations. Radio technology offered the possibility of creating an international communications system in which the United States would play a pivotal and not a marginal role. To convert that possibility into reality, however, called for two kinds of action: first, to prevent or at least impede Marconi access to American radio technology; and second, to create some entity, governmental or private, that would accept responsibility for protecting the American national interest in telecommunications.

* * *

When, in 1919, the victorious powers turned to the task of rebuilding the shattered world economy, the reconstruction of its communications system necessarily appeared on the agenda. The new technology of radio, however, was given little attention. It was the submarine cables that created most difficulty. This is at first sight surprising. What was difficult about repairing cables? It was not, however, the physical problems of repair that caused the trouble. What turned the submarine cables into a source of acrimonious debate at the Paris Peace Conference was a question of property rights. And underlying that question was the larger issue of how and by whom the international movement of information was to be controlled. The arguments at Paris over how to allocate the captured German cables made clear the distrust with which British hegemony over international electrical communications was viewed. And the attitudes and assumptions that underlay the American negotiating position were precisely those that were to determine American communications policy in the postwar years.

Wartime measures to disrupt German communications created the problem. The cables that, before 1914, linked Germany and Austria-Hungary to the outside world had been cut, most of them within a few hours of the declaration of war. Two of these cables had started from

Emden, continued from there to the Azores, and terminated on Long Island. These were both cut in the English Channel. One of them was landed at Penzance in England, the other at Brest in France.[8] In the latter part of 1916 the first of these cables was again cut by the British at a point about 600 miles east of New York and a new section was spliced on that ran to Halifax, Nova Scotia. The second cable, taken over by the French government, was also cut east of New York and its landing point transferred from Far Rockaway to Coney Island. By these actions Germany was deprived of direct cable communications with the United States.

In the Pacific, too, a German cable running from Shanghai to Yap and from there to Menado in the Celebes, together with a short cable between Yap and Guam, had been taken over by the Japanese. These had been vital links in German communications with China, Southeast Asia, and Australia. A German cable network in the South Atlantic was left undisturbed but its connections with Germany via the Canary Islands were severed. For a period early in the war Germany was able to communicate with the Western Hemisphere through a single cable between Monrovia in West Africa and Pernambuco, Brazil, with radio providing the connection between Berlin and Monrovia. This link was broken when a short section of the cable was cut and removed by the French Navy.[9] With this step the cable connections that had formerly linked the Central Powers with Africa, the Americas, and the Far East were finally severed.

The situation facing the delegations to the Paris Peace Conference, accordingly, was that extensive sections of the prewar German cable network had been expropriated by the Allies. The immediate issues were two: whether the cables that had been seized should be returned to their former owners; and, if not, how they should be divided up among the victorious powers.

These questions might conceivably have been decided by appeal to precedents in international law, if there had not lurked behind them issues that were much harder to deal with. From the American point of view,

[8] United States Congress, 66th Congress, 3rd Session, *Hearings* before a Subcommittee of the Committee on Interstate Commerce on S. 4301, A Bill to Prevent the Unauthorized Landing of Submarine Cables in the United States (1921) [referred to hereafter as *Cable Landing Hearings*], testimony of Undersecretary of State Norman H. Davis. There is some question as to whether the second cable was actually landed in France. It was not used for traffic during the war.

[9] United States, Department of State, *Papers Relating to the Foreign Relations of the United States: The Paris Peace Conference, 1919* [referred to hereafter as *Peace Conf. Papers*], 13 vols. (Washington, D.C., 1943), 4:647-48, statement of Admiral de Bon.

prime among these was distrust of British intentions and fear of the revival of British economic competition. British interests, as we have seen, owned the greater part of the world's submarine cable system and exercised great influence over the financing and construction of extensions to that system. Further, for historical reasons, the submarine cable network had by 1914 come to be laid out in such a pattern as to give to Britain itself and to certain of its overseas possessions the status of nodal points in the system. The network had, by that date, been extended so as to link all the major metropolitan centers of the world by rapid telegraphic communication. Key landing sites and relay points in that network, however, were under British control. British cable companies in many areas enjoyed exclusive landing rights, effectively preventing the entry of new and competitive cables. This state of affairs reflected the way the system had come into being. British capital and engineering had played key roles in building the system, and it had been built in such a way as to serve British commercial and strategic interests. The cables were laid where there was likely to be important traffic, and once laid they tended to generate important traffic, as business and military systems adjusted to take advantage of their existence. On the critically important route between North America and Europe, the crude fact of Britain's geographical position set her astride the flow of traffic. On a lesser scale the same situation existed with other nodal points. The tiny island of Yap in the Pacific, for example, acquired large importance in the peace negotiations because it was a major landing site and relay point for transpacific cables. The island had been occupied by the Japanese during the war. Any power that held Yap could intercept transpacific cable traffic and, if need be, cut the cable at will. Were the other powers content that Japan should have that ability?

Why was the question of who controlled cable landing sites important? Control of the cable itself surely did not necessarily mean control over the information that passed over the cable. Most of the time and in normal circumstances it did not. But circumstances were not always normal. Whoever controlled a cable did have preferential access to information moving over that cable and was in a position either to make use of it directly or to make it available to third parties. Control over the landing site or relay point, furthermore, meant that in time of stress or conflict the party in control could delay the transmission of information, or prevent its transmission entirely, or so distort its meaning by "errors in transmission" as to render it useless.

A submarine telegraph cable, in short, was not a neutral carrier. Control of the cable and of its termini conveyed an ability to control the movement of information—a control that in the vast majority of instances might never be exercised at all but which in particular cases or at particular moments could be of vital importance to the security of property or of nations. Considerations of this kind help to explain the widespread use of private cyphers. They explain, too, the sensitivity of governments to the question of where cables were landed and who controlled the relay points. Control of these nodal points meant the ability to interrupt, delay, or distort the movement of information through the system. An hour's delay in transmission, the omission or distortion of a word or a phrase, even leakage of the fact that a message had passed—these things could affect the fate of empires as well as of private fortunes.

Similar problems had arisen earlier in connection with written messages and institutional means had been developed to deal with them, at least in time of peace. That was the point of the Universal Postal Union and the conventions governing the transmission of letters and parcels. Long-distance radio was to present still another set of problems, but those were for the future. The issue that the First World War had brought forcefully to the attention of everyone concerned with long-distance communication—diplomats, naval and military staffs, propaganda agencies, private businessmen—was the vulnerability of the cable network around which they had built their systems of control. The submarine cables had made rapid transoceanic communication possible for the first time, but the price of that gain was a new kind of risk. A political, military, or business system built up around the existence of cables could not survive their sudden removal; and its security was jeopardized to the degree that its communications links were controlled by a competitive or potentially hostile power. Hence the new importance of cable landing sites and the unexpected acrimony that developed over the expropriated German cables. And hence too the new interest in long-distance radio, a technology that promised a measure of protection against the risks inherent in the cables and some relief from British control.

Britain had made effective use of her control of the submarine cables during the war. Germany was, in effect, subjected to a cable blockade. Loss of access to the New York money market meant loss of the ability to sell securities to raise foreign currency. Communications with German diplomats and intelligence agents depended either on the radio transmitter at Nauen or on circuitous cable routes, as for instance through Sweden. These alternatives were subject to British interception. The most dramatic example of this was the Zimmerman telegram. President Wilson, hoping to act as mediator between the belligerents, made American diplomatic cable facilities available to Germany in 1917 for communication with her legations and embassies abroad. Germany used this facility, in a coded telegram to its legations in Washington and Mexico, to propose an alliance with Mexico, offering the return of Arizona, New Mexico, and Texas to Mexico as quid pro quo. Intercepted by the British, published in the New York *Times*, and acknowledged as authentic by Germany, the telegram did much to bring the United States into the war.[10]

But in many less conspicuous ways British control of the cables facilitated the Allied war effort and impeded that of the Central Powers. Nor was this control limited to military or diplomatic matters. Interception of private commercial traffic provided information on neutral shipments and increased the efficiency of the British naval blockade. Attempts to corner the market in scarce commodities could be frustrated, with large gains to the British Treasury. Above all, Britain could control the flow of news from Europe to North America, and although this fell far short of an ability to manipulate American public opinion at will, over the long haul it had an undeniable effect on American sentiment.[11]

To what extent British authorities had intercepted American telegraph traffic, both before and after the United States entered the war, is not entirely clear. In 1921 Newcomb Carlton, president of Western Union, was questioned closely on the subject by a subcommittee of the U.S. Senate. Had Western Union turned over American government and commercial messages to British naval intelligence? Had the British government censored those messages? Had their contents ever been released to private parties? Carlton was reluctant to answer—"It puts my company in a very embarrassing position with the British government"—but admitted that Western Union had indeed, under pressure, physically turned

---

[10] Coates and Finn, *Submarine Telegraphy*, pp. 111-12.

[11] Interference with American overseas communications was not confined to the period of American neutrality and affected outgoing as well as incoming information. In March 1918 George Creel, head of the U.S. Committee on Public Information, complained to the military attaché in London that publications consigned to neutral countries and already approved by the U.S. Censorship Board were being thrown out of the mails by British censors; in the same month he complained to the State Department that, according to the U.S. minister at The Hague, President Wilson's speeches and messages, relayed through London, arrived so late and in such mutilated conditions that the newspapers could not use them. The London representative of the Creel Committee reported that U.S. propaganda material sent by radio ended up in the custody of the British Ministry of Information, which distributed or rejected material as it saw fit. The Creel Committee found it very difficult to get U.S. propaganda inserted in British newspapers and in the press of some neutral countries and held British censorship largely responsible. See James R. Mock and Cedric Larson, *Words that Won the War: The Story of The Committee on Public Information: 1917-1919* (Princeton, 1939).

over all telegraph messages entering and leaving Britain to the British authorities. Indeed, he believed that the company had no option but to do so, under the terms of its British cable landing license. He had been assured, however, that they would not be censored or even deciphered and, challenged by the British government to cite a single instance in which information had been improperly released, Western Union had been unable to do so. "We have investigated and are satisfied that during that period not a single message, commercial, diplomatic, or otherwise, has been actually handled by the Naval Intelligence Bureau, and that their contents are unknown to the British government because of that fact."[12] Carlton did, however, describe for the senators the standard procedure by which British naval intelligence, as a routine matter, picked up the sacks of in-and-out cables every day at the Western Union offices, stored them overnight in an Admiralty warehouse, and returned them unopened the following morning—all this, he was assured, only "as a matter of form," to avoid the appearance of discriminating against European countries whose telegraph traffic naval intelligence did wish to examine.

Whether Carlton himself was naive enough to believe this, or thought the senators were, is hard to decide. Others, however, had their suspicions less easily satisfied. Capt. F. K. Hill of the U.S. Navy, who had been naval attaché at Rio de Janeiro between 1917 and 1920, told the same subcommittee that he had received many complaints from American merchants that their messages had been delayed or turned over to British businessmen, and he presented a number of specific cases to the subcommittee in circumstantial detail. The Department of State received many reports that all cablegrams relating to trade with Scandinavian and South American countries passing through Britain were sent by the British censors to the Board of Trade and either detained or allowed to go forward as the board directed; and it was alleged that information gathered in this way was distributed by the Board of Trade among private British firms according to their special interests.[13] These allegations were vehemently denied by the British authorities, but dissatisfaction and resentment remained strong: delays and censorship seemed to go far beyond what military or naval security required. As was pointed out by Walter S. Rogers, Wilson's advisor on communications policy, a guarantee that the contents of a message would not be leaked, even if such a guarantee was credible, was hardly enough. "The complaint goes beyond the actual

[12] Cable Landing Hearings, testimony of Newcomb Carlton.
[13] Robert Lansing, War Memoirs (Indianapolis and New York, 1935), pp. 124-25.

use of the contents of a message. There is a distinct commercial advantage to a third party to know that two people are negotiating. The fact that messages are being exchanged between any two commercial houses is an intimation that trade is going on, so that the whole thing is very difficult to trace."[14]

Conclusive evidence that the British had made improper use of their ability to intercept traffic originating from a neutral power was, perhaps, hard to obtain. There was, however, a widespread and firm belief in the United States that they had done so. During the war this belief led the U.S. Shipping Board, for example, to devise its own cable code rather than rely on one of the usual commercial codes that the British could easily decipher.[15] After the war, distrust of British intentions, combined with a new appreciation of the power that control over international communications could bring, had a powerful influence on American policy toward radio.

✸   ✸   ✸

President Wilson received two memoranda on communications policy to guide him during the Paris Peace Conference. One was from his postmaster general, A. S. Burleson, whose organization had controlled the American domestic telegraph system during the period of hostilities. Burleson urged that the peace treaty should include a clause prohibiting any nation from granting exclusive cable landing rights in future. Existing agreements should be respected but in future "each country . . . will give to duly authorized corporations of other countries reciprocal rights to land and operate cables, erect [and] open stations, with the right and authority to arrange for and to operate lines exclusively for business in transit to interior or exterior points."[16] To support his case Burleson quoted from a recent issue of the London *Standard* which had argued against too speedy a relaxation of cable censorship on the ground that "the powerful weapon of cable control should not be wholly scrapped . . . the commercial interests of the country [Britain] demand that it should

[14] *Cable Landing Hearings*, testimony of Walter S. Rogers.

[15] Michael J. Hogan, *Informal Entente: The Private Structure of Cooperation in Anglo-American Economic Diplomacy, 1918-1928* (Columbia, Mo., 1977), p. 107.

[16] Cable, Burleson to Wilson (forwarded through Secretary Tumulty, 14 March 1919), in Ray Stannard Baker, *Woodrow Wilson and World Settlement*, 3 vols. (New York, 1922) 3: 425-26, quoted by permission of the copyright holder, Judith M. Macdonald.

in some form continue."[17] Britain realized the importance of the cables for world business, argued Burleson; it was time the United States did too. "The world system of international electric communication has been built up in order to connect the old world commercial centers with that world business. The United States is connected on one side only. A new system should be developed with the United States as a center." The goal to be sought was "an international comity" under which electrically conveyed information should have the same rights of free transit as people, ships, mails, and parcels.

The analogy between freedom of transit for telegraph traffic and freedom of navigation on the high seas was one to which Burleson could be sure Wilson would respond. So was his emphasis on equalization of trading opportunities. The key to Britain's domination of world cable communications was the system of exclusive landing rights; to prohibit them in the future would enable American capital and enterprise to compete on equal terms. The values implicit in this position were those Wilson had espoused in other connections, as for instance in antitrust policy.

The second and more detailed memorandum came from Walter S. Rogers, the "communications expert" attached to the American delegation. Dated 12 February 1919, it dealt both with the immediate problem of the German cables and with longer-range issues, such as the possibility of linking international control of cables and radio to the proposed League of Nations.

It was in a sense appropriate that Rogers should be Wilson's communications adviser in Paris, for the Wilson that most Europeans knew was largely a creation of Rogers and his staff. Since 1917 Rogers had been working for the Committee on Public Information, better known as the Creel Committee, the major U.S. wartime propaganda agency. Rogers had been the director of the Wireless-Cable Service, and in that capacity had been responsible for supplying the committee's network of foreign agents with U.S. news and propaganda. During 1918 and into the early months of 1919 Wilson's speeches and statements had provided much of the content for the CPI's foreign news releases; and if by the time of the Armistice (as the historians of the Creel Committee indicate) "the name of Woodrow Wilson and the general idea that he was a friend of peace, liberty, and democracy, were nearly as familiar in some of the remote places of the earth as they were in New York, St. Louis, or San Francisco," most of the credit belonged to Rogers and his writers. The adulation that greeted Wilson on his arrival in Europe and the impact

[17] Ibid., p. 426.

that his Fourteen Points had earlier made on the newspaper-reading public reflected the thoroughness with which they had done their job.[18]

The Creel Committee, in cooperation with U.S. naval and military intelligence, had been deeply involved in American censorship, but Rogers's section had little to do with that activity. Its responsibility had been foreign propaganda, and in carrying out that responsibility Rogers had become acutely aware of the fact that foreign censorship agencies, particularly in Britain, could frustrate his best efforts. Dependence on foreign communications facilities and foreign news agencies made direct access to public opinion at times impossible. Unlike Britain, with Reuters, and France, with Agence Havas, the United States in 1917 had no established news service of its own in Europe, South America, or the Far East. So one had to be improvised, not so much to gather information for American news media but rather to ensure that the story of the U.S. war effort and of the American conception of the postwar world should reach foreign news media in the form and volume that American authorities wished.

This was essentially what Rogers and his staff had been up to. It was partly a technical problem—finding communication channels that were available and that would do the job—and partly a semi-diplomatic one—dealing with foreign news agencies, wireless and telegraph administrations, and censorship bureaus. Since the submarine cables were congested with higher priority traffic, heavy reliance had been placed on radio, particularly on the capabilities of the high-power station at Tuckerton that the U.S. Navy had taken over from its German builders and re-equipped with a Federal arc. Daily news dispatches were sent out from Tuckerton to the French station at Lyons and, through the cooperation of Agence Havas and Agence Radio, relayed from there to Italy, Spain, Portugal, Holland, Switzerland, and other countries. The same dispatches were intercepted by Navy operators in Britain and passed on to the British press and eventually to Scandinavia. For the Pacific and the Far East, Navy operators in San Francisco sent news dispatches to Pearl Harbor and Guam; from that point they went by cable to China and Japan for distribution by the news agencies in those countries. The U.S. Navy station at Darien in the Canal Zone served as distribution point for Central America, while Rio de Janeiro, Buenos Aires, and other cities in South America were serviced by cables. The CPI bureau in Moscow depended mostly on interception of the Tuckerton–Lyons radio link for

[18] The quotation is from Mock and Larson, *Words that Won the War*, pp. 235-36.

its press material and instructions, supplemented by the transpacific cable, but communications with that office were always difficult.[19]

Roger's wartime experience gave him an intensive education in the control of information. His efforts and those of his staff had depended for their effectiveness, to the extent that they used the cables, on the cooperation of those who controlled the nodal points in the system, for at those points decisions were made as to what information should be relayed and what should not, what would be passed on to other media such as the newspapers and what would not, and what priorities should be given to different classes of traffic. That cooperation had not always been forthcoming, and Rogers had learned how helpless his agency was without it. He had learned, too, both the potentials and the limitations of long-distance radio; first to provide communications channels independent of the overloaded cables, but more importantly to move information to points that could not be reached by cable. The great technical weakness of "wireless" had always been thought to be the fact that its messages could not be kept secret; once transmitted from an antenna, they were broadcast indiscriminately to all who could receive them. The cables, in contrast, seemed to offer secrecy: interception of cable traffic required access to the cable itself. Roger's experience with the American propaganda effort showed him that the conventional virtues and limitations could be turned upside down. The great weakness of the cable as a propaganda medium lay in the fact that it was a point-to-point carrier; information passed only with the consent and cooperation of those who controlled the termini. The great potential of radio lay precisely in the fact that the information was sent "broadcast." Knowledge of Wilson's Fourteen Points reached European news media by radio, not by cable. Negotiations for an armistice were initiated not by cable telegrams, for in 1918 there was no cable to Germany, but by a call-up of the German radio station at Nauen. German radio transmissions had been monitored as a matter of course by the Allies all through the war; and the operators at Nauen had likewise been monitoring Allied transmissions. To establish communication was easy: a single call to POZ (Nauen's call sign) did the job.

Roger's advice to Wilson reflected this experience. His major theme was that the idea of a League of Nations implied not only "a central organization with power," but behind that "a world of people acquainted with each other." This called for elimination of the barriers to the flow of information. Technical development of cable and radio would help, but there was also a need for statesmanship. "Fraught with danger is a

---

[19] For a more detailed account, see ibid., pp. 239-41.

situation in which the commerce of some nations languishes through lack of means of communication, while the commerce of others is subventioned [sic] through control of communication facilities. And there must be direct, unhampered communication lest suspicion lurk that intermediaries profit by trade information passing through their hands."[20]

Getting down to particulars, Rogers argued that each nation should nationalize its radio facilities and cooperate to develop "a truly worldwide radio service." There was little reason to think that radio would in the foreseeable future render the cables obsolete. But there was a distinct danger that commercial exploitation of radio and "hit-or-miss competition" might cause capital to hesitate from financing extensions to the cable network. Government ownership of radio would eliminate that possibility. Desirable extensions to the cable system could be encouraged while at the same time provision could be made for the construction of radio stations even in far-distant countries "quite apart from possibilities of financial gain." Radio and the cable had each its own sphere: they were complementary and should act as feeders for each other.

Implicit here was the idea of a worldwide communications system in which the United States would play a much more central role than it had in the past. Cable extensions would to some extent reduce dependence on the British-owned system. Radio, however, offered a revolutionary new capability. The danger was that competition between the two modes would hinder the orderly development of each; hence Rogers's argument for government ownership of radio. The new technology had destructive as well as constructive potentials: to maximize the second while minimizing the first called for management by government.

Nationalization of radio facilities would remove one obstacle to the extension of the cable system. But more than that was necessary. It had been suggested, said Rogers, that the important cables of the world should be internationalized and placed under the control of the proposed League of Nations. But there were political difficulties to that idea, as well as a host of administrative and financial problems. Better to proceed along established lines, by reforming the system as it existed. Exclusive cable landing rights, for example, should no longer be tolerated; cable messages in transit through a country should not be subject to inspection in peacetime; unfair practices such as rebates and discriminatory tariffs should be prohibited; and other familiar abuses should be rectified. Finally, "the

[20] Baker, *Wilson*, 3:427-42, Document 63, memorandum on cables and radio, submitted to President Wilson by Walter S. Rogers, communications expert of the American Commission to Negotiate Peace, on February 12th, 1918. Quoted by permission of the copyright holder, Judith M. MacDonald.

great nations should commit themselves to encourage extensions and technical improvements."

Roger's suggestions echoed a well-established State Department policy favoring greater American ownership and control of submarine cables. But they also reflected the concerns of the large American telegraph companies, such as Western Union, and the pressure of American news media. Long-distance radio threatened investments in the submarine cables. Radio should therefore be brought under government ownership so that development of both cables and radio might proceed in an orderly manner. Developed, as they should be, as complementary and not competitive communications services, cables and radio could together give American media direct access to world sources of news and world markets for news. The opportunity existed, in short, to create for the first time a communications system that would be American-based and American-controlled but that nevertheless could be worldwide in its coverage.

What role could the German cables play in these plans? Clearly, said Rogers, they could serve as bargaining counters to extract concessions on other issues. But they were also communications facilities in their own right. To leave them in the possession of Britain and France would, he argued, not only penalize Germany but also injure the United States. The French cables, poor in quality and inefficiently operated, were not an important factor. All other cable communications between the United States and Europe now passed through Britain. That, for the United States, was the heart of the matter. American national interests were vitally involved. Those interests called for using the captured German cables to reduce dependence on the British-owned system. To return them to Germany was one possible solution, to place them under international control another. Both presented problems. But whatever was done, the cables should not be left in the possession of the nations that had seized them.

✷　✷　✷

Rogers and Burleson, in their advice to Wilson, could write bluntly about using the German cables to advance American national interests, but at the Peace Conference itself the issue had to be approached, initially, in more legalistic terms. Discussion began at a meeting of the Council of Ten on 6 March 1919.[21] Wilson was absent in the United States at the time; the American position was argued by Secretary of State Robert Lansing, supported by the chief of naval operations, Adm. W. S. Benson.

[21] *Peace Conf. Papers*, 4:227-29.

The draft terms of peace had included a clause stating that the German-owned cables should not be returned to Germany.[22] What was the legal basis for this, inquired Lansing? Were the cables to be taken over as an indemnity due from Germany or on some other grounds? Were they subject to capture and permanent retention in the same way as ships of war? That question had never been decided in international law. The cables could have been taken out of the ocean and destroyed, but that had not in fact been done. They had been left *in situ* and confiscated. Was this proper? The cables might be taken as an indemnity by agreement, but they could not be seized as a capture of war.

Balfour, for Great Britain, pointed out that there were really two questions to be answered. First, was it in accordance with international law that the captured cables should be taken permanently from Germany? And second, if the answer to the first question was in the affirmative, what should be done with them? If they were not to be returned to Germany, who should have them? Further discussion, he thought, was unlikely to be fruitful until the legal issues had been discussed by experts, and he suggested that they be referred to a special committee. This was agreed to, and the council moved on to other business.[23]

Lansing had at least made clear the American determination that title to the German cables should not pass without question to the nations that had seized them; but the comments of other delegates, particularly those of France and Japan, made it no less apparent that the nations which now held the cables had no intention of returning them to Germany. Referral to a special committee at least got the issue off the agenda. It did not, however, bring it closer to resolution. Three questions were put to the committee. Was it right under the principles of international law to treat enemy cables in the same way as enemy ships of war? That is, could they be captured or taken as prizes? Second, was it right for a government whose naval forces had seized an enemy cable to retain it as reparations? And lastly, if a cut or captured cable were diverted and landed in the territory of another nation, what powers did that nation have to control its use in the future?

Even on these narrowly defined legal questions, the jurists who staffed the special committee were unable to reach complete agreement. All members agreed that military necessity was a justification for the cutting of enemy cables (a question they had not been asked to address). And all were of the opinion that the answer to the third question depended

[22] Ibid., 3:941-42, Naval Clauses for Insertion in the Preliminary Peace Terms with Germany.
[23] Ibid., 4:226-28.

on the particular terms of the contract between the owner of the cable and the government of the territory in which it was landed. But on the second question there was disagreement. The British, French, and Japanese representatives held that the capture and confiscation of enemy cables were legally justified by "the general principle of the rights of capture of enemy property at sea." The United States and Italian representatives, on the other hand, held that this opinion was "not well founded" in international law. The property in enemy cables, they held, could not be "assimilated to property subject to capture at sea."[24]

With this inconclusive report before them, the Council of Ten again took up the issue on 24 March. By this time Wilson had returned to Paris. His presence, together with the fact that a committee of jurists had been unable to resolve the problem, resulted in a broadening of the discussion. The policy questions had to be faced squarely. Should the cables be returned to Germany? If not, who was to get them?

Wilson almost certainly intended to argue for some form of international trusteeship. Balfour, who began the discussion, tried to eliminate that line of argument at the outset. There were two questions, he suggested, one of which was relevant to peace with Germany while the other was not. The latter question was whether "world-arrangements" should be made for the regulation of submarine cables. That was an important question but it could well be postponed. The immediate question was whether Germany had any right over cables that had been cut or diverted. Had Germany any right to complain? Personally he felt sure Germany had not. The Allied governments had the right to appropriate cables in exactly the same manner as ships captured at sea. This was the issue that should be discussed; the question of general international regulation of cables could be postponed until "a more favourable occasion."[25]

This was unacceptable to the American delegation. The worst possible outcome would be one that left the former German cables in the possession of the British (and, to a lesser degree, the French and Japanese). Better than that would be to return the cables to Germany. But better still would be a form of international trusteeship that gave American interests broader scope. Balfour's position, if generally accepted, would give the quietus to these hopes. If the cables were considered as prizes of war, chances of prying them loose from British, French, or Japanese control were slim indeed. And the prospect of an international conference on the subject, after the peace treaty was signed, was not much of a consolation prize.

[24] Ibid., 4:460-61, Minutes of the Council of Ten, 24 March 1919.
[25] Ibid., 4:461, Minutes of the Council of Ten, 24 March 1919.

Wilson's position was a difficult one. Much had happened during his absence in the United States. Concessions had been made, by Colonel House, his principal aide, among others, that caused him deep anxiety. Relations with Lansing, his secretary of state, were strained. And his health was poor. "Anxious, confused, exhausted, ill, solitary"—so he is described by one recent historian.[26] Ray Stannard Baker, probably his most sympathetic biographer, suggests that Wilson had not yet had time, in the last week of March, to give the problem of the cables a thorough examination. And it would be understandable if, amid the press of other business, he found it difficult to consider the disposal of submarine cables an issue of the first importance. On the face of it, it was just one more example of that bickering over the spoils of war that threatened his hopes for a just peace.

Wilson labored under three difficulties: first, the strong British contention that seized cables should be treated, legally, just like seized ships; second, the fact that Britain, France, and Japan were already in secure possession of the cables and could not easily be dislodged; and third, on one critical point, a misunderstanding of cable technology. The first difficulty was serious enough. Submarine cable technology was far from new in 1919, but international law had not yet come to grips with its distinctive characteristics. The French, British, and Japanese delegations wished to assimilate cables to the status of prizes of war: like enemy ships, they had been captured and were now the property of the Allied powers. The American and Italian representatives refused to accept that definition of the situation but were unable to suggest an alternative.

On 24 March, when the discussion reopened, Lansing tried once more. Cables, he argued, were not like ships. There was a "very great difference" between the capture of ships at sea and the seizure of cables. Cables were "attached to a submarine region" that was not in the sovereignty of any nation. The cutting of cables was merely an expedient of war and it was wrong that such a cable should continue in the possession of the nation that cut it, after hostilities were over. "The basis of capture on the high seas was that the ship could be brought within the jurisdiction of the captor, where it could be reduced to possession. This could not be done with cables."[27]

What Lansing was advocating was, however, a legal innovation. Cables, he implied, were a special kind of property. They should be treated

[26] Kenneth S. Davis, *FDR: The Beckoning of Destiny, 1882-1928: A History* (New York, 1972), p. 564.

[27] *Peace Conf. Papers*, 4:459 ff., Minutes of the Council of Ten, 24 March 1919.

differently from other kinds of property with which international law was familiar. The trouble was that there was no legal precedent for this point of view; and no matter how vigorously the Americans might argue that the cables had not been "reduced to possession," since they still lay in their original positions on the bed of the ocean, it was easier and simpler to argue that they had indeed been captured and that they had been "reduced to possession" in the only way that, for cables, made sense. It was hard for Lansing to avoid leaving the impression that he was laboring to make a distinction where in fact there was none; and his legal niceties could make little headway against (for example) Lloyd George's bluff assertion that the right to take cables was just as strong as the right to take ships. The one carried communications under the ocean, the other on top of it. "He agreed that cables had not heretofore formed the subject of capture; but there had never been a war of the same kind before."[28]

Even if international law had spoken clearly on the point, the American position was a difficult one to sustain. Both Lansing and Wilson originally argued as if the German cables had merely been cut and left lying *in situ* in what Wilson called "no man's land," on the bottom of the ocean. In the case of the two North Atlantic cables, however, this was not the case: both had been diverted to new termini and integrated into new systems. The cable seized by the British was now an important link in communications with Canada, and any suggestion that it be returned to Germany and connected again to its New York terminus brought immediate objections from the Canadian representative, Sir Robert Borden. Much the same was true of the cables seized by Japan. It was not just a question of returning property that had been seized; it was a question of disrupting one communications system and rebuilding another.

Realization that the cables had not only been cut but also diverted seems to have dawned slowly on several members of the council. Balfour, on 24 March, was the first to call attention to the fact, pointing out that, although they had been discussing the cables as if they might be physically returned to Germany, this was an error. Whole sections of cable had been taken up and re-laid in new locations. The only way they could be returned to Germany in their *status quo ante bellum* was if the Allies were to re-lay them again, at their own expense. This, as the Italian delegate underlined, was hardly likely. To do so would not only be very expensive; it would also be "to admit that what had been done had not been right."

Wilson, even before this new element entered the discussion, had tried

[28] Ibid., 4:489, Minutes of the Council of Ten, 1 May 1919.

to broaden its scope. The cables were, he said, more than instrumentalities of war; they were also indispensable instruments of commerce, and their status as property had to be regarded "from a peace point of view." Steps should not be taken that might reduce their usefulness in reestablishing the ordinary course of trade. Further, the victorious powers expected Germany to pay heavy reparations. Germany could do this only by earning foreign exchange. The question of the cables, therefore, should also be looked at "from the German trade point of view." When the fact that the cables had been diverted was pointed out, he used the opportunity to revert to fundamentals. He was interested, he said, in seeing that there should be an entirely just peace, rather than that material advantage should accrue to any one country.

Arguments based on international law or on Wilson's high-principled hopes for a just peace must by this stage of the discussion have begun to seem somewhat irrelevant. Little help could be expected from either of those quarters. Wilson and Lansing seemed to be saying that the cables were still German property and should be returned to Germany. If that position could not be sustained, they were at a minimum unwilling to concede that those particular nations that had cut and diverted the cables should retain exclusive possesssion of them in peacetime. If they were indeed spoils of war, there was a question of how those spoils should be divided up. If that was the game, the United States wanted its share.

The question of allocation among the victors, however, could well be postponed until that later communications conference at which Balfour had already hinted. The immediate issue was the nature of the peace treaty with Germany; something had to be said in that treaty about Germany's claim to the cables. Balfour proposed a resolution that, on the face of it, seemed to come to grips with that question without prejudging other issues. It read: "The Treaty of Peace should not debar Germany from repairing at her own expense the submarine cables cut by Allied and Associated Powers during the war, nor from replacing at her expense any parts which have been cut out from such cables, or which without having been cut are now in use by any of those Powers." It was a shrewd resolution. Those present were clearly tired and frustrated by the complexities of the issue and anxious to move on to other business. When Baron Makino of Japan asked whether the resolution might not be interpreted to mean that all cables might be returned to Germany (which, as it turned out later, was precisely what the American representatives thought it meant), he was given short shrift by Clemenceau: the whole question was merely being referred to a Drafting Committee and would be reconsidered later. And this was true; the problems had not been resolved but merely shunted to a lower level.

The Drafting Committee did its best. By the last day of April the Council of Foreign Ministers had before it a draft that purported to convey the sense of Balfour's resolution. It stated that Germany was at liberty to repair at her own expense submarine cables that had merely been cut during the war and were *not* being utilized (hardly a controversial point); and that, in the case of cables that had been cut and later diverted, or that, without having been removed, were being used by any of the Allied or Associated Powers, Germany could replace them at her own expense. And, to make things clear, a complete list of the cables that had been removed or utilized by the Allies and therefore would *not* be returned to Germany was appended.[29]

Instant disagreement followed. To Lansing the draft did not carry out the purpose of Balfour's resolution at all. He had thought the resolution meant that the cables would be returned to Germany, subject to her making repairs at her own expense. Balfour could not agree. The general principle underlying his resolution had been that Germany might, at her own expense, restore her cables to their prewar state; but she could not make the Allies responsible for any damage done to them, nor could she ask them to restore any portion that had been removed, nor could she claim control of any cable set up by any of the Allies and composed of pieces of German cable. Very well, responded Lansing: if that were the principle, the question to be faced was one of allocating the spoils. The United States was not prepared to yield the line from New York to the Azores merely because the Allies had diverted it. As for the cable between Monrovia and Pernambuco, seized by the French, the United States would rather see it in German hands than ceded to any of the Allied powers.

That final remark, predictably, brought Admiral de Bon into the discussion, defending France's right to retain a cable that she had cut but not put to use. But the waters were already quite adequately muddied. Lansing and Balfour seemed to be in agreement on a general principle, but they were diametrically opposed on the practical implications to be drawn from that principle. Part of the problem arose from Lansing's insistence that German citizens still owned the cables.[30] And he refused to retreat from that position. Balfour and de Bon, however, continued to insist that the cables were spoils of war and that capture conferred title. The result was an impasse. The only action on which the foreign secretaries could agree was to refer the whole question once more to the heads of state. As if the situation were not already sufficiently complex, Lansing closed the discussion by giving notice of his intention to raise a

[29] Ibid., 4:645-55, meeting of the Council of Foreign Ministers, 30 April 1919.
[30] Ibid., 4:651.

question about the possible internationalization of the island of Yap—a move that triggered an immediate response from Baron Makino, the Japanese foreign minister. Japan, he warned, would have "a good deal to say" on that question.

The disagreement was acute—more acute, perhaps, than the participants had expected. Balfour's resolution of 24 March had seemed to catch the "sense of the meeting." Yet now all was discord again. Why? Part of the reason was that certain of the fundamental issues were not being faced directly, as perhaps they could not be. Lansing's insistence that the cables were not spoils of war might seem a rather forced legalism, but it was the only way to block Britain's clear intention to retain control of the cable diverted to Halifax and France's claim to the cable between New York and Brest. And the precedent set for the Atlantic cables would certainly apply to the Pacific also, where Japanese interests were at stake. Behind the veil of politely worded disagreement there were clear conflicts of national interest. And certainly everyone around the conference table understood this.

There was, however, something that they did not understand. This became embarrassingly evident when the Council of Ten convened on the following day. The first item on the agenda was the question of the cables. Balfour reviewed the history of the issue and explained his interpretation of his own resolution: all acts taken by the Allies in connection with the cables should stand, and Germany should have no claim to compensation; on the other hand, Germany would have a perfect right to reconstruct her cable system as it had existed before the war.[31] He asked the Council to imagine a German cable running from A to C through a point B. During the war the cable had been cut at B and connected with a new line running from B to D. To make the matter clear beyond all question, he drew a simple diagram:

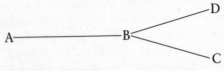

The American position was that the section A-B should now be restored to Germany, whereas the British, French, and Japanese representatives contended that, since A-B was now an essential part of the new line A-B-D, it should not be returned to Germany, but Germany should have a right to join up its old section B-C with a line laid from B to A.

[31] Ibid., 4:484, meeting of the Council of Ten, 1 May 1919.

Balfour's presentation was somewhat schoolmasterish but it did clarify the issue. Lloyd George at once wanted to know "whether the Germans would have the right to the use in common of the line A-B," and with that simple question one of the sources of misunderstanding became evident. No, replied Balfour: telegraph cables did not work that way. It was not possible for two separate systems to use the same cable. The cable from A to B had to be owned either by one of the Allies or by Germany. If it was returned to Germany, the Allies would have to spend a very large sum of money to lay a new cable between those points, or else the newly laid extension to point D would become useless. This would not be right: it was Germany who should bear the expense of reconstructing her lost cables: that is, of laying a new line from A to B. But, in any case, the section from A to B could not be shared: that was not technically possible.[32]

It was Wilson who was most embarrassed by disclosure of this arcane piece of technological lore, for it quickly became evident that the American negotiating position had been based on the assumption that the main line of the cable, which still lay on the ocean floor, could be shared. Clearly it was the cable from Emden to New York, now running from Land's End to Halifax, that was on everybody's mind. Wilson had assumed that all Germany would have to do to regain use of the cable was run new connections to this "main stem," by laying short sections of new cable from New York to the point of diversion at the western end, and from the English Channel to Emden at the eastern one. This was the point of his insistence that the cables that still lay on the ocean floor had not been "captured." He had assumed that it could still be used in common by Germany and by the Allies—thinking of it, perhaps, as a kind of highway over which more than one nation might send its vehicles. Now he was reduced to apologies. He had entered the discussion, he said, with "an unfortunate ignorance of technical details." He had agreed to the Balfour resolution only because he had been under "an erroneous impression, which was entirely his own fault." What he had assented to was not what, at that time, he had thought he was assenting to. But "the error was due to his ignorance."[33]

[32] Multiplexing techniques for passing several different messages over a single cable simultaneously were developed in the 1920s but were not available in 1919. In 1921 Western Union's proposed new line to Italy was to use the newly discovered "loaded" cable that allowed simultaneous transmission on four channels. See Hogan, *Informal Entente*, p. 119.

[33] *Peace Conf. Papers*, 4:485-86, meeting of the Council of Ten, 1 May 1919.

The apologies cannot have been easy for Wilson to make. The responsibility, of course, lay partly with those technical advisers who had allowed him to enter the negotiations in ignorance of what could and could not be done to re-lay cables or share their use. Balfour's position was not in fact correct. The captured cables could have been shared, if not by a technically sophisticated method like multiplexing, then by some simple arrangement such as permitting German authorities to use the cables between certain hours and British authorities at other times. Administratively tricky, perhaps, and demanding goodwill on both sides, but not impossible. But Wilson did not make that point; instead he retreated behind embarrassed confessions of ignorance. There had been much talk of international law, which in fact had nothing to contribute. There had been no talk at all about technology, which mattered a great deal. Now Wilson had to rebuild a negotiating position from scratch. The final objective remained the same: to prevent the captured and diverted cables from being used to reinforce British (and, in the Pacific, Japanese) control over international communications. But to insist that they be returned to Germany was no longer a feasible strategy.

Wilson's first reaction was to argue for some form of international control. Balfour, however, had already made the point that international regulation of submarine cables was a subject best left to a later conference—one that might deal with all forms of electrical communication, including radio. It was not immediately relevant to the drafting of a peace treaty with Germany. The immediate assignment was to draw up a clause for the treaty that would make it clear that the cables which the Allies had cut or diverted would not be returned to Germany. On that proposition it now appeared that all parties might agree. Even Wilson finally admitted that, if it were merely a question of literally returning the cables to Germany, the solution was easy, as the answer could only be in the negative.

And that was how it turned out. What was noteworthy about the discussion when the Council of Ten reconvened on 2 May was the way in which issues previously only hinted at were brought out into the open—particularly differences between the United States and Britain. Wilson introduced the draft resolution he had been asked to prepare. It began by listing the cables Germany would have to renounce—the same list as that to which Lansing had objected at the council of foreign ministers. It went on to state that the five Allied and Associated Powers should jointly hold these cables "for common agreement as to the best system of administration and control." And it closed by pledging the five powers to call, as soon as possible, an international congress to consider and report on all international aspects of telegraph, cable, and radio communication, "with a view to providing the entire world with adequate communication facilities on a fair, equitable basis."[34]

Only the second paragraph was controversial. Wilson defended it on grounds of basic principle. All the belligerents had a vital interest in the cables. All, therefore, should have a voice in their future control. It was not right to assign ownership of the cables to just one or two of the nations that had been partners in the war. They should be controlled by the five powers jointly. But Lloyd George would have none of this. He charged that the Atlantic cables were in fact "almost wholly in the hands of American monopolies" (meaning Western Union and Postal Telegraph). During the war Britain had captured one German cable and connected it with Canada. It was now the only Canadian state-owned line. President Wilson wanted to take that cable away from Canada and put it under international control. That was not the way to break monopolies. If America wanted to break American monopolies, the way to do it was to lay additional cables. That would cost between £700,000 and £800,000 for each cable, and whoever wanted to break monopolies had better be prepared to pay that sum. He failed to see the point of dispossessing Canada of something that had been legitimately captured in war and was now essential to her business success.[35]

Wilson protested against this reading of his intentions, but the game was already lost. And when Lloyd George proposed that they revert to a French proposal made the previous day—which was simply to inform Germany that the cables would not be returned but would remain the property of the Allied and Associated Powers—Wilson assented with what grace he could muster. His proposal, he said, did not differ in principle—a statement he must have found difficult to swallow. The only difference was that under his plan the cables would be vested in trustees during an interim period, until their final disposition could be determined. And so it was decided. All references to joint ownership and administration were struck out; and the crucial second paragraph was deleted, to be replaced by an innocuous amendment stating that the cables should "continue to be worked as at present without prejudice to any decision as to their future status."[36] A meeting on the following day tied up the loose ends. With minor changes Wilson's first paragraph, listing the cables that would not be returned to Germany, was approved for insertion in

[34] Ibid., 4: 493-94, meeting of the Council of Ten, 2 May 1919.
[35] Ibid., 4: 497-98.
[36] Ibid., 4: 499.

the peace treaty. Added to it was a clause stating that the value of those cables, less depreciation, would be credited to Germany in the reparations account. The second and third paragraphs were approved with merely verbal emendations.[37]

* * *

It had not been the most brilliant chapter in the history of American diplomacy. None of the objectives of the American delegation was achieved. Allocation of ownership rights to the captured cables was left for later diplomatic maneuvering. The idea of returning them to Germany was abandoned. And the broader policy objectives—nondiscriminatory pricing, an end to exclusive landing rights and so on—were left to a later conference.[38]

Did it make much difference in the long run? In the larger tragedy of the Versailles Treaty this was a minor matter. But even in the narrower field of international communications, the debate over the German cables was significant more for the attitudes it reflected than for its influence on the course of history. There were broader forces at work, compared with which the disposition of a handful of obsolescent cables was no more than a ripple on the tide of change. Some of these are worth noting. Consider, for example, some of the implicit assumptions that shaped American objectives.

In the first place, American attitudes and arguments during these discussions were dominated by a conception of Britain as controlling the world's submarine cable system, as if the fact of the matter were self-evident and as if everyone would agree on what "control" and "domination" meant in this context. Yet neither of these assumptions was valid. When Lansing argued that Britain should return the captured cable to Germany on the ground that she owned too many cables already, Balfour found it easy to retaliate by pointing out that the cable captured from Germany was in fact the *only* one that Britain controlled. Of the thirteen other transatlantic cables, each one was either owned by or leased to an American corporation. British interests, as Balfour saw it, had been frozen out by the powerful companies that controlled the land-line telegraph systems in the United States. Here were definitions of control and a

[37] Ibid., 5: 437-38, meeting of the Council of Four, 3 May 1919.
[38] For a summary of later developments, see G. H. Hackworth, *Digest of International Law*, 8 vols. (Washington, D.C., 1940-1944) 4: 272-79; Keith Clark, *International Communications: The American Attitude* (New York, 1931), pp. 123-68; and Hogan, *Informal Entente*, pp. 105-28.

perception of reality quite different from those Wilson, Lansing, and Rogers were using. Ownership of the transatlantic communications links, as RCA was later to discover in the age of radio, was an asset of limited value without control over the "feed" systems that distributed and collected traffic domestically.

Secondly, the American negotiators, as we can now see in retrospect, overestimated the strength of the British economy when hostilities ended. Their fears of a revival of British commercial imperialism, reasonable in terms of past history, were less reasonable in terms of Britain's limited ability to recover from the human and social costs of the war, to reequip her basic industries and find markets for them, and to rebuild the foreign investments that before 1914 had done so much to pay for Britain's imports. But few were thinking in those terms in 1919.

And thirdly, the American negotiators overestimated what was at stake in the disposition of the German cables. They allowed their resentment of British wartime censorship to color their expectations for the years of peace. Hence their insistence on communications with Europe that did not pass through Britain. But communications technology already offered an escape from that problem, as the antenna towers of Nauen, Bordeaux, Tuckerton, Sayville, and Arlington demonstrated. They spoke of Britain's domination of the prewar cable system and explained it in terms of who controlled the gutta-percha supply and who owned Telegraph Construction and Maintenance Limited. But what domination there was rested, at bottom, on Britain's lead in the technology of long-distance electrical communications. In 1919 the United States held that lead, not Britain. Technology would soon find substitutes for gutta-percha; American communications companies would prove fully capable of negotiating with their British counterparts as equals; and by the 1920s high-frequency radio would offer communications channels that, if they did not make cables obsolete, at least would serve to hold down cable rates and furnish direct access to points the cables did not reach.

* * *

Radio played no part in the arguments over the German cables. Wilson was the only one to refer to the subject, and he did so only to echo the conventional wisdom of the time. Wireless, he said, had not the same value as cables, since anyone could pick up wireless messages whereas cables possessed a certain degree of privacy, "depending on the good faith of the employees."[39] American postwar policy toward radio, how-

[39] *Peace Conf. Papers*, 4: 486, meeting of the Council of Ten, 1 May 1919.

ever, was to be decisively influenced by the very attitudes that had been displayed during the dispute over the German cables. Preconceptions formed in the context of the older technology were applied to the new. The result was that, in the immediate postwar years, the development of radio technology in the United States became politicized to a degree that had not been true before 1917.

What made this possible was the development of continuous wave radio. General Electric's alternators and the Federal Company's arcs gave the United States in 1919 a slight but nevertheless real margin of technical superiority in long-distance radio over its European rivals. The return of peace presented the opportunity of exploiting that margin of superiority in what American policy-makers took to be the national interest. There now existed the potential for building a truly reliable system of long-distance radio communications such as had not existed in the days of spark. And there also existed the possibility of carving out for the United States a role in that system more authoritative than it could ever have enjoyed with submarine cables. The new technology was there in 1919, proven in wartime and looking for markets in peace. And a new ideological context was there also, a context of nationalism and xenophobia more assertive than had been known before 1914. Technology provided the instruments by which nationalism could be made effective in action. Nationalism created the markets that the new technology required.

※   ※   ※

What radio could do in 1919 had been made clear to Wilson personally. During his voyages back and forth across the Atlantic he had kept in touch with Washington by radio; cables could never have offered that facility. And he had found time, during the course of the Paris negotiations, to set in motion a sequence of events that were to have a profound influence on American communications policy. This happened after a breakfast meeting with Lloyd George, in the course of which an aide brought the British premier a radio telegram, and some comment was passed about the probable importance of wireless in the postwar world. Wilson picked up the thought and sent word to his new director of naval communications, Adm. W.H.G. Bullard, then passing through Paris on his way to Washington, that he counted on him to keep a careful watch on American interests in radio. Conveyed to Bullard through Wilson's personal physician, Dr. Cary Grayson, this message was apparently not specific about what Wilson expected Bullard to do. But the fact that the President had intervened in the matter, that he believed there was a threat to American interests in radio, and that he expected the Navy to do something about it—these were to play an important role in later dealings between the Navy and General Electric.[40]

To the extent that the Wilson administration had a policy toward radio at this time, it was still the policy long advocated by Josephus Daniels: complete government ownership of both marine and long-distance stations, with private interests playing no role except in the design and manufacture of equipment. Not all of Daniels's staff officers supported this position; his new director of naval communications, for example, did not. But it was the official position of the department, and it was supported in principle by the Departments of State, Commerce, and Labor, by the Treasury, and by persons of influence such as Wilson's adviser, Rogers.

Chances that Daniels's objective might be achieved had never been great; they dwindled to the vanishing point in November 1918 when it became clear that the Republican Party would control the new Congress. But even had the elections turned out differently, the probability of success, if the issue had been presented to Congress as something requiring committee scrutiny, debate, and legislation, was not high. Wartime experience with government control of the telegraph system had left few people, whatever their party affiliation, enamored with the idea of government ownership of the means of communication. Legislation to give the Navy Department "exclusive ownership of all wireless communication for commercial purposes" had been introduced to Congress several times before 1917, and each time it had been tabled or allowed to die in committee. In the closing months of 1918 Daniels managed to get the hearings revived. His best efforts, backed up by testimony of other Navy spokesmen but vigorously opposed by commercial and amateur interests, were of no avail. On 16 January 1919 the House Committee on the Merchant Marine and Fisheries unanimously voted to table the bill (H.R. 13,159). With a more limited objective—naval control of the coastal radio system, for example—he might have been successful, but Daniels

[40] See below, Chapter Six; *Cable Landing Hearings* (1921), pp. 345-60, testimony of Owen D. Young; United States Senate, Committee on Interstate Commerce, 71st Congress, 1st Session, *Hearings* on S. 6, A Bill to Provide for the Regulation of the Transmission of Intelligence by Wire or Wireless [hereafter *FCC Hearings*] (Washington, D.C., 1936), pp. 1081-1220, testimony of Owen D. Young; W.H.G. Bullard, "Some Facts Connected with the Past and Present Radio Situation of the United States," *Proceedings* of the United States Naval Institute, 49 (September 1923), 1623-34; Owen D. Young, "Freedom of the Air," (as told to Mary Margaret McBride) *Saturday Evening Post*, 16 November 1929. There is no reference to this event in the diaries of Woodrow Wilson or of Cary Grayson, now in the custody of Professor Arthur Link of Princeton University.

wanted "the whole thing."[41] And most particularly he wanted the high-powered transoceanic stations. To return them to their original owners, as he saw it, would place them in the hands of foreign-controlled corporations, and that would allow foreigners to own "a military instrument within our borders."[42]

Direct legislative action by Congress, however, was not the only possible route to Daniels's objective. He could also hope to present Congress and the country with a *fait accompli*: a radio system already in Navy possession. In that event all that would be asked of the legislators would be acquiescence in the status quo and annual appropriations to operate a system the Navy already controlled. This was Daniels's "fall-back position," and in 1919 he was very close to achieving it. His sources of vulnerability were two: first, although the Navy controlled all radio stations in 1919, it did not own them all; and second, there were those in Congress still unwilling to see a Democratic secretary of the navy arrogate to the executive branch prerogatives that properly belonged to the legislative.

In 1914 there had been three important privately owned radio operating companies in the United States.[43] On the Pacific coast Federal Telegraph operated its service between San Francisco, Los Angeles, and Hawaii, with domestic links to Phoenix, Portland, and Seattle. In the Caribbean and Central America the United Fruit Company, through its subsidiary, Tropical Radio, operated an extensive network of land-based stations and relied on radio to control the movements of its famous "Great White Fleet." On both Pacific and Atlantic coasts the Marconi Wireless Telegraph Company of America operated a large number of shore stations for marine traffic; it provided most of the equipment and operators for radio installations in ships of American and British registry; and in addition it had begun construction, at New Brunswick and Belmar, New Jersey, of high-power stations to exchange traffic with sister stations at Towyn and Carnarvon in Wales. It also had just completed new sta-

[41] Stanford Hooper Papers (Library of Congress), Box 38, pp. 913 ff.: "There wasn't any excuse for the government owning it [the transoceanic business] at all . . . Mr. Daniels would never give in on that point; he wanted the whole thing."

[42] Hooper Papers, Box 1, survey of contributions of U.S. Navy to advances in radio, 1910-1923; author not identified but almost certainly Josephus Daniels.

[43] The U.S. Navy also handled commercial traffic under the provisions of the Radio Act of 1912, which permitted such operation in cases where a private commercial station within 100 miles of a naval station was not open for general public ship-to-shore service for twenty-four hours a day. The Navy made no charge for this service.

tions at Bolinas and Marshalls in California and at Kahuku and Koko Head in Hawaii, looking toward an eventual radio link to Japan, and at Marion and Chatham in Massachusetts for communication with Norway; these stations were still being tested when war intervened.

In addition the German Telefunken organization, through its American subsidiary, the Atlantic Communications Company, had constructed at Sayville, on the south shore of Long Island, N.Y., a high-powered station for communication with a sister station at Nauen, near Berlin. It would appear that the German authorities, fully aware that their cable network would be disrupted when war broke out, had counted on that link to maintain communications with the United States. Originally equipped with a Telefunken quenched spark transmitter and licensed by the Department of Commerce to operate with that equipment, Sayville in 1914 was reequipped with a 100 kilowatt Von Arco alternator and a more extensive antenna system. Using its spark transmitter, it had shown itself able to maintain twenty-four-hour communications with Nauen, though at slow speed. With its new alternator and antenna it was expected to do much better than that, but it was not yet offering a commercial service when war broke out in Europe in August 1914.[44]

A second high-power station intended for European communications was being built in 1914 at Tuckerton, New Jersey, by the German firm, Hochfrequenz-Maschinen Aktiengesellschaft für Drahtlose Telegraphie (usually known as HOMAG), nominally for the Compagnie Universelle de Télégraphie et Téléphonie sans Fil of France. Although completed just prior to the outbreak of war, it had never been turned over to French ownership, nor had it been granted a license to operate.[45] Potentially it was a powerful and useful station. Equipped with a Goldschmidt 100 kilowatt alternator and a large "umbrella" antenna, it should have been capable of reliable communications, given favorable propagation conditions, either with France or Germany.

This, then, was the private sector of the U.S. radio communications industry in 1914—the sector that, if Daniels was to attain his objective, he had to bring under Navy ownership and control. The Navy, of course, also had its own systems: a coastal system for marine communications, and a system of high-power stations for long-distance work. There were

[44] E. David Cronon, ed., *The Cabinet Diaries of Josephus Daniels, 1913-1921* (Lincoln, Neb., 1963), pp. 100-101.

[45] Call letters WGG were assigned to Tuckerton for purposes of testing, and these were listed in the U.S. call book for 1914, but no station owner nor wavelength was shown. See Thorn L. Mayes, "History of the Tuckerton Wireless Station" (Mimeo, 1979.)

forty-nine coastal stations in the Navy's system in 1914. The long-distance system at that time included, for transpacific work, stations at Cavite, Pearl Harbor, and San Diego; for coverage of Central America and the Caribbean, a station at Darien in the Canal Zone; and as "flagship" station for the whole system, a station at Arlington, Virginia, just outside Washington, D.C. These high-powered stations were all equipped with arc transmitters manufactured by the Federal Company.

When war broke out in Europe, President Wilson's overriding concern was for the preservation of American neutrality. To that end, on 5 August 1914 he prohibited the transmission of "unneutral messages" from radio stations within American jurisdiction and directed the Navy to enforce this policy.[46] The American Marconi Company, not without protest, opened its long-distance and coastal stations to monitoring by Navy inspectors. Most British telegraph traffic to and from the United States, of course, still passed via the submarine cables; use of radio for sensitive or confidential material was not necessary. The German-built stations at Tuckerton and Sayville presented more of a problem. Once the German submarine cables had been cut, these stations were indispensable links in German communications with the United States. It was hardly conceivable that they could both perform the functions for which they had been built and at the same time conform to Wilson's edict against unneutral messages.

Members of Daniels's staff, and in particular Lt. Stanford Hooper, of whom we shall hear more later, had expressed concern as early as 1913 about the ability of the strategically located Sayville station to monitor radio traffic between units of the American fleet. Recommendations that an officer be stationed permanently at Sayville as censor and that the station be required to use American operators exclusively had been rejected on the ground that there was no statutory provision allowing censorship in peacetime. The Radio Act of 1912, as it had passed the Senate, had included a provision requiring that all licensed operators of radio stations be American citizens, but this provision had been stricken by the House as being "a purely military feature of slight importance" and had not been restored in committee.[47] Now, in 1914, there was added to the Navy's earlier concerns an awareness of the ability of the

[46] Hooper Papers, Box 1, survey of contributions of U.S. Navy to advances in radio, 1910-1923; compare L. S. Howeth, *History of Communications-Electronics in the United States Navy* (Washington, D.C., 1963), p. 226.

[47] Hooper Papers, Box 1, Fleet Radio Officer, USS *Wyoming*, to Commander-in-Chief, 21 May 1913, on "Secrecy of Radio Communication"; and ibid., Bureau of Engineering, report, 2 June 1913.

Sayville station to monitor the movements of ships entering and leaving New York harbor and report that information to Berlin.

Authority to take over the Tuckerton and Sayville stations came in an Executive Order of 5 September 1914 that cited the desirability of taking precautions to insure that unneutral messages were not transmitted in secret code or cypher.[48] Five days later the Navy assumed control at Tuckerton and made the station available for communication, in plain language only, with shore stations in Europe. The only traffic actually exchanged was with Eilvese in Germany. Sayville proved somewhat harder to absorb, since it was nominally owned by an American-chartered corporation. The substitution of an alternator for the original quenched spark transmitter, however, provided the secretary of commerce with the necessary pretext to refuse an operating license, on the ground that the increase in power was equivalent to the construction of a new station and that "to grant a license for a new station erected since the war began with Germany, with German apparatus, avowedly under German ownership and control, communicating with stations known to be under the control of the Imperial German Government, would be an unneutral act." In July 1915, "in order not to leave the station idle," the Navy took over control of Sayville and put it into commercial service for traffic with Nauen—an operation that, with the German cables out of action, proved highly profitable, gross revenues of almost $1 million being earned in 1916.[49] Foreseeing eventual expropriation, the chief executive of the Atlantic Communications Company, Dr. K. G. Frank, tried to sell majority ownership in the company to the Swedish Government—a trans-

[48] Executive Order No. 2042, 5 September 1914.

[49] For the earnings, see Howeth, *History*, p. 226. There were many allegations that Sayville had been transmitting unneutral messages in cypher, despite Navy censorship, and phonograph recordings of Sayville traffic made by Charles Apgar, a New Jersey radio amateur, may have played a role in inducing the Navy to intervene. See Thorn L. Mayes, "The Sayville Wireless Station," (Mimeo, 1978), pp. 6-7. For the legal issues involved in the denial of a license, see James M. Herring and Gerald C. Cross, *Telecommunications: Economics and Regulation* (New York, 1936), pp. 241-42. The secretary of commerce had from the outset been reluctant to issue a license for Sayville, on the grounds that, although owned by a corporation chartered under the laws of the State of New York, it was German-controlled, and Germany did not allow American-controlled radio corporations to operate in Germany. The attorney general informed him, however, that under the terms of the Radio Act of 1912 he was without authority to refuse a license "if the applicant came within the class to which licenses were authorized to be issued"—a ruling that foreshadowed the eventual breakdown of the Commerce Department's licensing authority with the advent of popular broadcasting.

action that, he said, would involve only comparatively small sums and would ensure direct communications between Sweden and the United States if America entered the war. The negotiations were unsuccessful, however, and in 1917 title to Sayville was transferred to the Alien Property Custodian.[50]

Immediately upon the U.S. declaration of war, the Navy assumed control of the Marconi Company's stations in Massachusetts, New Jersey, California, and Hawaii and of the Federal Company's stations on the West Coast and in Hawaii. Ownership (as distinct from operating control) of the Federal Company's stations passed into the Navy's hands some ten months later when that company's patents and radio stations were acquired by the government under circumstances that we shall examine later. (See below, pp. 288-301.) Control of the shore stations handling marine traffic was also assumed by the Navy immediately after American entry into the war. The Navy at that time had forty-nine coastal stations of its own. There were in addition sixty-seven private coastal stations, most of them owned by the American Marconi Company. Some twenty-eight of these were closed as redundant on 7 April 1917; the remainder were integrated into the Navy's system.[51]

There remained a large number of radio transmitters on board ships of American registry. These were almost entirely owned by the American Marconi Company. In 1918 the U.S. Shipping Board authorized the purchase of all leased radio stations on vessels owned and controlled by the board, a move in which the Railroad Administration, which had sixty-three vessels under charter, quickly joined. This covered by far the

[50] Swedish Foreign Office files, K. G. Frank to F. A. Ekengren, Minister of Sweden, 22 April 1916; Ekengren to Frank L. Polk, counselor of the Department of State, 18 April 1917; Polk to Ekengren, 19 April and 30 April 1917. Frank was for a time placed under arrest, and some suspicion seems to have fallen on the Swedish engineer, Seth Ljungqvist, with whom he had been in communication. The German intent was clearly to keep open an indirect communications channel with North America by use of the Telefunken-equipped station at Karlsberg in Sweden. Outgoing traffic from Sayville was a matter of concern at the cabinet level in the United States; see Cronon, ed., *Cabinet Diaries of Josephus Daniels*, pp. 100-101 (28 June 1915). Two Telefunken engineers stationed at Sayville managed to make their escape to Mexico and built there at Chapultapec a station with which, from August 1918, they communicated nightly with Nauen. See Haraden Pratt to Ellery Stone, 3 August 1924, memorandum on "Telegraphic Communication Development in the Republic of Mexico" (Pratt Papers).

[51] U.S. Navy Department, Office of Naval Records and Library, Historical Section, Publication Number 5, *History of the Bureau of Engineering, Navy Department, during the World War* (Washington, D.C., 1922), p. 110.

greater part of the U.S. merchant fleet at the time. When the Navy's Bureau of Engineering opened negotiations for the purchase of the shipboard stations, however, the American Marconi Company balked. Officers of the company pointed out, not unreasonably, that their shore stations (which the Navy controlled but to which the company still held legal title) would be useless if the Navy bought the ship installations. If the Navy were going to take over marine radio, it should buy the shore stations also. Without great reluctance, the Navy acquiesced in this argument, and on 30 November 1918 (that is, after the Armistice had been signed) purchased the company's 45 coastal stations plus its 330 ship stations for the sum of $1.45 million, of which the Shipping Board provided $219,200, the Railroad Administration $141,200, and the Navy the rest. This purchase had not been authorized by the Congress, and Daniels was later to be threatened with impeachment for the act.[52]

By the end of 1918, therefore, the U.S. Navy had assumed ownership and control of all marine radio stations, comprising at the time 229 coastal stations and about 3,775 ship stations. All the high-power stations intended for long-distance communication were also under Navy control. Not all, however, were government-owned. The Federal Company's stations were now government property. Tuckerton and Sayville were in the hands of the Alien Property Custodian and it was highly unlikely that they would ever be returned to their original owners. That left the Marconi stations: Marion and Chatham in Massachusetts; Belmar and New Brunswick in New Jersey; Marshalls and Bolinas in California; and Koko Head and Kahuku in Hawaii.[53] These were still privately owned, though under Navy control. Of these the most important was the New Brunswick transmitting station, where an Alexanderson alternator had been installed at the Navy's request. This held the key to the resumption of commercial transatlantic service. Operational control could be retained until the national emergency was officially terminated, but after that the Marconi Company would take over once again.

Daniels's objective of total government ownership of radio did not command universal support among his staff. A policy of reducing and if possible eliminating foreign ownership, however, did. To both programs the Marconi stations in 1918-1919 posed the most immediate threat. It

[52] Ibid., p. 114; Clark Radio Collection (Smithsonian Institution), Cl. 100, v. 1, p. 305.

[53] The receiving station is mentioned first in each case, then the transmitting station that worked in conjunction with it. Marion and Chatham were intended for communication with Stavanger in Norway, Belmar and New Brunswick for communication with Towyn and Carnarvon in Wales.

became, therefore, a matter of settled policy in the Navy Department to impede in every way possible the Marconi Company's attempts to reequip its high-powered transmitting stations, and in particular New Brunswick. Behind that lay the more general goal of denying the Marconi organization the access to American continuous wave technology that it needed to modernize its system worldwide. And behind that again were the concerns that had animated Wilson and his advisers in Paris, particularly the wish to secure for the United States a wider and more autonomous role in international communications.

There were two sources from which the Marconi organization in 1919 could hope to secure high-powered continuous wave transmitters of proven reliability: the Federal Telegraph Company and the General Electric Company. An approach was in fact made to each of them, and in each case the Navy intervened to block the transaction.

*   *   *

The Federal Telegraph Company had received its first Navy contract in 1913. This was for the 100 kilowatt arc transmitter for Darien, first station in the Navy's high-power chain. Later contracts, after the success of that first installation, called for successively higher power levels: 200 kilowatts for San Diego; 350 for Cavite and Pearl Harbor; 500 for the new Annapolis, Maryland, station, commissioned in 1918; and two 1,000 kilowatt units for the Lafayette station near Bordeaux in France. Twin 2,000 kilowatt arcs were intended for the station to be built at Monroe, North Carolina, and there is no doubt that they could have been built; construction plans were cancelled, however, in late 1919. This represented a remarkable upward movement of power levels in a very short period of time and was eloquent testimony to the engineering and manufacturing expertise of the Federal Company.

Construction and commissioning of these arc stations provided the U.S. Navy with a long-distance radio communications network unparalleled in coverage at the time. It was also the most advanced technologically, largely because construction of the high-power chain had forced the Navy to commit itself, not just to continuous wave technology, but to a particular type of continuous wave generator. The result was forced-draft development of arc technology and the acquisition by the Federal Company of a body of design and manufacturing experience to be found nowhere else in the world. This had several consequences. First, it made the Federal Company an indispensable resource for the Navy Department as long as the construction program continued—that is, through the planning of the Monroe station in 1918-1919. Secondly, it convinced officers of the Navy's Bureau of Engineering that the Federal Company's patents and experience were national assets; information on arc design, and particularly on the design of the magnetic circuits, came to be regarded as close to official secrets.[54] And thirdly, since the performance of the Navy's transmitters and the identity of the firm that manufactured them were well known in the radio world, relations with the Federal Company became a matter of concern to the Marconi Company. Federal arcs, with their proven ability to make themselves heard over intercontinental distances, seemed precisely what was wanted to reequip the Marconi network. The prospect was particularly attractive in that Poulsen's basic British patent had expired and rights to the oscillating arc were open to anyone.[55] All that was required was access to the development and improvement patents—that is, to the work of Elwell and Fuller.

The Naval Radio Service, part of the Bureau of Navigation, was responsible for operating the Navy's radio system, but the procuring of apparatus was the responsibility of the Radio Division in the Bureau of Steam Engineering. Officers of that division, therefore, had the job of monitoring the Federal Company's manufacturing contracts. There were a few complaints about the difficulties of supervising from Washington work done in Palo Alto, and Federal's engineers occasionally balked at the rapid escalation of power levels that the Navy demanded.[56] In general, however, the arrangement worked smoothly.

For this much of the credit was due to Lt. Comdr. George C. Sweet, who was stationed at the Mare Island Navy Yard in San Francisco from

[54] Fuller, "Leonard Franklin Fuller," (Norberg interview), pp. 99-100; Fuller, "Design of Poulsen Arc Converters," IRE Proceedings, 7 (October 1919), 449-97. It is probably significant that Fuller's design ideas were not presented to the Institute of Radio Engineers until after the war was over, and that he did not submit his dissertation to Stanford and receive his doctoral degree until 1919.

[55] Poulsen's British patent (No. 15,599 of 1903) had been acquired by one Christopher Hage, later associated with the British and Overseas Engineering Syndicate, Ltd. It expired on 14 July 1914 and an application for extension was denied by the courts in 1921. See Radio Review 2 (3 March 1921), 206. S. G. Sturmey, in his Economic Development of Radio (London, n.d.), p. 24, states that the Marconi Company acquired the British rights to the Poulsen patents in 1915. The reference must be to the fact that the Marconi Company took over the obligations of the British and Overseas Engineering Syndicate in that year.

[56] Admiral Griffin urged the Federal Company in December 1916 to have a representative on the Atlantic coast "to look out for the details of deliveries, etc." This was later done. See Clark Radio Collection, Cl. 100, v. 2, Admiral Griffin to Federal Telegraph Company, 29 December 1916.

1914 to 1917 and served as liaison officer between the Navy and the Federal Company.[57] In this role he reported directly to the head of the Bureau of Steam Engineering, not to the commandant at Mare Island. Sweet had followed an unusual career in the Navy. He had caught diphtheria while in hospital in 1911 and later contracted a type of neuritis that made it impossible for him to get about readily aboard ship. Extended periods of sick leave were followed by appointment to Mare Island in 1914 on permanent shore duty and then in February 1915 by a medical review that found him "physically incapacitated for active service." He was placed on the retired list in March 1915. On 1 June of that year, however, he reported again for active duty at Mare Island. His biographer offers no explanation for this abrupt change in status; we can only presume that he was badly needed for that particular assignment.[58] It had, after all, a special importance: the Navy depended absolutely on the Federal Company for its arc transmitters, and those transmitters were vital for naval communications.

The first approach made by the Marconi organization to the Federal Company took place in 1917 and involved radio communications with South America. Federal, with State Department approval, had secured in 1915 a concession to build a radio station in Argentina, hoping to construct eventually a network in South and Central America to link up with its system in the United States. Nothing had been done, however, because of inability to raise the necessary capital. Meanwhile the British Marconi Company had acquired similar concessions in Brazil and other South American countries but, because of the war, was unable to do anything with them. The American Marconi Company, which might have been able to do something with its parent company's concessions, was prevented from doing so by opposition from the State Department and the Navy, the basis for this being the belief that American Marconi was controlled by its British parent and therefore could not be regarded as an acceptable instrument for the expansion of American communications interests in South America. The same objection had, incidentally, limited the Navy's willingness to purchase radio equipment from American Marconi.

The State Department was very anxious in these years to extend American radio coverage in Latin America, partly for commercial reasons and partly for propaganda purposes. Telefunken had shown considerable interest in the region before the war, as had British Marconi, but both

[57] Lillian C. White, *Pioneer and Patriot: George Cook Sweet, Commander, U.S.N., 1877-1953* (Delray Beach, Fla., 1963).
[58] Ibid., pp. 71-72.

of these organizations were now preoccupied by the war in Europe. There was, therefore, an opportunity for American interests. Federal, however, appeared unable or unwilling to carry the responsibility by itself, and there was no other acceptable American communications company to step into the breach. The result was an impasse and nothing was done.

These were the circumstances in which Edward J. Nally, vice-president and general manager of American Marconi, proposed the formation of the Pan-American Telegraph Company, to be owned three-eighths by British Marconi, three-eighths by American Marconi, and one-quarter by the Federal Company. On the face of it this hardly seemed a constructive idea, since majority control would still rest with the British firm and its U.S. subsidiary. One can, however, see the advantages from the Marconi point of view. It would remove the threat of competition in the South American market by a Federal Company backed by the Navy and the State Department. And it would make it possible to equip the stations planned for the new system with Federal arcs. The big obstacle was the conviction held by the Navy and the State Department that the American Marconi Company was controlled by British interests. As far as Nally was concerned, this was simply not true. British interests, including the British Marconi Company, did indeed own a majority of the voting stock in American Marconi, but they did not, he insisted, control its policies nor affect its identity as a truly American company. And in saying this, in 1917, he may not have been far from the truth, for by that time American Marconi had come to exercise considerable managerial autonomy and was financially independent of its British parent. The problem was to convince other people of what was, to Nally, a plain fact. His solution was to give verbal assurances that the British-held stock in American Marconi would not be voted.[59] What legal force such assurances could have is not clear, but they did the job. The Navy withdrew its objections to the consortium, and the Department of State gave its blessing.

The fruit of these labors was the Pan-American Telegraph Company, a joint venture of the two Marconi companies and Federal Telegraph. All South American concessions were to be pooled, and also the South American rights to all radio patents held by the three companies. On paper it was an impressive organization and, if its plans had been energetically pursued, it might well have served to offset German influence on South American communications, which was the prime objective at

---

[59] Hooper Papers (Library of Congress), Box 1, Hooper to Bastedo, 3 November 1917; Clark Radio Collection, Cl. 100, v. 2, pp. 75 and 79; compare Hogan, *Informal Entente*, p. 141.

the time. However, nothing was done. British support for the consortium was never more than halfhearted, in view of the concessions that had been made, at least nominally, to the insistence on "American control." The final blow came in 1918, when Secretary Daniels announced that the U.S. terminus for the South American system would have to be a government station—specifically, the "superstation" the Navy was then planning to build at Monroe, N.C.[60] This was not what the Marconi interests, British or American, had in mind, and serious planning to make the Pan-American Company into something more than a paper organization seems to have ceased at that point. The company lingered on for a few more years, to complicate RCA's planning for South American expansion, but it never became an operating organization.

What influence on these negotiations was exercised by the Federal arc patents and Marconi's need for a continuous wave transmitter is hard to estimate. It was taken for granted that the Pan-American stations would use arc transmitters, and it is clear that Marconi personnel would in that way have learned what they needed to know about the new technology. The evidence does not suggest, however, that this was a major factor in Nally's thinking, nor was the desire to keep the Federal patents out of Marconi hands a dominant element in Navy or State Department opposition at that time. The issue of American control of Pan-American was pivotal.

In the second approach to Federal, however, the issue of Marconi access to the arc patents was paramount. The matter came to the attention of the Navy Department in a rather strange way. Our knowledge of the circumstances is indirect at best and has to be pieced together from scraps of information. Lieutenant Commander Sweet, shortly after American entry in the war, had been relieved of his assignment at Mare Island and put in charge of the Shore Station Section of the Radio Division. His major responsibility was to oversee construction of the new high-power stations to be built at Annapolis and Bordeaux—a response to General Pershing's insistence on the need for reliable radio service between the United States and Europe in case the submarine cables should be cut by the enemy.[61] This was a demanding assignment for a man who only two years before had been declared incapacitated for active service, but Sweet probably knew as much about arc transmitters as any commissioned officer in the service at that time and certainly had more experience in dealing with the Federal Company than anyone else.

[60] Howeth, *History*, pp. 364, 530.
[61] Hooper Papers, Box 37; White, *Sweet*, pp. 79-80; War Department, Annual Reports, 1919, Vol. 1, Pt. 1, Report of the Chief Signal Officer, pp. 1013-16.

Shortly before Sweet left for France in July 1918 he was contacted by the New York sales representative of the Federal Company, a man named C. W. Waller, and introduced to Sidney F. St. John Steadman, legal counsel for the British Marconi Company. Steadman informed him, according to Sweet's later account, that he was on the point of concluding an agreement with Federal Telegraph for the purchase by the British Marconi Company of rights to all Federal's patents, and mentioned the sum of $1.6 million as the purchase price. Sweet's reaction was to go immediately to Washington and lay the matter before the assistant secretary of the navy, Franklin D. Roosevelt, and through him the secretary, Josephus Daniels. He urged that the Navy Department take immediate action to block the sale to Marconi, preferably by making an offer to Federal of at least the same sum.[62]

Sweet and Roosevelt had become personal friends, on a first-name basis, during one of the assistant secretary's visits to California, and there is nothing implausible about the idea that Sweet was able to get prompt access to the secretary and to present his case in the most effective manner possible.[63] It is a little harder, however, to accept the idea that this meeting with Steadman was the first time Sweet had heard about the proposed sale of Federal's patents, for he had kept up his California contacts and San Francisco newspapers had reported the imminence of a deal between Federal and Marconi since at least December 1917.[64] In that month Washington Dodge had become president of Federal.[65] Dodge was a highly respected citizen, a former tax assessor of the City of San Francisco and a vice-president of the Anglo and London-Paris National Bank. Press reports of his acceptance of the presidency stated without qualification that it "marked the consummation of a deal by which strong financial backing is assured to the Federal Telegraph Company"; that it was an "open secret" that during a recent visit to New York Dodge had concluded an agreement with the Marconi Company assuring a steady return from the Poulsen inventions; and that Federal had recently secured large

[62] Haraden Pratt to E. J. Simon, 1 July 1963 (Pratt Papers). Pratt, who had previously been employed at Mare Island, was serving as one of the Navy's civilian radio experts in Washington at the time of these events. He later became chief engineer of Federal Telegraph.

[63] For Sweet's friendship with Roosevelt, see White, *Sweet*, pp. 73, 105-107, 113-21.

[64] Confirmation that Sweet had not severed his personal ties with Federal Company personnel comes from Leonard Fuller (letter to the author, 12 December 1979).

[65] Foothill College, Federal Telegraph file, newspaper clippings for 8 December 1917.

contracts from European governments as well as from the government of the United States.[66] Much of this was, of course, hyperbole, intended to boost the price of Federal and Poulsen shares. But, to anyone who cast an eye on the San Francisco papers, it cannot have come as a shock to realize in early 1918 that Federal Telegraph and British Marconi were talking about a deal, or that Washington Dodge had assumed the presidency with the expectation that the deal would soon be concluded.

Some skepticism may be permitted, therefore, regarding whether news of the proposed deal between British Marconi and the Federal Company came to Sweet as a surprise; and it is possible that the facts did not entirely justify the air of urgent crisis that he conveyed to Roosevelt and Daniels. Without that atmosphere of urgency, however, they would not have acted as promptly as they did. Sweet heard from Steadman about the Marconi proposition on or about 5 April 1918. A draft contract providing for purchase of the patents and other property by the Navy was on Daniels's desk by the 16th of that month. It was signed on 15 May.

Though drawn up in great haste—Sweet, we are told, gathered up his brother, who was an attorney on loan to the State Department, and a stenographer, rushed them out to his apartment on Columbia Road, and had a draft agreement ready within about two hours after talking with Daniels—the contract appears, to a layman's eyes at least, to be a valid transfer of title from the Federal Telegraph Company and its parent firm, the Poulsen Wireless Corporation, to the Navy.[67] The point is important, as within a few years it was to be given a completely different interpretation. The Federal and Poulsen companies agreed, for the sum of $1.6 million, to "grant, bargain, sell, set over, transfer and assign" to the federal government all the real and personal property that the two companies owned, including specifically the patents and rights "known as the Poulsen Arc Patents and the Fuller improvements," together with their stations at Heeia Point in Hawaii, Los Angeles, Inglewood, San Diego, Ocean Beach, and South San Francisco in California, Lents in Oregon, and Phoenix in Arizona. Outstanding bonded indebtedness of the Federal Company, amounting to $500,000 at face value and secured by a mortgage, was to be cleared up before the sale became final. And

[66] Ibid.

[67] For the haste with which the contract was drawn up, see Haraden Pratt to Emil Simon, 1 July 1963 (Pratt Papers). Pratt was reporting to Simon the story as he had heard it from Sweet. A copy of the contract may be found in the Hooper Papers (Library of Congress) Box 2, file May-August 1918.

the Federal Company agreed to invest the purchase price immediately in Liberty Bonds.

A few years earlier $1.6 million would have been a lot of money to the Navy Department. The days of budgetary dearth had ended in 1916, however, when a $600 million appropriation signalled the beginning of rapid naval expansion, and by 1918 the Navy had more funds at its disposal than it properly knew what to do with. If there were any hidden reservations about purchasing the Federal Company's stations and patents, they were not occasioned by the amount of money involved. And Secretary Daniels was gratified, we are told, when Washington Dodge and his associates indicated that they would be happy to accept payment in Liberty Bonds (then selling at 95 and available in either registered or "coupon" form), not pausing to speculate as to why they would rather not receive a check.[68] As far as he and Roosevelt were concerned, it was a sound business transaction. The price was reasonable; otherwise why would the Marconi Company have been willing to pay as much? But, beyond this, the purchase advanced Secretary Daniels's grand strategy of government ownership. The Federal Company's stations were already under government control, but this contract transferred title.

Patents were a different matter. The Navy Department already had the use of the Federal Company's patents before the contract was signed. In that respect the contract gave the Navy nothing it did not already possess. But that, of course, was not the point of the transaction. The purpose of the contract was to deny the patents to the Marconi Company, not to make them available to the Navy. It was, as Lieutenant Commander Hooper later put it, "the first step to forestall the British control of the continuous wave."[69]

As far as the manufacturing and operating personnel of the Federal Company were concerned, the transaction at first made little difference. The Navy had already taken over the stations; the company was maintaining a skeleton service between San Francisco and Los Angeles by leasing lines from the Telephone Company. And for some years past the manufacturing operations of the company had tended more and more to overshadow its operating activities. Manufacturing for the Navy would presumably continue as long as the Navy needed arc transmitters—the sale would have no effect on that. And rights to sell transmitters to the Pan-American Company were specifically reserved. The only real options foreclosed to the company were its right to sell its patents to Marconi

[68] Emil Simon to Haraden Pratt, 30 May 1963 (Pratt Papers).
[69] Hooper Papers, Box 38, p. 660.

and to sell its transmitters for use in the United States or its dependencies without the permission of the secretary of the navy.

It seemed, on the face of it, an eminently satisfactory piece of business, certain to advance the major objective of the Navy Department and demonstrating its ability to act promptly and decisively in the national interest. Shortly after the signing of the contract, however, evidence began to accumulate that everything was not as it had appeared to be.

The first cracks in the structure appeared in California and they suggested that Washington Dodge had not taken all his associates fully into his confidence. Suit was filed against him on 14 January 1919 in the Superior Court of Marin County by T. C. Tognazzini and C. F. Leege, vice-presidents of the Anglo-California Trust Company, alleging that they and the interests they represented had sustained damages by the purchase of Poulsen Wireless shares, because Dodge had represented their value at $12 per share, exclusive of foreign patent rights and the sale of rights to the U.S. government, when in fact he knew at the time that, on such a basis, they were worth no more than $3.[70] This charge, however, was only the tip of the iceberg. Tognazzini and Leege, it soon appeared, were acting as representatives not only of local disaffected stockholders but also of Coleman du Pont, Lazard Frères, and certain other eastern capitalists. They had bought their Poulsen shares at the peak of a market flurry occasioned by news of the prospective deal with Marconi; they had bought them directly from Dodge at a time when he was liquidating his personal holdings; and he had failed, they alleged, to inform them of certain relevant circumstances associated with the sale of the company's property to the federal government.

What these circumstances were became evident at a meeting of the board of directors on 17 January. The attack on Dodge was led by Hiram Johnson, Jr., son of the distinguished and influential Progressive senator from California. Amid the storm of charges and countercharges certain facts were admitted by both sides. Chief among these was the fact that, of the $1.6 million paid by the federal government, only about $1 million had found its way into the treasury of the Federal Telegraph Company. The remainder had ended up in the custody of an organization called the Valencia Improvement Company. This, it turned out, was a corporation created and controlled by Dodge himself, in association with his brother-in-law, Nathan Vidaver (a New York attorney), C. D. Waller (eastern representative of Federal Telegraph, who had introduced the Marconi representative to Lieutenant Commander Sweet), and one Wil-

[70] *San Francisco Call and Post*, 17 January 1919 (Foothill College file).

liam D. Loucks.[71] The negotiations with the federal government had been carried on by the Valencia Improvement Company and, after those negotiations were successfully concluded, a commission of $480,000 was paid to that company by order of the board of directors of Federal Telegraph.[72] Those directors, however, or some of them, were not aware at the time that the Valencia Company was in fact controlled by their own chairman and his friends.

These bare facts were not denied by Dodge. He did, however, deny other charges, notably Hiram Johnson's accusation that the whole transaction was fraudulent, since he had concealed from his fellow directors the fact that the commission was to be paid, in effect, to him, and that the deal with the Navy had already been concluded at the time they voted to pay a commission to Valencia Improvement if the sale could be put through. These allegations were in addition to the original charges levied by Tognazzini and Leege, which essentially charged fraud in the sale of Poulsen shares to Coleman du Pont, Lazard Frères and other eastern investors.

Dodge tried to ride out the storm. He claimed, indeed, that honor impelled him to retain the presidency, as he had induced the eastern group of investors to buy more than 100,000 shares of Poulsen stock at pretty near the top of the market and therefore had to remain in office to protect their interests. Johnson, however, in what must have been a dramatic moment, produced telegrams from the eastern investors Dodge had named, stating that they were opposed to Dodge being continued in the management of the company and wished to make common cause with the disaffected California stockholders. This stripped Dodge of the last of his defenses, and he resigned forthwith, along with all but two of the members of the board. He died just over six months later.[73]

These events had not gone unnoticed in Washington. It would have been strange, indeed, if someone in the Navy Department had not wondered whether there had not been something a little odd about the affair from the beginning. Inquiries were set in motion, and attempts were made to recover some of the $600,000 that had so regrettably been diverted in transit from the Navy Department to the Federal Company's treasury, but nothing came of them. There were no indictments, no courts martial, and nobody's career was ruined. Nor was there any political fallout, which is remarkable. When, in 1919, Josephus Daniels was threat-

[71] *San Francisco Chronicle*, 22 May 1920.

[72] *San Francisco Call and Post*, 17 January 1919.

[73] Dodge shot himself on 21 June 1919 and died in St. Francis Hospital on 30 June.

ened with impeachment for misuse of public funds, the issue that caused congressional outrage was the $1.45 million spent to buy the Marconi shore stations, not the $1.6 million spent for the Federal Company's property.[74] Yet Daniels was far more vulnerable on the second issue than on the first.

What had the Navy got out of the deal? At the very least, ownership, one would think, of the Federal Company's patents. This certainly was what Daniels and the officers of the Radio Division believed at the time.[75] It was not, however, how the new management of the Federal Company preferred that the matter be interpreted. The new president of the company, after Dodge's resignation, was R. P. Schwerin. Formerly president of the Pacific Mail Steamship Company and then of the Associated Oil Company, Schwerin was recalled from retirement to salvage what was left of the Federal and Poulsen companies. He did not, at first, have much to work with: "Just a leased line service and a factory all on the wrong side of the U.S.," was the way one of his engineers expressed it.[76] The leased line service was almost lost when the Telephone Company withdrew its wire leasing privileges between California cities, and only a court injunction kept it functioning until a new radio communication system could be constructed.[77] The factory had been kept busy during the war filling orders for the Navy and the Emergency Fleet, but these disappeared with the cancellation of the Monroe station and the cutback in government spending. Schwerin's best hopes were to get Federal back into the radio operating business as quickly as possible and to find someone who was in the market for large arc transmitters.

He moved energetically on both fronts. By 1921 new stations had been built at San Francisco and Portland, Oregon, and three full-duplex circuits were in operation between the two cities. Service between San Francisco and Los Angeles was reopened in 1922-1923. No less important, in 1921 Schwerin successfully negotiated a $13 million contract with the gov-

[74] See *Congressional Record* (65th Congress, 3rd Session), 57, 2294-2309, debate on supplemental navy appropriation bill, 29 January 1919; compare Cronon, ed., *Cabinet Diaries of Josephus Daniels*, p. 372.

[75] Lt. Comdr. H. P. LeClair was head of the Radio Division at this time and Lt. Comdr. G. C. Sweet was in charge of the Shore Station Section. Captain D. W. Todd was superintendent of the Navy Radio Service. Lt. Comdr. S. C. Hooper, usually a prominent figure in the Radio Division, was on sea duty at the time. For the Navy's understanding of what the contract conveyed, see Director of Naval Communications to Hooper, 16 April 1918 (Hooper Papers, Box 2, file January-April 1918).

[76] A. Y. Tuel to H. Pratt, 29 July 1919 (Pratt Papers).

[77] Pratt Papers, Box 4, folder, "History of Mackay Radio and Tel. Company."

ernment of China to construct radio stations at Shanghai, Harbin, Canton, and Peking and open a radiotelegraph service with the United States.[78] The Pan-American Company might be moribund, but the China contract opened up once again the intriguing possibilities of transpacific expansion of which Elwell had dreamed.

To finance construction of the stations in China, however, Schwerin had to accept payment in Chinese government bonds that Federal undertook to market. And before turning over these bonds, the Chinese government required, in its contract, that Federal be able to show that it held full title to the patents on all the equipment to be installed. Here was a problem. Federal certainly intended to use arcs in its China transmitters, and those arcs were covered by the patents that the Navy thought it had bought in May 1918. Schwerin's solution was to claim that the contract signed in 1918 had not been a contract of sale at all; it merely gave the Navy a nonexclusive license to use the Federal Company's patents. And he sought from the Navy an agreement making it clear that ownership of the patents did reside with the Federal Company.

Schwerin's assertion was remarkable. True, the contract had been drawn up in haste, but nowhere in its language had there been any mention of licensing. Even more remarkable, when confronted with Schwerin's suggestion, the Navy raised no serious objection. On 19 March 1921 a new contract was signed, by which the Navy Department agreed to "resell, reassign, and retransfer" to the Federal Company all the rights conveyed in the contract of 15 May 1918. The Navy retained a nonexclusive, nonrevocable, and nontransferable license to all existing and future Federal patents, but there was no monetary consideration at all.[79]

How did Schwerin do it? In 1918, when the Navy first acquired the Federal Company's patents, it had no need for a license. It had been buying Federal arcs since 1913 and no question of licensing had ever been raised. Far less did it need to pay over a million dollars for the privilege. Patent restrictions played no role in government procurement anyway, as long as the war lasted. The argument that the first contract merely conveyed a nonexclusive license could not stand; and it is to be noted that the contract of 1921 spoke not of licenses acquired in 1918 but of the resale of rights purchased then. The fact remains, however, that Federal Telegraph in 1921 received back, at no cost whatever, patents for which it had been paid approximately $1 million three years before.

The most plausible explanation for this strange turn of events rests on

[78] F. J. Mann, "Federal Telephone and Radio Corporation, a Historical Review: 1900-1946," *Electrical Communication* 23 (December 1946), 377-405.

[79] A copy of the agreement may be found in Hooper Papers, Box 3.

the hypothesis that the arc patents were of very little value to the Navy in 1921 but of great value to the Federal Company in connection with its China concession. The major purpose of acquiring the patents in the first place had been to keep them out of the hands of the Marconi companies. But by 1921 these companies were no longer a significant factor in American radio. American Marconi had been absorbed into the Radio Corporation of America, a Navy-approved corporation created explicitly to safeguard American interests in radio. The structure of the industry, in short, had changed in such a way as to eliminate the rationale for government ownership of the patents. Why not then give them back? The State Department wanted to see Federal succeed in its China project. If regaining title to the patents would help, what objection could there be?[80]

The Navy Department's conduct with reference to the Federal Company's patents, examined in detail, can hardly be described as above criticism. Bureaucratically maladroit would be a charitable phrase to use. But what can be said about its efficacy in terms of its own objectives? In the short run it certainly did prevent the Marconi companies, British and American, from securing access to the particular form of continuous wave radio technology represented by the Federal Company's patents. Denial of access to these patents, however, was not the same as denial of access to continuous wave technology in general. The basic Poulsen patents were open to anyone. If higher-powered arc transmitters were required, Cyril Elwell had been in Britain since 1913, ready to place his skills and experience at the disposal of anyone who would meet his price. Breaking through the 30 kilowatt barrier may have been a problem to him in Palo Alto in 1913, but he was capable of learning from experience too. The 100 kilowatt transmitter he built for the British Navy in 1914 gave good service during the first Battle of the Falkland Islands and during the Gallipoli campaign. The Italian government in 1916, despite its loy-

[80] For an explanation along these lines, see Clark Radio Collection, Cl. 100, v. 1, pp. 287-287A, and compare Howeth, *History*, pp. 367-70. To supplement such an explanation it is necessary to add only that, according to the testimony of one well-placed observer, Schwerin was able to apply political muscle at just the right time. Haraden Pratt, who had served in the Navy as a civilian radio aide during the war, became acting chief engineer of the Federal Company in 1920. According to Pratt, Schwerin in 1921 got himself put in charge of the Republican Party's campaign fund for the Pacific Coast and exacted from the party's leaders a promise that, if the Republicans won the election, he would have the privilege of naming the next secretary of the navy. The second contract with the Federal Company was signed only a few weeks after Secretary Denby took office. See Haraden Pratt to Emil Simon, 1 July 1963 (Pratt Papers).

alty to Marconi, was glad to have him build an arc station at Rome that could maintain communications with the United States. And the first two stations in Britain's Imperial Chain, after the war, were to be designed and built by Elwell. The Marconi Company had no high opinion of Elwell, nor of the transmitters he built; but that did not mean that high-power arc transmitter technology was not theirs to purchase, whether or not the U.S. Navy held the Federal patents. Those patents represented a technological differential in arc technology between the United States and Europe and easy access to them would undoubtedly have enabled the Marconi companies to move faster in reequipping their stations once the war was over. But the differential was not a large one and to think of the Navy's actions in acquiring the Federal patents as denying to the Marconi organization access to continuous wave technology is a considerable overstatement.

Furthermore, the Federal arc was only one of the available technologies of continuous wave radio. There were also the Alexanderson alternators. Denying the Marconi organization access to those devices was to involve the Navy Department in an even more complex set of maneuvers. Purchase of patents might serve to keep the Federal Company's arcs in American hands but that strategy would not serve for the alternator. The differential advantage held by the United States in the case of that device rested less on the kind of information that could be embodied in patents than on the design skills and manufacturing expertise of the General Electric Company. GE, as the Navy well knew, had been negotiating on and off with the Marconi interests since 1915. In the early months of 1919 there seemed every reason to expect that a contract for the sale of alternators to the British and American Marconi companies would be signed in the very near future.

SIX

"An American Radio Company"

REGINALD FESSENDEN parted company with his Pittsburgh backers in January 1911. The National Electric Signaling Company survived, and its possession of the Fessenden patents was later to prove an important factor in the consolidation of interests that followed the formation of RCA (see below, pp. 460-61), but its hopes of becoming a communications company of major importance dwindled. With them went the possibility that NESCO might serve as the nucleus of an American organization able to challenge Marconi.

Fessenden's departure from NESCO also meant the end of his creative partnership with Ernst Alexanderson of General Electric. Up to this point NESCO had paid all development costs and served as the "natural outlet" for the alternators GE designed and built.[1] Who would underwrite development costs now? Where was the market? At one time the Telephone Company had shown interest, but that interest evaporated in 1907 with the shift in financial control from Boston to New York and J. J. Carty's redirection of the company's research program. For a line amplifier the Telephone Company turned in 1912 to Lee de Forest's audion. When in 1914-1915 it again ventured into radiotelephony, it relied on vacuum tubes, not alternators.

Alexanderson's work on high-frequency alternators at GE had enjoyed strong support from E. W. Rice, vice-president of the company, and from C. P. Steinmetz, head of what became in 1910 the Consulting Engineering Department. But the question of the market for such devices could not be evaded. Alexanderson himself was convinced that development work on the alternator should be continued, foreseeing a particularly promising

[1] Ernst Alexanderson, "Reminiscences" (Columbia University Oral History Collection), p. 18

future in radiotelephony. For a time after 1910, however, GE's corporate interest in the device slackened and Alexanderson was assigned to other projects.

There was available at this time a 2 kilowatt alternator, normally operating at 100 kHz, which had been designed for Fessenden and completed at the end of 1909.[2] This was a standardized, marketable device and could have been produced in volume if purchasers had been found. Several were in fact manufactured—ten were under construction in October 1911, with three already delivered to customers—but they were used for research purposes rather than for radio communications. One went, for example, to the Army's Signal Corps, where Maj. G. O. Squier used it for experiments on multiplex telephony.[3] Another went to Harvard, where Arthur Kennelly, working in collaboration with Alexanderson, undertook a series of experiments on the physiological tolerance of human beings to high-frequency currents; the results were reported in a joint paper in 1910.[4] Later Kennelly used a similar machine, on loan from NESCO, to investigate the "skin effect" resistance of conductors at radio frequencies.[5] Alexanderson himself used the alternator to measure the high-frequency properties of iron.[6] In a purchase that had important consequences for GE's research on vacuum tubes, two were bought by John Hays Hammond, Jr., for his research on radio guidance. Others were sold to Columbia University and to American Marconi, and inquiries regarding price and availability were received from the government of Japan. But these were all orders and inquiries for units to be

[2] E.F.W. Alexanderson, "Alternator for one hundred thousand cycles," AIEE *Transactions* 28 (1909), 399-412. By an ingenious modification of the windings, Alexanderson was later able to operate this machine at 200 kHz with no change in the rotor speed.

[3] George O. Squier, "Multiplex Telephony and Telegraphy by means of Electric Waves Guided by Wires," AIEE *Transactions* 30 (1911), 1617-65.

[4] A. E. Kennelly and E.F.W. Alexanderson, "The Physiological Tolerance of Alternating-Current Strengths up to Frequencies of 100,000 cycles per second," *Electrical World* 56 (1910), 154-56.

[5] IRE *Proceedings* 4 (December 1916), 523-74. The research was financed by AT&T as part of its program of investigations into the multiplexing of long-distance telephone circuits by the use of radiofrequency carrier currents. It is interesting to note that the commentator on Kennelly's paper, H. Zenneck, was "astonished" to note that, in 1916, the investigators had used a mechanically driven alternator instead of an oscillating audion. Kennelly's assistant responded that they had used the alternator because it was "immediately available."

[6] E.F.W. Alexanderson, "Magnetic Properties of Iron at Frequencies up to 200,000 cycles," AIEE *Transactions* 30 (1911), 2433-48.

used for experiments and testing. The 2 kilowatt alternator was a useful, even an exciting tool, but it did not find a ready market in commercial radio.

Alexanderson's hopes rested partly on higher power, partly on developing his alternators into a complete radio system. In a report for the Consulting Engineering Department in October 1911 he pointed out that GE was receiving "constant inquiries . . . from all parts of the world" about the 2 kilowatt alternators and should be able to sell a large number of them at a good profit. Meanwhile a 35 kilowatt machine, also undertaken originally for Fessenden, had stood in the shops half-completed for eighteen months. "I believe this business is sufficiently promising," continued Alexanderson, "that we should appropriate money ourselves to complete the machine without being dependent upon the National Electric Signaling Company."[7] But the 35 kilowatt alternator never was completed, despite Alexanderson's urgings. Some important development work in radio continued. There were radiotelephone tests, using the 2 kilowatt alternator, between Schenectady, Pittsfield, and New York City. Design and construction of the "magnetic amplifier," essential if high-powered alternators were to be modulated for voice transmission, made good progress. And Alexanderson worked closely with Langmuir and other members of GE's Research Laboratory on the design of vacuum tube amplifiers and receivers. But nothing was done to build higher-powered alternators. General Electric was clearly not willing to invest further time and money in building such machines until it had some assurance that there were outside purchasers ready to buy the product when completed.

But who besides scientists and experimenters would buy alternators between 1910 and 1914? NESCO was no longer in the market. The Federal Company was committed to its arcs. American Marconi and, while it survived, United Wireless, seemed content with their spark transmitters. The dearth of American operating companies meant that in the short run there was no market beyond the single-item orders from experimenters and foreign governments to which Alexanderson attached such significance. His machines were well spoken of in the scientific and engineering journals. Knowledgeable people agreed that, in principle, the days of spark transmission were numbered. But none of this added up to the substantial orders that would justify General Electric in designing and manufacturing a more powerful alternator.

Part of the difficulty lay in the fact that the alternators had never had

---

[7] Alexanderson Papers (Union College), folder 15, report for Consulting Engineering Department, 9 October 1911.

a chance to show what they could do under operational conditions—that is to say, as part of a traffic-handling radio station. The Federal arcs, in contrast, had a record of proven efficiency, both on that company's own circuits and in the Navy's Arlington tests of 1913. The Navy knew that the arcs worked, that they could make themselves heard through fading, static, and interference when a spark station could not, that they could stand up under arduous service conditions. No tests of the alternator from Schenectady or Pittsfield or New York, under carefully controlled conditions, with factory personnel standing by, could match that kind of evidence. But General Electric was not a radio operating company nor was it working in liaison with any such company, nor with the Navy. What was called for was a test of an Alexanderson alternator of higher power under service conditions. This meant siting it at a radio station where it could feed a large antenna, work in tandem with one of the more powerful stations in Europe, and handle intercontinental traffic for a substantial period of time. Not many organizations could offer such facilities.

Alexanderson had received an inquiry from the Telefunken company in 1911, possibly stimulated by the first published accounts of the Goldschmidt alternator under development in Europe. Nothing had come of it, and Telefunken went to other designers for the alternators it installed at Nauen and Sayville. With the outbreak of war in Europe and the possibility that submarine cable communications might be seriously disrupted, the construction of radio facilities in the United States that could take some of the traffic load off the cables and maintain communications if the cables were cut became an urgent matter. Alexanderson was quick to seize the opportunity. In September 1914 he contacted Rice, now president of GE, and urged completion of an alternator capable of transatlantic work, saying that such a unit was already 80 percent complete.[8] In the same month he reported to Steinmetz on the status of alternator development, stressing that GE now had a strong patent position covering a complete system of radiotelegraphy based on the alternator, and that design work on a 50 kilowatt machine capable of transatlantic communication was far advanced. GE's radio system, he argued, once in full operation, would prove superior to all others. But it had to be tested and its capabilities proved under actual service conditions. He suggested that it would be desirable to work out arrangements with an operating company such as American Marconi to install the system and put it to work.[9]

[8] Alexanderson to Rice, 11 September 1914 (Alexanderson Papers); James E. Brittain, *Alexanderson* (forthcoming) chap. 4, p. 32.

[9] Alexanderson to Steinmetz, 25 September 1914 (Alexanderson Papers), "Development of Apparatus for Radio Communication."

At General Electric there were, therefore, internally generated pressures toward cooperation with the American Marconi Company. Meanwhile pressures toward accommodation were also building within the Marconi organization. Plans to build a new high-powered station at New Brunswick, New Jersey, were announced in August 1914. At that time the Marconi engineers, true to their tradition, intended to use a synchronous spark. By the close of the year, however, that decision was being reconsidered. The turning point was an important one, marking the start of a long overdue shift. Possibly news of Telefunken's decision to install Von Arco alternators at Sayville and Nauen, and HOMAG's choice of a Goldschmidt alternator for Tuckerton, had something to do with it. But it may also be true that the American Marconi Company, under the leadership of E. J. Nally, was now beginning to play a larger role in Marconi decision-making, particularly in regard to the equipping of stations on American soil. The first tentative inquiry to GE came, not from Marconi headquarters in London but from F. M. Sammis, chief engineer of American Marconi, in December 1914.[10] He wanted to know whether GE could design and build an alternator of 200 kilowatts power and what its upper frequency limit would be. GE at the time had no such machine on the drawing boards, far less under construction. The jump from 2 kilowatts to 200 would have been a challenging one, and Alexanderson's response was cautious. He suggested that a better solution would be to equip the station with three 100 kilowatt alternators, one of which would normally be a spare while the other two could be run separately or in parallel.[11] Agreement was reached on price ($20,000 for the first 50 kilowatt unit, when it was completed, and $16,000 for later ones) and a contract was signed. But no such alternator was yet ready for installation, and when the New Brunswick station began operations it was with a Marconi synchronous spark.

Guglielmo Marconi visited Schenectady on 18 May 1915 and found time to look at the 50 kilowatt alternator then under test. He seems to have been impressed. Many years later Alexanderson recalled with pleasure how he and Marconi had watched the machine in action.[12] Marconi's comment on that occasion was perhaps open to misinterpretation: he said, "I can see you are doing good experimental work." The emphasis was probably on the word "experimental," with the implication that the

[10] F. M. Sammis to General Electric, 22 December 1914 (Alexanderson Papers).

[11] Alexanderson to Sammis, 24 December 1914 (Alexanderson Papers); Brittain, *Alexanderson*, chap. 4, pp. 34-35.

[12] Alexanderson, interview with James Brittain, 17 October 1972; see Brittain, *Alexanderson*, chap. 5, p. 6.

alternator was not yet ready for commercial service.[13] And Alexanderson, who described the setup as "a mess of instruments and wires and coils and everything," would have agreed with him. The machine could not yet be run at full speed (which implied that it could not attain its intended frequency of operation), nor could it be kept in operation for more than a few minutes, as the cooling system was incomplete. What Marconi and his engineers saw at Schenectady was a prototype under development. The power output was much less than they wanted and the critical evidence of operation under service conditions was still lacking.

Nevertheless, the future course of events must at this point have seemed predictable. GE needed a working arrangement with one of the operating companies and a market for its alternators. Marconi needed a high-powered continuous wave transmitter, and that need was now urgent. One of Marconi's engineers had admitted as much to Alexanderson during the visit to Schenectady: the Marconi Company, he said, either had to get alternators or get out of business.[14] There were pressures on both sides. If GE did not make a deal with Marconi, to whom would it sell alternators? If Marconi did not acquire them from GE, what other source was there?

Already, however, there were signs of hesitation from some elements in General Electric. What exactly was the company getting itself into? Charles Neave, patent counsel for GE, warned Vice-President Owen D. Young in May that "the Marconi Company is not strong financially."[15] A. G. Davis, head of GE's patent department, took the opportunity in a survey of Marconi's probable requirements to inform Rice that "our people have a disagreeable impression of the American Marconi Company and its business methods," and recommended strongly that GE should not enter into any exclusive arrangements, particularly with respect to the magnetic amplifier (which could be used with arc transmitters as well as with alternators). Alexanderson too had reservations. He wanted to be reassured that the Marconi Company would make proper use of the equipment he had built, and he did not want Marconi buying some items and not others. "His idea is that the thing which he has produced is a complete sending station."[16]

Certainly what was in the offing was no routine business deal. The Marconi organization was on the verge of committing itself, for the first

[13] Alexanderson, "Reminiscences," p. 23.
[14] Alexanderson to F. C. Pratt, 12 April 1915 (Alexanderson Papers).
[15] Charles Neave to O. D. Young, 14 May 1915 (Young Papers, Van Hornesville, N.Y.).
[16] A. G. Davis to Rice, 5 June 1915 (Young Papers).

time in its history, to an outside supplier, and to a technology that up to this point it had repudiated. General Electric was about to add an important new product line, acquire a major new customer, and begin to play a key role in the communications industry. A certain hesitancy on both sides was to be expected.

Nally, writing for American Marconi, described his conception of a desirable future relationship between his company and GE in a letter to President Rice in early June 1915. The Marconi companies, he stated flatly, owned the controlling patents on wireless communication. On the average they had to replace their equipment, to keep up with advances in the art, about once every five years. There was therefore a substantial manufacturing business to be done. General Electric was in a position to take this over. The development of the alternator, in conjunction with Marconi's existing patents, held out the prospect of a significant enlargement of both short- and long-range radio. This called for an efficient and experienced communications company. Upon the success of the communications company would depend the success of the manufacturing company that supplied it. "It is our view," wrote Nally, "that the Marconi Company is the logical concern to control exclusively the apparatus as well as the operating end of the business, while your company can secure the benefits which accrue from being the sole manufacturers of wireless equipments."[17]

Whether this glittering prospect of dual monopoly originated with Nally or reflected the views of the British parent company is not clear; one suspects the former. There was no assurance, as GE executives soon learned, that the two corporations spoke with one voice. E. P. Edwards, assistant manager of GE's lighting department, had been keeping an eye on the market for radio equipment since 1908. "There seems to be a decided feeling between the British and American Companies," he wrote to Vice-President Anson Burchard, "placing them in the position of competitors rather than allies."[18] It was a shrewd observation: the interests of British Marconi and those of the American company that it nominally controlled did not always coincide. Nally wanted to see General Electric in the role of exclusive supplier to American Marconi partly because the relationship would help remove the stigma of foreign control. Godfrey Isaacs and the officers of the British Company, on the other hand, were primarily concerned with the need to reequip their long-distance stations, to meet the competition from Telefunken (if it survived the war), and to place their organization in a position to contract for construction and

[17] Nally to Rice, 4 June 1915 (Young Papers).
[18] E. P. Edwards to Anson Burchard, 27 October 1917 (Young Papers).

operation of the British Imperial Chain. Both groups, however, wanted more than merely alternators: they wanted exclusive rights to the device. That was coming to be the sticking point as far as GE was concerned. A large and assured market was attractive; but to cede monopoly control of the innovation to Marconi was not.

If the Marconi companies had signed a firm contract for the purchase of alternators in 1915, the agreement would almost certainly have called for General Electric to grant exclusive rights. Opposition to that idea had not yet crystallized at GE and the certainty of volume sales to the world's leading operating company would have been hard to resist. There were two main reasons why such an agreement was not signed. In the first place, the 50 kilowatt alternator was still under development while work on the 200 kilowatt machine had not begun. GE was in no position to give guarantees of performance or delivery dates. In the second place, financing the transaction would have been difficult for Marconi. The sales agreement would have called for a large initial deposit and payment in installments as the work proceeded. These payments could have been made only by drawing on the assets and reserves of American Marconi, for British reserves of gold and dollars were coming under severe pressure by the summer of 1915. Even to liquidate Marconi-owned American assets would probably have required the permission of the British Treasury, and there was no assurance that such permission would be granted.

General Electric and British Marconi did sign an agreement on 2 July 1915 but it did not call for any immediate purchases or payments.[19] The document recited certain patents owned by GE that related to the Alexanderson alternator, and it listed others that were owned or controlled by the Marconi Company and were, it alleged, involved in the utilization of Alexanderson's system.[20] It proceeded to state that the machine, when perfected, was to be known as the "Alexanderson-Marconi Alternating-Current High-Frequency Generator." (What Alexanderson thought of these concessions to the Marconi name is not recorded.) The heart of the agreement, however, lay in its fourth and fifth clauses. These bound the General Electric Company to manufacture alternators solely for sale to and use by the Marconi Company (the only exceptions being sales to Western Electric for use in wireless telephony and sales to the U.S. gov-

[19] Brittain (*Alexanderson*, chap. 5, p. 6) writes that this agreement was signed within a few days of Marconi's visit to Schenectady. The date on the copy in the Young Papers, however, is that given in the text.

[20] These Marconi-controlled patents were in fact NESCO patents on which Marconi at that time held licenses. See Edwards, "Analysis of Proposal," 7 February 1918 (Young Papers).

ernment), and it bound the Marconi Company to purchase any alternators it might require solely from General Electric. Provision was made for terminating the agreement on three months' notice if, in any calendar year after the alternator had been "brought to an operative and successful commercial standard," the Marconi Company did not submit orders amounting in the aggregate to 5,000 kilowatts of power. If GE failed to bring the alternator to such a standard before 1 January 1917, the agreement could be cancelled on sixty days' notice. In the meantime the Marconi Company was at liberty to purchase or use any device that scientific research might show to be superior. It agreed, however, to purchase GE's alternators exclusively as long as they were "more simple, efficient, and generally superior," and to use its influence to induce all other Marconi companies to do likewise.[21]

On the face of it the agreement was impressive. With the specified exceptions, it linked Marconi and GE together as exclusive purchaser and exclusive supplier of the new device. And the scale of operations looked substantial. If the 200 kilowatt alternator planned for the future were taken as the standard, the agreement seemed to envisage the purchase of twenty-five such devices each year for the term of the contract. One wonders where the Marconi Company intended to locate even as much as one year's output. When, after the formation of RCA, GE got the big 200 kilowatt alternators into volume production, the total number manufactured was only twenty.

On closer examination, however, it is clear that the agreement was hedged around with so many qualifications and included so many escape clauses that in effect it amounted to little more than a statement of intentions. GE was not committed to manufacture even one of the larger machines, nor were any penalties stipulated if it failed to do so. It was committed to attempt to bring the alternator to "an operative and successful commercial standard," but no more than that. Nor was the Marconi Company committed to purchase any. The language merely provided that if it failed to buy a certain minimum number, the agreement would be null and void. If any superior device turned up in the interim, Marconi was free to purchase it. The agreement was, in fact, contingent and hypothetical. If Marconi bought alternators, it would buy only from GE. If GE manufactured any, it would sell only to Marconi (unless the Navy or AT&T submitted orders). Apart from the new formality, the situation was essentially unchanged.

From Alexanderson's point of view, the most immediate benefit of the agreement was the right it gave GE to "arrange for practical wireless

---

[21] Young Papers, Box 72, "Heads of Agreement," 2 July 1915.

tests between Marconi stations at the expense of the General Company," provided that they did not interfere with commercial service. This opened up for the first time the possibility of full-scale operational tests. The opportunity was not lost. Most of 1916 was taken up with tests at Schenectady. But in January 1917 an informal agreement was reached between Edwards of GE and Nally of American Marconi to install the alternator at the New Brunswick station, with GE defraying all expenses and American Marconi providing the facilities. Installation was complete and the machine ready for service by the end of March. In April, when the United States entered the war, the Navy assumed control of the station and temporarily suspended operations. Tests finally began in May 1917.

* * *

There were, inevitably, a few anxious moments. Twice in March 1918 the drive belt slipped off the alternator's oil pump and the machine came to an abrupt and disconcerting halt. Fortunately no damage was done to the bearings. Alexanderson took appropriate action: on the larger 200 kilowatt alternator then under construction at Schenectady the oil pump would be driven directly from the drive shaft.[22] But in general the alternator performed magnificently—a fact obvious not only to GE and the Marconi Company but also to the Navy, which was running the station, and indeed to anyone who heard its clear, penetrating, and distinctive signal on the air. How much of this was due to the alternator itself and how much to other improvements that Alexanderson and his co-workers made to the station is difficult to judge, nor is the question very relevant. The multiple-tuned antenna, the magnetic amplifier, the new vacuum tube speed regulator—these were as much part of Alexanderson's system as the alternator, and the New Brunswick installation made it possible to give that system a full-scale operational test for the first time. By the end of March 1918 it was clear that it had passed with flying colors.

One consequence was a new solicitude on the part of the Navy. The Radio Division had been keeping an eye on alternator development for some time. In 1915 Lt. Stanford Hooper, then planning the first stations in the high-powered chain, had seen the 50 kilowatt alternator under test at Schenectady and had been highly impressed: "It was very ingenious and it ran just like a watch, just as smooth as could be."[23] But his invitation to General Electric in 1916 to install a 200 kilowatt alternator

[22] Young Papers, "Report on Operation of New Brunswick Station," 20 March 1918.

[23] Hooper Papers, Box 38, p. 801.

in the new San Diego station had been turned down, even when the Navy offered to assume all risks if the machine failed to meet contract specifications. General Electric, he was told, preferred not to do business that way. Now in 1918, however, the Navy actually had an alternator in service and under its control. It was clearly underpowered for transatlantic traffic—"a boy doing a man's job" was how Hooper described it—but it was rendering excellent performance. Though rated at only 50 kilowatts output, in terms of its ability to provide twenty-four-hour transatlantic service even in adverse propagation conditions it was superior both to the 100 kilowatt arc at Arlington and to the Marconi 350 kilowatt synchronous spark originally installed at New Brunswick. In fact for a time, when winter storms levelled the antennas at Sayville and Arlington and the Goldschmidt alternator at Tuckerton burnt out, Alexanderson's alternator at New Brunswick carried almost all transatlantic radio traffic. For the first time the Navy's allegiance to the arc for high-power work began to weaken. The alternator was more stable in frequency, its signal was purer, it radiated no backwave, and it could be keyed at faster speeds. As priced by General Electric, alternators cost no more than arcs of equal radiated power and, on the evidence, they seemed completely reliable.[24]

The Navy's reaction was twofold. First, it sought alternators of higher power. And second, it began to intervene in GE's plans to sell alternators commercially—that is, to Marconi. The first objective presented no great problem. Alexanderson was already working on a 200 kilowatt alternator when the 50 kilowatt machine was installed at New Brunswick. Minor modifications were made in design but essentially it was a matter of increase in scale. This first 200 kilowatt alternator was completed at Schenectady in May 1918 and placed in service at New Brunwick in September. From that date until February 1920 it carried the bulk of radio traffic with Europe, earning an impressive reputation for reliable and trouble-free service. It became the prototype for the standard 200 kilowatt machines that GE manufactured and sold after the war.

Installation of this second alternator at New Brunswick accentuated an already anomalous situation. The station was owned by the American Marconi Company; it was being operated by the Navy; but its two transmitters were the property of General Electric. The announced intent of the Navy Department, following Secretary Daniels, was to take the station under government ownership. But even those officers of the Radio Division who did not accept that objective were determined that the

[24] See, for example, Hooper to Sweet, 13 September 1918 (Hooper Papers). For the higher keying speeds (up to fifty words per minute), see Hooper to Sweet, 24 January 1919 (Hooper Papers).

Pl. 12: Alexanderson's 200 kilowatt alternator
(upper half removed to show rotor).
Source: General Electric Company

alternators installed during the war would not become the property of
American Marconi when hostilities ended. Lieutenant Commander Hooper
(promoted to that rank in 1916), who was rapidly emerging as the key
man in the Radio Division, put the matter bluntly: the Navy was determined, he said, that the alternators would never become the property of
any company that was not "100 percent American."[25]

For neither of the alternators at New Brunswick had General Electric
yet received any payment, from the Navy or from American Marconi,
nor had any sales agreement been signed. In the hope of clarifying the
situation, E. P. Edwards and Alexanderson went to Washington in June
1918 to discuss terms of payment with Lieutenant Commander Sweet.
They received instead a lecture on the Navy's plans for General Electric.
Sweet told them that the Navy would sign a purchase contract for the
alternators only if GE was willing to include in it a clause binding the
company not to sell alternators to anyone in the U.S. or its possessions
without government consent. And he made it clear that this restriction
would apply not only to war conditions but also to the period after the
war. He asserted that government control of radio was in principle the
best policy, but since that could probably not be achieved, GE and the
Navy should enter into a "business arrangement" by which the same
objective could be accomplished. He told them that the government had
recently purchased the Federal Telegraph Company "lock, stock, and

[25] Hooper Papers, Box 37 (transcript of tape-recorded memoirs).

barrel" and Federal was now permitted to manufacture for other customers only with government permission. General Electric, he said, should
accept a similar restriction.

Edwards was not impressed. "I told him as I understood his proposition
he was asking us to do for nothing what the Federal Telegraph Company
had agreed to do in consideration of the payment of three million dollars,
and that I thought there was no prospect of our subscribing to such a
one-sided arrangement."[26] Whereupon Sweet informed him that, if GE
took that attitude, there would be no contract. Edwards had an inflated
idea of the price the Navy had paid for the Federal Company's assets,
and his later report to his superiors may have exaggerated the bluntness
of Sweet's proposition. But about its general thrust there was no doubt
at all, nor was there any doubt that Edwards had rejected it out of hand.[27]

This presented General Electric with a problem to which there was no
easy solution. Whether Sweet spoke for the Navy Department might
indeed be questioned; what kind of "business arrangement" between GE
and the Navy he had in mind was vague; and how the Navy proposed
to exercise in peacetime the powers it held during the war emergency
was not explained. But GE had been given a clear warning that the Navy
intended to intervene to prevent the Marconi companies from acquiring
alternators. This was a threat to the only commercial market for the
machines that had yet shown itself.

Navy Department intentions, as expressed by Sweet, imposed one set
of constraints on General Electric's behavior. Distrust of Marconi intentions, however, was no less limiting. During the period when the 200
kilowatt alternator was under construction at Schenectady and its 50
kilowatt predecessor was being put through its paces at New Brunswick,
the attitudes of influential GE executives toward negotiations with the
Marconi companies became progressively stiffer. Every day it became
clearer that the alternator would do everything that had been claimed
for it. And every day there was greater reluctance to contemplate any
agreement that would give the Marconi companies exclusive rights to
the device.

The issue came to a focus over the terms of sale—a matter that, given
greater confidence on both sides, could easily have been negotiated. The
question was whether GE should give the Marconi companies an exclu-

[26] E. P. Edwards to Anson Burchard, 26 June 1918 (Young Papers).

[27] See also Edwards to Young, 8 July 1918 (Young Papers, Box 75). Edwards
suggested that the policy put forward by Sweet was probably of his own devising
and that, if the situation were called to the attention of Secretary Daniels, he
would not subscribe to it.

sive license or merely a general license that would leave GE free to sell to other buyers. Should it sell the alternators for a flat price? Or should it sell them for a somewhat lower price, but charge in addition a royalty based on the revenue that alternator-equipped stations would generate? An exclusive license and outright sale would give the Marconi companies control over how the alternators would be deployed. A general license and provision for royalties, on the other hand, would give GE a continuing stake in the machines, even after they had passed under the operational control of Marconi—a stake that would not only allow GE to share in the profits that the alternators generated but also enable the company to influence Marconi policy on where the alternators would be located and how they would be placed in service. A royalty clause should not have been a strange idea to Marconi executives: it was fully consistent with the leasing policy they had themselves followed in the early days of the Marconi system.

There had always been uncertainty over what price to charge for the alternators. Since the device was new and since there was, in the short run at least, only one buyer and one seller, there was no going market price. The closest competitive device was the Federal arc, but Federal sold only to the Navy. Up to the start of work on the 50 kilowatt alternator Fessenden and NESCO had paid all development costs, so there was no need for GE to be concerned about recovering those expenses. What then would be a reasonable price to charge Marconi?

A. G. Davis, of GE's patent department, raised this question as early as June 1915 but was unable to suggest a good answer. The Marconi Company had quoted a price of $75,000 for one of its own 100 kilowatt spark sets. The Goldschmidt alternator of 100 kilowatt output was selling at "something between fifty and one hundred and fifty thousand dollars." Poulsen in Europe was getting $20,000 for an arc set rated at 100 kilowatts but actually delivering 40 to the antenna. Alexanderson regarded all cost data for the alternator as "fictitious" but thought that there probably were in the 50 kilowatt alternator "about two thousand dollars' worth of labor and about two thousand dollars' worth of material." Davis thought that GE should charge at least $25,000 or $35,000 for that machine but clearly he had little confidence in his own figures.[28] The tentative agreement that GE signed with the British Marconi Company in July 1915 included a sliding scale of charges according to which $500 per kilowatt would have been charged for the first 3,000 kilowatts, $450 per kilowatt for the next 1,000, and $400 for the next 1,000 after that.

[28] A. G. Davis to Rice, 5 June 1915 (Young Papers).

The first 200 kilowatt alternators, on that basis, would have sold for $100,000 each.[29]

By September 1917, however, Edwards was urging that GE should quote a flat price of $127,500 per set, assuming that Marconi bought not less than ten sets of 200 kilowatts each and that the sets were sold outright with no royalty. Edwards's price included not only the alternator itself but also the driving motor, magnetic amplifier, antenna inductances, a photographic receiver capable of receiving messages at several hundred words a minute, and all the other equipment necessary to make up a complete transmitting station except the antenna itself and the buildings. If ten sets were sold on this basis, Edwards estimated, it would wipe out GE's unliquidated development costs and yield a total profit of $502,730. The price per unit was not unreasonable, he thought, in view of the fact that the Federal Company had recently got a price of $125,000 from the Navy for the 300 kilowatt arc set.[30] Arc transmitters were rated by input power and never operated at more than 50 percent efficiency, so that a 300 kilowatt arc would at best deliver 150 kilowatts to the antenna—less than Alexanderson's new machine. The price Edwards suggested was therefore very close to the price of the nearest equivalent transmitter available at the time. The price actually quoted to American Marconi for the first 200 kilowatt machine in November 1917 was, however, $127,000 f.o.b. Schenectady—plus (and it was an important plus) a royalty payment "on some mutually satisfactory basis."[31] No such basis could in fact be found, which was why negotiations broke down and GE ended up installing the machine at New Brunswick at its own expense.

Uncertainty over price was one reason to charge royalties. If, in the eyes of General Electric, the income-earning potential of the alternators was higher than it was in the eyes of the Marconi Company, a base price plus royalties provided a means whereby the two companies could still do business. General Electric would, in effect, share the financial risks of the innovation, as it was accustomed to doing with many of its other customers, such as local public utilities and transit systems. But there was more to it than this. Underlying GE's growing insistence on royalties instead of outright sale was a lack of confidence in the intentions of the Marconi Company. Once Marconi controlled the alternator—and the demand for exclusive rights implied control and not just ownership—

[29] Young Papers, Box 72, "Heads of Agreement," 2 July 1915.

[30] Young Papers, Box 72, "Cost Data & Price Recommendations, Alexanderson Radio System," enclosed in Edwards to Burchard, 20 September 1917; ibid., Edwards to Burchard, 3 December 1917.

[31] Edwards to Nally, 21 November 1917 (Young Papers).

would it exploit its potential to the full? And would it promptly accept and introduce the improvements that Alexanderson and his colleagues had in mind? Or would it adopt a more cautious and conservative policy, one amounting perhaps not to suppression of the device, but tantamount to exploiting its possibilities less energetically than GE thought desirable? Anxieties on this score were not unreasonable considering the technological conservatism the Marconi organization had shown in the past.

A contract that included a royalty clause would at the very least enable GE to monitor the use made of the alternators and, at best, give it some legal leverage with which to insist on vigorous development. It would be in a position to look over the shoulders of the Marconi Company, so to speak, and would have some basis in equity to complain if development seemed to lag. This is probably the major reason why negotiators for the Marconi organization objected to the very idea of royalties. Otherwise a royalty clause would have been an attractive element in the contract, implying a lower initial price and less drain on cash reserves. American Marconi did, after all, have a clear opportunity to buy the first 200 kilowatt alternator in 1917; the stumbling block was not price but royalties. On that issue Nally would not yield.

The idea that the Marconi Company, if left to its own devices, might enter the age of continuous wave transmission somewhat cautiously was not a fanciful one. Self-interest alone might have suggested moderation. The British company and its subsidiaries had a large investment in spark technology; in fact, installation of new spark equipment continued all through the war years. Marconi first tried out his timed spark discharger on the Clifden–Glace Bay circuit in 1915. The first full-sized unit was set up at the Carnarvon station and tested there in 1916. The unit built for the station at Marion, Massachusetts, was delivered late in 1917 and was ready for test in the year following. These events came after, not before, Marconi first expressed interest in the alternator.[32] There was in 1917 no public evidence at all that the Marconi organization had turned its back on spark technology nor that, in the view of Marconi engineers, the last possibilities of spark had been exhausted. This may have been related to the fact that reception of continuous wave telegraphy required a type of receiver—the heterodyne or at least some form of mechanical "tikker"—that Marconi did not possess. Neither, of course, did General Electric—a fact that was hinted at only occasionally in GE's internal memoranda.[33]

[32] Clark Radio Collection, Cl. 100, v. 2, unsigned comments on Clark's history of naval radio.

[33] But see, for an exception, Young Papers, Box 72, Edwards to A. G. Davis,

Uncertainty about Marconi's commitment to the continuous wave and to the alternator in particular became a major factor in the negotiations with the arrival in October 1917 of a letter from H. M. Hobart, one of GE's consulting engineers temporarily in London. Hobart urged in the strongest possible terms that the Marconi Company should not be allowed to get control of the Alexanderson system. "Please accept and act upon my assurance," he wrote, "that when I see you I can satisfy you of the importance of my advice that the G.E. Co. retain <u>complete</u> [in original] control and develop it themselves."[34] The warning was taken seriously. A cable was promptly sent to Hobart instructing him either to cable the full reasons for his opinion or to return to Schenectady immediately.[35] And Edwards was concerned enough to write to the New York office suggesting that negotiations with Marconi be held up until further information arrived. "Knowing Mr. Hobart," he wrote, "we can easily believe that his reasons are good, and further than he would have cabled them if he thought the cable would go through without causing trouble."[36]

Hobart got back to Schenectady in January 1918 and made clear his reasons then. They amounted to a conviction, based on discussions with "over a dozen of the best wireless people in England," including men associated with the Post Office, the Admiralty, the Army, and the railways, that the Marconi Company could not be trusted to support progress in radio technology. In almost all cases, reported Hobart, discontent with the Marconi interests was very much in evidence. "The Marconi people appear to have become very unpopular owing to their policy of minimizing the importance of engineering progress and applying their activities almost exclusively to financial operations." Their fundamental patents had almost all expired. Belatedly, they were now beginning to realize that they could not rely on the old spark method much longer. They had acquired control of the Poulsen arc in Britain but had not exploited it. Now the Alexanderson alternator had come along and "they probably realize that in order to retain any technical prestige they must acquire

---

1 November 1917, "Hetrodyne [*sic*] Reception"; Edwards pointed out that all long-distance radio systems were virtually dependent on heterodyne reception; the Alexanderson system could be so modified as to generate a modulated wave and thus dispense with heterodyne reception but this would require about twice the power for a given distance, which was obviously a serious handicap. He recommended that GE should either buy the heterodyne patents from NESCO or secure an exclusive license under them.

[34] Hobart to Alexanderson, 10 October 1917 (Young Papers).
[35] Cable, Pratt to Hobart, 30 October 1917 (Young Papers).
[36] Edwards to Burchard, 30 October 1917 (Young Papers).

this alternator and put it forward in place of the Spark Method." Whether they would use it energetically was quite another question. The only reason the Marconi Company held such a dominant position in Britain was that it had bought out all its more enterprising competitors. "As soon as these competitors were bought up, all progress in developing their systems is stopped." The implication was clear: if Marconi got control of the alternator, progress in developing that system would stop also.[37]

Hobart suggested, in effect, that GE should develop its own radio system in competition with Marconi. Nobody showed much enthusiasm for that idea at the time. From late 1917 onward, however, lack of confidence in the technical dedication of the Marconi Company became a recurrent theme in GE correspondence and memoranda. Edwards in particular stressed the problem. In February 1918 he said that Hobart's view represented "the concensus [sic] of opinion held by every one with whom I have talked," including members of a visiting French scientific commission, the U.S. Navy Department, and representatives of the Italian government.[38] In June of that year he characterized Marconi's patent policy as being designed "not for the purpose of developing the art but for the purpose of creating a monopoly."[39] Nobody in General Electric doubted that the Marconi Company wanted to buy alternators, and several of them. But what did they want them for? As mere symbols of their dedication to technical progress, as Hobart suggested? Or as the key component in the postwar reconstruction of the world's radio communications? To the executives at General Electric, it made a difference.

In the circumstances negotiations with the Marconi Company could not go smoothly. The provisional agreement of July 1915 was followed in March 1916 by an exchange of letters between Rice of GE and Steadman, who handled most of the negotiations for the Marconi interests. Both parties agreed that the time was not ripe for the signing of a formal contract, since the alternator had not yet reached a stage of development that would justify GE in giving guarantees of performance and delivery dates, nor the Marconi Company in expecting them.[40] Each company committed itself to make no engagement with any third party that might interfere with their entering into a definite contract when conditions became more clear.

[37] Hobart to Alexanderson, 30 January 1918, enclosed in Edwards to Burchard, 7 February 1918 (Young Papers, Box 72).

[38] Edwards to Burchard, 7 February 1914 (Young Papers).

[39] Edwards to Pratt, 7 June 1914 (Young Papers).

[40] Steadman to Rice, and Rice to Steadman, 9 March 1916 (Young Papers).

This understanding soon came under strain. After the successful tests of the 50 kilowatt alternator in 1917 GE began to receive inquiries from other possible purchasers, some of them phrased in urgent terms. Understandably, GE again broached the subject of a contract with Marconi.[41] Marconi insisted that GE should adhere to the "stand-still" agreement of the previous year.[42] In the eyes of the Marconi Company GE was not free to sell alternators to any other party, since that would jeopardize the exclusive rights that the contract would convey when finally signed. GE executives, however, were finding the very idea of an exclusive contract with Marconi increasingly unattractive, as their confidence in the alternator grew and as inquiries began to arrive from other quarters. An Italian military mission, in the summer of 1917, asked GE to quote prices on alternators of various sizes. A French mission was ready to give a firm order—or so GE was led to believe—until it was dissuaded from doing so by the U.S. Navy Department. The Swedish Telegraph Administration was known to be actively interested. And in October 1917 GE's Australian subsidiary sought quotations for two alternators, one to be located at Port Darwin in the Northern Territory and the other at Apple Cross in Western Australia.[43]

To all of these inquiries General Electric, at Marconi insistence, had to give an equivocal response. They did suggest to the company's officers, however, that there might be a sizable market for the alternator independent of the Marconi Company, and resistance to the idea of an exclusive contract began to stiffen further. By August 1917 Edwards was urging that GE "depart radically" from the form of agreement contemplated previously. GE, he argued, should sell its apparatus "at a normal profit, licensing its use by the Marconi Company on a royalty basis, charging so much per word for all messages sent." Under such an arrangement the Marconi Company would benefit from any future improvements GE might make, and GE would be stimulated to make improvements.[44] A. G. Davis agreed. The Alexanderson system, he believed, represented a remarkable improvement in the radio art. "We do not feel that we should sell outright apparatus which would produce such an extraordinary benefit to the Marconi people; rather that we should make a general agreement to give them the benefit of our skill and inventions

[41] Edwards to Rice, 24 July 1917, quoting a cable from Nally to Isaacs: "They [GE] are ready to discuss definite proposal and desire conclude matters at an early date." (Young Papers, Box 72.)

[42] Steadman to Rice, 4 June 1917 (Young Papers, Box 72).

[43] Edwards to Burchard, 27 October 1917 (Young Papers).

[44] Edwards to Burchard, 16 August 1917 (Young Papers).

. . . and to sell them apparatus . . . at a fair price, the understanding being that we should receive a royalty of a certain percentage of the gross income of each station using any of our ideas, patented or unpatented." Certainly no exclusive licenses or exclusive sales agreements should be considered: at most GE should promise, if the Marconi Company promptly submitted a definite order, to give its machines priority over any orders placed by others. Farther than that the company should not go.[45]

Steadman arrived in New York for negotiations with Rice and Burchard in mid-September 1917. Faced with a refusal to sell alternators outright, a firm demand for royalties, and an insistence that he either submit a definite order or free GE from its obligations, he made no progress. For more than a year thereafter there were no official discussions between the two companies. By no coincidence, it was during this period of stalemate with General Electric that the first rumors began to circulate about a deal between the Marconi Company and Federal Telegraph. General Electric, for its part, regarded all previous agreements with Marconi as having been cancelled.[46] Arrangements for the installation first of the 50 kilowatt alternator at New Brunswick and then of the 200 kilowatt unit were made with the American Marconi Company, not its British parent, and they did not involve sale of the machines (although some GE executives were concerned lest Marconi, by possession and use, might acquire some kind of equity right to them).

Fundamentally, however, the situation remained unchanged. The pressures that had brought General Electric and the Marconi companies into negotiation between 1915 and 1917 were still present when, with the signing of the armistice, thoughts turned to peacetime markets. Marconi could not postpone much longer the modernization of its high-power stations and, with access to the Federal arc patents foreclosed by Navy intervention, the alternators were the only possibility remaining open. General Electric still had not earned a dollar from its larger alternators; hopes for sales to the Navy had been dashed by Sweet's insistence on exclusive rights; civilian markets would have to be found. Nor was there much indication in 1919 that either side had changed its bargaining position or redefined its corporate interests. The Marconi group still wanted exclusive rights to the alternator and did not want to pay royalties. General Electric was determined that royalties would be paid and that there would be no exclusive contract. And underlying that polarization was distrust: the recurrent suspicion on the part of the engineers

[45] Davis to Burchard, 21 August 1917 (Young Papers).
[46] Cable Landing Hearings (1921), testimony of Owen D. Young, p. 332.

and executives of General Electric that the Marconi companies could not be trusted to use the alternators as they could and should be used.

*　　*　　*

Negotiations between General Electric and the two Marconi companies were resumed at the end of February 1919 and continued through March with slow progress being made. By the last week in March Nally and Steadman had made clear their intention to purchase alternators in quantity. They were prepared to place immediately a firm order for at least twenty-four alternators of 200 kilowatt capacity each, fourteen for the American company and not less than ten for the British.[47] The base price for each unit (including not only the alternator but also the power switchboard, pumps, exciter, speed regulator, magnetic amplifier, and telegraphic signalling system) was to be $125,000 f.o.b. Schenectady. All charges were to be billed to, and paid by, the American company.[48] There was no explicit provision for an exclusive license, but GE did undertake to ship the first unit within eight months from the date of the contract and the second by the end of twelve months. Thereafter they expected to produce at the rate of two complete sets of equipment each month. It was clear, therefore, that if a contract with Marconi were signed in April 1919, it would be the summer or fall of 1921 at the earliest before any other customers could expect their orders to be filled. By that date the Marconi companies would have their transmitters in place and operating.

So much was relatively straightforward. Nally and Steadman had made it clear in the course of the discussions that they wanted exclusive rights to the alternators. GE for its part had ruled that unacceptable, but (as Davis later expressed it) "it was understood by all present that the Marconi Company might place an order with us which, in view of our limited capacity for producing these machines, would monopolize our output for a considerable time."[49] Where there had been no meeting of minds was on the old issue of royalties. GE executives were agreed on this point, though they differed as to whether royalties should be based on number of words of traffic handled or on the gross revenue of each station. The first proposition submitted had been for a royalty of 7 ½ percent of the gross revenue of "each station unit equipped for sending" (of which there

---

[47] Davis to Young, 13 May 1919 (Young Papers).

[48] Edwards to Nally, 7 April 1919 (Young Papers). This letter was not in fact sent, but does report the state of negotiations at the time.

[49] Davis to Young, 13 May 1919, p. 3 (Young Papers).

might be more than one at each station). Later in March the figure was reduced to 5 percent. Neither was acceptable to Nally, and it became evident that he and Steadman either had been given instructions from which they could not deviate or that it was a matter of nonnegotiable principle. Instead, Nally made an oral offer of $1 million cash in addition to the normal purchase price in lieu of royalties.[50] The significance of the issue from the Marconi point of view could hardly have been demonstrated more forcibly.

This offer was not accepted by General Electric, A. G. Davis in particular advising against it. By the end of March something very close to stalemate had been reached. On the 31st of that month Edwards went so far as to recommend to Burchard that General Electric should make an approach to Western Union or the Postal Telegraph Company, since it seemed to be impossible to come to terms with Marconi.[51] Alexanderson also argued against any further concessions. The 5 percent royalty, he asserted, could easily be paid from the operating economies that the alternator would yield, quite apart from any new traffic generated. Nally's estimates of future revenue, he thought, were far too low. The earning power of each alternator would prove to be at least $300,000 a year; a 5 percent royalty on the total would yield GE an income of $370,000 annually. The expenses of Alexanderson's own department for radio development had been averaging $200,000 a year for the last two years. If GE was to maintain its position in the field, it was only reasonable to assume that it would have to double that rate of expenditure by the time it had twenty-five high-power stations in operation. The royalty income, therefore, was essential. If the Marconi Company was not willing to recognize that logic, Alexanderson would interpret it to mean that they were not serious in proposing to establish a bona fide partnership and community of interests. And he concluded, in words unusually strong for a man who normally kept his emotions to himself, "Rather than looking forward to the mental agony of having moral responsibility for such a large undertaking without a business basis for making the same a continued success, I should personally prefer to have the Marconi Company adopt some other system and leave us free to develop our system on its own merits."[52]

Any account of these negotiations, therefore, that depicts General Electric as being on the point of concluding an agreement with Marconi in the spring of 1919 is wide of the mark. The executives and engineers at

[50] Ibid., p. 2.
[51] Edwards to Burchard, 31 March 1919 (Young Papers).
[52] Alexanderson to Burchard, 4 April 1919 (Young Papers).

Pl. 13: Alexanderson inspecting a 200 kilowatt alternator.
Source: General Electric Company

GE who had been closest to the negotiations were, during the first week of April 1919, the ones least sure of a successful outcome and most inclined to look elsewhere. Lacking any alternative, they would of course have worked out a contract of some kind. If, however, an alternative strategy appeared on the horizon, it was certain to receive an interested welcome.

Vice-President Burchard of GE sailed for Europe on 5 April, to meet pressing obligations and incidentally talk with Godfrey Isaacs of British Marconi. He left Edwards in charge, with instructions to draw up one final counterproposal to submit to Nally and Steadman. As drafted by Edwards, the proposal contained nothing new in principle. It stipulated that a firm order would be placed immediately for twenty-five of the 200 kilowatt alternators and associated equipment, for a price per unit of $125,000, or a total of $3,125,000, of which 15 percent was to be paid in cash with the order. On the thorny issue of royalties there was a modification in detail but no retreat on the principle: the percentage was to start at 1 percent on the first $100,000 of gross revenue per "station unit equipped for sending" each year, and rise to the full 5 percent with the fifth $100,000. And a maximum royalty of $25,000 per year per station unit was stipulated. Delivery dates were specified; and GE undertook to guarantee performance "equivalent to that demonstrated in actual practice at the New Brunswick station."[53]

Apart from the introduction of a sliding scale of royalties there was nothing in this proposal that was new. Nor was there any reason to suppose that this minor change would be enough to overcome the determined opposition to the payment of royalties that Nally and Steadman had shown for several years previously. Strange things can happen in close bargaining, however, and sometimes a minor concession by one party can get stalled negotiations moving again. Whether something like that would have happened in this case must remain conjectural, for Edward's draft was never in fact shown to Nally and Steadman and all copies except one, which Owen Young of GE kept in his personal records, were destroyed.[54] Instead, on 9 April, the British and American Marconi

[53] Edwards to Nally, 7 April 1919, attached to Edwards to Young. 9 April 1919 (Young Papers).

[54] In his last letter to Nally before leaving for Europe, Burchard assured him that GE's counterproposal, though delayed, would be in line with Nally's suggestions, although GE would not be satisfied with the very low royalty rate Nally had proposed (Burchard to Nally, 4 April 1919, Young Papers). This is the first suggestion that Nally was prepared to consider royalties of any kind. Nally also insisted (writing in 1940) that at the last conference with GE, with Steadman not present, it was stipulated only that, if the Marconi companies gave GE a large order for alternators, GE would not sell them to any other company until after the Marconi order had been completed. See Clark Radio Collection, Nally to G. H. Clark, 16 August 1940. For Young's orders to destroy all copies of the counterproposal, see Edwards to Young, 9 April 1919, and Young to Edwards, 10 April 1919 (Young Papers).

companies were formally notified that all negotiations regarding the sale of alternators were terminated. No explanation was given at the time, but the reason for the sudden change in strategy soon became apparent. Once again, the U.S. Navy had intervened, this time not with an offer to buy patents, such as had been made to the Federal Company, but with a proposal that promised both a market for the alternators and freedom from the Marconi insistence on exclusive control.

Two chains of events, taking place at different levels in the official hierarchy, converged to bring about the Navy's belated intervention. The first began modestly enough. On 25 February 1919 Rear Admiral Griffin, chief of the Bureau of Steam Engineering, sent to General Electric a letter requesting that Ernst Alexanderson's services be made available to visit the naval radio station at Sayville and make a report on a speed control system for the high frequency alternator in use there. The request probably seemed to Griffin almost routine. Speed control (and therefore frequency control) had always been a problem with the Sayville alternator. This was a Von Arco machine that generated its fundamental frequency at 8 kilohertz, doubled to 16 kilohertz, and then doubled again to give a radiated wavelength of 9,400 meters. This ingenious frequency-doubling system meant that the speed of the alternator rotor itself was much less than in an Alexanderson alternator; but it also meant that any irregularity in the speed of the alternator had a multiplied effect on the frequency of transmission. The Atlantic Communications Company, original owner of the station, had paid a substantial bonus to the local power company to keep the line frequency constant, but the Navy could not or would not do this and acute problems of frequency stability were the result. Alexanderson's alternators, in contrast, were renowned for their excellent speed regulation. If anyone could stabilize the Sayville alternator, Alexanderson could—assuming, of course, that his talents and General Electric's patented speed-regulating system were made available to the Navy.

If Griffin's letter had been received while the war was in progress, Alexanderson would have been on his way to Sayville at once. But now the more calculating days of peace had arrived, and there were those at General Electric inclined to question the Navy's right to get something for nothing. Sweet's uncompromising insistence on the Navy's right to veto commercial sales of the alternator was, after all, still fresh in the corporate memory. A. G. Davis in particular read much significance into the request. The Navy clearly intended, in his judgment, to get the benefits of Alexanderson's system—the magnetic amplifier and speed regulator in particular—without paying for them. This should be resisted. Now that the war was over, he argued, the government should be treated no differently from any other customer. The sooner matters were back on a straight business basis, the better; and that meant refusing to install any of GE's devices at Sayville, or Annapolis either for that matter, without appropriate compensation.

The task of answering Griffin's letter fell to Owen D. Young, vice-president and general counsel for General Electric. Young had joined the company in 1913, shortly after an antitrust suit filed in 1911 had been settled by a consent decree on the government's terms; his major area of responsibility had been governmental relations and what was referred to generally as "policy." Little known outside the electrical and public utility fields when the United States entered the war, and known within those industries mainly for his legal work with local power and transit companies, he had spent most of his time since 1917 working on problems of material supply—negotiating over "priorities" as a kind of general liaison officer with government agencies and private suppliers.[55] He was a skilled negotiator, a master of accommodation and reconciliation, and by 1919 he had emerged as one of GE's key executives.

Young's first inclination was to accede to Griffin's request. He saw no reason why it should not be met. "Personally, I should prefer to cooperate with the government and meet the license question directly with the feeling on their part that we are trying to be helpful rather than antagonistic."[56] But this was far too mild for his colleagues, who "descended on him in protest" and insisted that he take a harder line.[57] The result was a letter that, going far beyond the immediate business at hand, put the Navy on notice that General Electric had interests of its own to protect.

Young began by acknowledging Griffin's request for Alexanderson's services and assuring the Navy of GE's desire to cooperate with the government in any reasonable way. He expressed confidence that the Navy would not ask the company to do anything "inimical to our commercial interest" and reviewed some of the contracts for radio equipment that had been successfully concluded in the past. He continued: "At the same time we are in active negotiations with the British and American Marconi Cos. for the sale to them of a substantial number of our high-

[55] See Josephine Young Case and Everett Needham Case, *Owen D. Young and American Enterprise: A Biography* (Boston, 1982). Outdated by the Cases' biography but still worth reading is Ida Tarbell, *Owen D. Young: A New Type of Industrial Leader* (New York, 1932).

[56] Young to Davis, 25 March 1919 (Young Papers).

[57] Case and Case, *Owen D. Young and American Enterprise* (Boston: David R. Godine Inc., 1982; quoted by permission), p. 175.

power radio equipments with the necessary accessories . . . including a license to those two companies to utilize our system commercially on a royalty basis." And he concluded by suggesting that it would be very helpful if, in the near future, an opportunity could be found to talk the situation over fully, so that the company could be in a position to furnish such equipment and advice as the Navy might require, and at the same time "retain a reasonable protection of the commercial interests of the General Electric Co."[58]

The letter was carefully phrased, but it did not move arrangements for Alexanderson's visit to Sayville one step closer to realization. That, in any event, was a matter that could have been handled by subordinates. The true purpose of the letter was quite different: it was intended to give the Navy Department formal notice that the General Electric Company had been negotiating with the Marconi companies for the sale of alternators and intended to come to a final agreement on that subject unless the Navy did something to stop it. If the Navy intended to intervene, it would have to act quickly. That was the message Young's letter conveyed. It would be going too far to say that it positively invited the Navy to act but certainly it carried a warning that General Electric was not an agency of government but existed to earn a return on its capital and intended to act accordingly unless the Navy did something to prevent it.

The letter was sent, significantly, not to Admiral Griffin or his deputy but to Acting Secretary of the Navy Franklin D. Roosevelt, in charge of the Navy Department while Josephus Daniels was in Europe. Roosevelt replied on 4 April, asking GE not to conclude any agreement with the Marconi companies until after a conference with representatives of the Navy.[59] He suggested 11 April as a suitable date and Young agreed. What might have happened at that conference must remain a matter for speculation, since it never took place—not because either party was reluctant but because, at lower levels in the two hierarchies, events were moving along a parallel course but at a faster pace.

The officer in charge of the Radio Division of the Navy's Bureau of Steam Engineering at this time was Lt. Comdr. Stanford C. Hooper.[60]

[58] Young to F. D. Roosevelt (draft), 29 March 1919 (Young Papers); the letter is reproduced in its entirety in FCC Hearings (1929), p. 1108. Admiral Griffin was out of the country by the end of March; normally, however, the correspondence would have been handled by his deputy.

[59] F. D. Roosevelt to Young, 4 April 1919 (Young Papers).

[60] Stanford Caldwell Hooper was born in Colton, California, in 1885. His father was a railroad agent for the Southern Pacific and later a banker in San Bernardino. Hooper was originally nominated to the Naval Academy as an al-

This was no mere temporary assignment for Hooper. He had been involved with Navy radio almost from the day he received his commission as ensign in 1907 and was completely dedicated to that branch of the service. First placed in charge of the Radio Division in 1913, he had been associated with it ever since, except for a short period of sea duty in 1917-1918. In that capacity he had borne major responsibilities for the procurement of radio equipment and, inevitably, he had come to know many of the individuals who worked for the Navy's civilian suppliers. With some of them he had become friends. Among these was Edmund P. Edwards of General Electric. Edwards's somewhat anomalous title was assistant manager of GE's lighting department; in fact he was in charge of the marketing of radio equipment and, as we have seen, he had been very active, along with Anson Burchard, in the negotiations with the Marconi companies.

Hooper and Edwards understood each other and shared many of the same values. To Hooper, Ted Edwards was "a good American" who wholeheartedly supported the Navy's efforts to keep American radio from falling under foreign control.[61] To Edwards, Hooper was the man who, more than any other single individual, had built naval radio into the impressively efficient communications system it was by 1919. He was also, of course, a very useful man to know in the Navy Department, just as, for Hooper, Edwards was a useful contact at GE. They kept in touch regularly, by face-to-face meetings, by correspondence, and by telephone. Through Edwards, Hooper was continuously apprised of the progress of negotiations between GE and the Marconi companies.[62]

Hooper's major concerns during the winter months of 1918-1919 were two. First there was the immediate problem of what to do about the Alexanderson alternator at New Brunswick. Whatever happened, he was determined that this machine should not fall into the hands of the Amer-

---

ternate on 29 May 1900. The cadet who actually entered the academy in that year from that congressional district was dismissed. Hooper applied to take his place, was renominated in the year following, and was appointed a midshipman on 31 August 1901. On graduation in 1907 he ranked 52nd in a class of 108. For further biographical information, see Hooper Papers (Library of Congress), Boxes 4, 37, 40, and 44; L. S. Howeth, *History of Communications-Electronics in the United States Navy* (Washington, D.C., 1963), pp. 113-14 and *passim*; and National Archives, U.S. Navy, Bureau of Navigation, files relating to naval cadets.

[61] Hooper Papers, Box 37 transcript of tape-recorded memoirs.

[62] Ibid., and Box 3, Hooper to Edwards, 28 April 1921; Clark Radio Collection, "Radio in War and Peace," pp. 26, 37A and C1. 100, v. 2, Hooper's comments on Clark's manuscript.

ican Marconi Company. Second there was the longer-term problem of how to keep control of American radio in American hands. Secretary Daniels's strategy of government ownership seemed certain to fail, and Hooper had never fully supported it anyway. But if radio were turned back to private hands, those hands would almost certainly prove to be the American Marconi Company—a far cry indeed from Hooper's dream of a corporation that would be "100 per cent American."[63]

But if government ownership was improbable, and the American Marconi Company unacceptable, what alternatives were there? Hooper could think of only two: General Electric or AT&T. But the Telephone Company, after its long-distance radiotelephone tests in 1915, had confined itself strictly to the wired telephone business. General Electric was a more likely candidate, but GE had shown no public interest in the traffic-handling side of radio. It was a radio manufacturing company (among its many other interests), not a radio operating company. Still, it was a possibility. In January 1919 Hooper thrashed the whole matter out with a fellow officer, Lt. Comdr. E. H. Loftin, head of the patent board in the Radio Division. To strengthen the Navy's position he arranged for the Navy to acquire legal title to the station at Sayville; and, at Loftin's suggestion, he acquired from the Alien Property Custodian on behalf of the Navy title to all the German radio patents seized during the war. Meanwhile, to Edwards, his friend at Schenectady, he confided that he had a "big proposition" in mind for the General Electric Company. He asked him to keep the matter secret and believed that he did so. But Edwards, who was indeed Hooper's friend but who was also a loyal employee of General Electric, made sure that his superiors were informed.[64]

In mid-February 1919 Hooper asked for and got an interview with Secretary Daniels. He explained to him the problem of the alternator at New Brunswick and told the secretary that, according to his sources of information, GE and the Marconi companies would probably reach an agreement very soon if nothing were done to prevent it. Daniels asked his advice. Hooper replied that the first essential was to appoint an officer of flag rank, or at least a senior captain, as director of naval communications. He himself was only a lieutenant commander, and in any case he was in charge of materiel and had no jurisdiction over the communications work of the department. Todd, the previous DNC, had been

[63] Hooper Papers, Box 37.

[64] Young Papers, Box 72, Edwards to A. W. Burchard, 16 August 1917, and A. G. Davis to Burchard, 21 August 1917; Box 75, Edwards to Young, 8 July 1918; J. and E. Case, *Young*, pp. 173-76.

sent to the Baltic to take command of the *Pittsburgh*, but no relief had been appointed and the officer temporarily in charge was junior to Hooper. If the Navy Department was to take a firm stand with General Electric and with Marconi, its spokesman had to be someone with rank and authority.

Daniels, according to Hooper's later account, asked for suggestions and Hooper named Adm. W.H.G. Bullard, the only available senior officer with radio experience. Bullard had been superintendent of the Naval Radio Service from 1912 to 1916 and before that had originated the department of electrical engineering at the Naval Academy.[65] He was at the moment in command of U.S. naval forces in the Adriatic but his only immediate responsibilities were to oversee execution of the naval clauses of the peace treaty with Austria-Hungary and to take part in the work of the commission investigating conditions in Fiume. He could be spared from those duties. Daniels concurred, and a memorandum went off at once to the Bureau of Navigation directing that Admiral Bullard be appointed director of naval communications, with instructions that he stop off in Paris on his way home for further instructions.[66]

Bullard did indeed spend some time in Paris on his way back to Washington. While there he spoke with Daniels, primarily to make it clear that he could not support the secretary's policy of government ownership, and with his predecessor in office, D. W. Todd, who had been temporarily detached from his command for duty at the Inter-Allied Radio Conference.[67] He also spoke with President Wilson's personal physician, Dr. Cary Grayson; and it is possible—the evidence on the point is contradictory—that he may have talked briefly with President Wilson himself. As a result of these conversations Bullard came back to Washington carrying a message that he believed embodied the president's wishes. Steps should be taken to conserve American rights and resources in the field of radio, and those possessing such rights should not part with them but use them to establish an international communications system for the United States. This was the message that was later conveyed as a "state secret" to Owen Young (see below, pp. 340-41); and it was on this message that Young relied in 1929 in suggesting to skeptical senators

[65] W.H.G. Bullard, "Some Facts Connected with the Past and Present Radio Situation of the United States," United States Naval Institute *Proceedings* 49 (October 1923), 1630.

[66] Clark, "Radio in War and Peace," pp. 38-39; L. S. Howeth, *History of Communications-Electronics in the United States Navy*, (Washington, D.C., 1963) pp. 353-54.

[67] Clark Radio Collection, Cl. 100, v. 2, p. 39, Hooper's comments on Clark's manuscript; Howeth, *History*, pp. 353-54.

that the Radio Corporation of America had been formed as a result of presidential initiative.[68]

In some later versions of these events Bullard is depicted as having carried from President Wilson a message to be delivered specifically to Owen Young, requesting him, as a patriotic duty, not to sell alternators to the British.[69] Hooper's account of the matter is more mundane but, even if we discount any tendency to play up his own role in the matter, it has a ring of plausibility about it. According to his recollections, Bullard arrived in Washington about 3 April.[70] Hooper gave him a few days to unpack and get settled. Then, in company with Lieutenant Commander Sweet, he called on Bullard to bring him up to date. He told him essentially the same story as he had earlier told to Secretary Daniels. He described "the alternator situation." He pointed out the importance of the machine already in place at New Brunswick. And he made it clear that, if the GE–Marconi negotiations were allowed to proceed on course, it was well-nigh certain that a binding contract would be signed within a matter of days. If that happened, he argued, one of the supreme achievements of American radio engineering would become the exclusive property of a foreign corporation. For, even if the contract with GE did not guarantee Marconi exclusive rights to the alternator, it was certain that filling the Marconi contract would keep GE's production facilities fully occupied for several years to come. By the end of that period the Marconi companies would undoubtedly have won the position of world dominance in radio communications that they had sought and almost achieved before 1914.

We may be sure that Hooper did not understate his case. He felt very strongly about it; and, through Edwards, he knew that the signing of a contract with Marconi might be only a few days off. Owen Young's letter of 29 March to Roosevelt had of course reached the Navy Department by this time, and Roosevelt had replied suggesting a conference on 11 April. At that level GE's intentions and the Navy's response were now matters of record. But Hooper needed no letter from Young to tell him what was going on, nor had he any intention of waiting until 11 April.

Bullard's response, as Hooper later recorded it, was one of interest and gratification. Interest, because he had been confused and uncertain before; gratification, because now he understood what he was supposed

---

[68] FCC Hearings (1929), testimony of Owen D. Young, p. 1089.

[69] FCC Hearings (1929), testimony of Cary T. Grayson, pp. 1174-76; "Freedom of the Air," by Owen D. Young, as told to Mary Margaret McBride, Saturday Evening Post, 16 November 1929.

[70] Hooper Papers, Box 37, transcript of tape-recorded memoirs.

to do. Matters had not been made entirely clear to him, it would appear, in Paris. He told Hooper about Wilson's breakfast with Lloyd George: how an aide brought a radiogram to the British prime minister and Lloyd George had let drop some remark about the importance of radio; and how this had triggered in the president's mind the thought that something would have to be done to protect American interests in that field. Cary Grayson, at Wilson's request, had talked to Bullard about it, but it had all been quite indefinite. Bullard continued (as Hooper later recorded his recollection of the conversation): ". . . maybe that is what the president sensed was needed after he heard Lloyd George make this remark to the young officer, because Dr. Grayson came to me and said that the president instructed him to tell me something very important about the radio situation and told me to do something about it. Neither I nor Dr. Grayson knew what it was. I had no inkling of what it was until I just heard your story. So maybe they fit together."[71] The president had been vague as to details, said Bullard; all he said was that there was "something important" in the radio situation, and that Bullard was to find out what it was and then act on it. Wilson had had a clear idea of what was involved when he first heard Lloyd George's remark, but the original thought slipped his mind and "left merely the impression that there was an important idea there for the U.S. and that he wanted it looked into."[72]

Whether Bullard had met and talked with Wilson personally in Paris is not certain; it is quite possible that he did.[73] But it is clear, if Hooper's account is to be trusted, that any instructions Bullard received, whether from Wilson directly or through Grayson, were cast in general terms. Wilson was deeply concerned, at this point in the negotiations, with problems of postwar communications policy. The memoranda he had received from his advisers had laid heavy stress on British "domination" of the submarine cable system and the new potentials opened up by radio.

[71] Ibid. The conversation between Wilson and Grayson took place between 16 and 21 March. Bullard's appointment as director of naval communications was effective the end of February. See Hooper to Dodd, 18 February 1919 (Hooper Papers) and Howeth, *History*, p. 354n. Lloyd George had, of course, particular reasons to appreciate the political significance of radio; his own career had been nearly ruined a few years before by ill-advised speculation in Marconi stock.

[72] Hooper Papers, Box 45, manuscript "History of Radio."

[73] Hooper at one point quotes Bullard as saying, " That just fits in with something the President said to me when I saw him in Paris." (Hooper Papers, Box 45, manuscript "History of Radio") In his tape-recorded memoirs, however, he describes the message from Wilson as having been conveyed to Bullard through Grayson (Hooper Papers, Box 37). I have been informed by Professor Arthur Link that Grayson's diary, in his custody, contains no reference to this event.

To convey to his new director of naval communications, either directly or through a trusted intermediary, a general injunction to look out for American national interests would be entirely natural.

This is, however, a far cry from any specific instructions concerning alternators or General Electric or Owen Young. It is possible, of course, that Bullard received more detailed and specific briefings from Secretary Daniels or from Lieutenant Commander Todd. But, if so, it is hard to understand his later statement that his talk with Hooper in Washington was the first time that the impending sale of the alternators to Marconi was brought to his official attention. Instructions received from the secretary of the navy or from the outgoing director of naval communications would hardly fall into the category of "unofficial." The evidence suggests, therefore, that Hooper's account is essentially correct: Bullard was not informed of the state of negotiations between Marconi and General Electric until after his arrival in Washington. Nor did he think of the situation as critical and requiring immediate intervention until Hooper drove home the point. Grayson in Paris had given Bullard no more than a warning and a puzzle: Hooper supplied the key.[74]

✸   ✸   ✸

Hooper, however, was not the only individual anxious to inform Bullard of the impending contract with Marconi, nor was he alone in believing that the situation called for immediate action. For more than two years the Swedish government had been asking the U.S. Navy to find out whether its Telefunken-equipped station at Karlsberg could be heard regularly and reliably in the United States. Roosevelt at first said that

[74] Compare Hooper Papers, Box 37: "He didn't even know what it [the problem] was and I had the key to the lock." Bullard's own account of these events makes no mention either of Wilson or of Hooper but states clearly that the GE–Marconi negotiations were brought to his official notice "a few days after my arrival in my office"—a statement he could hardly have made if the negotiations had been discussed with him in Paris. See Bullard, "Some Facts Connected with the Past and Present Radio Situation in the United States," United States Naval Institute *Proceedings*, 49 (September 1923), whole number 247, 1630. Howeth (*History*, p.355) says that Bullard "must have" discussed the situation with Acting Secretary Roosevelt, but this is purely speculative, as is his earlier assumption that Wilson and Daniels had discussed an American-controlled operating company while traveling together to France on the U.S.S. *George Washington* (ibid., p. 354). For other accounts of the episode, see *Cable Landing Hearings* (1921), pp. 345-60, testimony of Owen D. Young, and "Admiral Bullard: The Director of Naval Communications in a New Role," *Wireless Age* 8 (February 1921), 13-14.

there would be no problem. But W.A.F. Ekengren, the Swedish minister in Washington, was skeptical—"In earlier cases he [Roosevelt] has been found to promise more than he could later fulfill"—and, as it turned out, with good reason.[75] Not until January 1919 were arrangements concluded; when the tests were carried out, they showed the Karlsberg transmitter to be quite inadequate.[76] Naval experts in the United States told Ekengren the solution was higher power. A Telefunken quenched-spark station of 60 kilowatts power might be good enough for communications within Europe but transatlantic work was a different proposition.

The Alexanderson alternator looked as if it might offer a solution. Preliminary negotiations for the purchase of a single alternator were opened on 14 January 1919. General Electric held out hopes that a reasonable price would be quoted and that delivery and installation might be completed within eight months. At the same time, however, Alexanderson warned the Swedish representatives that discussions were going on with the Marconi Company and that, if they reached a successful conclusion, sales price and conditions of sale would probably be quite different.[77]

What most concerned the Swedish Foreign Office and Telegraph Administration was that their radio communications with the United States should be completely free from British control.[78] Experience with the submarine cables before and during the war had left bitter memories. Freedom from British control of radio meant, as a practical matter, freedom from control by the Marconi Company. Swedish suspicions of British influence, and specifically of the Marconi companies, had found ready sympathy in the U.S. Navy Department.

Ekengren was informed on 5 April 1919 of the state of negotiations between General Electric and Marconi, and learned for the first time that Marconi hoped to obtain exclusive rights to the alternator, or at the very least to submit an order large enough to tie up GE's manufacturing capacity for the indefinite future.[79] This, of course, would spell disaster for Swedish plans. At the most Sweden was in the market for a single machine, while Marconi intended to order twenty-five. Even if GE held

[75] Swedish Royal Telegraph Administration files, Ekengren to Rydin, 13 March 1918. I acknowledge with thanks the indispensable assistance of my friend Kaye Weedon in locating and translating these sources in the archives of the Swedish Foreign Office and Royal Telegraph Administration.

[76] Swedish Foreign Office files, Ekengren to Foreign Minister, Stockholm, 23 January 1919.

[77] Swedish Foreign Office files, Seth Ljungqvist to Ekengren, 5 February 1917.

[78] Swedish Royal Telegraph Administration files, Gosta Lindman to Rydin, 22 March 1919.

[79] Swedish Foreign Office files, Lindman to Ekengren, 5 April 1919.

out against the demand for exclusive rights, it was hardly to be expected that they would hold up the Marconi order to satisfy what had never been more than a tentative understanding with Sweden. Control of the alternator by the Marconi companies would give the quietus to Sweden's hopes for an independent radio facility. Marconi would undoubtedly install one of the new alternators at Stavanger in Norway and that station, working with its sister station at Marion, Massachusetts, would command radio traffic between the United States and northern Europe.

Ekengren needed no one to spell out these implications for him. He immediately "rushed to the Navy Department" and laid his case before Bullard, pointing out "the danger of Marconi through his manoeuvers achieving the object of excluding them from the right to the Alexanderson machines."[80] He found Bullard gratifyingly well-informed on the subject, for Hooper had already done his work. In fact, action had been taken that very day—action designed to safeguard what the Navy regarded as American national interests but which also, incidentally, served to keep alive Swedish hopes.

That action involved a telephone call from Hooper to Edwards, who had just returned from New York to Schenectady to prepare, as instructed by Anson Burchard, the final draft of a contract with the Marconi Company.[81] Hooper's request was simple: would Edwards hold up the final signature of the contract; and would he arrange a conference in the immediate future, to include at least Bullard, Hooper, Edwards, and if possible E. W. Rice, president of GE, to discuss the implications of the contract and also the plan that Hooper had mentioned to Edwards on several previous occasions. There was no mention of any correspondence between Owen Young and Roosevelt, no reference to any conference already scheduled for 11 April. Hooper had the initiative and he wanted action at once. Edwards obliged. The conference was set up for Tuesday, 8 April, in the New York offices of the company.

Edwards phoned Bullard on 7 April to confirm final arrangements.[82] That evening Bullard and Hooper took the train from Washington to New York. The men sat in the smoker for two hours while Hooper went over the situation yet again and explained what needed to be done. The admiral listened patiently, took it all in, and at the end said he understood,

[80] Swedish Royal Telegraph Administration files, Ekengren to Rydin, 10 April 1919. Compare John W. Hammond, *Men and Volts: The Story of General Electric* (New York, 1941), p. 375.

[81] Edwards to Hooper, 20 November 1920 (Hooper Papers, Box 3). Edwards gave the date of the telephone conversation as "the 5th or 6th of April." Compare Hammond, *Men and Volts*, p. 376.

[82] Hammond (*Men and Volts*, p. 376) gives the date of the conference as 5 April, but this is an error. See below, n. 87.

he approved, and he would present it the next day as it had been presented to him.[83]

Stanford Hooper went to some pains, in later years, to make sure that his role in the formation of the Radio Corporation of America was properly recorded and appreciated. George Clark's ambitious history of naval radio (never, unfortunately, completed) was written, he tells us, partly because Hooper wanted "the exact truth" to be known regarding the birth and growth of RCA; and the chapter dealing specifically with that topic was written from Hooper's dictation and revised several times to incorporate Hooper's suggestions.[84] But Hooper was not alone in believing that he had played a pivotal role. "The original visit of Admiral Bullard and yourself to us was the beginning of the Radio Corporation," wrote Owen Young to Hooper in 1920.[85] Two years later Young was even more direct: ". . . you were the initial and inspiring influence which caused the whole setup and Admiral Bullard and I were merely agents of execution."[86] Young had pragmatic reasons, beyond personal respect, to recognize Hooper's contributions. RCA in 1922, facing massive public resentment over its legal monopoly of vacuum tube production, had much to gain from the Navy's goodwill and from public recognition that the corporation had been formed at the Navy's initiative. Young would hardly have been so explicit, however, if his own knowledge of the events had not confirmed what it was expedient to say.

✴   ✴   ✴

Ida Tarbell, the historian, in one of her more enthusiastic moments, referred to 8 April 1919, as "a date in the history of radio which takes

[83] Clark Radio Collection, Cl. 100, v. 1, "Radio in Peace and War," by George H. Clark (typescript), chap. 12, p. 42.

[84] Clark Radio Collection, Cl. 100, v. 2, draft of a letter from G. H. Clark to Elmer Bucher, 31 March 1947. Clark also submitted his draft of the relevant chapter to David Sarnoff of RCA, but Sarnoff refused to read beyond the first page and sent it to the Legal Department for approval. See Clark Radio Collection, Cl. 100, v. 2, David Sarnoff to Horten Heath, 22 August 1940. In a further letter to Bucher dated 5 April 1947 Clark said that he had enclosed "a more complete, and accurate, statement of the origin and censorship connected with that chapter." This statement, unfortunately, has not come to light. Clark's typescript and large collection of documents were used extensively by Howeth in preparing his *History of Communications-Electronics in the U.S. Navy* and by Rupert Maclaurin for his *Invention and Innovation in the Radio Industry.*

[85] Young to Hooper, 8 May 1920 (Young Papers).

[86] Young to Hooper, 6 June 1922 (Young Papers).

about the same place as the Fourth of July in the history of the country."[87] Nationalism did indeed characterize much of what went on that day in the New York offices of General Electric. There was also, however, a fair measure of hardheaded calculation.

The conference met in the company's offices at 120 Broadway. There were present, for the Navy, Bullard and Hooper; and for General Electric, President Rice, Vice-President Young, A. G. Davis, C. W. Stone, and E. P. Edwards. Bullard, it would appear, had met none of the GE executives before. Hooper knew Edwards well, and possibly Davis, but not the others. Conspicuously absent from the meeting were Vice-President Burchard, who was on his way to Europe, and GE's chairman of the board, Charles Coffin, still a powerful figure in the affairs of the company.

Bullard made the first presentation for the Navy. He began by telling of his experiences in the Far East, where foreign cables provided the only means of rapid communication with the outside world, and described how American business messages had often been held up—the inevitable result, he suggested, of dependence on communications controlled by foreigners. Then he launched into a plea for an American radio service controlled by Americans. He explained particularly the danger to American national interests that would result if the Alexanderson alternator were sold to any foreign government or foreign private company, arguing that possession of the machine gave Americans a chance to control long-distance communications in a way that had never been possible with the cables. And he set forward his dream of "a policy of wireless doctrine similar to the greater Monroe Doctrine, by which control of radio in this continent would remain in American hands." Specifically, he urged the officers of General Electric, as patriotic Americans, to refuse to sell alternators to either the British or the American Marconi companies. Experience with GE's radio system at New Brunswick had proved it to be the best in the world; the essential thing now was to keep control of that system in American hands. Failure to do so would "fix in British hands a substantial monopoly of world communications."[88]

[87] Tarbell, *Young*, p. 130. There has been considerable confusion about the date of this conference. Young gave the date as 5 April (*Cable Landing Hearings*, p. 333 and *FCC Hearings*, p. 1101), Hooper as 7 April ("Keeping the Stars and Stripes in the Ether," *Radio Broadcast* [June 1922] 131), and Bullard as 6 April ("Some Facts," 1631), while Howeth (*History*, p. 355) heroically resolved the problem by asserting that there were two conferences, one on 7 April, the second on the day following. The date in the text is authenticated by GE correspondence; see Edwards to Hooper, 20 November 1920 (Hooper Papers) and Davis to Young, 30 June 1919 (Young Papers).

[88] Davis to Young, 13 May 1919 (Young Papers); *Cable Landing Hearings*

It was an effective presentation, and Bullard was allowed to proceed without interruption. When he had finished Owen Young began questioning him, in a manner that made Hooper, who had so far been silent, somewhat edgy, for Young's questions were searching and Bullard, who had "learned his lesson" only the evening before, had difficulty handling some of them.[89] Soon, however, it dawned on Hooper that, far from trying to break down Bullard's argument, Young was trying to develop it. As he later recalled, "I became an admirer of his from that moment."[90] Several telling points were made, however. Young pointed out to Bullard that GE's business was to build and sell apparatus to its normal customers, and it was not in the habit of entering into competition with those customers. A very large amount of money had been spent on developing the alternator and the Marconi companies were, practically speaking, the only large customers in sight. Negotiations with the Marconi group had already gone very far. If they were suspended now, how could GE hope to realize any return on the investments it had already made? Certainly the officers of GE, said Young, were responsive to the considerations of patriotism that Bullard had mentioned; on the other hand, there were commercial considerations to be borne in mind.[91]

This, it would seem, was as far as discussions went that morning. Bullard had made his plea, based essentially on simple patriotism, a particular conception of the American national interest, and the same image of Britain as dominating the world's long-distance communications as had inspired Wilson's arguments in Paris. General Electric had responded with sympathy and understanding, but the need to protect the company's interests had also been underlined. What action, if any, the company would take was at this stage very much in doubt. No explicit suggestion had yet been made of the formation of a new radio operating company—one to which GE could sell its alternators, one that would be "American controlled" and able to safeguard American interests in radio. The problem had been posed: if GE did not sell its alternators to Marconi, to whom could it sell them? But no answer had been suggested. Bullard's

---

(1921), testimony of Owen D. Young, p. 333; Clark, "Radio in Peace and War," chap. 12, p. 42; Bullard, "Some Facts," p. 1631.

[89] The quotation is from Clark, "Radio in War and Peace," p. 43. Compare Clark Radio Collection, Cl. 100, v. 2, draft of letter, Clark to Elmer Bucher, 31 March 1947: "As to the chapter on the Formation of RCA, all said therein was dictated by Admiral Hooper. . . ."

[90] Hooper Papers, Box 37, transcript of tape-recorded memoirs. This transcript was apparently never proofread by Hooper. Rice is referred to throughout as "Wright," Coffin as "Proffitt," and there are many similar errors in transcription.

[91] *Cable Landing Hearings*, testimony of Owen D. Young, p. 333.

later account of the discussion makes it sound as if the formation of such a company was part and parcel of his morning presentation.[92] Hooper, however, is quite explicit that this was not so, and that the omission was deliberate: he wanted the suggestion to come from the company, not from the Navy.[93] Bullard's plea had been essentially negative: he was appealing to General Electric not to do something. Beyond the reference to a Monroe Doctrine for radio, he had made no positive suggestion.

The meeting broke up for lunch. As they were taking the elevator up to the top floor of the building, Rice expressed his regrets that Coffin, chairman of the board, had not been able to be present that morning. Perhaps he would be able to drop in after lunch.[94] One reason for the apparent inconclusiveness of the morning's discussion became clear: no decision would be taken without Coffin's knowledge and consent.

During the course of the luncheon break, Bullard took Owen Young to one side, out of hearing of the others, and confided to him "as a state secret" the fact that President Wilson personally had asked him to dissuade General Electric from selling its alternators to the Marconi companies. The president had become convinced, said Bullard, that future world preeminence would be determined by three factors: oil, transportation, and communications. Britain had long been "mistress of the seas" and was still supreme in transportation. Through her control of the cables, she still dominated international communications. Radio was now beginning to threaten those cables. If Britain won control of radio she would own two of the three essentials for world dominance. This had to be prevented. The president had therefore asked Bullard to say that he hoped that General Electric would not transfer control of the alternator to any other country, and particularly not to the Marconi companies. And Bullard warned Young that GE's decision might change the course of world history.[95]

Only three days before, Bullard had told Hooper that he had "no

[92] Bullard, "Some Facts," pp. 1631-32.
[93] Clark, "Radio in War and Peace," p. 43.
[94] Ibid., p. 44.
[95] This account rests on Young's testimony, and in some cases on his testimony as reported by others. The assertion that Bullard imparted the information privately to Young as a "state secret" rests on an interview with Young reported in Gleason Archer, *History of Radio to 1926* (New York, 1938), p. 163n, and is accepted by Josephine and Everett Case in their recent biography of Young. Our account of what Bullard said rests on Young's description as reported in the "Freedom of the Air" article and in his testimony at the *FCC Hearings* in 1929. Bullard's statement, as reported by Young, clearly echoes the memoranda submitted to Wilson by Walter S. Rogers (see above, pp. 263-67).

inkling" of what Cary Grayson had been talking about in Paris until Hooper explained it to him. He had described the president as vague as to the details and convinced only that there was something important in the radio situation that needed looking into. Later, in a lecture to the U.S. Naval Institute, he was to state explicitly that the negotiations between GE and Marconi had been brought to his official attention only "a few days after my arrival in my office."[96] But now, on 8 April, these vague instructions and warnings from Paris had been transformed into a "state secret" to be imparted to Owen Young alone, and a secret, moreover, that referred to a specific business transaction and a specific piece of machinery. Where did Bullard get the idea that there was any secrecy involved? From whom was the information to be kept secret? From the British government? From Congress? Or from American news media? Why did he choose Young as a confidant? Why not Rice, who was after all president of the company? Why not share it with the select group of executives with whom he had been talking that morning? Could they not be trusted with the information as readily as Young—a man he had never met before?

Young was impressed, as undoubtedly Bullard hoped he would be. He left the group and went next door to pass the information on to Coffin, who thereupon joined the meeting as coffee was being served. Young told no one else about Bullard's message except his wife, and not until ten years later did he speak of it in public.[97] His discretion was certainly admirable, as was his sensitivity to the fact that "for diplomatic reasons the head of the nation could not openly show his hand in the matter."[98] On the other hand his reported behavior suggests a certain credulity that is completely out of character, for Young was a veteran of many negotiations and not in the habit of accepting facts without verification. True, it is difficult, more than half a century later, to appreciate the deference that Bullard's uniform, rank, and invocation of presidential authority evoked in 1919. But one wonders whether Owen Young might not have probed a little more deeply. And the suspicion cannot quite be eliminated that perhaps he felt it was not in the company's interest to do so, in view of what seemed to be emerging from the discussions.[99]

[96] Bullard, "Some Facts," p. 1630.

[97] Case and Case, *Young*, chap. 11, p. 179.

[98] Archer, *History of Radio*, p. 163n, interview with Young, 5 February 1937.

[99] Eric Barnouw, the historian of broadcasting, has referred to the president's message as Young's "diplomatic trump card" in the negotiations that followed. The Cases have correctly pointed out that it was a card he did not feel free to play; and perhaps it is as well that he did not try to do so, for its authenticity

Hooper, of course, knew nothing of what had gone on between Bullard and Young, and, when Coffin joined the group, he nudged the admiral and told him, "Here's your chance—tell *him* about it."[100] Bullard was a little slow to get the idea—hardly surprising, in view of what he had been up to—so Hooper, who had been uncharacteristically silent up to this point, seized the initiative and, for Coffin's benefit, repeated in his own words Bullard's plea to suspend negotiations with Marconi. But what had been only implicit in the morning's presentation became the heart of the afternoon's discussion. It was not enough merely to deny the alternators to Marconi; it was also necessary to create an American-controlled radio operating company to which the alternators could be sold and which would be strong enough to deal with the Marconi companies as an equal. If such a company were created, the Navy Department would do all that it properly could to see that it was given a monopoly of long-range radio communications. The Navy, said Hooper, controlled valuable radio patents—those acquired from the Federal Company as well as the German patents acquired from the Alien Property Custodian—and these might be made available to such a new company. And the government's help could certainly be counted on when it came to securing concessions in foreign countries.

This was the essence of the "big proposition" that Hooper had been hinting at in his correspondence and conversations with Edwards for weeks past.[101] What he wanted was (in his language) "a real and proper American Radio Company."[102] Neither he nor Bullard supported Daniels's policy of government ownership. The policy they advocated opened up a new alternative, that of a private corporation with government-sanctioned monopoly privileges. Daniels's objective of a "single hand" in control of radio would be achieved as easily by such a private agency as by the Navy Department. Political objections would be much less. And the overriding objective of eliminating foreign control could be assured.

But the proposal also opened up alternatives for General Electric. The idea that GE might become or create a radio operating company was in

---

as a "state secret" might not have withstood examination. Not even Cary Grayson, testifying to the event in 1929, suggested that there was anything secret or confidential about the message. See Eric Barnouw, *A History of Broadcasting in the United States, Vol. 1, A Tower in Babel* (New York, 1966), chap. 1; Case and Case, *Young*, p. 181; *FCC Hearings*, testimony of Cary Grayson, pp. 1174-76.

100 Clark, "Radio in War and Peace," p. 44.

101 Edwards to Hooper, 20 November 1920 (Hooper Papers, Box 3).

102 Hooper to Young, 11 December 1920 (Hooper Papers, Box 3).

itself not new. Hobart had suggested it in 1917. Alexanderson, despairing of reaching agreement with Marconi, had repeated the suggestion only a few days earlier. What Hooper and Bullard added was the assurance of Navy cooperation and endorsement. Beyond this, however, they offered an acceptable and legitimate escape route from the negotiations with Marconi—negotiations which had proved very difficult and for which few of the leading executives and engineers at GE could now muster much enthusiasm. And, thanks to Bullard and Hooper, this alternative could now be seized, not in the name of corporate self-interest, but as a patriotic act, a decisive step taken to liberate American communications from foreign influence and preserve for American benefit the products of American genius.

It was an attractive prospect, couched in the language of nationalism.[103] But it also held out more pragmatic benefits. These were what A. G. Davis emphasized in his report on the proceedings.[104] Bullard had promised, according to Davis, that if GE would deal exclusively with the new company, the Navy would do everything in its power to make the enterprise a success. For its own use the Navy proposed to retain the coastal radio service; but long-distance radio would be the field for the new company and in that field it would have a commercial monopoly, if Navy could so arrange matters. Any patents usable in that field that Navy held would be turned over to GE and the new company, prov only that the government received in return such licenses as it neede carry on government business in its own stations.

These were the essential elements in what was to become, within weeks, the draft of a formal contract between GE and the Navy partment. No final decision on the matter was taken on 8 April could one be. There were serious questions about the powers

[103] Especially for those who could forget that Fessenden, Steinmetz, and anderson had all received their education in foreign countries.

[104] Davis to Young, 13 May 1919 (Young Papers). Davis attributes the p solely to Bullard; there is no mention of Hooper's having made any inde contribution, nor of his having participated actively in the discussion, eithe or after lunch. Bullard, in his lecture to the Naval Institute, does not even Hooper's presence. The account in the text is based largely on Hoop memoranda and tape-recorded memoirs and on Clark's history of the which was written from Hooper's dictation. Hooper's own published does not stress his personal role but refers merely to "the Navy represer See S. C. Hooper, "Keeping the Stars and Stripes in the Ether," 125-: is no doubt, however, that the conference of 8 April was Hooper's ide he gets the date wrong in his article). If it had not been held, presun already scheduled for 11 April would have been.

government and of the Navy Department in the matter, and there were others in General Electric to be consulted. But Coffin assured the two officers that they would have their answer within a few days.[105] For a decision on at least one of the issues they did not have to wait long. Conversations with the Marconi representatives had been suspended by GE on 7 April.[106] Edwards, still a reliable channel of communication, telephoned Hooper to give him the news. GE had terminated negotiations with the Marconi companies: the alternator agreement would not be signed.[107]

✻ ✻ ✻

Much of the talk at the conference had been about a particular transaction—the sale of the alternators—but the heart of the matter was really the creation of a new institution: the radio corporation. Termination of negotiations with Nally and Steadman did not mean that alternators would not be sold to foreigners. Quite the contrary: installation of a 200 kilowatt alternator at Marconi's Carnarvon station began in 1920 and was completed in 1921, with General Electric rendering the fullest cooperation every step of the way. Other foreign installations followed—Poland in 1923, Sweden in 1924. Japan had a 400 kilowatt alternator, built in that country from GE blueprints, on the air by July 1922. If the purpose of the Navy's intervention was to prevent transfer of the new technology to foreigners, it cannot be said to have succeeded.

Likewise, if General Electric believed after the conference of 8 April that it was about to establish a unique partnership with the federal government in the field of radio, one that would grant it exclusive privileges and a legal monopoly, these hopes too were disappointed. A contract with the Navy was indeed drawn up but it was never signed. And even if it had been signed the privileges it conveyed were much less than those that General Electric had been led to believe were obtainable. A new corporate entity was created; it did become an instrument of national policy in telecommunications; but its relation to the federal government was not that which Bullard and Hooper had described.

[105] Clark, "Radio in War and Peace," p. 44. Hooper, in his memoirs, describes the meeting as having ended with Coffin saying, "Oh, well, we must report whatever you recommend." (Hooper Papers, Box 37) Apart from the inherent implausibility of any responsible board chairman acting in this way, without consulting his colleagues, it is inconsistent with what is known of Coffin's personality.

[106] Young to C. W. Stone, 7 April 1919 (Young Papers).

[107] Clark, "Radio in War and Peace," p. 46.

It had been agreed at the conference that representatives of GE would go to Washington in the near future to consult with Bullard, Hooper, and Nagle, assistant solicitor of the Navy Department, about the legal issues involved. Davis and Edwards accordingly arrived in Washington on Thursday, 10 April. They were shown the draft of an agreement between the Navy Department and General Electric prepared by Lt. Comdr. E. H. Loftin, Hooper's colleague and friend.[108] Davis, reading it with a professional lawyer's eye, found it less than satisfactory. Discussion revealed no basic disagreement, however, and Nagle agreed to prepare a second draft if he were given all the necessary information. The document went through several further revisions, but by early May a final version was available.[109]

It was intended that this document would be submitted to Congress for ratification, since the new company was to have a federal charter. It was, therefore, carefully phrased, to withstand exacting scrutiny, and any summary runs the risk of serious distortion or omission. The version now in Stanford Hooper's papers in the Library of Congress appears from internal evidence to be the last draft of all, prepared on 3 May, and since it is a rather unique document, it is reprinted as an Appendix to this book.[110]

The contract had three major provisions. First, it divided the radio spectrum into a "government field" and a "company field." The "government field" included short-range communication with ships and airships, using the shorter wavelengths, plus long-range communication for government messages only. Essentially, this reserved the coastal radio network for the government and provided channels for the government-owned long-distance stations on the longer wavelenths. Notable to the modern eye is the identification of long wavelengths with long-distance transmission and the shorter wavelengths with shorter-range work; this of course reflected the assumptions about radio propagation commonly accepted at the time. The "company field" was to include land stations using the long waves for commercial purposes; privately owned ship stations using any wavelength; and coastal radio stations using the short

[108] Clark Radio Collection, Cl. 100, v. 2; Davis to Young, 10 April and 13 May 1919 (Young Papers).

[109] Davis to Young, 13 May 1919 (Young Papers).

[110] The internal evidence referred to is the deletion of all references to monopoly privileges and the inclusion of a clause giving the company title to government-owned radio patents and not merely a license to them. See Davis to Young, 13 May 1919 (Young Papers).

waves—except that no such coastal stations were to be established as long as the government continued to operate its system of shore stations.

A. G. Davis summed up this section of the contract in a letter to Young on 10 April. The Navy, he said, divided radio into three classes: low-power shortwave communication; medium-power medium-wave communications between ships and shore; and high-power long-wave stations for long distance work. In the first class they wanted complete freedom, in the sense that they did not wish to interfere with the right of any experimenter or amateur to do whatever his patents gave him a right to do. The second class they wanted to be a government monopoly. The third they wanted to see handled as a monopoly under government control by a private corporation formed for the purpose, except that they also wanted freedom to use the high-power government stations for government traffic. And this is what the contract provided for, except that the word "monopoly" was not used and provisions were made for private ownership and operation of radio stations on board ship.[111]

Secondly, the contract committed the new company to the construction and operation of "a chain of wireless stations intended to constitute a high grade international system of communication"; and it committed the government to provide and permanently assign to the company channels in the radio spectrum adequate for this purpose, and to protect these channels as far as possible from interference. This was, in a sense, the heart of the matter. The preamble had recited the desirability of such a radio system and had asserted that an essential part of its value was that "the American part of it" be wholly controlled by American interests, while other parts were to be controlled by American interests "as far as possible." The new company was to be the instrument for achieving those goals. The assurance that adequate frequencies would be permanently assigned to the company, and that the government would protect those frequencies from encroachment, was all that remained in the contract from earlier insistence on monopoly rights and exclusive privileges.[112] In practical terms this might have been sufficient for a while, although the right of any agency of government to allocate frequencies had not been constitutionally established at the time and this section of the contract might not have survived a serious legal challenge.

Paralleling these assurances of government support and protection were provisions to ensure that the company would remain free from foreign

[111] Davis to Young, 10 April 1919 (Young Papers).

[112] Notice, however, the inclusion of a clause explicitly prohibiting the secretary of the navy from authorizing the Federal Telegraph Company to engage in long-distance work, as that company might well have wished to do.

control. Both the company itself and General Electric were required to guarantee that they would not furnish radio apparatus for long-distance work to any other party without the consent in writing of the Navy Department. This consent would not be withheld, provided that the purchaser were "a loyal American citizen or an American corporation" and guaranteed that it would remain free from foreign domination. Nor was the new company or GE to enter into any working arrangement involving radio with any foreign company or government without written Navy consent. Further, the charter of the new company was to include restrictions to make sure that the voting power of 80 percent of its capital stock would remain at all times in the hands of U.S. citizens; and only U.S. citizens were to be employed as operators at the company's stations. Lastly, a representative of the government, designated by either the secretary of the navy or the president, was to have the right to attend all meetings of the board of directors to present the government's views, but without voting power.

Thirdly, the contract provided that GE and the government would exchange rights to the radio patents each controlled, and the GE would release the government from any claims for damages that stemmed from unauthorized use of its patents in wartime.[113] What the Navy had to offer in this category were, first, the patents on arc transmitters that it had bought from the Federal company in 1918; and second, the German Telefunken patents that had been seized in 1917 and later acquired by the Navy from the Alien Property Custodian. The Federal patents were of considerable commercial value, though whether the new company would have used them is another question. The Telefunken patents were an unknown quantity. Nobody seemed to have examined them thoroughly, and Hooper, when the purchase was first suggested, had half-seriously offered "$5 for the lot." The final purchase price was a nominal $1,500.[114] The General Electric patents that the Navy was primarily interested in were the Alexanderson patents on the alternator, its speed

[113] There was an asymmetry in the exchange of patents. GE granted to the government nontransferable exclusive licenses for the use of its radio patents in the "government field," but the government granted to GE, not exclusive licenses, but title to the patents themselves. The difference arose because the government was unwilling to permit GE to sue in the government's name, and because of doubts as to the government's power to grant such licenses. See Davis to Young, 13 May 1919, p. 11 (Young Papers).

[114] Clark, "Radio in War and Peace," p. 33. Later litigation over this transaction, including the acquisition of the Sayville station and the stock of the Atlantic Communications Company, raised the total, with accrued interest, to almost $7 million.

regulator, the magnetic amplifier, the multiple-tuned antenna, and the "barrage" receiver; but the vacuum tube patents taken out by Langmuir and his fellow workers in the GE Research Laboratory were probably also an important consideration.

On balance, the exchange of patent licenses would have benefited the Navy more than GE. There was nothing in the Telefunken patent portfolio that GE needed or, as far as we know, used. It did need the Fessenden heterodyne patent which, as Hooper later pointed out, was probably worth more than all the Telefunken patents combined; but that was still owned by NESCO. The Federal arc patents would have been indispensable to any company that did not already have a continuous wave transmitter of its own, but any communications company created by GE would use the alternator. Control of the Navy's arc patents was significant primarily because it meant that GE could deny them to any potential U.S. competitor.

No less significant than what the contract included is what is left out. And particularly interesting are clauses contained in earlier drafts that do not appear in the final version. Dropped at the last minute, for example, at the insistence of Acting Secretary Roosevelt, was language stating that General Electric had formed the new company "conforming to a suggestion made by the Government."[115] Bullard, Hooper, and all the GE executives urged Roosevelt to reconsider his opposition to this clause but he would not budge, belatedly sensitive to the fact that "the Government," in the sense of the Cabinet and particularly his immediate superior, Secretary Daniels, had never sanctioned any such action in the first place. Dropped also were all references to the word "monopoly" that had been bandied around so freely over the luncheon table on 8 April. An early draft dated 11 April had noted that "The Navy Department will recommend to Congress and urge before Congress a special charter for such corporation, giving it the sole right in the United States ... for high power communication."[116] And a later version, dated 22 April, committed the government to cooperate with the company in securing concessions in foreign countries and pledged that it would use its best efforts in all proper and legitimate ways "to the end that the Company or the Radio Corporation shall have exclusive control in the United States for the term of this agreement except in the field reserved

[115] The deletions are clearly indicated on pp. 2 and 3 of the copy in the Hooper Papers. For Roosevelt's role, see Davis to Young, 13 May 1919, p. 12 (Young Papers). Bullard, Hooper, Loftin, Young, and Davis all urged Roosevelt to reconsider his opposition to the inclusion of these clauses.

[116] Clark, "Radio in War and Peace," p. 48.

by the Government."[117] The charter was also to convey, if the company requested it, "the sole right ... for effecting radio communication by long waves." These were the clauses that, once signed by the secretary of the navy and ratified by the Congress, would have given the new company a legally enforceable monopoly.

None of this language survives in the final draft. The only provisions in that version of the contract that would tend to give the new corporation monopoly privileges were the clauses providing for the permanent allocation of frequencies, for defense against interference, and for transfer of title to the arc patents. These were important privileges, without doubt; but they were a far cry from a government-sanctioned monopoly, written into a federal charter approved by Congress.

It was not the Navy representatives who suggested that these references to monopoly and exclusive privileges be removed. They, apparently, were perfectly willing to go before Congress and defend such explicit language, convinced as they were that radio was a natural monopoly and that the only way to negotiate with foreign administrations was to have a single bargaining agent speak for the United States. Removal of these clauses that spoke of monopoly and exclusive rights came at the insistence of General Electric. While Davis and his assistants were in Washington, Owen Young had been in New York, looking into the patent situation, taking initial steps toward acquiring rights to the heterodyne patent, and "interviewing bankers." He also consulted William Gibbs McAdoo, former secretary of the treasury and now a prominent New York attorney. Presumably the conversation revolved around the constitutional and political issues raised by the contract. Young put Davis in touch with McAdoo's partner, Joseph P. Cotton, and, as Davis later reported it, "As a result of his suggestions and the consideration which we all of us had been giving to the matter, we decided to make some pretty substantial changes in the contract in the direction of removing the covenants giving us a monopoly."[118]

They were probably wise. To go before Congress with a document that contained so many references to exclusive rights would have been asking for trouble. And there was a real question whether the new corporation would need to have monopoly rights written into its charter. Its power would be based on technological superiority and on the unique manufacturing capabilities of General Electric, not on any act of the United States Congress. Attempts to secure monopoly by charter would have aroused all the familiar antitrust hostility and, as long as the com-

[117] Ibid., p. 50.
[118] Davis to Young, 13 May 1919, p. 11 (Young Papers).

pany confined its activities to long-distance international communications, would not have increased its income-earning power one iota. Talk of monopoly was heady stuff; but in a document intended to be laid before the public, best to downplay the idea.

This was not the only way in which the understanding between General Electric and the Navy, arrived at in such a flush of self-congratulatory patriotism on 8 April, was beginning to come apart at the seams. The federal bureaucracy, strained by the war but even more strained by the absence of its leading figures at the Paris Peace Conference, was beginning to pull itself together. Actions that at one time had seemed justified by a commonsense appeal to the national interest now began to seem questionable. The frontier between initiative and insubordination, always difficult to draw in any bureaucracy, began to shift, and formal lines of authority and responsibility acquired an importance that earlier had been overlooked.

Up to this point Hooper and his fellow officers in the Radio Division had been acting as if they were free to follow policies of their own. Josephus Daniels, their nominal superior, had staked out his position, but by the spring of 1919 everyone knew that it was not politically defensible. Alternatives had to be explored and, with Daniels absent in Europe for much of 1919, it was easy to ignore the fact that he was still secretary of the navy. The man effectively in charge of the Navy Department in Daniels's absence was his deputy, Franklin D. Roosevelt, and Roosevelt was not about to check the initiatives of his staff officers in their defence of the national interest as they saw it. Unlike Daniels, who was never popular with his officers, most of whom put him down as a stubborn, plodding landlubber, Roosevelt was a young man who knew the sea and who seemed to have the verve, dash, and decisiveness that Daniels lacked. Since his first days as assistant secretary, officers had got in the way of taking matters to him for action that properly should have been decided by his chief; and sometimes they deliberately waited until Daniels was absent and Roosevelt was acting secretary before bringing an issue up for action. If, on occasion, this led Roosevelt to the very edge of insubordination, it was not a practice he discouraged. A nod, a significant wink, an offhand remark that the secretary was likely to be out of the country in the near future—these were enough to get the message across.[119] The crucial intervention in the negotiations between

[119] For the relationship between Daniels and Roosevelt, see Kenneth S. Davis, *FDR: The Beckoning of Destiny, 1882-1928, A History* (New York, 1971), pp. 305-33 and p. 395, and Frank Freidel, *Roosevelt: The Ordeal* (Boston, 1954), pp. 19-21. Roosevelt was the youngest of the fifteen assistant secretaries of the

General Electric and Marconi had taken place during Daniels's absence and without his knowledge. Now, however, his return was imminent and even Roosevelt began to move with caution.

Warning of the problems that lay ahead came when Roosevelt referred the draft contract to the Munitions Patent Board for an opinion, presumably because of the proposed patent exchange between GE and the Navy. It was a prudent step, but in other circumstances might well have been omitted. A Mr. Ewing, formerly assistant commissioner of patents, was now chairman of the board; and although described as being very ill with a dislocation of the spine, "in great pain a large part of the time," and anxious to be relieved from duty, Ewing nevertheless was inclined to dig in his heels. He said that, as long as he was chairman, he wanted to act on the contract intelligently; he was inclined to criticize it; and he wanted to talk to Bullard and Roosevelt before taking action.[120] A minor piece of bureaucratic obstinacy, perhaps, and easily overridden if enough pressure was applied. But, in the meantime, a problem.

In the meantime, too, Hooper had been remembering his lines of authority. A few days after the 8 April conference Young came to Washington and asked for an interview with Bullard and Hooper. He wanted to know more specifically what the two officers had in mind when they talked about a "partnership" between the government and General Electric. GE would have to have a guarantee of some sort, he said, before it could raise the capital and launch the new company. At which point Hooper "remembered a pressing engagement in his own office" and left the room, feeling that this was dangerous ground and Young was getting too far ahead of him.[121] Belatedly, too, Hooper recalled that the policy he had been so energetically pursuing had never been approved by his chief in the Bureau of Engineering, Admiral Griffin, nor by the chief of naval operations, Admiral Benson, far less by the secretary of the navy. These minor lapses from bureaucratic propriety, it turned out, were not easily remedied.

It had originally been planned, following a suggestion made by Nagle, that as soon as a mutually acceptable draft of the contract had been worked out, Admiral Bullard and one of the GE executives would go to Europe and lay the whole matter before the president and Daniels. At a

navy who served between 1890 and 1936 and served the longest term of any of them: seven years and five months.

[120] Davis to Bullard, 10 May 1919 and Davis to Young, 13 May 1919, p. 13 (Young Papers).

[121] Clark, "Radio in Peace and War," p. 46; the phrase in quotation marks is, according to Clark, from "Notes of Admiral Hooper."

conference on 24 April, however, it was learned that Daniels would soon be returning to the United States—so soon, in fact, that even if someone left Washington immediately it was unlikely that they would be able to meet the secretary before he sailed. The alternatives were, therefore, either to get the contract signed immediately or wait until Daniels's arrival.

Roosevelt said that he was willing to sign the contract immediately as acting secretary: there was no need to wait until the secretary's return. He had already discussed it with Bullard and thoroughly approved, though of course the Departments of State and Commerce would also have to give their endorsements. Bullard in the meantime had drafted a cable to be sent to Admiral Benson, then in Paris with the Peace Commission, asking for authority to proceed. Roosevelt advised him not to send it in that form, since asking for authority might cause long delay. Better just to send the cable "for information." He made some changes in the wording, signed the cable himself, and sent it off.[122]

Roosevelt's intention was clearly to present the secretary, when he got back to Washington, with a *fait accompli*. And in this he had the support of General Electric. On 10 May, with Daniels already en route to the United States, A. G. Davis wrote to Bullard about the problem with Mr. Ewing and the Patent Board and took the opportunity to underline the desirability of settling the matter before Daniels returned.[123] It was not to be expected that Daniels would react with pleasure when he learned of what had been going on while he was in Europe—far less that he would be willing to sign a document that clearly was designed to create a private monopoly, no matter what care had been taken to expunge that word from its language.

Hopes for getting the contract signed before Daniels's return depended on keeping him in ignorance of what was in the wind. And this might have been possible if Stanford Hooper, with commendable if somewhat tardy prudence, had not thought it best to cable his superior, Admiral Griffin, a brief description of the contract and a request for instructions.[124] Griffin, on receiving this message, took it at once to Daniels. And Daniels immediately cabled back instructions that the matter should be held in abeyance until his return. General Electric was so notified on about 12 May.

[122] This account is based on Davis to Young, 13 May 1919 (Young Papers). Clark, "Radio in Peace and War," p. 58, confirms that Roosevelt was prepared to give his "willing signature" to the contract.

[123] Davis to Bullard, 10 May 1919 (Young Papers).

[124] Clark, "Radio in Peace and War," pp. 58-59. Clark dates this cable as having been sent "on the very eve" of Roosevelt's signature of the contract.

One historian of these events calls attention at this point to Hooper's "strict adherence to discipline."[125] A less friendly critic might ask why he waited so long. Be that as it may, once the information reached Daniels, the game was essentially over and it is idle to speculate on what might have happened if Hooper had been able to restrain his response to discipline's imperatives a day or two longer. He might have been able to win Griffin's support. Certainly he tried: a long memorandum of 22 May spelled out the benefits likely to accrue to the government from the contract, depicting the General Electric Company in terms that its own advertising department could not have bettered. "Its resources are almost unlimited, its business relations strong and universal, its production facilities without peer, its high power radio apparatus the best to date, its patent position very strong, and its loyalty unquestioned."[126] Daniels, however, could not be won over by such eloquence. He called Bullard in immediately upon his return to Washington and went over the contract with him. Without denying that it contained items of value, he said flatly that he was opposed to the appearance of monopoly that it presented.[127] At a conference with GE officials on 23 May he repeated the point, adding that he still believed government ownership to be a necessity, though admittedly he saw little hope of convincing Congress of the fact. In any case, he doubted that he had the authority to sign such a contract without the express consent of Congress, and he intended to do nothing further about it until he had conferred with Cabinet colleagues and with leaders of the House and Senate. And there the matter rested.[128]

The contract was not yet entirely dead. An attempt was made to attach a rider to the Naval Appropriations bill giving the secretary authority to act. Nothing came of it, however, in the absence of any strong pressure from Daniels. And as the weeks passed the idea of a contract, of exclusive privileges, and of a federal charter slipped gradually into the background.

[125] Ibid., p. 58.

[126] Hooper Papers, Box 2, file May-August 1919, memorandum for Chief of Bureau of Steam Engineering.

[127] Clark, "Radio in Peace and War," p. 59.

[128] *Report* of the Federal Trade Commission on the Radio Industry in Response to House Resolution 548, 67th Congress, 4th Session, 1 December 1923 (Washington, D.C., 1924), p. 3; Clark, "Radio in Peace and War," p. 60, quoting from Nally's diary. Nally had received a firsthand report of the meeting from Young. Howeth, official historian of naval communications electronics, criticizes Daniels for "failure to support his subordinates." (*History*, p. 358) Why the secretary of the navy should be expected to "support" subordinates who pursued policies that he had never authorized and that ran diametrically counter to his own is not explained.

There were other possibilities. The concept of an American-based communications company, created in the first instance by General Electric but designed to include eventually all American corporations with a substantial interest in radio, had taken firm hold. If the abortive negotiations with the Navy had done nothing else, they had at least alerted General Electric and others to the existence of a major commercial opportunity. The need to reequip the world's radio systems was still present. The technology was available, although its ownership was dispersed. The problem was to create an organization that, within the political constraints imposed by the Navy, could integrate ownership rights to the technology and bring it to bear on the job to be done.

SEVEN

The Formation of RCA
Part I: Washington and
New York

LATE in 1919, when the General Electric Company was beginning to contemplate seriously the possibility that, with or without a contract with the federal government, it might take the initiative in forming a new American radio company, Owen Young received from his staff a fifty-page document entitled "The Marconi System." It contained an analysis of the Marconi corporate empire—the network of enterprises throughout the world with which Marconi's Wireless Telegraph Company Ltd. of Great Britain was affiliated. Included were important operating companies, such as the parent firm and its American subsidiary; firms holding potentially valuable concessions, such as those in Argentina and Spain; and formally independent companies in which it was known or suspected that the British firm held an interest. Information was given on financial structure and dividend payments; and in each case there were listed the names of the directors and their business connections. In the case of the Marconi subsidiaries these lists had a certain sameness about them. Guglielmo Marconi always appeared first, closely followed by Godfrey Isaacs, managing director of the British company since 1910; then there followed the representatives of the local investors in each country, usually prominent financiers, industrialists, and well-connected attorneys.[1]

---

[1] Young Papers (Van Hornesville, N.Y.) Box 71, report on "The Marconi System." The report also contained information on the Federal Telegraph Company and the Poulsen Wireless Corporation, presumably because of their affiliation with Marconi interests in the Pan-American Wireless Telephone and Telegraph Company, and on the Gesellschaft für Drahtlose Telegraphie m.b.h. (Telefunken), presumably because of its patent pooling agreement with Marconi

This network reflected the strategy of expansion that the Marconi organization had pursued before 1914. In its original form this strategy had called for the establishment of corporate subsidiaries in all countries where there seemed prospects for profitable radio traffic or for the sale of radio equipment. In these subsidiaries Marconi personally, the other directors of the parent firm, and that firm itself in its corporate capacity held a substantial though not necessarily a majority interest. Each subsidiary held an exclusive license to use Marconi patents in its territory, and in normal circumstances it obtained its equipment exclusively from its British parent. In countries such as France and Germany, where well-entrenched private corporations controlled radio, the British Marconi Company, instead of establishing subsidiaries, sought alliances through corporate treaties and sometimes the purchase of the stock interest. In France, for example, the Marconi Company held a substantial equity interest in the Compagnie Générale de Télégraphie sans Fil (TSF); and in Germany after 1910 Marconi and the Gesellschaft für Drahtlose Telegraphie m.b.h. (usually known as Telefunken) were joint owners of the company that handled all German merchant marine radio, and in addition had agreed to pool their patents. In countries such as Norway or Sweden where government agencies controlled radio, the Marconi Company sought to become the preferred supplier of equipment, and bid aggressively on contracts for the construction of new stations.

The United States was in some ways an anomaly. The American Marconi Company was formally a subsidiary of the usual type, with a majority of its stock held outside the United States and a sizable block held by the parent company. But since it controlled the American end of the important transatlantic radio circuits, as well as a large coastal system and the elements of a transpacific circuit from San Francisco to Hawaii and Japan, it occupied a distinctive position in the Marconi system and by 1918 its management had come to enjoy a considerable degree of autonomy. Though clearly the dominant firm in American radio, it had never been entirely free from competition, both from other private radio companies and from the U.S. Navy; and, as we have seen, it was not regarded with favor in government circles, partly because of the taint of foreign control, partly because it was reluctant to sell its equipment outright, and partly because it consistently opposed even the smallest element of government control.

In its original form, too, the Marconi strategy had called for non-

---

and joint ownership of the German marine radio company, the Deutsche Betriebsgesellschaft für Drahtlose Telegraphie (DEBEG). See W. J. Baker, *A History of the Marconi Company* (New York, 1971), p. 133.

intercommunication with other radio systems. Marconi stations, in other words, would communicate only with other Marconi stations, on land or on shipboard. Intense political and diplomatic pressure and two international conferences had finally broken down that element of Marconi exclusiveness, but the vision that inspired it remained intact.[2] This was a vision of a worldwide system of Marconi-controlled and -affiliated companies, each of them nominally a national entity but in fact linked to the British parent by stock ownership, interlocking directorates and a common technology. And even though legally required, for the safety of ships at sea, to accept maritime traffic of non-Marconi origin, the normal expectation was that in intercontinental traffic Marconi stations would communicate only with other Marconi stations, the whole forming, in its ultimate extension, a system truly imperial in scope.

When people spoke of the Marconi "monopoly" of radio, it was this vision of a worldwide Marconi radiocommunications network that they had in mind. It was never, of course, a true monopoly. In marine radio the Marconi companies met stiff competition almost from the beginning. The legal requirement for intercommunication between systems blunted the major weapon the Marconi group might have used to consolidate its originally strong position in that field. Intercontinental radiocommunication was just beginning to look as if it might become commercially feasible in 1914; but in that sphere also it was clear that the Marconi companies were going to meet vigorous competition, not only from German and French systems on the transatlantic circuits but also from the Federal Company in the Pacific. If, through aggressive patent litigation, the Marconi group had been able, internationally, to prevent the entry of new firms or to eliminate them after entry, its long-run goal of dominating radiocommunications might have been feasible. In certain cases, such as the elimination and absorption of United Wireless in the United States, this strategy worked; in others it attained partial success. But there were too many variations in radio technology, too many gaps that the ingenuity of Marconi engineers and lawyers never quite plugged, for it to be totally effective. Patents expired, foreign courts were unsympathetic, new methods and devices were being introduced constantly. The technological leadership that the Marconi enterprises had once been able to

---

[2] John D. Tomlinson, *The International Control of Radiocommunications* (Ann Arbor, Mich., 1945), pp. 11-44, and Susan Douglas, "Exploring Pathways in the Ether" (Ph.D. diss., Brown University, 1979), chaps. 5 and 8. The international conferences referred to took place in 1903 and 1906, largely on German initiative. In the United States legislation to require intercommunication was not passed until 1910.

claim was short-lived. Already shaky by 1914, by 1919 it had gone; and with it went any hope of basing a world monopoly of radiocommunications on exclusive control of the necessary technology.

Distrust of Marconi intentions, however, remained strong, and not only in the ranks of the U.S. Navy. It was well known that the Marconi interests, under the vigorous leadership of Godfrey Isaacs, hoped to construct the British government's "Imperial Chain" of radio stations. That phrase and its associations triggered all the old hostilities that had clustered around Britain's alleged "domination" of the submarine cables. In reality the Imperial Chain was only part of Marconi's grand design, and certainly not its central feature. There were contracts and prestige to be won if the government made it a Marconi project. And strategically it had its importance, though mostly as a backup system for the imperial cable network. Commercially, however, prospects elsewhere were at least as attractive. North America, South America, the Pacific, the Far East— these were the areas likely to generate new traffic, and also the areas where the cables were most vulnerable to competition. Nor were relations between the British government and the Marconi Company as harmonious as outsiders, particularly Americans, assumed. Any reading of Hansard or of the company's annual reports would have disproved that belief. When and if the decision was taken to go ahead with the Imperial Chain, it was by no means a foregone conclusion that the contracts would go to Marconi. The first two stations in the system, in fact, at Leafield and Cairo, were arc stations, designed and built by C. F. Elwell, formerly of the Federal Company.[3] As the outside world viewed the situation, however, the aspirations of the Marconi companies and those of the British government were all of a piece.

Fears of Marconi monopoly and fears of British imperialism fed on each other. Both were, if not baseless, at least exaggerated. Few realized at the time the full damage that four years of war had done to the fabric and morale of British society; few appreciated the structural weaknesses that afflicted the British economy; and few understood the centrifugal forces that were reshaping the British Empire and redefining the ties that bound colonies and dominions to the mother country. In the same way, few appreciated in 1919 that the Marconi system of companies was largely a paper tiger. It looked formidable, as for instance in the report that Young's staff prepared for his guidance. In the short run, however, its technological position was weak. Before 1914, confident in its technical supremacy, it had failed to mount a sustained research program. Now, in 1919, it found itself caught short in the shift from spark to the

[3] *Radio Review* 2 (October 1921), 509.

continuous wave. The only high-power transmitter it could call its own was the timed spark, a prewar design that most observers considered obsolete and that was not a true continuous wave machine. Hopes for a return to prewar eminence hinged on the acquisition of the necessary technology from outsiders, and from the United States in particular.

This was an urgent short-run problem. The critical issue was how to reequip the Marconi long-distance stations in the immediate postwar years. This demanded quick access to high-power continuous wave generators. If this immediate crisis could be handled, the longer-run prospects for the Marconi companies were not necessarily gloomy. Everything depended on how successful Marconi scientists and engineers were in recapturing technical leadership. There was much to be learned about the new technology of the vacuum tube and about the mysteries of long-range propagation. In those realms the Americans held no unchallengeable lead. European receiving tubes were at least the equal of anything American manufacturers were turning out in 1919; and British high-powered transmitting tubes available at the end of the war had no counterparts in the United States.[4]

Knowledgeable scientists and engineers knew that the future lay with the vacuum tube. To Marconi executives in 1919, however, this was small consolation. It would be several years before these new possibilities in electronics could be brought to fruition. In the meantime the technology would have to be borrowed, for it was dangerous to wait. New radio stations were going to be built in the postwar years. Once they were established as reliable communications channels, once traffic agreements had been worked out with sister stations in other countries, it would be very hard to displace them or move them off the frequencies on which they had settled. Who was to build these stations? Who would equip them? Who would operate them? What role would the Marconi companies play in these altered circumstances? How were they to adjust to the new and more strident forms of nationalism and the new sensitivity to foreign control over communications that the war seemed to have bred? The older vision—an international system of Marconi companies

[4] Hooper Papers, Box 3, Report to Chief of Bureau of Engineering on visit to Europe, 19 August to 18 September 1920 (dated 4 October 1920). Hooper visited in England the radio station at the Air Ministry field at Croydon and the Naval Research Laboratory at Portsmouth. "I was permitted to see nearly all their tests and apparatus, including the 10 KW. valves which are far ahead of anything in the United States." Alexanderson, while on a visit to England in 1921, was instructed by cable to obtain full information on high-power tubes and circuits and to ship sample tubes to Schenectady as soon as possible. (Pratt to Alexanderson, 29 January 1921, Alexanderson Papers, folder 16)

dominated by the parent firm in London—seemed less than ideally suited to the new realities.

* * *

The return of peace and the shift in technological leadership called for a new corporate strategy on the part of the Marconi Company. The same was true for General Electric. What role should the corporation play in the future development of radio? That depended on what GE's management thought the future of radio was likely to be. In 1919 that future included no conception of popular broadcasting. No one in General Electric, and only a very few individuals outside the company, thought in those terms. Radio still meant primarily radiotelegraphy: marine communications and transoceanic traffic in Morse code in competition with the submarine cables. Radiotelephony—the transmission of the human voice—had been proved technically feasible, but no one was sure what its commercial future might be.

General Electric had a certain stake in exploiting the markets that this conception of the future of radio suggested. That stake, however, was not a large one, and it would be an error to imagine that decisions with respect to radio were in any sense critical to the company's fortunes in 1919. GE had a highly diversified line of products; radio equipment was only a small part of the total. In 1919 it would have been quite possible for the company to withdraw completely from the manufacture of radio apparatus with no perceptible effect on its net earnings. Neither the public reputation of the company nor its profitability depended on radio.

There were, however, two particular areas in which the company had invested substantial amounts of capital, talent, and prestige: high-frequency alternators and vacuum tubes.[5] In neither of these product lines was its position altogether secure. The Alexanderson alternator was an excellent piece of equipment for long-distance radio on the very low frequencies. But there were other types of alternator available, almost as good, and arc transmitters were preferred by some for their greater flexibility in use. In vacuum tube research General Electric had made important advances and had built up substantial manufacturing capacity; but its patent position was vulnerable to challenge, particularly by that

---

[5] E. P. Edwards estimated on 21 June 1919 that total expenditures on the development of the alternator to that date were $550,000, none of which had yet been recovered. (Edwards to Young, 21 June 1919, Young Papers, Box 72) No analogous estimate of unrequited investment in vacuum tube development seems available.

formidable adversary, AT&T. And in both these areas the market was highly uncertain. The military demand for vacuum tubes evaporated with the end of hostilities. No comparable civilian market was in sight. As for the alternators, the Navy's intervention had deprived GE of its prime customer; and even if another purchaser could be found—or created—how many of the machines could be sold anyway? Twenty to twenty-five at most, for channels on the very long wavelengths where the alternator was at home were limited in number. Alexanderson himself thought that there could be room for only twelve "first class transmitting stations" between 10,000 and 20,000 meters, unless major improvements were made in directional antennas and selective receivers.[6]

Prudence, therefore, might well have suggested in 1919 that the probable gains to GE from expanding its role in radio manufacturing were not very large. In the absence of some major new vision of what radio was, what it was for, and what it could do, prospects for growth were modest. For intercontinental traffic radio would always have to meet the entrenched competition of the cables, and it was an open question whether the new technology could carve out a market share for itself in peacetime. For domestic communications the wired telephone and telegraph systems, mature in their techniques and controlled by powerful corporations, provided an efficient service that seemed to leave little room for wireless. There were some areas of the world, to be sure, where the submarine cable network was inadequate. There long-distance radio could play an important role. For marine communications radio would always be vital. And for intercontinental traffic there was value, both strategic and commercial, in the redundancy that radio could provide. Cables could be cut, wires could blow down, but radio might still get through. There were some functions, in short, that only radio could perform, or that it could perform more efficiently than wired systems.

These functions, however, were all marginal to the world's existing communications systems. Were they enough to support an economically viable radio manufacturing industry? If so, on what scale? Were the probable markets large enough, and would they grow fast enough, to justify a continuing commitment by General Electric? Might it not make more sense to leave the business of supplying the equipment needs of the industry either to smaller and more specialized firms, such as had provided radio apparatus before the war, or else to the Telephone Company, well situated as it was to integrate the new technology of radio into the

---

[6] E.F.W. Alexanderson, "Transoceanic Radio Communication," IRE *Proceedings* 8 (August 1920), 264. Alexanderson believed, however, that this number could be vastly increased by continued engineering efforts.

existing structure of the communications industry? AT&T already held strong patents in radio, many of them overlapping or interfering with patents held by GE. Was there room in radio manufacturing for both companies? Patent litigation would certainly guarantee lifetime salaries for stables of lawyers on both sides, but cooperation might be cheaper. Why not leave the communications field to the Telephone Company and reserve power engineering and lighting to GE? In those older fields Westinghouse could be counted on to provide more than enough competition.

Several factors worked against the adoption of any such strategy. Not least among them was an emerging corporate tradition. Since its founding in 1892 General Electric had learned that the key to corporate growth lay in aggressive exploitation of the opportunities that new technology offered. Its executives had come to understand that the survival of the organization depended on developing new products and new markets to offset those that it would inevitably lose through obsolescence or the expiring of patents. The introduction of new products and the opening-up of new markets had become the norm of corporate policy, not the exception. This was why the General Electric Research Laboratory and the Engineering Laboratory existed; it was why men like Steinmetz, Whitney, Coolidge, Langmuir, and Alexanderson had come to General Electric, and why they stayed there. Systematic research and development was no occasional luxury for General Electric; it was the corporation's life insurance. The scientists and engineers who worked there knew that they had the support of a well-established corporate tradition. Part of that tradition was the assurance that they would be given the resources necessary to carry out their projects in a relatively nonbureaucratic environment. Another part was the confidence that, if they created something that looked both interesting and useful, the corporation's commercial departments would not permit any likely markets to remain unexplored. For the scientists in the company this may have been a secondary consideration, but not so for the engineers. A man like Alexanderson was not interested in building ingenious but useless gadgets.[7]

[7] On the origins and development of research at General Electric, see Kendall Birr, *Pioneering in Industrial Research: The Story of the General Electric Research Laboratory* (Washington, D.C., 1957); John Broderick, *Willis Rodney Whitney* (Albany, 1945); John Winthrop Hammond, *Men and Volts: The Story of General Electric* (Philadelphia and New York, 1941); E.F.W. Alexanderson, "Reminiscences" (Columbia University Oral History Collection); James E. Brittain, "C. P. Steinmetz and E.F.W. Alexanderson: Creative Engineering in a Corporate Setting," IEEE *Proceedings* 64 (September 1976), 1413-17; Leonard Reich, "Radio Electronics and the Development of Industrial Research in the Bell System" Ph.D. diss., Johns Hopkins University, 1977), especially chap. 6; Reich, "Irving

Vigorous exploitation of new electrical technology was, then, a well-established policy at General Electric. It was part of the corporate culture. In this particular case it was reinforced by the fact that, under the pressure of wartime needs, new technology in radio had already been brought to a relatively advanced stage of development. In vacuum tube research there was still a long way to go. Some tough problems had still to be solved. But the difficulties had been identified and overcoming them was only a matter of time and talent. In the laboratory Langmuir and his fellow workers knew where they were heading; in the factory most of the problems of quantity production had been solved.

As for the Alexanderson alternator, here GE had a fully developed and tested machine. The marriage of power engineering and radio technology had proved highly fruitful. There might be room for minor improvements—better grades of iron, perhaps, could be used, and it might be possible to cut air friction losses by running the armature in a partial vacuum—but in all essentials the alternator in 1919 was a perfected device.[8] It was ready to be manufactured and sold in quantity. Furthermore, if it was to be sold, it should be sold soon, for there were clear indications that its market life might be short. The alternators themselves did not lack durability: once set up there was no reason why, with proper maintenance, they could not be kept running indefinitely.[9] But it did not take much perspicacity nor did it require access to arcane sources of information to realize that within a few years the alternator would be obsolete. Vacuum tubes would be the continuous wave generators of the future. And who was in a better position to understand this than GE's scientists and engineers at Schenectady, who themselves were working on the frontier of vacuum tube technology? There was, therefore, some

---

Langmuir and the Pursuit of Science and Technology in the Corporate Environment," *Technology and Culture* 24 (April 1983), 199-221; George Wise, "A New Role for Professional Scientists in Industry: Industrial Research at General Electric, 1900-1916," *Technology and Culture* 21 (1980) 408-29.

[8] Alexanderson reported to Young in September 1922 that French engineers had gone further than the Americans in certain details of alternator design, such as the use of a vacuum to reduce air friction and "the use of better grades of high frequency iron than it was possible for us to obtain when our alternators were manufactured." (Young Papers, Box 3, "Forecast of New Developments and Notes on Europe," by E.F.W. Alexanderson, 28 September 1922)

[9] An Alexanderson alternator installed by RCA at Warsaw remained in service until destroyed by German forces in the closing phases of World War II. See Society of Wireless Pioneers, *The Old Timer's Bulletin* 21 (December 1980), 25. The alternator at Grimeton in Sweden is still regularly put in operation for test purposes; it is believed to be the only machine of its type still serviceable.

urgency attached to the marketing of the alternator. General Electric had perhaps two or three years to find a market for the device, but not more.[10]

The growth prospects of the radio industry as it was understood in 1919, before the advent of popular broadcasting, were, in short, quite modest. It is doubtful whether, on a strictly commercial basis, they would have justified GE in undertaking any major new investments either in alternator development or in vacuum tubes. Lieutenant Commander Hooper, who had been in charge of a vacuum tube development board during the war, recalled that immediately after the Armistice the "Company men" on the board unanimously said that they could not continue developing transmitting tubes because "they saw no future use for such tubes except the Military, and this was too far distant to interest them."[11] Only a special appropriation and a requisition for three new types of transmitter, he believed, kept tube development going at that time. What was true of tubes was true of most other types of radio equipment; until civilian markets developed there was little reason for GE to involve itself any more deeply in radio manufacturing and development than it already had.

With the alternator, however, General Electric had come very close to tasting success—commercial, and not merely technical, success. How the

[10] Between 1920 and 1922 hesitation on the part of Swedish officials to contract with RCA for a high-power station was due in part to uncertainty as to whether vacuum tubes or the alternator should be used. Alexanderson, who was in immediate charge of the negotiations, recommended the alternator but advised RCA's management to be ready to offer a tube-equipped station if that should be insisted on. See Alexanderson to Sarnoff, 23 November 1920 (Alexanderson Papers, folder 16) and Wireless Age 10 (October 1922), 57. A 100 kilowatt transmitting tube, designed by Western Electric, was available by October 1922, the critical breakthrough being W. G. Houskeeper's solution of the problem of getting a reliable glass-to-copper seal. In the same month RCA successfully tested a tube transmitter, using six 20 kilowatt pliotrons, on its regular circuits to Great Britain and Germany. RCA officials on that occasion assured the press that the alternators would not be immediately superseded, although the eventual adoption of tubes was inevitable. See Wireless Age 10 (October 1922), 60, and (November 1922), 55.

[11] Hooper Papers, Box 40, manuscript "History of Radio," and Box 3, commendatory letter to be attached to Hooper's record. Compare Hooper Papers, Box 1, "Radio Developments, 1910-23": ". . . the company representatives . . . advised the chairman, after the Armistice, that their managers would not permit them to attend further sessions of this Committee because they saw no future value to the vacuum tube, and could not spend any funds in that direction in the future."

negotiations with the Marconi companies would have worked out if Bullard and Hooper had not intervened is an unanswerable question; clearly there were still unresolved differences. After that intervention, however, it was possible to assert, and to believe, that only government action had prevented the deal from being consummated. And this became in fact the accepted RCA version of the affair.[12] The appeal to GE had been made on grounds of simple patriotism, garnished by an invocation of presidential authority. And there is no reason to suspect that GE's management were anything but sincere in responding to it. What made such a response possible, however, was their expectation that, through some kind of contract with the federal government, an alternative customer for the alternators would be found. In the absence of that assurance Bullard and Hooper might well have expended their eloquence in vain. GE had been given assurances—vague in detail but well understood in principle by all who heard them—that an American-controlled organization would be formed to acquire and use the alternators, and that this organization, in return for its public responsibilities, would enjoy a special status in relation to the federal government, particularly in the allocation of scarce frequencies.

There was, however, a problem, which became distressingly evident as the spring of 1919 merged into summer. The naval officers who had offered these comfortable assurances were now unable to implement them. The reason for this unhappy state of affairs was that, in a no doubt laudable desire to protect the national interest as they saw it, they had exceeded their authority. The policy they were promoting was not merely unauthorized by their civilian chief, the secretary of the navy, but inconsistent with the policy that he had publicly advocated and with which he was identified: government ownership of radio. It was true that in April 1919, when the appeal was made to General Electric, it required no great political wisdom to predict that Daniels's policy was unlikely to prove acceptable to Congress. That had become evident by mid-December of the previous year, when the House Committee on Merchant Marine and Fisheries tabled the Alexander bill. [13] But no approach to

[12] For only one example, see the article by Gen. James G. Harbord, then president of RCA, in Harvard University, Graduate School of Business Administration, *The Radio Industry, the Story of Its Development, as told by Leaders of the Industry* (Chicago and New York, 1928), pp. 87 ff.

[13] House of Representatives, 65th Congress, 3rd Session, *Hearings* before the Committee on the Merchant Marine and Fisheries on H.R. 13159, A Bill to Further Regulate Radio Communication, December 12, 13, 17, 18, and 19, 1918 (Washington, D.C., 1919). Testimony by Navy representatives at these hearings seems to have struck several members of the committee as arrogant and officious,

Congress for approval of a contract between the federal government and General Electric's proposed radio company was possible without Daniels's endorsement or, failing that, decisive leadership from the White House. Neither of these was forthcoming. In the circumstances the assurances given to General Electric were hollow indeed.

What was the company to do? Daniels seemed prepared to carry the draft contract around in his hip pocket indefinitely rather than go to Congress and get the matter resolved one way or the other. Delay might well mean that the company would find itself out of the radio manufacturing business by default, rather than by decision. There were, too, elements of personal pride involved. Owen Young in particular had reason to feel frustration and annoyance. He had made the negotiations with the Navy his special concern; ever since Bullard and Hooper had successfully made their appeal to GE's top management, it had been understood within the company that "the ball was Young's to run with."[14] And he seems to have taken up the challenge with enthusiasm. His previous work for GE had brought him rank and salary but not much independence of action, and the prospect of carving out a role as architect and builder of a major new American corporation was an attractive one. Alexanderson too was deeply and personally involved in the fate of the machines he had done so much to create. And A. G. Davis, GE's patent counsel, who had long harbored reservations about the Navy's intentions, was not inclined to let time slip by at the company's expense. More was involved, in short, than the fate of a corporate investment. Individual aspirations and ambitions were also at risk.

* * *

Matters reached the critical point in June 1919. On the 13th of that month J. W. Elwood, one of Young's assistants, held a conference with Hooper in Washington in which he pointed out that, since a preliminary meeting with Daniels on 23 May, General Electric had heard nothing from either the secretary or his staff. In the meantime the company was unable to sell its apparatus or to make a contract with the government. Could Hooper tell him what was going on?

---

while opposition from civilian scientists, amateurs, and radio companies was vehement. The Navy's case finally boiled down to the assertion that radio had to be controlled by a single authority if interference was to be prevented. This was an argument for spectrum allocation and stricter technical standards for the licensing of transmitters, not for government ownership.

[14] Josephine Young Case and Everett Needham Case, *Owen D. Young and American Enterprise: A Biography* (Boston, 1982), chap. 11, p. 178.

Hooper replied that Daniels was not willing to take any action in the matter until he was sure of support, first from Admiral Benson, chief of naval operations (then in Europe), and second from the House Committee on Merchant Marine and Fisheries. He had already been severely criticized by that committee for paying an exorbitant price for the Federal Company's patents and he was not prepared to go before it with this new project until he was sure the committee would support and sponsor it. In the meantime Admiral Bullard had tried to get the secretary to act, but his reception had "practically amounted to a rebuff" and he did not feel that he could approach Daniels again. The only way to get at the secretary, Hooper felt, was if Young were to communicate with him, either by letter or in person.[15]

Since Admiral Benson was known to be a strong supporter of government ownership, the news that Daniels was waiting for his advice was hardly encouraging. Young did not follow the suggestion that he might communicate with Daniels personally. On 16 June, however, he wrote to Bullard, summarizing the chronology of past events and, in his final paragraph, hinting that, unless signs of movement became evident soon, the company might feel free to reopen negotiations with Marconi. "I assume that the Secretary is making progress in the matter as rapidly as possible, and when he is prepared to do so, that he will proceed further with the plan or advise me that the request which has been made upon us to suspend our regular business in the sale of machines is no longer operative."[16] When this mildly phrased missive yielded no results, pressure mounted within General Electric for more positive action. A. G. Davis pointed out, at the close of a long letter to Young on 30 June, that "the opportunity we had before the Navy Department intervened in this matter may be slipping away from us" and recommended that "at an early date we should either be told that the Department endorses Admiral Bullard's statements . . . or that it disagrees with his views and that we are free, as a manufacturing company, to sell these devices to our natural customers."[17]

But the signal Davis asked for never arrived. What information General Electric did receive from the Navy Department came by way of Lieutenant Commander Hooper. And what Hooper had to say seemed more calculated to sharpen the crisis than to resolve it. A meeting with Young on 15 July produced a remarkable exchange of views. The immediate issue was the alternator at New Brunswick, still nominally the property

[15] Young Papers, Box 75, "Memorandum" by J. W. Elwood.
[16] Young to Bullard, 16 June 1919 (Young Papers, Box 75).
[17] Davis to Young, 30 June 1919 (Young Papers, Box 72).

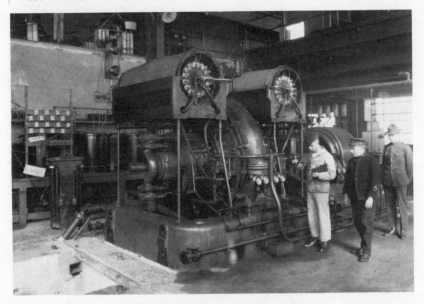

Pl. 14: The military presence: GE's first 200 kilowatt alternator installed at
New Brunswick, N.J.
Source: General Electric Company

of General Electric. This single machine had carried the bulk of transatlantic radio traffic in the closing months of the war; since the Armistice
it had provided Woodrow Wilson with direct radio contact with his staff
in Washington. Soon, however, with the imminent ending of the state of
war emergency, the New Brunswick station would have to be returned
to its owners, the American Marconi Company. Did this mean that the
Marconi Company would get the use of the alternator?

Certainly not, said Hooper: the Navy intended to remove the alternator
from New Brunswick and store it at the Brooklyn Navy Yard. As for the
second 200 kilowatt alternator, on order for New Brunswick, that should
be held at Schenectady. But if this happened, Young pointed out, all
commercial radio communications with Europe would be brought to a
standstill. Hooper cheerfully agreed, adding for good measure that there
was not another station in the country that could take over the job. Well,
asked Young, did the Navy intend to replace the old apparatus that the
Marconi Company had previously used? No, said Hooper, but the Marconi Company could put it back into use if they saw fit. Was it suitable
for transatlantic communication? No: and probably there was no way
it could be made suitable.

These eventualities Hooper was apparently prepared to contemplate with equanimity. And, to cap the discussion, he informed Young that, in the event of a dispute over vacuum tube patents, if the Navy Department were pressed by the government's Patent Board to say on which side its interests lay, it would have to reply that they lay with Western Electric, rather than with General Electric, since the Telephone Company had promised, if the dispute were resolved in their favor, to give the government a nonexclusive license and to forgive any back royalties that might be payable. To which Young could only reply that, if the Navy Department took that attitude, the proposed contract between the department and General Electric would be automatically cancelled.[18]

The conversations between General Electric and the Navy Department which had begun in such a mutually congratulatory flush of patriotic virtue on 8 April probably reached their coolest in this conference a little over three months later. Hooper was clearly threatening to make resumption of civilian control over radio communications with Europe impossible. Why did he do this? It was not because of any foot-dragging on the part of General Electric that negotiations had ground to a halt. Responsibility lay with the Navy Department and its civilian chief. What did Hooper hope to gain by threatening to remove the alternator from New Brunswick?

What he hoped to gain was precisely the objective he had sought for years: American control over American technology.[19] That was why he and Bullard had blocked the sale of alternators to Marconi; and that was the purpose behind the proposed federal contract. Now, in July 1919, it was becoming clear that no such contract would ever be signed. The danger was that, lacking a contract, General Electric would reopen negotiations with Marconi on a normal commercial basis (as A. G. Davis was already urging). This was why the alternator at New Brunswick and its sister about to be shipped from Schenectady had both a symbolic and a practical importance. Defeat of the Navy's efforts to keep these machines out of the hands of the Marconi companies would almost inevitably be followed by resumption of normal commercial relations with General Electric. The Marconi Company would be the operating com-

[18] Young Papers, Box 75, "Memorandum" dated 15 July 1919 by J. W. Elwood. The Young Papers (Box 71) also contain Elwood's handwritten notes on this meeting. They indicate a somewhat more forceful tone by Young than appears in the final typed version.

[19] Compare Hooper Papers, Box 37 (memoirs): Hooper was determined that the alternators at New Brunswick (both the 50 and the 200 kilowatt units) would not go to any company that was not "100 percent American."

pany, as it had been before the war; General Electric would manufacture the equipment. When commercial radio service resumed between North America and Britain, Marconi companies would control both ends of the circuit. But now they would have what, before the war, they had not had: equipment adequate to do the job.

Removal of the alternator from New Brunswick would not in itself prevent such an outcome. The threat to do so, however, served clear notice on General Electric and the American Marconi Company that the Navy would go to extreme lengths to prevent it. The Marconi Company might indeed resume possession of the New Brunswick station; but, if it did, it would not find the alternator there. And the warning that the Navy Department might make its influence felt in patent disputes was a not too subtle reminder to General Electric that, though hostilities had ceased and military contracts were only a fraction of their former size, official favor still counted for something.

Hooper's objective was to eliminate American Marconi as a factor in American radio. The most effective way to do this was to make sure that it was absorbed into the new radio corporation that GE intended to create. The threat to remove the alternator from New Brunswick was intended to apply pressure in that direction. General Electric probably needed little urging. Talk of a new radio corporation to protect and represent American interests was heady stuff. The hard fact remained, however, that no one in GE knew anything about radio as a business, however sophisticated they might be in radio technology. Expertise was needed that could not be acquired overnight; and the prospect of taking over responsibility for handling long-distance radio traffic with completely inexperienced personnel was not an attractive one. The people who had the experience and the expertise were in the American Marconi Company: men like E. J. Nally; his young assistant David Sarnoff; W. A. Winterbottom, the company's traffic manager; Roy Weagant, an engineer of talent and ingenuity; and the company's cadre of veteran telegraph operators. There, already in existence, were the organization and personnel that the new company would need. Not to be forgotten, too, were the facts that American Marconi owned the transmitting and receiving stations that would be returned to civilian control, and that, if radio traffic were to be passed across the Atlantic and Pacific, working arrangements would have to be made with the companies or government bureaus that controlled the other end of the circuits. In the case of Great Britain, that was the Marconi Company.

There was, therefore, good reason why General Electric should show interest in acquiring the personnel and property of the American Marconi Company. And there was good reason for the Marconi interests to be more than casually interested in talking to GE. Their coastal stations had already been sold to the government; they were out of that line of business, presumably, for the indefinite future. The odds against being permitted to resume operation of the high-power transoceanic stations seemed uncomfortably high. Even if Daniels's program of government ownership were defeated, the Navy had made it painfully clear that every possible obstacle would be put in the way of acquiring new equipment, while Bullard and Hooper had left no doubt about their determination that no foreign-owned or foreign-controlled company would ever again be permitted to operate a radio station on American soil. Nally's diary makes it evident that, although in April 1919 he had heard only rumors about the formation of a new company and knew nothing of the proposed contract, he had few illusions about the future. The only way out he could see at that time was for the American Marconi Company to buy out its British stockholders, thus validating its claim to be a truly American enterprise.[20]

This was no idle dream. The British company at this time, Nally believed, owned just under 500,000 shares in its American subsidiary, or about 25 percent of the total.[21] These shares were selling on the American market at $4.25 to $4.50 per share and in London at from 20 to 30 shillings.[22] To buy out the equity of the British company would therefore have required something like $2.25 million, if the shares could have been acquired at the New York price. At the end of November 1919 the American Marconi Company held among its assets $1,237,500 of U.S. Government bonds and a total of $2,875,321 in high-grade corporate bonds. It had no bonded debt, mortgages, or notes of any kind.[23] The funds to buy out the parent company's equity could have been found— if British Marconi had been willing to sell on those terms. This would

[20] E. J. Nally, diary entry for 25 April 1919, as transcribed in Clark, "Radio in Peace and War," chap. 12, pp. 53-54. The original of Nally's diary has not been located.

[21] Nally to Sweet, 6 February 1918 (Clark Radio Collection). Nally's first figure was 318,986 shares, but on 25 February he amended this, on receipt of a cablegram from England, to 494,826 shares. The total authorized common stock was 2 million shares of a par value of $5.00.

[22] Young to John Gray, 1 November 1919 (Young Papers). The quotations given by Young referred to the period "when we began to negotiate with Mr. Nally"—i.e., May 1919.

[23] Young Papers, Box 71, auditor's report on Marconi's Wireless Telegraph Company of America, as of 31 November 1919. The company's total assets were $10,667,395.28 and its total liabilities only $590,233.52, leaving a net worth of more than $10 million.

not have purged American Marconi of all foreign ownership, since some 954,000 shares were held by individuals in other countries, but it would have eliminated the parent company's equity and it might have been sufficient to neutralize the Navy's principal complaint.[24]

Owen Young at General Electric had another strategy in mind. He wanted American Marconi to form the operational core of GE's new radio company—a conception at variance with Nally's vision of an autonomous American Marconi, freed from its British connection. Little could be done to advance Young's plan until Nally and his fellow directors could be induced to accept the idea that the only viable future for their organization lay in accepting absorption into a larger entity. In this endeavor Young had the full support of Stanford Hooper at the Navy Department.[25] But to convince Nally required patience and tact. Nally wanted American Marconi to survive, but with a greater degree of independence than before. Young's strategy implied its disappearance.

Young and Nally held the first in a series of conferences on this issue in New York on 12 May 1919.[26] While the talk ranged over a variety of topics, what chiefly concerned Nally was the new radio company he had heard about. He understood why the Navy Department did not favor his company, although he insisted that British Marconi had never tried to influence its policies. But what was this new company supposed to do? Establish new radio stations in foreign countries? That, he suggested, would be difficult and expensive. And what would its relation be to "the other people in the United States who are in the wireless field"?

Young, for his part, depicted General Electric as virtually helpless— "practically from a commercial standpoint, paralyzed." He asked Nally to understand that General Electric had such wide-ranging relations with the various departments of government that "even an office boy or a clerk" could write or telephone the company to go on or not to go on with a contemplated plan, and it would have no choice but to obey. Specifically, GE could be stopped from negotiating with another company by anyone in any department in Washington, even though that department was without authority to present a constructive alternative.

Nally may not have found this picture of a General Electric Company

---

[24] Young Papers, Box 71, analysis of the list of stockholders of the Marconi Wireless Telegraph Company of America.

[25] Hooper, in his memoirs, claims credit for initiating negotiations between American Marconi and General Electric. See Hooper Papers, Box 37, and Clark, "Radio in Peace and War," chap. 12, pp. 53-54.

[26] Young Papers, Box 72, notes by J. W. Elwood on a luncheon held at the Bankers Club, New York, 12 May 1919 (dated 14 May 1919).

at the beck and call of any government flunky entirely convincing, but he did not press the point. There was a more important issue. What, he asked, would be the attitude of the government in Washington if the British interest in the American Marconi Company were to be bought out by a syndicate, and that syndicate then sold the stock to the new company that General Electric proposed to establish? Would that change the government's attitude? In particular, "would the government do business with the American Marconi Company if the British did not have a share interest in the Company?"

What Nally was suggesting was not, it should be noted, the absorption of American Marconi into GE's proposed new company but rather an arrangement by which GE, through a newly created subsidiary, would take over the equity interest that British Marconi currently held. American Marconi would continue to exist as the operating company. Young's response was guarded. He could not say what the attitude of the government to such an arrangement might be, and he again emphasized that GE could act only under the direction of the government. But, he added, "In this particular case, I recognize the right of the Government to stop us from negotiating with you, but I am almost inclined not to recognize their right to force us into competition with you."

It was a nice answer, committing Young and GE to nothing but nudging the door a little wider open toward a mutually advantageous accommodation. There was some further talk about how an arrangement of this kind would be regarded under the Sherman Act, but the next step was clearly up to Nally. Neither he nor Young could do much more without knowing whether British Marconi would sell its interest in its American subsidiary.

Young had stressed several times in the course of this conversation that a request from even a low-level government official was equivalent to an order. General Electric, he implied, had so much at stake in its relations with the federal government that to steer a course differing from that endorsed by a federal department was unthinkable. Did he really believe this? Or did he say it to put the best possible face upon the embarrassing fact that GE had precipitously withdrawn from business negotiations that were being carried on in good faith, at the behest of individuals who now, it turned out, could not make good upon the assurances they had offered? If, in May 1919, the American Marconi Company found itself in a distinctly awkward position, so did General Electric. And both companies had the Navy to thank for their predicament. By mid-May it was evident that the Navy was not going to extricate them from it.

The next meeting between the two men occurred a week later.[27] Nally in the meantime had cabled Godfrey Isaacs confirming his belief that the negotiations with GE had been broken off because of interference from Washington and asking him to come to the United States at once. Isaacs, unfortunately, could not arrive until about 20 June. This, Young suggested, was too late: matters could not be held up until then. Nally agreed, adding that he had discussed the situation with his board of directors and executive committee. They were in favor of cooperating with GE in working out a plan of action acceptable to Washington, and they understood that this would probably necessitate taking over the interest of British Marconi in the American company. How to finance the transaction was still an open question. Young made light of the problem. Purchase of the British interests, if they could be obtained, represented such a small part of the financing required for the new company that it could be handled as part of the same package and "through the same set of bankers." Nally said that was exactly his idea.

From this point on, General Electric and the American Marconi Company were agreed in principle on the general objective of purchasing the British company's financial interest in its American subsidiary and transferring it, directly or indirectly, to General Electric. Two vital issues remained. The first was the status of the proposed federal contract, the second the willingness of British Marconi to sell its equity interest at a reasonable price. Neither issue proved easy to resolve.

On the contract, a conference with the Navy Department in Washington on 23 May produced only equivocation.[28] At a second meeting on 21 July Daniels, meeting with Bullard, Hooper, Nally, and Young, was still fretting over what he called the monopolistic features of the charter, his authority to grant exclusive rights to government-owned patents, and the need for further consultation with Secretary of State Lansing and the president.[29] It was probably after this latter conference that General Electric in effect gave up hope that the contract would ever be submitted to Congress, far less approved, and decided to push ahead anyway.

Stanford Hooper, in his memoirs, takes credit for this decision.[30] As he later recalled, after the meeting broke up no one seemed to know what to do. Young was discouraged, upset, and disconcerted by Daniels's

27 Young Papers, Box 72, "Memorandum of Conference with Mr. Nally," 19 May 1919.

28 Davis to Young, 26 May 1919 (Young Papers, Box 75).

29 Davis to Young, 23 July 1919 (Young Papers, Box 75).

30 Hooper Papers, Box 37 (transcript of tape-recorded memoirs).

attitude, and Hooper made a point of walking down the corridor with him alone, "because I knew what he was going to say to me." Young said that he didn't see how General Electric could go ahead if the secretary would not approve the charter. Hooper had his answer ready. "Mr. Young," he said, "you don't need that charter. I think you are better off without it to tell the truth. You have got the patent[s], you have got the alternators, and you will have the radio stations if you take over what there are in this country, and the thing to do is go ahead. No other company can get ahead in less than two years and you will be in full, worldwide operation by then. Certainly nobody will put money into a competing company for many years to come and by that time you will be well established." Young, according to this account, agreed immediately. "That is just exactly what we are going to do. That settles that."

Young, apparently, acceded readily to Hooper's suggestion. Had he perhaps already discussed such a possibility with his colleagues? Hooper's logic was to some extent compelling. General Electric's principal current asset in the radio field was the alternators—not so much the machines themselves as the capability of building them and integrating them into a highly efficient system of long-distance low-frequency radio. This was, however, a wasting asset. Delay was costly, and if GE was going to produce and market the alternators, it would have to do so soon. There was a strong case to be made for getting a head start on the competition.

On the other hand, giving up the proposed federal charter was not without its costs, although some of them were not to become evident until years later. What GE gave up was not so much the government-held patents, which were of debatable value, but rather the clear and explicit recognition of public purpose. The radio company that General Electric created, which we know today as the Radio Corporation of America, might indeed claim to be a chosen instrument of national purpose, formed at the instance of the federal government and dedicated to defending the American national interest. And, in the telecommunications field at least, this claim had substance. But the federal charter that would have sealed this claim to special status was lacking. The explicit privileges that the charter would have given—the preferred claim on frequency allocations, for example—might indeed have proved very valuable. The charter itself, however, would have been worth more. Without the charter, RCA's claim to special status and a special relation to the federal government lacked explicit validation. It became something for the advertising and public relations departments to work on, a suitable topic for speeches, articles in popular periodicals, and, when necessary, congressional testimony. But, when the corporation later came under public attack for its alleged monopolistic practices, these were not the secure defense that a federal charter would have been.

On the purchase of the British interest in American Marconi, Young and Nally made steady progress. Nally at first tended to drag his feet, on the ground that he and his fellow directors knew nothing about the terms of the proposed contract with the federal government and thus could not know what they might be committing themselves to. He was inclined to talk in terms of a holding company, which would purchase the British equity, hold the necessary patents, and then license the various interested operating companies, including American Marconi. This Young declined to consider. On instructions from Assistant Secretary Roosevelt, he refused to show Nally a copy of the proposed federal contract, but he did make it clear that General Electric would either have to become an operating company itself or else set up a new company to do the operating. In such a company, he assured Nally, the personnel of American Marconi would play an essential role. "I had always assumed that the General Electric Company was the best fitted to do the manufacturing. As far as operating is concerned, we know nothing about it and we ought to take advantage of your experience."

Nally raised no objection to that idea, but he did worry about who was going to be in control. The American shareholders in the American Marconi Company were certainly willing to enter into an arrangement with GE, but only if they were to have a controlling interest in the operating company or "at least the final say in an operating company." This was rather more than Young could accept, but he did assure Nally that "it ought to be in your hands to operate the company should it be formed"[31] This seems to have given Nally the reassurance he needed. By 17 June he was able to state definitely that he and his fellow directors "would like to put the deal through on fair terms, provided that it was definitely understood that no one interest should control the new company."[32]

The major question mark remaining was the attitude of Godfrey Isaacs, representing the British firm. Would he sell? If so, at what price? What was the American Marconi Company worth anyway? As Young had indicated to Nally, raising the capital to buy out the British interest was only a part of the larger problem of how the new corporation should be financed. But it was an important part, because the equity interest held by the British firm represented control, at least in the eyes of the federal

[31] Young Papers, Box 72, "Memorandum" by J. W. Elwood on luncheon held on 2 June 1919.

[32] Davis to Young, 17 June 1919 (Young Papers, Box 72).

government. Furthermore, if American Marconi were to be absorbed into a new radio corporation, Isaacs's approval was essential, and not only because of his influence over the foreign stockholders who would vote on the transaction. There were longer-run issues involved. If the new American radio corporation were formed, it would have to do business with the Marconi Company all over the world. Traffic agreements would have to be worked out, territorial spheres of influence arranged, and possibly some exchange of patents negotiated. It was essential, in short, that when the negotiations were over and the dust had settled, the new American radio corporation and the Marconi Company should be able to do business with each other in a reasonably amicable way.

* * *

The most immediate issue was the worth of the American Marconi Company and how its acquisition should be financed. Owen Young was an old hand at this game. It was how he had made a name for himself, first as legal counsel for the prestigious firm of Stone & Webster in Boston, later with General Electric. In its operating characteristics radio was not much like the electric street railways and public utility systems with which Young had been familiar before the war. From the legal and financial points of view, however, the problems were not very different. A new consolidation was to be formed. What were its component parts worth? What would be its real assets? What was its potential earning power? How much money was it worth risking in it? And in what form should the money be raised? These were familiar problems to Young. He knew what questions to ask, what information was needed, what kind of assurances GE's directors would need, and what GE's bankers would want to know. And this was the arena in which he had developed and perfected those skills of negotiation and compromise for which he was well known[33]

The first draft of a possible agreement between General Electric and the American Marconi Company was prepared by Nally and discussed at a luncheon meeting with Davis, Young, and L.F.H. Betts, counsel to American Marconi, on 13 June 1919[34] Nally stated that his company was worth $12 million on the basis of its physical and financial assets, goodwill, claims against the U.S. and British governments and holdings in other companies (such as Pan-American). This figure was somewhere between its market value as tested by the price of its stock in the United

[33] Ida Tarbell, *Owen D. Young: A New Type of Industrial Leader* (New York, 1932), esp. chaps. 4 and 5.

[34] Davis to Young, 17 June 1919 (Young Papers).

States and that value as tested by the price of its stock in London. It had 28,000 shareholders, and over half the stock was owned abroad. How much of it was held by the British company he did not exactly know. Large amounts were held by London brokers, probably in the interests of British Marconi, and how much would eventually turn up in Isaacs's hands would depend, he thought, on how favorable a trade General Electric offered. Probably Isaacs would end up holding about a quarter of the total.

Looking toward the future, Nally accepted that General Electric would do all the manufacturing for the new company, and to that end American Marconi's plant at Aldene, New Jersey, valued at $750,000, would be sold to GE.[35] He estimated that the new company would require twelve alternators to begin with, and GE could presumably turn these over in exchange for stock, on the basis of $127,000 each or a total of $1,524,000 in all. He thought it would be wise to bring into the consolidation "what is left of the Federal Company" and also the patents of the International Radio Company (formerly NESCO) as well as the de Forest patent on the oscillating audion; but the heterodyne patent, he thought, was not really indispensable. General Electric's patent rights in the field of long-wave radio, he thought, were worth $5 million plus a royalty of 5 percent on all business. Mention was made of the fact that the original charter of American Marconi contained a covenant binding the British Marconi Company to turn over to it the U.S. rights to all its inventions forever; this would presumably be transferred to any company that bought American Marconi and might prove very valuable.

Neither Young nor Davis found anything substantial in this to object to, though Young suggested greater reliance on bond financing. Acquire the assets in exchange for preferred and common stock, he suggested, and raise whatever outside money might be needed through the sale of bonds. An amended version of Nally's proposal dated six days later reflected this suggestion[36] It proposed that a radio corporation be formed with a Delaware charter. Any remaining objections to foreign ownership could be taken care of by providing that 20 percent of the stock could be voted no matter who owned it, while the remaining 80 percent could be voted only by American citizens. The charter should allow the company to sell, but not to manufacture, radio apparatus and devices. Funds

[35] The typed text of Davis's letter states: "Mr. Nally fully understands that the American Company is to do all the manufacturing." The context makes clear that this is an error in transcription; it was accepted from the beginning that GE would be responsible for the manufacturing.

[36] Young Papers, Box 72, "Wireless Program," 19 June 1919.

should be raised by an open-ended series of convertible debentures, starting at $15 million and capable of increase at any time to not more than 60 percent of the net worth of the company. In addition, there should be issues of common stock and cumulative preferred stock, with equal voting power; these would be used to acquire the patents and other radio rights of General Electric and its subsidiaries, plus the assets and goodwill of the American Marconi Company. Preferred stock to the amount of $3 million would also be issued to General Electric when it acquired the interests of British Marconi in its American subsidiary. It was suggested that an attempt should be made to acquire the patents of the Federal Company and the equity interest held by Federal and British Marconi in the Pan-American Company; that an agreement should be reached with British Marconi by which GE would furnish all the radio apparatus that the Marconi companies throughout the world might need, with the restriction that it not be used in the Western Hemisphere; and that possibly an attempt should be made to bring AT&T and Western Electric into the combination by offering them a stock interest in exchange for their patent rights. The proposed agreement with the federal government was dismissed in a couple of lines. If it could be obtained, well and good; if not, the consolidation would go ahead without it.

If the new company were set up on this basis, Nally suggested, its financial structure might take the following form:

Debentures	$15,000,000
Preferred Stock:	
To GE for patents, less the Aldene factory	$4,500,000
To GE for twelve alternators	$1,500,000
Reserved for purchase of British Marconi's holdings in American Marconi	$3,000,000
To Marconi individual stockholders	$9,000,000
To Federal Company	$ 250,000
For other patents, etc.	$ 750,000
Total Preferred	$19,000,000
Common Stock:	
To GE for patents	$5,000,000

To Marconi stockholders for
patents $5,000,000
To International GE for patents $3,000,000
Reserved for other patents $3,000,000
Total Common $16,000,000

Of the common stock, General Electric and American Marconi would each contribute to their bankers, to compensate them for marketing the debentures, the sum of $1,875,000. This would leave GE holding $3,125,000 of the common stock. Combined with the International Company's holdings of $3,000,000, General Electric would own 38.3 percent of the common stock and 31.6 percent of the preferred.

An unfriendly critic, looking at Nally's financial plan, might wonder how a total capitalization of $35 million could be justified on the basis of earning assets that, by Nally's own estimates, were worth much less than that. He had said that American Marconi was worth $12 million. Presumably this was represented by the $12 million of preferred stock in the radio corporation designated as reserved for the purchase of British Marconi's holdings and the shares of individual stockholders. Why then the additional allocation of $5 million of common stock "to Marconi stockholders for patents"? In the same way, he had estimated that GE's radio patents were worth $5 million plus a royalty, and $4.5 million of radio corporation preferred stock ($5 million minus the price of the Aldene factory) had been set aside for their purchase. Why then an additional $5 million to GE, plus $3 million to International GE, in common stock "for patents"? Financing in this style was precisely what critics of corporate mergers had in mind when they talked about watered stock. Exactly what earning assets was the common stock in the new radio corporation supposed to represent?

Owen Young's papers give some insight into this question. A set of handwritten notes, under the cryptic title "Trade Marconi," suggests the rationale behind GE's allocation of common stock. Under the heading "Method," the notes read: "We put in money against *Pref*, are to have common based on theory ordinarily on intangibles—here against theory buying out British—Excess cash over par = our right to common.[37] The common stock, in other words, represented not so much earning assets or "goodwill" in the ordinary sense but rather GE's stake in the control

[37] Young Papers, Box 71, notes headed "Trade Marconi" (no date; probably by J. W. Elwood). The following line in the notes is particularly intriguing. It appears to read: "Easier—51-49—in *loot*." I have tried to read the final word in some other way, but without success.

of the company—its assumption of the role formerly played by British Marconi. Much the same was true of the $5 million of common stock which, by Nally's plan, would go to former stockholders in American Marconi; and the same pattern was to be followed later when AT&T, Westinghouse, and United Fruit joined RCA. The common stock of RCA in fact never paid dividends throughout the 1920s, and this should have occasioned no surprise.

Justification for the enlarged capitalization proposed for the new radio corporation, both in Nally's plan and later, rested on the presumption that the income-generating power of the consolidation would be larger than that of its component parts, functioning separately. This was not an unreasonable belief for anyone who understood the problems that faced both the Marconi companies and GE, in the radio field, in 1919-1920.

The way in which RCA was actually set up and financed did not follow Nally's plan in detail. He (and Young) had proposed to rely rather heavily on bond financing. This part of the plan soon dropped out of sight and RCA began its corporate life without any bonded debt whatsoever. Relations with British Marconi were not in fact structured as Nally had suggested they might be. And he had explicitly called for the new corporation to establish its own radio stations in Europe, estimating that at least three would be required. This glossed over the probability that British Marconi and its European allies would regard such action as an invasion of their territory. The essentials of his plan, however, were adhered to rather closely in the negotiations that followed.

Young had GE's comptroller, C. E. Patterson, go over the books of American Marconi on 1 July.[38] In the time available no complete audit was possible, but Patterson's impression was that the balance sheet Nally had provided did reflect the probable true value of the company's assets. The fixed investment had been set at $4,860,000. This, Patterson thought, was probably conservative. Three-quarters of the total represented land and buildings at the costs of 1913-1915; after four years of double-digit inflation they could not be replaced in 1919 at anything like the same figures. The Aldene plant was a substantial modern factory, well-equipped and probably worth more than its book value. The electrical apparatus, on the other hand, was subject to substantial depreciation and part of it was obsolete. There were, on the other hand, no obsolete or slow-moving inventories; any stock on hand was fully covered by existing orders. The largest item among the "receivables" was a regular account with the U. S.

[38] C. E. Patterson to Young, 1 July 1919, report on Marconi Wireless Telegraph Company of America (Young Papers, Box 72).

government, and the list of investments seemed perfectly satisfactory. Patents and goodwill had formerly been set at $3 million, but the amount for goodwill had been largely written down the year before and the figure currently on the books—$2,100,000—seemed reasonable. There were reserves of $680,000 against fixed investments and patents and a surplus of $2,130,000. As for earnings, net profits had risen from $260,000 in 1916 to $618,000 in 1917 and $712,000 in 1918; the 1918 figure represented a rate of return of about 7 percent on the $10,000,000 capital stock. Dividend policy had been highly conservative: in fact, only three dividends had ever been paid—2 percent in 1916, 5 percent in 1917, and 5 percent in 1918. The only note of caution that Patterson sounded was with regard to depreciation: in view of "the rapid development of the art," the amount set aside on that account seemed too low. Taking one thing with another, however, the findings were reassuring. GE would be safe, its comptroller concluded, in taking the "sound value" of American Marconi as $12 million, or $6.00 per share of capital stock.

It was a responsible and informative report and undoubtedly gave Young part of what he needed to know. All the comptroller could look at, however, was the record of the past: the properties that had been acquired, the amounts spent on them and the income they had earned. By the standards of his profession, this was entirely proper. The value of American Marconi as a going concern, however, depended less on these items than it did upon an uncertain future, and it was the shape of that future that Young had to discern. The question to which he had to find some kind of an answer was how much the properties of American Marconi would be worth if they were integrated with other things of value—specifically, General Electric's technological capacity, its borrowing power, its reputation, and its standing with the federal government. But the principal finding of GE's comptroller was reassuring none the less: there was nothing in American Marconi's past history, as reflected in its debts and other liabilities, that threatened to interfere with Young's plans for its future.[39]

There were two main problems with Nally's scheme for financing the new company. One was its heavy reliance on bond issues. Young had originally suggested this, and we may surmise that the suggestion reflected

[39] Quite the contrary, in fact. The current liquidity of the company was very high. The asset side of the balance sheet included $5,230,000 in high-quality liquid assets ("receivables," bonds and notes, loans and cash), while liabilities included $2,810,000 in surplus and reserves and only $1,030,000 in loans and accounts payable. From that point of view American Marconi was a bird ready for plucking.

his prior experience with the refinancing of street railway companies and public utilities. It implied, of course, a burden of fixed interest charges against future earnings. In the case of a radio company, future earnings were highly uncertain and the prudence of bond financing was therefore questionable.

The second problem concerned who was going to control the new company. Under Nally's plan, General Electric and its allied interests would hold 38.3 percent of the common stock and 31.6 percent of the preferred (47.4 percent after buying out British Marconi's holdings). And in normal circumstances this would have been more than enough to give voting control. To the individual stockholders in American Marconi, however, the plan called for the allocation of 31.25 percent of the common stock and 47.4 percent of the preferred. Since, share for share, common and preferred stock would have equal voting rights, the possibility existed that the American Marconi stockholders, if they made common cause, could control the new corporation, at least until GE received its additional allocation of preferred stock to compensate it for buying out British Marconi. Even then, GE would hold less than 50 percent of the combined common and preferred. This could prove important during the critical period when the new corporation was being organized and the first board of directors elected. The arithmetic was uncomfortably close from GE's point of view.

Much depended upon whether and on what terms British Marconi would sell its interest in its American subsidiary. If that could be arranged, and if the other stockholders in American Marconi consented to the arrangement, then the new radio corporation could become a reality. Until then, all understandings had to remain hypothetical.

A draft agreement of 12 July committed the two companies to work together to achieve the purchase. This was followed by three further agreements on 25 July. The first, which remained in escrow to be executed only if the new corporation was chartered before 1 January 1920, cancelled all wartime claims between the two parties. The second recorded their consent to the substance of the third or main agreement, which was to be executed as soon as the consent of a majority of the stockholders of American Marconi had been obtained. In this main agreement General Electric undertook to establish a corporation under the laws of the State of Delaware to be known as the Radio Corporation. To it GE and International GE would assign their patent rights in radio devices, and it would become the exclusive marketing agent for any radio equipment they might manufacture. The agreement also recorded GE's intention to purchase British Marconi's holdings in its American subsidiary, and the new Radio Corporation's intention to acquire all the assets of the American Marconi Company (except certain "reserved assets"), including its share of the Pan-American Company. If it proved impossible to reach an agreement with British Marconi, the agreement could be terminated by either party on or before 1 January 1920.

The agreement also laid out the financial structure of the new corporation, and although the structure of RCA, when it finally took shape, differed substantially from that indicated in this document, it nevertheless gives some insight into the assumptions of the two contracting parties before serious negotiations with British Marconi began. There was now no mention of a bond issue. Two and a half million shares of 7 percent preferred stock were to be issued, with a par value of $10, along with three million shares of common with no par value. Of the preferred, GE was to be allocated less than one-tenth, while of the common it would receive two-thirds. These amounts would, of course, change after GE purchased British Marconi's holdings in its American subsidiary, the exact amount of the change depending on how many shares in American Marconi the British firm turned out to hold.[40]

There was clearly some hard bargaining ahead, particularly on the price to be paid for the British company's holdings. What Young was working toward was an arrangement that would transfer control with the least possible cash outlay. General Electric would buy the shares in American Marconi held by its British parent for cash, at a price close to or perhaps slightly above the current market rate. Shares in American Marconi held by individuals, however, would be exchanged for common and preferred stock in the new radio corporation. The uncertainties were two: first, whether Godfrey Isaacs was willing to sell the British company's holdings, and if so at what price; and second, whether a majority of the individual stockholders of American Marconi, once the British interest had been bought out by GE, would approve the sale of their company to the radio corporation. Buying out the British interest was the first essential; but that by itself would not give GE enough votes to carry a stockholders' meeting of the American Marconi Company. A sufficient number of proxies would have to be obtained from stockholders overseas. If that could be done, the American stockholders could probably be persuaded to consent, or they could be outvoted.

The Navy was formally notified of this plan at a meeting on 21 July. The primary purpose of this conference, as far as Nally was concerned, was to persuade Daniels to rescind his order to remove the alternator from New Brunswick. This was the threat that Hooper had held over American Marconi, and one may well wonder whether Nally and his

---

[40] Young Papers, Box 72; Case and Case, *Young*, chap. 11, pp. 185-77.

fellow directors would have been as ready to negotiate with GE without it. Daniels agreed to leave the alternator in place, though he would not commit himself to returning the station to private hands until peace was formally declared. Young seized the opportunity to present the plans for the new company, making it explicit that he was doing so for the secretary's information only, not for approval. Daniels, characteristically, grumbled about the monopolistic features and whether the promoters might not "over-capitalize the Company or rob the public in some other way." Explanations of the proposed capital structure were given. They probably did not lay the secretary's populist fears to rest, but at least he refrained from outright condemnation.

Davis and Young also used this occasion to get informal clearance from the Navy for a number of future actions. They were clear in their own minds that Nally should be president of the radio corporation. They had promised him from the beginning that, in an operational sense, the organization and personnel of American Marconi would form the core of the enterprise. They respected the man for his character and for his intimate knowledge of the communications business. And they knew that, when the time came for the stockholders of American Marconi to vote on its absorption into the new company, endorsement by Nally was essential. The proxies would be needed.

Already, however, the Navy Department had made it clear that they did not care for the idea. Nally was closely associated with the past policies and attitudes of American Marconi, and there was a large legacy of distrust. And behind the Navy's objections lurked the larger question of how the new corporation would be publicly regarded. As a subsidiary of General Electric? Or as the old American Marconi Company in a new guise? It was not a trivial question, nor one that could be taken care of by a couple of press releases. If Nally were to be president of the radio corporation, there would have to be substantial non-Marconi representation on the board of directors and, as chairman of the board, an individual strongly identified with strictly American interests. The conference of 21 July did not settle this issue.

Young and Davis also told Bullard that it would probably be necessary or desirable to sell some alternators to British Marconi. Here was an issue that might indeed have caused problems. Had not the proposal to sell the machines to the British provoked the Navy to intervene in the first place? Now, however, with an American radio corporation in the offing, Bullard's reaction was mild. He hoped the company would not sell the alternators "all over the world." No, replied Davis, "only in places where we think it proper." And, continued the admiral, if they sold alternators to British Marconi, they should be sure to get a traffic agreement as part of the deal. Yes indeed, said Young; and he expressed his confidence that the interests of the radio corporation would prove to be "identical with the interests of the Government." Whatever trade Nally and GE might make in London, he was sure it would be one that the government would find satisfactory. Bullard acquiesced.[41]

Clearly the climate was changing. Daniels's program of government ownership was making no headway, although neither he nor his assistant secretary had abandoned it.[42] The aggressive self-confidence with which Bullard and Hooper had blocked the alternator sale had dissipated, to be replaced by Young's easy assurances that the interests of the government and those of the new radio corporation were identical. The British would get their alternators after all. So indeed would any other customers to whom the new company saw fit to sell the machines, for Davis's promise that they would sell alternators only where they thought it "proper" was no real constraint at all.

What made the difference was not merely the end of hostilities, the shrinking of Navy budgets (and with them Navy hubris), and the return to "business as usual." It was also the confidence that there would shortly appear on the scene a new entity, an American-controlled corporation that could be held responsible for the formulation and execution of American radio communications policy. This was part of the price that British Marconi was to pay for access to American technology. The technological balance had shifted, and so had the balance of economic power. There was no question but that in the future British Marconi and the new American radio corporation would work very closely together. Corporate self-interest would ensure that. But, whether as collaborators or as antagonists or as an uneasy combination of the two, they would deal with each other as equals, not as parent and subsidiary.

[41] Davis to Young, 23 July 1919 (Young Papers, Box 75).

[42] On 14 August 1919, Assistant Secretary Roosevelt, testifying at an executive session of the subcommittee on radio of the Senate Naval Affairs Committee, still argued strongly in favor of government ownership. He admitted that the second-best alternative was "to have the government back a private corporation giving that corporation a monopoly" but went on to say that such a monopoly would soon become controlled by British Marconi. Admiral Benson, testifying before the same sub-committee, was convinced that control of radio by the Navy Department was "the best way to exclude British domination of worldwide communication and to further American trade throughout the world." See Young Papers, Box 75.

EIGHT

The Formation of RCA Part 2: London and Jersey City

OWEN YOUNG may well have felt confident when negotiations with the British Marconi Company began.[1] After all, what options did Godfrey Isaacs really have? The American Marconi Company was helpless as long as it faced the hostility of the Navy Department. The probability that it would ever again be able to function in long-distance radio was negligible. It had sold its system of coastal stations. It had been denied access to new high-power equipment. All it had left was the radio rental service to ships. In 1919 American Marconi was little more than "the wreck of a business."[2]

On the other side of the scales, an alliance with General Electric promised many benefits. It is true that British Marconi could have used other types of high-frequency alternator—it owned rights to the Goldschmidt machine—or it might have adopted the Poulsen arc.[3] But General Electric's alternator was the best device of its kind available. It was part of an integrated transmitting system. And, perhaps most important, it was backed up by GE's proven manufacturing ability. Given the orders, the Schenectady shops were ready to start work immediately.

And why should Isaacs object to the idea of a new American-controlled radio corporation? He had the reputation of being a realist. What was essential was that there should be an organization in the United States capable of managing the American end of the transatlantic radio circuits.

[1] Compare Josephine Young Case and Everett Needham Case, *Owen D. Young and American Enterprise: A Biography* (Boston, 1982), pp. 180-87.

[2] Paul Schubert, *The Electric Word: The Rise of Radio* (New York, 1971; copyright 1971 by the New York Times Company, quoted by permission), p. 166.

[3] See *Radio Review* 2 (3 March 1921), 206, and Cyril F. Elwell, "Autobiography" (Elwell Papers, Stanford University Library, Box 1), p. 83.

It was not essential that it be controlled by British Marconi. Competent cooperation was all that was really needed. And if, along with cooperation in traffic-handling, there could also come some kind of patent-sharing agreement and an intercorporate treaty to minimize the chances of open conflict in other parts of the world, so much the better. British Marconi was used to dealing with government telegraph bureaus. If the Americans preferred a hybrid more to their peculiar tastes—a private corporation with vaguely defined public responsibilities—that was their business. And if that corporation wished to purchase Marconi's American subsidiary, and was willing to pay Isaacs's price, why should he object? The American firm had never been much of a moneymaker anyway; its sale would not hurt the income-earning ability of the British parent. The contrary was more likely. Funds for rebuilding and reequipping the Marconi system were short, at least until claims against the British and U.S. governments could be fully settled.[4] A transfusion of cash from across the Atlantic would be timely and helpful.

But if Young could feel confident that a deal would eventually be worked out, he could not afford complacency about the terms of the transaction. Isaacs had other suppliers to whom he could turn. If the worst came to the worst he could let the U.S. government expropriate his American stations and then sue for damages—he had some experience in seeking recompense from governments. At the very least he was certain to be a hard bargainer. He would want every penny he could get for American Marconi. He would want it in cash rather than securities. And he would want in addition a clear understanding with the Americans over postwar spheres of influence.

What did General Electric have to offer in exchange? Essentially three things: a fair price for American Marconi; mutually advantageous cooperation in reconstructing the world's radiocommunications system; and immediate access to American radio technology. Of these three, the last two were intangible and it was to be expected that Isaacs would try

[4] The profit and loss accounts of the British Marconi Company between 1915 and 1923 are summarized in Hiram L. Jome, *Economics of the Radio Industry* (Chicago and New York, 1925), p. 39. See also the *Economist*, 9 August 1919, for complaints by Isaacs that he had not yet been able to arrive at any settlement with the postmaster general with respect to interpreting and monitoring services performed during the war. Damages to the amount of £590,000 were, however, awarded by the British government late in 1919 as compensation for the seizure of Marconi stations in 1914. A circular distributed to Marconi stockholders in 1919 referred to the need for new funds to finance the reequipping of old stations and the construction of new ones. (See *The Wireless World* 8 [January 1920], 620.)

to minimize their value. A. G. Davis in particular anticipated an attempt to "talk down" the worth of GE's radio system and, before sailing for England to begin negotiations, he enlisted Alexanderson's aid in preparing counterarguments.

The long memorandum that Alexanderson prepared in response throws much light on what GE's engineers thought they were selling. He admitted without equivocation that no single feature of GE's system was indispensable. The Telefunken or Latour alternator could be used instead of Alexanderson's. Relays could be used for telegraphy instead of the magnetic amplifier. Other types of speed regulator could be devised. And so on for all the particular elements of the system. Nevertheless, even if substitutions were made, "it would yet remain our system." "The strength of our situation is the fact that it is a complete and new transmitting and receiving system which does not infringe the patents which have heretofore dominated the radio situation." It was in fact "the only system which is complete and controlled by one corporation." The patents governing its components would, he was confident, be sustained by the courts if challenged. But the value of the system depended as much on "the engineering and manufacturing organization which is behind it" as upon the patents themselves. The whole system, complete and integral, was "a representative accomplishment of the G.E. Co.," and the fact that it had been developed in a relatively short time was proof that "we are in a position to maintain the lead that we have acquired."[5]

Alexanderson was pinning his case on two assertions: first, that GE's system was complete and invulnerable to challenge on grounds of patent infringement; and second, that GE, having attained leadership in radio technology, would keep it in future. GE, in other words, could offer in the short run an operational radio system, ready for immediate use; and in the longer run, the creative potential of its engineers and scientists. The one promised the Marconi interests prompt reequipment of their long-distance radio circuits; the other promised them insurance against technical backwardness in future, at least vis-à-vis the Americans.

Godfrey Isaacs was probably as well aware as anyone of the value of

[5] Alexanderson to A. A. Buck (GE Patent Department), 1 August 1919 (Young Papers, Van Hornesville, N.Y., Box 72). It should be noted that Alexanderson regarded his system as "complete" despite the fact that it included neither the de Forest audion patent nor the Fessenden heterodyne patent, both essential to a state-of-the-art receiving system. Alexanderson regarded his magnetic valve detector as just as good as the heterodyne; his magnetic amplifier, he claimed, though not as efficient as the de Forest amplifier, was "sufficiently effective so that the same practical results can be obtained if sufficiently large receiving antennas are used."

what GE had to trade. The problem would not be to convince him but to get him to admit it at the bargaining table. This called for special skills. GE needed to send shrewd traders to London, men who would know a bluff when they saw one and, if necessary, could do a bit of bluffing themselves. They would have to match Isaacs at his own game.

Owen Young himself did not go to London to bargain with Isaacs. In view of his known talents as a negotiator and the prominent role he had played in GE's dealings with the Navy and American Marconi, this seems strange. It may be, as his biographers suggest, that family considerations kept him at home and that he preferred to hold himself in reserve against contingencies.[6] And possibly too he realized that any provisional agreement reached in London would have to be "sold" to the senior officers of General Electric and to the Navy Department. An individual somewhat removed from the day-to-day bargaining was needed to play that role. In any event, Young stayed close to GE headquarters at 120 Broadway in New York City. A. G. Davis went to England on the *Aquitania* at the end of July as the company's representative, and with him went E. J. Nally, general manager of American Marconi. Already in Europe on other business were Vice-President Anson Burchard of GE and Gerard Swope, until recently head of Western Electric's foreign operations, now president of International General Electric.[7] For staff assistance and legal advice Davis would be able to draw on the resources of the British Thomson-Houston Company (BTH), GE's exclusive licensee in the British Isles.[8]

Young stayed in touch with the negotiations by frequent exchange of cables, sent in code to reduce the possibility that information might be leaked to other interested parties.[9] Letters provided fuller information at

[6] Case and Case, *Young*, p. 187.

[7] See David Loth, *Swope of G.E.* (New York, 1976), chaps. 6, 7, and 8.

[8] The British Thomson-Houston Company was formed to exploit the Thomson-Houston arc lighting patents in the British Isles. Though said to be financed largely from German sources, it was staffed by American engineers from the General Electric Company and used American equipment. In 1896 it was reorganized to operate GE's street railway traction patents. Note that the British General Electric Company was completely independent of the General Electric Company in the U.S. and of British Thomson-Houston. See I.C.R. Byatt, *The British Electrical Industry, 1875-1914* (New York, 1979), pp. 33-34.

[9] These cablegrams, unfortunately, were frequently so garbled in transmission that it was difficult for anyone in New York to follow what was going on or to make constructive suggestions. For examples, see Young Papers, Box 72. Nally in September 1919 described the transatlantic cable service as "demoralized and sadly congested." See Clark Radio Collection, extract from Nally's diary, 6 September 1919.

intervals. On-the-spot conduct of negotiations, however, was in the hands of Davis; all Young could do was give advice and set limits to bargaining concessions. Now, Davis at this time was a widely respected figure at GE and, as head of the patent department, well-equipped to handle any of the technical issues that might arise. But he was not really a member of top management. His elevation to a vice-presidency did not take place until after his return from England. And the question may be raised whether GE did not send to England a man underranked for the job. Davis was a highly competent executive but he did not carry the bureaucratic or personal authority of Godfrey Isaacs, managing director of British Marconi. And one wonders what impression was made on Isaacs when he found himself bargaining, not with one of GE's senior executives, but with the head of its patent department. It is possible that Paris was not the only location where American innocence was on display that year.

* * *

August 2, 1919, found Davis and Nally established in London, ready to begin negotiations but held up by the August Bank Holiday. Davis cabled to Young immediately, suggesting that early deliveries of alternators would turn out to be the controlling factor and asking for an estimate of the best possible delivery schedule.[10] On 5 August, still before meeting Isaacs, he sent off a long letter that raised several interesting issues. He requested, first, an estimate of the value of the Tuckerton station on Long Island, owned by the Compagnie Générale de Télégraphie sans Fil. According to Nally, British Marconi owned half of the stock of the Compagnie Générale and the other half was owned by French Marconi and allied interests. If so, operation of that station in the United States had been a violation of the contract and patent rights of the American Marconi Company, and American Marconi would have a valid claim against its corporate parent for part or all of the $400,000 that the U.S. government had recently paid the Compagnie Générale for its use of Tuckerton during the war. Possibly, suggested Davis, that claim

[10] Davis to Young (cablegram), 2 August 1919 (Young Papers, Box 72). On 5 August Young replied that, with an expenditure of $75,000 for additional facilities, GE could ship one machine a month from 1 November to 1 March and two a month thereafter. He took the opportunity to warn Davis that no machines could be shipped that would impair the effectiveness of the American company without highly unfavorable reactions on public opinion, thereby encouraging government ownership. See Young to Emmons, Pratt, and Edwards, 5 August 1919 (Young Papers, Box 72).

could be traded off in connection with the purchase of Tuckerton by GE's new radio corporation.

Nally's inside information on the Marconi system was clearly beginning to pay off for his new allies in General Electric. He also warned Davis that Isaacs was likely to attach great importance to the adaptation of the Alexanderson system to radiotelephony. Davis asked advice from Young as to what this would cost, noting that if patent conflicts made it difficult to manufacture the necessary vacuum tubes in the United States they could always be made by British Thomson-Houston in England. To Davis's mild distress it turned out that Nally, and indeed all the Marconi people, thought that GE had nothing to offer but a transmitting system and attached little importance to the successful operation of Alexanderson's "barrage" receiving system during the summer static of 1918. Clearly Davis had a job ahead of him if he was to convince the Marconi group that GE had more to offer than just the alternators.

What seems to have most concerned Davis, however, in this brief pause before serious trading began, was uncertainty over government policy, both in Britain and in the United States. He wanted to know whether Congress had adjourned, and he asked Young to forward—"not, of course, by cable"—any information he had as to the possibility of Josephus Daniels's recommendations being passed. On the other side of the Atlantic, the British government had announced just a few weeks earlier that it intended to equip the first two stations in the Imperial Chain with arc transmitters, and it was beginning to look certain that the Marconi Company would have no role in that venture. "The more I learn here," wrote Davis, "the more I appreciate that the British Marconi Co. is very much disliked by the authorities here." And this left him worried about giving that company exclusive rights in the British Empire. "We ought not to give the British Marconi Company any exclusive rights until we can be satisfied that it is in a position to exploit them."[11]

The first conference with Isaacs began on the afternoon of 5 August and was continued on the following morning.[12] It at once became clear that two issues would dominate the negotiations: definition of the spheres of influence of British Marconi on the one hand and of the Radio Corporation on the other; and the price to be paid for British Marconi's holdings of American Marconi stock.

On the first issue, Isaacs opened by insisting that, whatever arrangements might be made for the rest of the world, the whole of the British Empire should be the exclusive territory of the British Marconi Company;

[11] Davis to Young, 5 August 1919 (Young Papers, Box 72).
[12] Davis to Young, 6 August 1919 (Young Papers, Box 72).

in that area no other company (including, of course, the Radio Corporation) should have the right to make use of GE's radio patents. Davis responded by saying that Canada was too close to the United States for that to be possible, and when Isaacs admitted that GE might have "special interests" in Canada, the matter was temporarily set aside. GE's counterproposal was made on the morning of 6 August. Davis argued that the rights to GE's radio patents in the British Empire were of enormous value. To give them up would be, in effect, to grant the Marconi Company complete freedom from Radio Corporation competition in that whole area. If that were to be done, GE would insist on "substantial special consideration"—something on the order of £7 million sterling.[13] For its part, GE wanted the Radio Corporation to have as its exclusive territory the whole of the Western Hemisphere, and this would of course require British Marconi to turn over to the American corporation its holdings in the Canadian and South American Marconi companies. Not surprisingly, Isaacs was not prepared to give up these territories without an argument, any more than GE was prepared to concede Canada, and for the moment an impasse resulted.

What price GE should pay British Marconi for the stock it owned in its American subsidiary was, of course, a separable issue; but in this kind of negotiation all issues were related. Davis told Isaacs that, before they learned of his desire for exclusive rights in the British Empire, they had intended to offer him the market price of the stock in New York on 12 May, which was about $4.50 per share. However, in view of Nally's urgent representation that Isaacs could not be expected to sell the stock for less than the cash cost of the shares he had purchased (and apparently there had been some recent buying for cash), Davis was prepared to offer par ($5.00). But this figure represented what GE would have offered if there had been no question of exclusive rights in the British Empire. If those rights were to be part of the deal, GE could not be expected to pay so much.[14]

Isaacs at this point objected vehemently. The stock that day, he said, was selling in London at between 32 and 33 shillings. How could he look his stockholders in the face if he sold their American Marconi shares for 20 shillings (as he put it) when it was selling on the market for 32?

[13] This figure was based on a claim that British Marconi had made against the British government for the value of the Imperial Chain business in the same territory.

[14] Davis also offered to buy British Marconi's interest in the Pan-American Company at cost. Isaacs rejected this on the ground that he would be unable to justify it to his stockholders.

To which Davis replied that 20 shillings was not the right figure—23 shillings was more like it, at the exchange rate prevailing on 12 May—and, in view of all the circumstances, he thought it was a fair and proper offer.[15]

Isaacs remained unconvinced, but instead of continuing to haggle over valuation he turned the discussion onto a new track. Why, he asked, was it necessary for British Marconi to sell its shares at all? Why could it not hold an interest in the new Radio Corporation? The unpopularity of the American Marconi Company in America could not be due to the fact that British Marconi owned some of its stock, since it had already promised the U.S. government not to vote that stock. The real difficulty, he argued, was that the Americans wanted to have a radio corporation that was free to operate over the whole world, or at least a large portion of it, instead of being confined to the U.S. and its possessions, as the American Marconi Company was. That was understandable; but if a corporation with such wider powers did come into existence, there was surely no reason why British Marconi could not hold two or three million dollars' worth of its stock.

What Isaacs was suggesting was the kind of minority participation in ownership that the Marconi Company had used successfully on other occasions, as for example with the Compagnie Générale. Davis thought the suggestion interesting enough to report to Young. What Isaacs was working toward, he surmised, was an agreement that would allocate to British Marconi the British Empire, and to the Radio Corporation all or substantially all of the Western Hemisphere, leaving the rest of the world open to the competition of both companies, each being licensed under the patents of the other. And it would involve British Marconi holding some stock in the Radio Corporation. This last proviso, as Young well knew, was the sticking point. If Isaacs thought it was only a small matter, that was because he underestimated Washington's insistence that the new radio corporation should be completely free from British influence.

Davis also reported Isaacs's reaction to several related but smaller issues. He had denied that British Marconi held any *direct* interest in the Tuckerton station, but he seemed "distinctly disturbed" when it was

[15] Isaacs was closer to the correct figure than was Davis. For the week ending 19 May 1919 American Marconi stock was quoted on the New York stock exchange at a high of 4-3/4 and a low of 4-3/8. The sterling exchange rate for demand funds in New York was then 4.68-3/4. At that exchange rate a price of 4-3/4 translated into 20 shillings and 3 pence, not 23 shillings. More serious than the arithmetic, however, was the fact that New York and London differed so widely in their evaluation of the stock.

pointed out that Tuckerton had been operated in violation of American Marconi's exclusive rights. Davis and Nally had, for the record, entered a claim for the entire amount of Tuckerton's wartime earnings. They also quizzed Isaacs about the British government's decision to adopt arc transmitters for the Imperial Chain, but Isaacs pooh-poohed the matter. Large-scale use of the arc in future, he said, was an impossibility. In fact, he understood that the British government would instruct its delegates to the proposed international radio convention to insist that radio stations should emit nothing but pure waves, and that the arc could never do. Davis thought this was a significant admission: if that was what Isaacs thought about arcs, and if the Marconi Company wanted high-powered transmitters in the near future, they would have to use alternators.

Isaacs closed this first meeting by asking for twenty-four hours in which to draft a new proposal based on the idea of British Marconi owning some stock in the Radio Corporation.[16] In the meantime the British Marconi Company held its annual stockholders' meeting. Whether as a result of conversations at that meeting or for some other reason, when Isaacs again met with Davis and Nally late in the afternoon of 7 August, no mention was made of an equity interest in the Radio Corporation. Instead, Isaacs presented a set of demands that represented a significant hardening of his position. Because of the lateness of the hour, little discussion could take place that day, and Isaacs was unable to meet with the Americans on the day following. Davis took the opportunity to confer with Vice-President Burchard and representatives of British Thomson-Houston.

If Isaacs's new offer was intended to represent a movement toward compromise, that was certainly not evident to Davis. For the American Marconi stock, Isaacs now demanded £2 per share, payment to be made in dollars at a rate of $4.20 to the pound. This was substantially more than any figure mentioned at previous meetings. On the matter of territory, he proposed to pay General Electric £175,000 for the exclusive rights to use its radio patents in the British Empire in contrast to the £7 million that Davis had suggested as a basis for negotiation. In South and Central America, which Davis had hoped would become exclusive Radio Corporation territory, Isaacs proposed cooperation. He would sell half of British Marconi's stock in the Argentine Marconi Company for £85,000, and its interest in the Pan-American Company for what it had originally cost ($185,000).[17] In Brazil and perhaps other South and Central Amer-

---

[16] Davis to Young (cablegram), 8 August 1919, and letter of the same date. (Young Papers, Box 72)

[17] The principal asset of the Argentine Marconi Company was a perpetual

ican countries new corporations would be formed to construct and operate radio stations under Radio Corporation management, with stock ownership and all capital investments divided equally between the British and U.S. interests. These companies would have exclusive rights to handle all radio traffic with Europe and North America. In effect this would give the Radio Corporation exclusive operating control in South and Central America while reserving to British Marconi a share in ownership. A similar arrangement was suggested for China. British Marconi and the Chinese government were joint owners of a new company which, according to Isaacs, had been granted a legal monopoly of all radio communications in that country, and this company would construct and operate the necessary stations. The Radio Corporation, however, could have a monopoly of radio communications between China and the United States. As regards Canada, Isaacs was willing to offer the Radio Corporation equity participation in Canadian Marconi; there were, however, temporary difficulties involving the high market price of that company's stock and the absence of substantial tangible property, and he thought the matter could be postponed.[18]

Davis and Nally reacted to this proposition in quite different ways. Nally was much encouraged and felt that substantial progress had been made, primarily because Isaacs now recognized for the first time the right of the Radio Corporation to do business outside the United States. Davis had never thought there was any question about that issue, but Nally was of a different opinion. As he saw it, the Radio Corporation was proposing to inherit the rights and privileges of the American Marconi Company. That company had never enjoyed any rights outside the United States and in fact was "morally if not legally bound" to confine its activities to that country.[19] Here was a potentially disruptive issue. In Nally's eyes Isaacs had made a valuable concession. As far as Davis was concerned, Isaacs had conceded nothing.

Argentine franchise of uncertain validity. British Marconi owned about two-thirds of the stock, the remainder apparently being owned by the Pan-American Company.

[18] This description of Isaacs's proposal is based on Davis to G.E. Company (cablegram), 18 August 1919. The text of this cablegram is often cryptic and several passages are followed by question marks, indicating garbled transmission or uncertainty in deciphering.

[19] Davis to Young, 8 August 1919 (Young Papers, Box 72). Davis was prepared to concede that, if the Radio Corporation did business outside the United States, it might lose its rights in future inventions of the Marconi Company, but he also thought it an important question how far the Radio Corporation should be bound by any obligations or understandings between British and American Marconi.

There was much else that Davis could not but think unsatisfactory. The issue of radio rights in China, for example, could hardly be disposed of by a facile reference to a new concession. It was unlikely that the Radio Corporation—or the Department of State—would long be content with an arrangement which gave British Marconi the exclusive right to build and operate radio stations in that country while restricting the American role to handling transpacific traffic with the United States.[20] Isaacs's design for Central and South America also was an oversimplification, if only because it ignored German interests in that area. This was easy to do in 1919, to be sure, but the fact remained that before the war German radio had established a firm foothold in South America and Mexico and no plan that ignored the interests of that country, and to a degree the interests of France also, could prove adequate in the long run.

Nevertheless, on these territorial questions there was at least a basis for further negotiation, and it is easy to see why Nally thought the offer a good one and was seriously disappointed when Davis reacted to it with anger and resentment.[21] In Nally's eyes Isaacs was offering the Radio Corporation—an entity that at this time did not even exist—a generous partnership. British Marconi would have its exclusive sphere of influence, in which it could use the products of American technology but be free from American competition. The Radio Corporation would have similar exclusive privileges in the United States, joint privileges in Central and South America and possibly Canada, plus the prospect of exclusive rights

[20] Young stated in congressional testimony on 11 January 1921 that, although Isaacs had represented to GE that the Chinese government had granted to a company known as the Chinese Communications Company (owned half by Marconi and half by the Chinese government) a legal monopoly of radio in China, nevertheless since then Japanese interests had been allowed to construct a station in China and he believed that an American firm also had asked for a concession. He had therefore informed British Marconi that, if China was open to others than the Marconi-affiliated company, RCA must be at liberty to take such steps as might be necessary to protect Chinese communications with North and South America. See *Cable Landing Hearings* (1921), p. 347. It is not clear whether Young thought that Isaacs had misrepresented the situation to Davis, or that the Chinese government had misrepresented it to Isaacs.

[21] J. M. Glaster to Young, forwarding translation of cablegram from Davis (Young Papers, Box 72). This cablegram suffered from many errors in transmission, leaving some features of the proposition unclear. It would be interesting to know whether Isaacs met privately with Nally between 5 August and this meeting on 7 August. As will appear later, when Nally left England on 3 September he felt assured of the presidency of RCA. This assurance may have been given by Isaacs before the meeting of 7 August.

to radio traffic between the United States and China. The rest of the world would be "neutral territory," in which both British Marconi and the Radio Corporation would be free to compete, each having free use of the other's patents. To Nally this must have seemed an entirely acceptable design for the reconstruction of the world's commercial radio-communications. It allocated to American interests a sphere of activity much larger than they had ever enjoyed before. And it was a design in which his organization, the engineering and operating personnel of American Marconi, would play an essential role, albeit under a new corporate title. Nally was neither a lawyer nor a financier; his whole background had been in traffic-handling. Here was a plan that would get civilian radio traffic moving again.

Davis thought otherwise. Isaacs's whole proposition he considered "exceedingly disappointing and undoubtedly bluff." He insisted that full details be cabled to Young. Isaacs thereupon withdrew his commitment to the offer, which upset Nally but probably confirmed Davis in his suspicions. The prime sticking point for him was the price to be paid for American Marconi. The territorial allocations could be haggled over at leisure, but American Marconi was after all what General Electric was proposing to buy; and although a difference of a few shillings per share might not seem a large enough issue to block the conclusion of an otherwise satisfactory deal, in the aggregate it amounted to a sizable sum. And we may suspect also, from the tone of his letters, that he did not fancy being hustled by Isaacs. He cabled Young on 18 August that he intended to make a counteroffer of 30 shillings per share. This would more or less split the difference between his original offer of 5 August and what Isaacs was now demanding, and would not be too far from the price currently quoted for the shares in London. If Isaacs accepted that price, Davis estimated that the payment required from General Electric for the whole deal, including purchase of the shares and all other parts of the transaction, would be about $2.15 million. If Isaacs stuck to his price of £2 per share, it would be about $2.9 million.[22]

* * *

Differences of opinion also arose between Davis and Vice-President Anson Burchard. These were of lesser gravity but still required attention. Since Burchard was in England, with Swope at his elbow, it was both proper and expedient for Davis to consult him regularly, and Young duly

[22] Davis to GE Company (cablegram), 18 August 1919 (Young Papers, Box 72).

advised him to do so. But Burchard and Swope, having other business, were not always available when needed; and when they were consulted their perspective was not quite the same as Davis's. This was especially true with regard to GE's international operations, a matter on which Swope, as the new president of International General Electric, was naturally sensitive. The problem was that the British Thomson-Houston Company held exclusive rights to the manufacture and distribution of all General Electric products in British territory. Now, however, General Electric was proposing to sell directly to the British Marconi Company. On the face of it, this would be a breach of contract with British Thomson-Houston, and compensation would be necessary. What form should this take? One possibility was a substantial cash payment by GE, and this is what Burchard advocated.[23] Another was to arrange for British Thomson-Houston to manufacture some or all of the alternators that British Marconi wanted to buy. This was Davis's preferred alternative. Isaacs, who was not above muddying the waters when it was in his interest, proposed at one point to turn over to British Thomson-Houston a large interest in British Marconi's factory and manufacturing business, presumably as a form of indirect compensation for infringement of the company's exclusive rights, but this offer was soon withdrawn.[24]

General Electric and British Marconi, it is clear, found direct negotiations complicated by the fact that each had to take cognizance of exclusive rights granted to a foreign subsidiary. In the matter of the Tuckerton station, British Marconi had at least indirectly trespassed on the rights of its American affiliate, and Isaacs had to agree to "adjust" the matter with the French so that American interests would be protected. With regard to British Thomson-Houston, Owen Young intervened quickly to nip the problem in the bud, cabling Davis on 18 August that British Thomson-Houston should transfer its radio rights to the new Radio Corporation, with compensation to be adjusted directly between GE and BTH. The Radio Corporation in turn should pay GE for the British rights, the amount to be adjusted on Davis's return and if necessary submitted to arbitration.[25] This took care of the immediate problem. Young also took time to smooth any ruffled feelings that the difference of opinion might have left. "Under existing conditions here," he cabled to Burchard on 22 August, "we all believe best way to handle radio rights is through new radio corporation and we hope business will be reasonably profit-

[23] Davis to Young (cablegram), 18 August 1919 (Young Papers, Box 72).
[24] Davis to GE Company (cablegram), 18 August 1919 (Young Papers, Box 72).
[25] Young to Davis (cablegram), 18 August 1919 (Young Papers, Box 72).

able." He added, however, that it was "not essential to trade with Marconi Companies" although with some misgivings he thought it desirable to do so if a reasonable price could be arranged. They would not want the deal to go through, he added, unless Burchard thought it reasonable and desirable; but they were confident that a trade could indeed be consummated.[26]

To Davis on the same date Young sent a cable intended to clear the ground for a final settlement with Isaacs.[27] To remove any possibility of misunderstanding (for several of Davis's cables had arrived full of errors in transmission) he spelled out the terms of agreement as they were understood in New York. Territorial spheres of influence were clearly regarded as of prime importance. The new Radio Corporation should have exclusive rights to all GE and Marconi radio patents in United States territory, while British Marconi could enjoy the same exclusive rights in the British Empire, except that for Canada a special agreement would have to be drawn up restricting Canadian stations to Canadian business and giving the new Radio Corporation the opportunity to acquire a substantial interest in Canadian Marconi if it so desired. For South America a new company should be formed to handle all transoceanic traffic. The Radio Corporation should own a majority of its stock and operate its stations. In the case of Central America, Mexico, and the West Indies the Radio Corporation should have licenses to all Marconi patents, exclusive if possible. And in China it should have such rights as would enable it to control absolutely all radio communications with the United States. The rest of the world should be "neutral territory," in which each company would license the other, except where one party or the other had already acquired exclusive rights that would have to be honored.

There were related matters to be taken care of. The New York office understood, for example, that in the case of Brazil the new corporation's rights were to be limited in some way. They were not sure exactly how, but they did regard Brazilian rights as very important. That would have to be clarified. Also, it should be made clear that the Radio Corporation would acquire British Marconi's interest in the Tuckerton station, and British Marconi must agree to cooperate in securing a satisfactory contract with the Compagnie Générale for the exchange of traffic between France and the United States. Satisfactory agreements would have to be signed by British Marconi and the Radio Corporation for the exchange of traffic between all existing and future stations, with all radio traffic for the United States funnelled exclusively to the Radio Corporation and

[26] Young to Burchard (cablegram), 22 August 1919 (Young Papers, Box 72).
[27] Young to Davis (cablegram), 22 August 1919 (Young Papers, Box 72).

all traffic to "Marconi territory" directed to Marconi stations. And, finally, it was regarded as highly desirable that British Thomson-Houston should have a substantial interest in the radio manufacturing business.

If, under each of these headings, agreement could be reached with British Marconi in the terms specified, and if it was satisfactory to Burchard and Davis, Young and his colleagues undertook to approve the deal. The total cost of the whole transaction, excluding only acquisition of the Tuckerton station, they estimated at about $3 million.

None of the uncertainties mentioned in Young's cable proved difficult to resolve. On 2 September Nally and Emil Girardeau of the Compagnie Générale signed an agreement transferring Tuckerton to American Marconi in return for a cash payment and an annual rent (the rent obligation to be cancelled if the French government ever chartered a competing company). Rights to the Goldschmidt alternator and an exclusive license to all TSF patents for the next twenty-five years accompanied the transfer.[28] In effect the agreement guaranteed freedom from French competition in the United States for the indefinite future and sealed an alliance with the corporation that controlled radio in France and her colonies. The Brazilian issue turned out to involve the marine radio business in that country, in which British Marconi had an interest that it did not wish to give up. The problem was resolved by agreeing that the Radio Corporation or its Brazilian subsidiary would be free to compete in Brazilian marine radio but without using Marconi patents. With this exception—and excluding Colombia, where British Marconi had recently acquired a concession—the Radio Corporation and British Marconi were to act together in South America through a jointly owned corporation, each supplying half the necessary capital, but with the American organization exercising operational control. In Mexico, Central America, and the Caribbean both companies would nominally be free to compete, but it seems to have been tacitly assumed that British Marconi would probably not do so.[29] As for the potentially troublesome matter of British Thomson-Houston, an agreement was hammered out and accepted in principle by Davis and

[28] Young Papers, Box 72, "French Agreement," 2 September 1919. American Marconi agreed to pay to the French company one million francs upon delivery of the station and a further million francs at the end of 1920; thereafter it was to make fifteen yearly payments of at least $30,000, increasing after 1923 to 12 percent of the net profits derived from the French circuit.

[29] Young Papers, Box 72, "Advantages of Proposed Arrangement with General Electric Company," 18 September 1919. In Mexico, Central America, and the West Indian archipelago, the new U.S. corporation would have the right to use Marconi patents, but British Marconi acquired no reciprocal rights to use General Electric patents.

Nally which provided that the Radio Corporation would buy from BTH "at least as many Alexanderson alternators and accessories as are bought of the Radio Corporation by the British Marconi Company for use in British Thomson-Houston territory," but added the significant qualification, "provided that . . . the quality, deliveries, and prices after allowances are made for freight insurance exchange etc., are as favorable as those of the General Company." The legal niceties were observed, in short, and compensation would be paid, but in fact the alternators would be shipped from Schenectady.[30]

Young's cable set a figure of $3 million for the estimated cost of the whole transaction, excluding only the Tuckerton station. By this time it was known that British Marconi owned precisely 364,826 shares in its American subsidiary. If Isaacs's asking price of £2 per share had been met, at his stipulated exchange rate of $4.20 to the pound sterling, the cost would have been $3,064,538.40. Offset against this would have been the sum Isaacs had agreed to pay for the British Empire rights to GE's radio patents: £175,000 or $735,000. The net cost to GE would therefore have been $2,329,538.40.[31]

Did Isaacs get the price he wanted? Not quite. The Preliminary Agreement between General Electric and British Marconi that was signed on 5 September provided for a net payment by GE of $2,212,738. Add to this the price British Marconi paid for the British Empire rights to GE's radio patents, or $735,000, and we get a figure of $2,947,738 for what the shares alone would have cost. This represents a price of just under $8.08 a share, and this is what GE paid to buy out British Marconi. A good price, one must admit, for shares of $5.00 par value on which only three dividends had ever been paid and which had been quoted in New York only a few months before at $4.50.[32] But a little less than the $8.40 Isaacs wanted. (If, as Young's biographers suggest, the purchase of the British Empire patent rights is regarded as an "additional consideration" and not part of the same transaction, the price per share would be just

[30] Davis to Nally, September 1919 (written from Patent Department, The British Thomson-Houston Company Ltd., 83 Cannon Street, London). (Young Papers, Box 72)

[31] At least one participant in the proceedings got his arithmetic wrong. See the document entitled "Things to be Done" in Young Papers, Box 72 (probably by E. J. Nally): "For the American Company's shares, the Radio Company, through General Electric Company, paid at the rate of £2 per share, less £175,000 allowed for General Electric patents and rights in the British Empire, all computed on the basis of $4.20 to the pound sterling, or $2,212,738.00."

[32] See *Wireless Age* 6 (January 1919) and (February 1919); dividends of 2 percent had been paid in 1914 and of 5 percent in 1917 and 1919.

under $6.06—still an eminently satisfactory figure from Isaacs's point of view.)

Did General Electric and the Radio Corporation it was about to create get good value for the money? This was to be, after all, a real transfer of funds, not a paper transaction. Private stockholders in American Marconi would be offered RCA common and preferred on a share-for-share basis, but British Marconi was to be paid off in hard cash. What made this necessary was the insistence that the last vestiges of control had to be eliminated. Owen Young went to some pains to hammer home the point that the complete extinction of British Marconi's ownership rights was a sine qua non. On 22 August he cabled Davis: "Government situation here looks more favorable on the assumption that we will take out British interests and be able set up promptly satisfactory system for American business. . . . There will be sharp reaction against Marconi interests if program fails and may encourage government ownership to the extent of prejudicing our independent program."[33] And on 2 September he cabled in similar terms to Burchard: "Washington situation would make it undesirable for us to acquire any interest American Marconi if British Marconi retained any stock in American company."[34]

This insistence inevitably weakened Davis's ability to drive a hard bargain. When Isaacs demanded £2 a share for British Marconi's holdings in its American subsidiary, he was not even implicitly asserting that every share of common stock in that enterprise should or could command such a price. If that had been his claim, it would have set his valuation wildly at variance with that of the market, as well as with appraisals of the net worth of the company made by responsible auditors.[35] What he was asserting was that the particular shares that his company happened to hold were worth that much, and the outcome of the bargaining indicates that he was able to make the assertion stick. The shares of American Marconi held by the British parent company could command a premium price precisely because they were the shares that represented foreign control. They *had* to be acquired. The British Marconi stockholders, under the efficient stewardship of Godfrey Isaacs, collected the rent that American nationalism created.

Davis had gone to England believing, along with his fellow executives

---

[33] Young to Davis (coded cablegram), 22 August 1919 (Young Papers, Box 72).

[34] Young to Burchard (cablegram), 2 September 1919 (Young Papers, Box 72).

[35] Note, however, from a clipping headed "Woram and Co.—Stocks and Bonds" in Young Papers, Box 71, the prediction in mid-1919 that stock in American Marconi "is apparently headed for the $20 level."

at GE, that British Marconi's need for American radio equipment was urgent, and that, to obtain such equipment, Isaacs would be willing to sell his company's stake in American Marconi at no more than the going market price per share. General Electric was willing to take over American Marconi because it needed a nucleus of experienced operating personnel around which to build its new radio company, because the physical assets of the company might prove useful, and because it wished to eliminate a source of potential competition. But there was no urgency about it. A new radio operating company could be formed without necessarily involving American Marconi. Skilled personnel could be hired as necessary. And much of American Marconi's equipment was obsolete and would have to be replaced. As a competitive threat in 1919 it did not look formidable.

Godfrey Isaacs's ability to drive a bargain shows in the way he was able to turn these preconceptions around. By the last week of August 1919, he was acting and talking as if General Electric had no choice but to meet his terms. He would buy a few alternators, certainly; something had to be done to get the transatlantic circuits working until high-powered transmitting tubes became available. He would pay a reasonable price for the exclusive rights to GE's radio patents in the British Empire. And he would cede to the new company a share of the radio business in South and Central America and the Caribbean, where the commercial and political interests of the United States were evident. But what the Americans seemed to regard as indispensable was the elimination of the last vestigial elements of foreign control. They wanted a truly American radio corporation. Very well: Isaacs would not stop them. But he would make them pay.

*　　*　　*

General Electric's New York office thought the agreement a good one. "Very much pleased with your London trade," Young cabled to Davis. "It was an excellent piece of work and we will try to put our end through here."[36] Nally too felt that the negotiations had gone well and the future looked promising. "Provided our Government does not interfere and push its doctrine of Government monopoly to the end that it receives congressional authority," he noted in his diary, "I feel certain that the proposed new organization, which I am to head, to which I have given so much of my time and my thoughts, will revolutionize communication and open

[36] Young to Davis (cablegram), September 1919 (Young Papers, Box 72).

up a bigger and broader field for human endeavour than was ever dreamed of."[37] What Davis thought is not recorded.

The Preliminary Agreement signed on 5 September said nothing about territorial spheres of influence. That, along with other matters requiring careful legal specification, was relegated to the text of a Principal Agreement that was signed by British Marconi and delivered to Nally in escrow, to be executed by the Radio Corporation when it was formed. The Preliminary Agreement was concerned with what had to be done in the period of transition, before the Radio Corporation was chartered and before British Marconi's equity interest in the Marconi Company of America was transferred to GE. It referred to the action contemplated as a merger. General Electric undertook to form a corporation to exploit its radio inventions, and to merge that corporation with the American Marconi Company if—and it was to prove an important "if"—the stockholders of that company consented to such a merger before 1 January 1920. Proxies for the 364,826 shares that British Marconi owned in its American subsidiary were to be given to E. J. Nally, to enable him to vote those shares for the merger. Since the British company had, some years before, assured the American government that it would not vote its shares in American Marconi, this action could not be taken until special consent had been secured, and this Nally undertook to do. The British company also undertook to urge all others who held shares in American Marconi to give their proxies to Nally or vote them personally for the merger. If the merger was approved, the British company agreed to transfer its shares to the ownership of General Electric, and General Electric in return agreed to transfer the sum of $2,212,738 to the account of British Marconi. Any dividend payments or other benefits received by British Marconi from its American subsidiary in the interim would be duly reported and deducted from the payment due from GE; and British Marconi agreed not to acquire, prior to 1 January 1945, any shares or interest in American Marconi or the Radio Corporation without the latter's written consent. Finally, British Marconi agreed, pending the completion of the merger, not to purchase or order any high-frequency alternators except from General Electric.

The Principal Agreement of 21 November, a much longer and more carefully drafted document, was an agreement, not between British Marconi and General Electric, but between British Marconi and the Radio Corporation of America, which by this time had received its charter from the State of Delaware (and which we may hereafter properly refer to as

[37] Clark Radio Collection, extract from Nally's diary, 6 September 1919, on S.S. *Aquitania*.

RCA). By this date, too, RCA had signed an agreement with General Electric which in essence transferred to RCA licenses for radio purposes under all the patents and inventions that GE and its subsidiaries owned or controlled, bound GE not to sell any radio devices except to RCA (or the U.S. government), and required RCA to purchase all its radio equipment from GE, as long as the latter was able to supply it with reasonable promptness. As far as radio matters were concerned, therefore, British Marconi from this point on was doing business with RCA.

The terms of the Principal Agreement followed closely the understandings worked out between Isaacs and Davis, particularly with respect to territorial rights. It defined with great precision those areas of the world in which the British Marconi Company on the one hand and RCA on the other would enjoy exclusive use of GE and Marconi patents. These were (with carefully specified exceptions) the British Empire in the case of British Marconi, and the Western Hemisphere (again, with stated exceptions, particularly regarding Canada) in the case of RCA. The rest of the world was classified as "neutral territory" and in this area both companies would be free to seek business, with patent rights being shared through a jointly owned company called Shielton Limited. Here too, however, there were exceptions. China, for example, which was excluded from neutral territory and reserved for the time being to British Marconi; and certain European countries, such as Holland, Spain, France, and Italy, in which the British company or its affiliates already enjoyed exclusive concessions. In Central America, including the West Indies and other Caribbean islands, RCA received a nonexclusive license to use Marconi patents, but no corresponding license was granted to British Marconi to use GE patents. For South America there was to be formed a new corporation, provisionally entitled the South American Communications Company, in which the contracting parties would each hold stock to the amount of $2 million. Of the nine directors of this company, five were to be named by RCA and four by British Marconi; and the "sole control and management" was to be entrusted to officers nominated by RCA. To this company British Marconi agreed to sell its holdings in the Pan-American Company, the Argentine Marconi Company, and the Marconi Company of Brazil. In return British Marconi was to receive the sum of $880,761, to be credited against its subscription to the capital stock. The South American company was to be granted an exclusive license under GE and Marconi patents for communication across the Atlantic and Pacific oceans and the Caribbean Sea, as well as between South America, Cuba, and Central America, but only a nonexclusive license for communication within South America itself. RCA agreed to arrange for the company to erect within three years a station suitable for communication with Great Britain and to sell it alternators for that purpose.

With these territorial rights specified, British Marconi agreed to make the Alexanderson alternator and its accessories purchased from General Electric its standard high-frequency alternator (not its standard high-frequency *apparatus*—the key word was changed in draft) for long-distance communication (over 2,000 miles). RCA in turn agreed to sell alternators and other radio devices made under its patents exclusively to British Marconi for use in its exclusive territory, and nonexclusively for use in neutral territory. For use within the British Empire the price to British Marconi was to be the cost to RCA plus 2.5 percent; for use in neutral territory the cost plus 10 percent; and for use in China the cost plus 25 percent. Identical markups were to apply to any radio equipment that British Marconi might sell to RCA. And both parties agreed not to sell long-distance equipment to any outside party (except their own governments) if such a sale would imply violation of the exclusive rights created by other clauses of the agreement.

Any attempt to summarize a document of this kind runs the risk of introducing error, either by oversimplification or by omission of some apparently trivial phrase that, in later litigation, might prove of critical importance. Enough has been said, perhaps, to indicate what the contract gave to each of the parties involved—what rights and expectations it created that were important for future action. To British Marconi it gave assurance that the transoceanic radio circuits would promptly be reopened for traffic, under civilian management and with a type of equipment capable of reliable and remunerative performance. This was of major significance. The very first article of the agreement, in fact, referred back to the contract of 1902 between the British and American Marconi companies, calling for the establishment of a "continuous and high-grade wireless service" between Britain and the United States, and recognized that RCA had acquired both the privileges and the obligations of that original contract. Beyond that, limits had been set on American aspirations. Within the British Empire the Marconi position was secure—at least against American competition, if not against the British Post Office. Even in Canada a successful holding action had been staged. In the European countries where Marconi companies held concessions and exclusive privileges, these were left intact. In China a direct challenge was avoided, at least for the immediate future. It was in South America if anywhere that ground had been given, and even there, though American interests would certainly play a much larger role than before, British Marconi was assured of a share of the profits and, perhaps more important, of control over the long-distance circuits between South America and Europe.

For the time being, at least. Partitioning the globe in this grand manner made the contract an impressive document, reminiscent of diplomatic treaties and even of papal bulls. At bottom, however, it was only a contract between two private corporations, temporarily possessed of rights that they had reason to consider valuable. The Germans, given a year or two to regroup, would be active in rebuilding their formerly strong position in South America. And, where China was concerned, the Federal Company was certain to want part of the action. There was always the possibility, too, that technology might spring a few surprises. Technology had created the values that were being traded; it could also destroy them.

General Electric, for its part, got most if not all of what it wanted. Above all it got markets for its alternators. The ground had been cleared for the creation of that American-controlled radio corporation that had been the prime objective of policy ever since Bullard and Hooper had appealed to the directors on 8 April. And, though Davis for one would have preferred to pay less, the price was within the limits contemplated before negotiations began. The Radio Corporation, furthermore, now had clear opportunities open to it. Its first objective would be to get the transatlantic circuits working again under civilian control. This meant inducing the Navy to hand over the New Brunswick station and its alternator, and that in turn implied defeat or abandonment of the Navy's drive for government ownership. With British Marconi out of the picture, such a political objective seemed obtainable.

That taken care of, further markets for the Alexanderson radio system could be found and exploited. In the United States such stations as Tuckerton would have to be reequipped and new ones built, particularly to handle the anticipated new traffic with Europe, South and Central America, and the Far East. Isaacs would want alternators for his territory—how many was not exactly clear, but at least two in the immediate future for the transatlantic station with others probably to follow in Europe and the British Empire. And there were good market possibilities in "neutral territory"—Poland and Sweden, for example, both countries known to be in the market for up-to-date communication facilities and not tied to the Marconi system by concessions or exclusive contracts. And stations would also have to be built and equipped for the projected South American company.

All this meant orders for the Schenectady shops and a busy schedule for Alexanderson and his assistants. For it would be necessary to move fast and get the new equipment installed and working before Telefunken competition revived, before high-powered tube transmitters started cutting into the market for alternators, and before the international conferences planned for the future began allocating frequencies in the very-low-frequency range of the radiofrequency spectrum that alone was suitable—or so the experts said—for long-distance transmission.

And what of the American Marconi Company? Were there no regrets for its imminent demise, for the severing of the links that had connected it to its corporate parent? Quite the contrary. For several years the connection with British Marconi had been an inconvenience rather than an advantage, the source of the Navy's hostility and of innumerable difficulties in Washington. If that connection were once cut and the resources of General Electric combined with the talents and experience of American Marconi, there seemed no limit to what might be accomplished. "We have found," reported President John W. Griggs of American Marconi to his stockholders, "that there exists on the part of the officials of the Government a very strong and irremovable objection to your Company because of the stock interest held therein by the British Company. This objection is shared by the members of Congress to a considerable extent. . . . [We] are satisfied and convinced that in order to retain for your Company the proper support and good will of our own Government it is necessary that all participation in its stock, as well as in its operation, on the part of any foreign wireless company must be eliminated."[38]

But there was a little more to it than that. Few of the men involved in the management of American Marconi seem to have thought of the shift in ownership as marking the extinction of their company. And the same was true of the operating personnel. By this they meant more than just the fact that, for reasons of convenience, the corporation would continue to exist as a legal entity, so that claims against the government and private firms for unlicensed use of patents could be suitably pursued in the courts. They meant that it would continue to exist as a living and functioning organization, although under a new name and in a new alliance with the corporate resources of General Electric. The essence of the matter, as they saw it, was that British Marconi's property interest in their company was to be acquired by General Electric, and a new company was to be formed to which American Marconi would transfer most, though not all, of its assets. From then on, the business of wireless communication and the sale of wireless devices would be conducted through the new company, but the organization that handled the communications and did the selling would be in essence the same organization as before.

This perception did not entirely correspond to the realities of the situation, but from the point of view of the operating personnel it was not far wrong. And when voiced by the executives of American Marconi it

[38] "To the Stockholders of the Marconi Wireless Telegraph Company of America," by John W. Griggs, president, 22 October 1919.

was not merely a tactic to induce the stockholders to approve the sale, although it conveniently served that end. The rhetoric also reflected a real sense of organizational continuity. There is, for example, among the papers of Owen Young, a document entitled "Advantages of Proposed Arrangement with General Electric Company," dated 18 September 1919. It is unsigned but the author is almost certainly E. J. Nally.[39] Its substance is a comprehensive recital of all the benefits that absorption into RCA was expected to bring: an end to government opposition, access to new sources of capital and new types of equipment, possession of Tuckerton and alliance with the Compagnie Générale, new opportunities for profitable business in Latin America and in the sale of equipment to the British company, and so on. But the language is also revealing: it is full of what "we" are going to do, and what "we" are now assured of. That "we" is not the Radio Corporation: when that organization is referred to, it is so designated by title. The first person plural refers to the owners and officers of American Marconi.

What the agreement gave to American Marconi, accordingly, was assurance of continuity—not continuity of title, but continuity of organization. And, for a company that at the end of the war had seemed on the verge of disintegration—all its stations either in the hands of the government or shut down, and confronting in official Washington a stony hostility that no amount of argument seemed to soften—that kind of continuity was of no small value. It represented a substantial contribution to the strength of RCA, for it transferred to the new corporation the loyalty as well as the skills and experience of executives like Nally, Sarnoff, and Winterbottom, of engineers like Weagant, and of scores of men farther down the hierarchy. Decades later, in a much larger and more diversified RCA, it was still possible to spot the "old Marconi men." They all knew each other and they looked after each other. Much of the tremendous authority that Sarnoff later commanded as chief executive of RCA certainly derived from his personality; but much too stemmed from the fact that he had come up through the ranks of the old Marconi Company. He symbolized and personified the continuity between the old organization and the new.

* * *

In the short run, however, the Marconi heritage brought problems as well as talents. There was a question of public relations. And there was the related question of who was to run the new company. It was politically

---

[39] Young Papers, Box 72, "Advantages of Proposed Arrangement with General Electric Company," 18 September 1919.

vital that the radio corporation be publicly identified as a new, independent, and completely American organization. But it was also expedient, if only to reassure the American and foreign stockholders who would be voting on the sale of American Marconi, to make it clear that the new organization would have competent management—competent, that is, in radio, not in the arts of high finance and corporate mergers. Were these objectives compatible? The first newspaper stories on the merger emphasized the role of General Electric. When the New York *American*, the *Sun*, and the *Knickerbocker Press* reported the merger on 23 October, they referred to the formation of RCA as a "General Electric Deal." The *Sun*'s headline read: "American Radio to Absorb Marconi Co." But, although the text of the story emphasized that the new company would be "exclusively American," it also referred to the merger as an "alliance" and stated flatly that Edward J. Nally of the American Marconi Company would be RCA's first president.[40] That was enough to raise official eyebrows.

Who told Nally that he was to be the first president of RCA, and when, is not clear. Young had assured him, before formal negotiations began, that as far as operating was concerned GE intended to take advantage of American Marconi's experience; but that was far short of a promise that Nally would be the new corporation's chief executive. By 6 September, however, returning on the *Aquitania* with the negotiations in London successfully concluded, Nally was writing confidently in his diary about the glowing prospects of "the proposed new organization, which I am to head." Was this the price that GE had agreed to pay in return for his cooperation in the negotiations with Isaacs? Or was it perhaps the price, or part of the price, that Isaacs had exacted for his agreement to sell the British company's interest in American Marconi? Or was it a much simpler matter: everyone agreed that Nally was the best man for the job?

Young was clear in his own mind that, as soon as RCA was set up and functioning, he intended to move out of the organization and return to GE. Coffin and Rice were expected to retire in the near future. There would be a changing of the guard at General Electric, with a new chairman of the board and a new president. Young knew nothing about radio beyond what he had picked up in the last few months, and the evidence

[40] Newspaper clippings in Young Papers, Boxes 71 and 72. In an earlier story on 4 September 1919 the New York *Times* had stated that "the wireless company will in no sense be a subsidiary of General Electric." This stands in sharp contrast to the claim by Davis, in a private memorandum to Young, that the creation of RCA was "a matter of the formation of a mere book-keeping corporation . . . plus a consolidation of that Company with another corporation." (Davis to Young, 18 October 1919, Young Papers, Box 71) In fairness to Davis it should be added that he was discussing GE's tax liability.

of his correspondence and memoranda indicates unambiguously that, in the autumn of 1919, he did not think of his future as lying with RCA. The decision, however, was taken out of his hands, and by none other than Godfrey Isaacs.

Isaacs had agreed to sell his company's interest in American Marconi if a majority of that company's stockholders approved of the merger with the Radio Corporation. And he had promised, in addition, to use his influence to persuade all foreign stockholders to cast their votes in favor of the merger. It was important that he follow through on that promise. The consent of a majority of the stockholders could not be taken for granted. The 364,826 proxies that Nally held from British Marconi fell far short of a majority, and even if the votes of all the American-resident stockholders without exception were added to Nally's total, the resulting majority would be exceedingly slim. The distribution by country of residence in September–October 1919 was as follows:

Country	Number of Shares
England	1,049,964
Ireland	116,513
Scotland	3,845
France	5,571
Holland	7,600
Italy	1,206
Portugal	700
Germany	1,100
Switzerland	1,550
Belgium	12,125
Canada	16,925
Other foreign countries	880
U.S.A.	650,918
Total	1,968,897

The balance of the 2,000,000 shares issued and outstanding was made up of various small holdings of less than 50 shares each, the majority of which were believed to be in the United States.[41]

If all the known American stockholders voted in favor of the merger, and Nally cast the British company's proxies likewise, there would be a total of 1,015,744 favorable votes, or 50.78 percent of the total. It was hardly a comfortable majority. A few thousand abstentions or negative votes—from disgruntled or uninterested American stockholders, for ex-

[41] Young Papers, Box 71, "An Analysis of the List of Stockholders of the Marconi Wireless Telegraph Company of America."

ample—could swing the decision the other way. This would be embarrassing to say the least. It was imperative to secure more foreign proxies. That, however, required not just Isaacs's passive consent but his determined initiative.

Unfortunately, Isaacs was already showing himself to be far from the congenial, cooperative ally that Young would have liked. The first sign of trouble came early in September and it concerned Nally. Davis wrote to Young on 8 September quoting at length a letter from Isaacs. If he was going to urge the British stockholders to approve the trade, Isaacs wrote, he had to be fully satisfied with the arrangements made for the conduct of the new company's business. He could not feel assured of this unless he knew that Nally would be in charge. Isaacs continued: "I therefore hope and rely upon Mr. Nally being the First President, and my being able to be advised definitely upon this point before communicating with the European stockholders."[42]

Isaacs would not have written in this vein if Nally had already been assured of the presidency by Young or Davis. If, therefore, Nally felt certain of the presidency when he sailed back to the United States, and the excerpt from his diary quoted above indicates that he was, it must have been because Isaacs told him so. Isaacs could do this because he knew he held the trumps. Young and Davis were faced with a virtual ultimatum. Either Nally became president, or Isaacs would let it be known among the European stockholders that the deal did not have his full support.

Whether Nally would have become president in the absence of this pressure we shall never know. He was highly respected by the American stockholders and his nomination certainly helped to clinch their support no matter what Isaacs said or did. But the European proxies were needed also. As it was, there was no alternative but to do as Isaacs indicated. Late in October the news was released—significantly, to the Navy Department first of all—that Nally was to be president of RCA. The reaction in Washington was immediate.

The nature of that reaction is best conveyed by a letter that Young wrote to Nally on 23 October.[43] It cannot have been an easy letter to write. Since word got out that Nally was to be president and operating head of the new company, Young reported, there had been "very grave opposition" in Washington, in different quarters, to the whole enterprise. The source of the opposition lay in the belief that, if Nally was to be in charge, that indicated that "the Marconi Company was going to take

[42] Davis to Young, 8 September 1919 (Young Papers, Box 72).
[43] Young to Nally, 23 October 1919 (Young Papers, Box 71).

over the General Electric interests rather than the General Electric interests, through the new company, taking over the Marconi Company." This criticism was heard not only in the executive departments of government but also in influential places in Congress. If the new company was to start right, it was vital to correct that view. "All the arrangements with the British Marconi Company, advantageous as they seemed to Washington on the theory that the trade was made between the General Electric interests and the British Marconi, are viewed with suspicion when looked at from the angle that they are to be executed and carried out by the old Marconi organization."

Young had done his best, he wrote, to counter these suspicions. He had explained that General Electric had no intention of merely turning over its interests to the old Marconi organization and then paying no further attention to radio. Nally's selection as president did not mean that. It meant merely that, after careful consideration, General Electric thought that he was, by virtue of his experience, the man best qualified for the job of executing the policies the new company would follow. But in the determination of those policies General Electric would play a large role.

To make this perfectly clear, Young stated his belief that "in the ultimate set-up of the new organization, there would be a chairman of the board who, while he might not represent the policies of the General Electric Company, would at least be so impartial that he might be said truly to represent the policies of the new radio corporation, which would be the composite views of the old Marconi Company, of the General Electric Company, of the Government, and of American business interests broadly." This statement, Young reported, had quieted opposition temporarily, but it could and would break out again unless the matter was carefully handled.

None of this, Young emphasized, involved any reflection on Nally personally. But they had to face facts as they found them. Serious prejudice against the American Marconi Company was a fact. It was vital to make it clear that RCA was "really a new set-up" and not merely a continuance of Nally's old organization.

Concretely, Young made two suggestions. First, they should consider very carefully the selection of a chairman of the board. "He should be a man who, for the time being at least, will command the entire confidence of the public and of Washington." And second, they should also consider the selection as vice-president of the company of a man who was "familiar with the public service business and skilled in the handling of public relations." Nally could not be expected to fulfill all the duties that would be imposed on him as president and handle public relations as well. They would need a specialist for that.

Whether Young knew it or not, he had argued himself into a new job. He had also made sure that Nally would be the first president, for in the circumstances there was no way Nally could decline the position, and no way in which General Electric could refuse to offer it to him. Yet Nally must have accepted the inevitable with mixed emotions and with the suspicion that his tenure in the position would not be a long one.[44]

Final decisions on the top leadership of RCA were taken by the end of October, after conversations between Coffin and Rice of GE and Griggs of American Marconi. Nally would be president. Young would be chairman of the board—with the understanding that if, before the time of election, he could suggest a satisfactory alternative, he might withdraw in his favor, and that he might do the same at any later time. "Mr. Young does not intend to retain this position permanently," Davis informed Nally. There would initially be only eight directors, four named by GE and four by American Marconi. A ninth would be appointed whenever three of the four directors representing either side asked for such an appointment, and it was hoped that this ninth director, if satisfactory to everyone, might replace Young as chairman of the board. Griggs would be RCA's general counsel, and Alexanderson, though remaining with GE, would be appointed consulting engineer. As for the vice-president in charge of public relations whom Young had suggested, it was agreed that there should be such a functionary, that he should be a lawyer, and that he should have a title high enough to enable him to speak with authority, as either vice-president or assistant to the president; but no names had been put forward.[45]

There remained only the last essential requirement: approval by the stockholders of American Marconi. The directors of that company had already endorsed the merger on 22 October. November 20 had been set for the stockholders' meeting. There was not much time left for the European proxies to arrive, and in that connection Isaacs was still raising

---

[44] In December 1922 RCA announced that Nally was leaving the presidency to become the corporation's managing director of international relations, with his base of operations in Europe. Hooper remarked, in one of his less pleasant moments, that this would at least enable him to be closer to "his master's voice." The next president of RCA was Maj. Gen. James G. Harbord, formerly chief of staff of the American Expeditionary Force. Admiral Bullard, who had been regularly attending RCA board meetings as "government representative," is said to have been a disappointed candidate.

[45] These details are from Davis to Nally, 31 October 1919 (Young Papers, Box 71).

difficulties. In a series of letters and cables that Young and Davis found progressively harder to understand, Isaacs complained about changes he thought had been made in the understandings arrived at in London. He expressed concern over whether the stockholders were being treated fairly. He demanded details of when and how stock in RCA would be offered to the public. And he worried about how American Marconi shares were behaving on the stock market. Little of what he said made sense to the people at General Electric.

Unfortunately, GE had no representative in Britain at the time to straighten the matter out, and John Gray, legal counsel to British Thomson-Houston, was pressed into service to remedy what looked like a serious breakdown in confidence. To him Davis dispatched a cable on 1 November that raised the old specter of government ownership if the RCA project failed, and urged Gray to impress on Isaacs that his co-operation was indispensable. "Daniels fight for government ownership only checked by proposed arrangement probably impossible meet it if arrangement fails . . . proxies coming in regarded as only salvation American Marconi Company . . . contract requires Isaacs recommend proxies . . . his cable hopelessly indefinite not understandable dont explain difficulty . . . rely on you to straighten situation. . . ."[46]

On the same day he sent off a long letter to Gray that eloquently expressed the same sentiments of anxiety and exasperation. Isaacs's cables, wrote Davis, were "indefinite and utterly ununderstandable." He had agreed in the contract to promote the merger in every way possible and in particular to help get proxies. Nothing had happened to justify failure to do this. Reaction to news of the proposed merger had been most favorable. Even preliminary rumors had sent American Marconi stock up on the market and President Griggs's official announcement had driven it as high as 7 1/2, and this despite the publicity surrounding Secretary Daniels's campaign for public ownership. "The judgment of the American market then has been . . . that this proposed deal has added about 75% to the value of American Marconi shares."

Did Isaacs perhaps think, Davis continued, that American Marconi could get a better deal in some other way? If so, he was sadly mistaken. If the deal with GE fell through it was most unlikely that the company would ever get its stations back from the Navy, and Daniels's drive for government ownership was almost certain to succeed. The stations themselves were, in any event, obsolete. The only one with modern equipment was New Brunswick, and only GE's vigorous intervention had induced the secretary of the navy to rescind his order to get the alternators out

[46] Davis to Gray (cablegram), 1 November 1919 (Young Papers, Box 71).

of that station. There was, in short, no future for American Marconi except through the agreement already worked out with General Electric.

Davis made his points forcefully and yet, throughout the letter, there is a note of honest puzzlement. He really did not understand what was bothering Isaacs. "I cannot understand what Mr. Isaacs is doing nor what the situation is abroad. I hope there is no misunderstanding. . . . Mr. Isaacs knew the whole story when we were over there. . . . If Mr. Isaacs would only tell me what he wanted me to do we would do it if we could and if it was anything which was in accordance with the plans agreed to by him. . . . He knows more about the London stock market in radio than any man in the world, and if there was something wrong with the deal . . . he should have told us about it when we were over there. The whole thing is to me an absolute mystery. . . ."[47]

Nally, who should have been in a position to know, thought that Isaacs was speculating in American Marconi shares. He certainly had plenty of opportunity to do so between the time the Preliminary Agreement was signed and the time it became public knowledge.[48] The market quotations in London and New York were in fact higher at the end of October than they had been for a long time, but Isaacs may well have wanted them driven up even further before the merger became a reality. This would not have been difficult. A few judicious press releases, a few roseate forecasts of future earnings would have done the trick. Davis and his colleagues at GE, however, wanted no part of any such financial trickery; deliberate encouragement of speculation, they thought, would be very unwise. The circular that President Griggs of American Marconi had issued to the stockholders had been deliberately drafted to avoid a flare-up in the price of the stock.[49]

If speculative profits were not what was on Isaacs's mind, it was hard to see what was bothering him. True, certain changes had been made in the plans for the merger, but they were technical in nature and for the convenience of the stockholders. American Marconi's New Jersey factory, for example, originally to be sold to RCA, was instead to be kept and leased to GE, but with the option of selling it to GE at any later time at the price originally agreed upon. Young in particular expressed concern that the factory not be merged into the GE organization in such a way as to make it difficult to separate it in the future.[50] But it was hard to see anything potentially harmful to the stockholders in that. As for the

[47] Davis to Gray (letter), 1 November 1919 (Young Papers, Box 71).
[48] The *Economist* carried its first story on the merger on 1 November 1919.
[49] Davis to Gray (letter), 1 November 1919 (Young Papers, Box 71).
[50] Davis to E. P. Edwards, 12 December 1919 (Young Papers, Box 71).

financial side of the affair, the original plan had been to issue to the stockholders of American Marconi one million shares of RCA common stock of no par value and one million shares of RCA preferred of $10 par value. Each holder of two shares of American Marconi common ($5 par value) would have received one share of RCA common and one share of RCA preferred. It turned out, however, that several American Marconi stockholders held odd numbers of shares; so, for convenience, it had been thought better to split the preferred into shares of $5 par value, so that each shareholder of American Marconi common could get, in the exchange, an integral number of RCA preferred. In the aggregate, two million shares of RCA preferred of $5 par value would be issued to the American Marconi stockholders instead of one million of $10; and the corresponding change had been made for RCA common. What it amounted to was that each holder of one share of American Marconi common would end up holding instead one share of RCA preferred and one share of RCA common.

And that was all there was to it. It was purely a matter of arithmetic and meant no dilution of the original stockholders' stake in the company. As for an issue of RCA shares to the general public, none had ever been planned. There would be no attempt to raise capital from the investing public. This did not mean, of course, that no RCA stock would be traded on the market. General Electric had no intention of selling any of the RCA common or preferred that it would acquire in exchange for the American Marconi shares it had agreed to purchase; but the private stockholders were free to do so as and when they pleased.

John Gray's intervention seems to have had some effect. To make assurance doubly sure, Young cabled to Admiral Bullard early in November asking him to issue a statement making it clear that, in his judgment, the merger was desirable and stockholders should vote for it. This may have been done at Isaacs's suggestion, for it was public knowledge in Europe as well as in America that the U.S. Navy had other plans for the future of radio. Bullard obliged; in view of his past role in the affair he could hardly do otherwise. "I am entirely in accord," he wrote to Young, "with the proposition that the British Marconi Company should vote its stock in favor of the proposed transfer. . . ."[51] The letter was forwarded to Isaacs as soon as received. By the time it reached him (probably between 14 and 17 November) there was very little time left.

Was there ever any real danger that the stockholders would not approve the merger? In the case of the American stockholders, probably not. Those

[51] Young to Bullard (cablegram), 5 November 1919, and Young to Isaacs, 7 November 1919 (Young Papers, Box 71).

who felt pessimistic about future prospects had ample opportunity to dispose of their holdings on a rising market between the time when the merger plans became known and the date of the vote. There was no rival group bidding for the company. But to count on the American votes plus the proxies Nally held from British Marconi meant, as we have seen, relying on a fraction of a percentage point for a majority. The foreign proxies were needed, and could not be taken for granted. Many if not most of the foreign stockholders looked to Isaacs for leadership, and his confidence in the Americans was clearly paper-thin. It was not unreasonable for Young and Davis to be concerned.

After all these anxieties the outcome was almost an anticlimax. By noon on 14 November a total of 1,001,729 proxies had been received. Of these 750,126 were from Britain, including 365,000 from the British Marconi Company. The remainder were American. At the stockholders' meeting in Jersey City on 20 November, with thirty-five people present, 1,192,092 shares were voted in favor of the merger and only 6 against it. After Nally had read the resolution approved by the board of directors a motion for unanimous approval was made, and it was so voted.[52]

＊　＊　＊

What exactly had the stockholders of the American Marconi Company done? Their action did not bring the Radio Corporation of America into existence. That occurred when the company received its charter under the laws of the State of Delaware on 17 October 1919. Nor did it affect the so-called "Radio Agreement" of 20 November by which RCA became the exclusive sales agent for all GE radio equipment and GE the exclusive supplier of radio equipment to RCA. Nor had they been asked to approve directly the terms of the Preliminary Agreement between British Marconi and General Electric, nor the Main Agreement between British Marconi and RCA, although in general terms the provisions of these agreements had been made known to them.[53] In the narrowest sense, indeed, all they had done was approve the terms of a sales contract—a contract by which the American Marconi Company conveyed to RCA all its assets and property except the New Jersey manufacturing plant and its unsettled claims against the government and private firms for patent infringement, in return for two million shares of common stock in RCA and two million

[52] Young Papers, Box 71, memoranda from A. H. Morton to Young, 10 November, 14 November, and 20 November 1919.

[53] These agreements are conveniently reprinted in the Federal Trade Commission's *Report on the Radio Industry* (Washington, D.C., 1923), pp. 116 ff.

shares of cumulative preferred stock. There were, of course, a few an-
cillary details—provision for a cash dividend of 25 cents per share, for
payment to RCA of the first $500,000 from settlement of patent claims,
and for the lease to GE of the New Jersey factory—but in a technical
sense that was the essence of the matter.

That action, however, was the critical move that let all the other elements fall into place. On 20 November the Main Agreement with British
Marconi was signed. This finally cut the link between American Marconi
and its British parent by transferring ownership of British Marconi's
equity to General Electric. Assured now of title to American Marconi's
property, RCA was transformed from a mere paper entity into a major
communications company, a corporation with important responsibilities
for reconstructing and reequipping the world's long-distance radio circuits. In particular RCA now bore the responsibility for asserting and
defending American national interests in radio. In effect if not by formal
charter (for the proposed "Navy Contract" had never been approved by
Congress), it was the chosen instrument of American telecommunications
policy. There had been, it is true, in the discussions leading up to the
assumption of this role by RCA, an occasionally rather shrill and strident
note of chauvinistic nationalism. But the other side of the coin was a
new confidence in American radio technology and in the ability of American business to manage that technology and profit from its use.

What kind of corporation was RCA at this stage of its existence? It
was, of course, different things to different people. Financially its structure was simple. Its capital stock consisted of five million shares of preferred stock of a par value of $5.00 per share. This stock was to receive
preferred dividends of 7 percent per annum, and these were to be cumulative after 1923. There were also five million shares of common stock
of no par value. The preferred stock was supposed to represent, in a
general way, tangible property while the common represented patents
and goodwill. Share for share, preferred and common stock had equal
voting power. From the point of view of the individual shareholders, or
at least those of them who were well-informed about RCA and its prospects, the common stock was held in the hope of future capital gains
while the preferred was held for income. No one expected RCA to pay
dividends on its common stock for many years to come; indeed, the
struggle in the beginning was to make sure it earned enough to start
paying dividends on the preferred in 1923. There was no bonded debt,
and the corporation began its life with no fixed interest notes or obligations—a prudent strategy in the circumstances.

Of the preferred stock, 235,174 shares were issued to General Electric
upon the formation of RCA, along with two million shares of common

stock. The preferred represented the funds that GE had expended to buy out British Marconi; the common reflected the value of GE's radio patents, present and future.[54] In addition, it was agreed that GE would accept RCA preferred stock in payment for the alternators and other equipment it would supply, at a price of $127,000 per alternator. There were to be twelve of these machines in the initial purchase order, and for them GE was to receive a total of 304,800 shares of RCA preferred (the equivalent of $1,524,000 at par value).

To the former stockholders of the American Marconi Company (excluding of course the British Marconi Company) there were allocated two million shares of RCA common and two million of RCA preferred. Many of these shares were held by people who were not United States citizens. To limit the possibility of foreign influence, RCA's charter provided that not more than 20 percent of the stock could be voted by foreigners, and to this end foreign stockholders received "foreign share certificates" instead of the usual preferred and common stock.

As a functioning business unit, RCA in the closing months of 1919 was an anomaly. It was, on the one hand, exclusive sales agent in the field of radio equipment for General Electric. On the other, it was a communications company. As radio sales agent for GE, however, it had at that time only a single customer in sight (apart from its own equipment needs), namely the British Marconi Company. And as a radio operating company it had no stations, since those previously operated by the American Marconi Company were still in the hands of the United States Navy. As far as manufacturing facilities were concerned, RCA had none; nor did it have any independent research facilities.

What RCA had to sell was, in terms of hardware, the Alexanderson alternator and its associated equipment. Other product lines would soon

[54] There is a problem of evidence here. Young stated at the *Cable Landing Hearings* of 1921 (p. 335) that GE agreed to contribute "something over $3,000,000" to RCA, against which it was to receive preferred stock at par. This implies that GE should have received at least 600,000 shares of RCA preferred. According to the Cases, 235,174 shares of RCA preferred were issued to GE on the formation of RCA. Added to the 364,826 shares of American Marconi acquired from the British company (each converted into one share of RCA preferred and one share of RCA common), this would give the expected 600,000 shares of RCA preferred. I have therefore used the Cases' figure in the text. According to the FTC *Report* of 1923, however, GE received only 135,174 shares of RCA preferred on the formation of RCA, and this figure also appears in the Main Agreement between RCA and American Marconi of 20 November 1919 as it is reprinted in the Appendix to that *Report* (p. 118). See Case and Case, *Young*, pp. 186-87.

become important—vacuum tubes in particular—but in 1919-1920 the Alexanderson system was what counted. British Marconi had the exclusive right to this system within "Marconi territory," which as we have seen meant essentially the British Empire, those European countries in which Marconi-affiliated companies held exclusive concessions, and perhaps China. But it was not, of course, required by contract to buy any alternators at all. For the shorter distances (less than 2,000 miles) it could rely either on vacuum tube transmitters, which for intra-European traffic were by this time quite adequate, or on advanced spark equipment, still very common for marine use. Alternators would be used for long-haul traffic. By 1919-1920 it was becoming clear that the British Marconi Company would play no role in the Imperial Chain as the British government then conceived it. For that system the Norman Committee, which reported to Parliament in 1920, recommended tube transmitters operating through relay stations not more than 2,000 miles apart; it had dismissed, for what it deemed good technical and economic reasons, the possibility of using arcs or alternators.[55] The Marconi Company, after submitting in 1919 a large and ambitious proposal that was rejected, had refused even to testify before the committee, so there was clearly no market for alternators to be found in that direction.

That left the Marconi long-haul circuits. As far as revenue was concerned, the most important of these was the transatlantic service from Britain to North America. Reopening this circuit would require reequipping the Marconi station at Carnarvon in Wales, or building a new one; this would call for two alternators (one on standby). The circuit from northern Europe to North America depended on the Marconi-equipped station at Stavanger, Norway, operated by the Norwegian Telegraph Administration. In 1919 this station was still using a timed-spark transmitter; modernization could not long be delayed. As for the rest of Europe, the French had their new arc station at Croix d'Hins, near Bordeaux, built by the U.S. Navy, and the station at Lyons, equipped with a Bethenod alternator and arcs designed by Elwell. It was unlikely that either the Compagnie Générale or the French government, if it elected to play a direct role in radio, would buy American equipment. The Germans, through Telefunken, would certainly use their own alternators or tube transmitters. For the Low Countries, Italy, and Spain, the prospects in 1919 were still uncertain. The Poulsen arc was, after all, free from patent restrictions by this time, and Cyril Elwell was available to build low-cost arc transmitters of proven reliability for anyone willing

---

[55] *Report* of the Imperial Wireless Telegraphy Committee 1919-1920, The Right Hon. Sir Henry Norman, chairman (London, 1920).

to pay his fee, as in fact he did for the British Admiralty, the Italian government, and numerous others. Sweden and Poland were "neutral territory," and thus open to direct competition with RCA, as well as other suppliers hungry for orders such as Telefunken. In the Far East prospects for Marconi expansion were uncertain. For Imperial traffic, any Marconi circuits would have to meet the competition not only of the submarine cables but also of the projected government-owned Imperial Chain.

What all this added up to was that British Marconi's requirements for Alexanderson alternators would probably be quite limited, even ignoring the possibility of rapid development in tube transmitters. When negotiations had first been opened with GE, the talk had been of possibly ten alternators for British Marconi. By the end of 1919 a conservative estimate might have cut that figure in half, and even that would have proved too high. The surviving record leaves much to be desired, but it appears that GE never sold more than two alternators to the British Company.[56] In view of the role that prospective sales to Marconi had played in the

Pl. 15: Twin 200 kilowatt alternators at the Marconi Carnarvon station.
Source: Science Museum, South Kensington

[56] Attempts to locate relevant records of the American Marconi Company for this period have been unsuccessful.

formation of RCA, that outcome is ironic. If the contracts for the Imperial Chain had gone to the Marconi Company, the situation would have been entirely different; but, for the moment at least, that possibility had disappeared.

This drastic attenuation of the Marconi market for alternators gave added importance to the acquisition and modernization of RCA's own stations, to the expansion of its radio circuits, and to the vigorous exploitation of market opportunities in countries where it was free to compete for business. Events moved quickly. RCA opened its doors for business on 1 December 1919. On 1 March 1920 the Navy relinquished control of the high-powered stations and RCA took over. This was a remarkably quick transition and testifies both to the efficiency of the operating personnel whom RCA inherited and to the way in which the drive for government ownership, once formidable, lost its force in Congress and in the Navy Department once the specter of foreign control was banished.

The stations that RCA took over immediately included the transmitting station at New Brunswick, New Jersey, which operated in partnership with a receiving station at Belmar, New Jersey; a transmitting station at Marion, Massachusetts, with a complementary receiving station at nearby Chatham; the station at Tuckerton, New Jersey, acquired from the Compagnie Générale, which also operated in conjunction with Belmar; on the West Coast, the former Marconi transmitting station at Bolinas, California, near San Francisco, with its complementary receiving station at Marshalls, about thirty miles away; and in Hawaii a transmitting station at Kahuku, with a complementary receiving station at Koko Head. The Navy retained control of the former German-owned station at Sayville.

Of these facilities the only ones with reasonably adequate facilities in 1919 were the New Brunswick–Belmar system, where the 200 kilowatt Alexanderson alternator carried the load, and Tuckerton, where the Navy had installed a Federal arc. Marion and Bolinas both had Marconi timed-spark apparatus.[57] The circuit with Great Britain, between New Brunswick and Carnarvon, was opened for commercial business on 1 March 1920. This was essentially a continuation of the operation that the Navy

[57] The Marconi-built stations at Marion, Stavanger, and Carnarvon were all originally timed spark stations. According to Haraden Pratt, who worked on a similar transmitter at Bolinas (with, as he recalled, "a copy of Steinmetz's 'Transient Phenomena' at my elbow"), "it worked pretty well and to listen to it at a receiving station it did sound like an undamped wave except for some side mushy sounds and it generated a whole lot of strong harmonics." (Pratt to Lloyd Espenschied, 12 July 1963, Pratt Papers).

had conducted in the closing months of the war and since the Armistice. On the same date the circuit between California and Hawaii was opened, but without new equipment.[58] A circuit with Norway was opened on 17 May 1920, between Marion and Stavanger, after the American station was reequipped with alternators. Marion also took responsibility for handling traffic with Nauen, in Germany, beginning on 1 August 1920, and on 15 December a circuit was opened with France, using the New Brunswick transmitter. Tuckerton was intended to serve as the American transmitter for the French circuit, but installation of an alternator there was delayed by problems with electricity supply to the site, and Tuckerton was not brought into operation until early 1921.

Efforts expended in getting these circuits into operation had as one of their objectives the generation of income-earning traffic. Results were encouraging. Operations up to 31 December 1920 showed, probably to no one's surprise, a deficit of $45,728.44, but by the end of the following year this had been converted into a profit from operations of over $400,000 and Young was beginning to feel hopeful that by 1923 the corporation might be able to earn enough to pay dividends on its cumulative preferred stock.[59] Over one million words of paid traffic were handled in December 1920, and the trend was sharply upwards on all circuits then open— reflecting, probably, not only the congested condition of the submarine cables but also the fact that radio rates per word were significantly less than by cable.[60]

There was, however, an element of haste and improvisation about all

[58] According to Espenschied, the Marconi station in Hawaii was never able to maintain a commercial service with Japan. "Try as the Marconi boys would on Hawaii, they could not 'make' Japan. Compared to the c.w. stations of the Germans on the East Coast and those of the Navy on both coasts (arcs) and with Federal arcs, the Marconi spark stations of the 'Imperial Chain' were white elephants." (Espenschied to Pratt, 18 July 1963, Pratt Papers) The Marconi station in Hawaii was never, of course, officially part of the British Imperial Chain. According to Young, RCA's transpacific service handled 210,653 paid words in December 1920, before the equipment was modernized. This statistic may refer, however, to traffic between California and Hawaii only. See *Cable Landing Hearings*, testimony of Owen Young, p. 351.

[59] Young to Hon. Herbert Hoover, Secretary of Commerce, 6 March 1922 (Young Papers, Copy Book 802); Young to Hon. Eliot Wadsworth, Assistant Secretary of the Treasury, 10 May 1922 (ibid.).

[60] *Cable Landing Hearings*, testimony of Owen Young, pp. 329-31. The radio tariff from New York City to Great Britain was 18 cents per word for ordinary commercial messages while the cable rate was 25 cents per word. For press traffic the radio rate was only 5 cents per word. (See *Cable Landing Hearings*, testimony of W. A. Winterbottom, RCA Traffic Manager, p. 339)

these installations. Their purpose was not only to earn revenue but also to demonstrate to foreign administrations and to the American public that an aggressive and technically competent organization was now running American radio. For its long-term needs, RCA was counting on its ambitious plans for a new "Radio Central" to be built on a ten-square-mile site just east of Port Jefferson on Long Island. This was intended to be the most powerful radio station in the world, and an impressive demonstration of the capabilities of American radio technology. With its giant twelve-spoked antenna system and five alternators—one intended for South American service, the second and third for supplemental service to France and Germany, the fourth for traffic with Italy, and the fifth for communication with Poland—this was to be a transmitting complex capable of laying down a signal of commercial quality anywhere in the world, twenty-four hours a day, three hundred and sixty-five days a year, under any conditions of static and interference. Four of the alternators were in place and one span of the antenna system completed by November 1921. If it had ever been completed according to its original conception, this mammoth project would indeed have demonstrated GE's alternator-based technology in its ultimate form. By 1923, however, developments in tube transmitters and the discovery of new modes of radio propagation threatened to make that conception obsolete and further work on the project was suspended.[61]

In the meantime, the export market was not neglected. The years after 1919 saw, indeed, a scramble among the industrialized nations to acquire new long-distance radio facilities if they did not already possess them, to modernize them with continuous wave equipment if they did. By no means did RCA have this field to itself. The Germans had their own alternator systems; so did the French; by the early 1920s Japan had entered the competition; the British were aggressive in pushing the virtues of tube transmitters; and there were always the arcs—easy to build, simple to maintain—as a technical alternative. The proven efficiency of GE's alternator, however, had become something of a byword in international radio circles. In a sense the machine advertised itself, for New Brunswick's radio signal—stable, consistent, free from the mush and harmonics that were making arcs unpopular—could be heard by anyone with a suitable receiver. David Sarnoff's Commercial Department at RCA exploited these advantages vigorously, with strong support from Alexanderson. Contracts were signed for the installation of alternators with the government of Sweden in 1921 and with Poland in the following year. And lower-

[61] Elmer E. Bucher, "A History of RCA" (Sarnoff Research Center Archives), chap. 10, pp. 240–41.

powered stations were sold to the governments of Venezuela, Mexico, and the Philippines.[62]

Of all the radio alternators that General Electric built in these years—and there seem to have been, in total, about twenty of them—only one survives today. This is the machine at Grimeton, in Sweden, and in view of Alexanderson's origin and ancestry it is hard not to see something appropriate in this. It is still maintained in prime condition, almost as it left the shop floor in Schenectady in 1921. And about once a month it is started up and carefully brought to full operating speed.[63] The Warsaw alternator was destroyed in the closing phases of World War II. As for the rest, in Europe and the United States, they were scrapped long ago, to be replaced by the versatile vacuum tube. The same is true of the arc transmitters, once the last word in continuous wave technology; the only specimens surviving today are in museums.[64]

This, then, was RCA in 1919-1920, as far as its formal organization and functions were concerned. To the individuals involved in its future, however, it was much more than this. Like most organizations, it served as a vehicle for a multitude of hopes, ambitions, and fears. There were those who were disaffected, who thought that RCA was either superfluous or potentially pernicious. Even Major Gen. George Squier of the Army Signal Corps was ungallant enough to indicate in congressional testimony that the Army did not entirely agree with the way the Navy had handled the matter, and that he personally saw no reason why GE should need a separate corporation to sell its radio equipment since it already had an efficient sales organization of its own. And some radio operators thought it made no difference, that the radio business was really going on much as it had before. Consider, for example, the sentiments of E. T. Quinby, outward bound on a tramp steamer in 1919, who was informed via the radio station on St. Paul Island in the Bering Sea of the name of his new employer and, returning to New York almost a year later, took the opportunity to check in personally. "Imagine my pleasant surprise to discover Jim Sawyer and Jack Duffy ensconced amid mahogany desks and green plush carpet, doing business as the RCA Marine Supervisors, along with practically the entire staff of former de Forest–Marconi of-

[62] Ibid., pp. 243-44.

[63] For information on the history and present condition of the Grimeton alternator I am grateful to Kaye Weedon, who visited the station in 1979 and provided me with many photographs.

[64] Early models of Federal arc transmitters may be seen in the Foothill College Electronics Museum, Los Altos, California.

ficials. They had all become High Priests in the new Cathedral of Commerce. . . ."[65]

Comdr. Stanford Hooper observed the same continuity of personnel, but not with unalloyed pleasure. He thought the housecleaning had not been thorough enough. To complaints by Young in 1921 about the difficulty, despite Hooper's optimistic forecasts, of making money in long-distance radio, Hooper replied caustically that he had not chosen RCA's management. His private belief was that, if RCA got into real financial trouble, Congress would come to the rescue; but he did not share that thought with Young. Most Navy officers who had anything to do with radio, however, thought the new corporation a decided improvement—better, in the opinion of some, than government ownership would have been. Captain D. W. Todd, for example, commanding the USS *Pittsburgh* off the Dalmatian coast in February 1920, sent his congratulations to Nally: "From the time when the Marconi Company opposed *any* radio regulation . . . it has been very difficult for many of us who have had to do with radio matters in the Navy to deal sympathetically with that company. . . . Now that the Department has a strong, real American company to deal with, I feel sure that progress will be rapid, and the United States has a good chance of leading all countries in the development and use of radiotelegraphy.[66]

This "strong, real American company," however, struck some observers as potentially dangerous. What they feared, of course, was the concentration of power that it implied—the concentration that had in fact been deliberately created. Even Hooper worried about the monopolistic position that RCA soon acquired in the supply of radio equipment to the Navy, particularly after Westinghouse, AT&T, and United Fruit contributed their patents to the RCA pool (see below, pp. 432-79). A single organization to represent American radio in dealing with the outside world was a concept of which he heartily approved. For naval procurement, however, he would have preferred competition.

How to reconcile these conflicting requirements was to prove a recurrent problem. Already in 1921-1922 the Federal Trade Commission had launched an investigation of the radio industry, and of RCA in particular, triggered by complaints that General Electric had set up the Radio Corporation as a "bogus independent" and was trying, by the use of tying contracts and price discrimination, to acquire a monopoly in the man-

[65] E. T. Quinby to Lee de Forest, 6 November 1950 (Emil Simon Papers, Bancroft Library, Box 4). The Cathedral of Commerce was the name given to the Woolworth Building in New York City.

[66] Dodd to Nally, 14 February 1920 (Clark Radio Collection).

ufacture and sale of radio apparatus. Public resentment at the scarcity of radio tubes, after the phenomenal explosion of interest in popular broadcasting, added fuel to the fire and led to the passage by the House of Representatives, though not by the Senate, of a bill that would have denied radio transmitting licenses to any individual or corporation that sought to monopolize radio communication, directly or indirectly, through the control of the manufacture or sale of radio apparatus "or by any other means."[67] Any such legislation, if rigorously enforced, posed a lethal threat to RCA. Young and others might argue that, in dealing with foreign governments and foreign radio corporations, a monopoly was essential, while in domestic affairs it was either nonexistent or at worst transitory. Nevertheless, here was a vulnerability that had scarcely been anticipated when RCA was formed and that was to harass the corporation for many years to come.

To Young and his colleagues the RCA that existed in the closing months of 1919 was only a beginning, the mere skeletal lattice around which they intended to build a vastly more ambitious enterprise. Hooper, when the idea of a "truly American radio corporation" was first canvassed, had urged Young to make sure that it was a consolidation of all the principal interests in radio. What he had in mind was mostly a consolidation of patents, for the wartime immunity from prosecution that had enabled the federal government's suppliers to ignore the risk of litigation ended when the state of emergency ended. It was important that some means be found to integrate into usable systems the fragments of knowledge that the advance of the radio art had generated. These fragments were represented by patents; ownership of these patents was widely diffused; and there was great uncertainty over which were truly basic, which the courts might sustain, and which were mere paper claims that could safely be ignored. Consolidation was clearly called for. Some peacetime analogue had to be found for the umbrella of protection that wartime necessities had provided. This was a job for RCA. Created to defend American interests internationally, it could also serve a domestic function as the organizational framework within which the particular bits and pieces of knowledge that made up American radio technology could be integrated.

Such a conception was highly acceptable to Young. The organization he served—the General Electric Company—had itself been created in the first place to reduce conflicts over patents, by consolidating in the hands of a single entity the Edison and Thomson-Houston patents for electric

[67] FTC *Report*, p. 9.

light and power.[68] The same job now needed to be done in radio. Already signs of integration were evident. Despite the persistence of individual claims by inventors like de Forest and Armstrong, three major clusters of radio patent rights seemed to be emerging. One was the Marconi–General Electric complex over which Young himself presided. Another was in the hands of the American Telephone and Telegraph Company and its affiliates. And a third was being rapidly assembled by Westinghouse, as that corporation belatedly scrambled to establish a position of strength in radio—a position from which it could either move to establish its own operating company or alternatively bargain for a major role in any consolidation of interests that might emerge. None of these clusters was complete. Despite what engineers like Alexanderson might claim, none of them provided a basis on which could be constructed, without the high probability of extended litigation, an operating radio system that made full use of what was known to be feasible.[69]

For each of the corporations involved, some form of consolidation or pooling of radio patent rights seemed expedient. Even if it did not entirely eliminate the costs and uncertainties of litigation over patents, it at least promised to reduce them. And there were signs that each of them realized this. Gerard Swope, for example, while still with Western Electric, had sounded out his opposite numbers at General Electric as far back as 1918 about a possible unification of interests in radio. The Telephone Company, having staked its claims in radio by acquiring the de Forest audion patents and by its successful transatlantic radiotelephone tests in 1915, seemed disposed for the time being to confine itself to its traditional field of wired communications. If so, it might be persuaded to grant licenses for radio use to some organization that explicitly intended to confine its activities to radio and showed no inclination to invade the land-line business. If these two large organizations could be induced to cooperate, the smaller individual patent claims could be swept into the pool without too much trouble or expense.

The possibilities were there. RCA offered the means to explore them. And, in the person of Owen Young, there was a man who by personality and prior experience was disposed to move in that direction. Young did

[68] See Harold C. Passer, *The Electrical Manufacturers 1874-1900: A Study in Competition, Entrepreneurship, Technical Change, and Economic Growth* (Cambridge, Mass., 1953), pp. 321-29.

[69] General Electric, for example, although it could use the Fleming two-element vacuum tube by virtue of its alliance with Marconi, had no rights to the de Forest triode in any of its circuit configurations, and it had no claims to the Armstrong regenerative or superheterodyne circuits.

not care for conflict. He sought always the irenic solution. When conflict seemed to confront him, he looked for the underlying community of interest. And, time and again, he found it. This was the talent that had made him so valuable to General Electric and that, within two years, was to make him chairman of that company's board. RCA provided another field for his abilities.

NINE

Expansion and Integration

CONTINUOUS wave radio was the technological matrix within which the American communications industry was recreated after World War I. The most conspicuous event in that process was the birth of the Radio Corporation of America. RCA was formed to oversee the deployment of the alternator, the absorption of American Marconi, and the assertion of American independence in international radio. Those were its primary tasks, and it was toward performing them that its executives first devoted their energies and attention. The question remained open whether this restructuring of the industry had gone far enough. There were other matters on the agenda: the diffusion of the vacuum tube, the perfecting of radiotelephony, the new craze for entertainment broadcasting, and the opening up of the high frequency spectrum. What further restructuring would be called for?

The problem was that the technical, political, and social environment in which RCA functioned after 1919 was in continuous flux. It was problematic whether an organization so recently established, and one with such a special sense of its mission and of its technical function, could transform its structure and its sense of purpose quickly enough. If technology, the market, and public expectations could have been kept in stasis, as they were in 1919, RCA would have presented few management problems, its structure could have remained simple, and its role in the communications industry would have been straightforward. Its orientation when first created was almost exclusively toward the outside world—to Europe, Latin America, and the Orient. Its technical base was radio-telegraphy, and its major device the alternator. Its corporate affiliations were uncomplicated: it was essentially sales agent and operating company for General Electric. And it looked to the Navy Department, rather than the Departments of Commerce or Justice, for its mandate and for endorsement of its conduct. All this was to change, and with disconcerting speed, as the social, political, and technical parameters within which the organization had been brought to life began to shift.

* * *

It was in connection with vacuum tubes that signs of movement first became apparent. This should have surprised nobody. RCA and GE had acquired rights under the Fleming diode patent by the absorption of American Marconi, but rights to the de Forest triode patents remained under the control of the Telephone Company (and de Forest personally). AT&T, for its part, could not manufacture triodes without infringing the Fleming patent, and even if, by waiting a couple of years, it would see that patent expire, there were still unresolved questions of infringement in the past, the Arnold/Langmuir interference in the future, and a multitude of lesser interferences concerning details of design and construction. A move of some kind toward exchange of patent rights was to be expected.

On 3 January 1920 Captain A. J. Hepburn, acting chief of the Navy's Bureau of Engineering, sent a letter to AT&T, Western Electric, and General Electric. He began by referring to "numerous recent conferences" held in connection with the radio patent situation and stated the bureau's belief that all interests would best be served by "some agreement between the several holders of pertinent patents" whereby the market could be freely supplied with tubes. In the past, Hepburn wrote, the reasons for an arrangement of this kind had been monetary, but now the safety of life at sea was involved. "Today ships are cruising on the high seas with only continuous wave transmitting equipment. . . . Due to the peculiar patent conditions which have prevented the marketing of tubes to the public, such vessels are not able to communicate with efficiency except with the shore and, therefore, in cases of distress it inevitably follows that the lives of crews and passengers are imperilled beyond reasonable necessity." It was "a public necessity" that all ships be able to procure tubes without difficulty, since they were "the only satisfactory detectors" for receiving continuous waves. Hepburn closed by asking for a "speedy understanding" among the parties involved, saying that his letter should be considered "an appeal, for the good of the public, for a remedy to the situation."[1]

[1] Hooper Papers, Box 3, A. J. Hepburn to AT&T and Western Electric, 3 January 1920. A similar letter was sent to General Electric. The text is conveniently reprinted in Josephine Young Case and Everett Needham Case, *Owen D. Young and American Enterprise: A Biography* (Boston, 1982), pp. 209-10.

This letter has been often quoted in full or in part by historians of American radio and is conventionally taken as representing the Navy's second major attempt to bring order to the industry. Its author was Lt. Comdr. Hooper, though the signature was that of his chief.[2] Hooper, typically, did not underestimate its importance. In a covering letter to F. B. Jewett of Western Electric he underlined the fact that the tube situation was "very seriously handicapping the radio art" and urged him to help in "strengthening the efforts to get together, without the long delays, such as must needlessly follow fighting a matter out in court." And to Owen Young he depicted it as a matter second in importance only to the formation of RCA itself. "Next to the formation of an American Radio Company . . . the clearing up of the tube situation follows as greatest in importance."[3]

It would be naive to think that the bureau's letter took its recipients by surprise. We do not know exactly who had attended the "numerous recent conferences" referred to in the first paragraph, but a reasonable guess would include A. G. Davis and E. P. Edwards of GE and their opposite numbers in Western Electric and Westinghouse.[4] Proposals for joint action by GE and the Telephone Company to divide up the field of electrical communication had been in the air for some time, and they had gone beyond vague generalities. In May 1918 Gerard Swope (then still with Western Electric, AT&T's manufacturing subsidiary) had proposed that Western Electric and General Electric cooperate to form a radio company that they would own jointly. Western Electric, he had suggested, could then have wired telephone communications as its exclusive field of activity while General Electric would have "the X ray, light, and power fields." To the new company would be allocated "all other applications and uses," including specifically "all vacuum tube work . . . as well as all wireless work." In that field it would hold exclusive licenses to all patents owned by its parent firms.[5] This memorandum had been accepted in principle by President H. B. Thayer of AT&T and by George E. Falk, the corporation's general counsel, and it had been discussed with Davis of GE. Within General Electric, Hooper's friend and confidant, E. P. Edwards, had warned of the company's weak position in vacuum tube patents as early as March 1916; and in November 1918

<hr>

[2] For the authorship, see Clark Radio Collection, Cl. 100, v. 2, Hooper's comments on Clark's history of radio. Compare Case and Case, *Young*, p. 209.

[3] The covering letters are in Hooper Papers, Box 3.

[4] Cf. Case and Case, *Young*, p. 210.

[5] Young Papers, Box 95, Swope to Young, 28 February 1920, enclosing "Memorandum" of 29 May 1918; and ibid., Davis to Young, 5 January 1920.

he urged the necessity for "a conference of the principal parties interested
... in order that all of us may be in a position to advance the art to the
best of our individual ability and reap the benefit of our efforts." Edwards
specifically suggested that the Navy Department might be "a good medium" through which to work.[6]

These and similar initiatives were swamped by more urgent concerns
during the absorption of American Marconi and the formation of RCA.
They reemerged in 1920 in a somewhat altered context. General Electric
was now in a much stronger position, since it had access to the whole
arsenal of Marconi radio patents and in particular the patent on the
Fleming diode. On the other hand, AT&T still controlled the de Forest
triode, and without rights under the triode patents GE's tube research
was likely to prove largely abortive, as far as commercial exploitation
was concerned. Then, too, RCA was now functioning as an operating
communications company. General Electric had a large stake in its success, and therefore in working out some kind of rapprochement with the
Telephone Company that would avert disruptive competition by indicating each company's appropriate fields of activity.

The bureau's letter of January 1920 served as the cue for action. Would
they have acted without it? Almost certainly. Nevertheless, the convenience of having the Navy initiate the process of corporate treaty-making
was apparent. Once again the national interest, as defined by the Navy
Department, could be invoked to rationalize a restructuring of the industry. Who could argue against the safety of life at sea?

The suggestions made in the bureau's letter were in themselves quite
modest. The bureau did not propose a wholesale cross-licensing of radio
patents; nor that AT&T become a shareholder in RCA; nor that RCA,
General Electric, and the Telephone Company get together to work out
a convenient allocation of corporate "territory;" nor that any other company be brought into the syndicate. All that the letter asked for was some
arrangement by which "the market can be freely supplied with tubes,"
and it defined "the market" very narrowly, specifying detector circuits
in marine radio receivers. Ships of the Emergency Fleet had mostly been
equipped with arc transmitters, and to receive signals from these transmitters you had to have at least a regenerative receiver, and that meant
vacuum tubes. Some marine operators got their tubes from the same
semi-legal sources as the amateurs did—O. T. Cunningham's "audio-
trons" were widely used. Some found they could buy excellent tubes in

[6] Young Papers, Box 75, Edwards to F. C. Pratt, 12 November 1918, enclosing
memorandum on "Patent Situation," Edwards to Burchard, 30 March 1916.

Japan.[7] Most, however, still used crystal detectors, stable and predictable in performance and perfectly adequate for spark signals but not for continuous wave telegraphy. There was, therefore, an emerging problem in this area that could be solved only by making vacuum tubes more readily available. And that indeed called for some cooperative action by the firms that owned the relevant patents: AT&T and General Electric. But, as regards civilian use, there was no immediate urgency in the matter. What most concerned the Navy was that its own needs for vacuum tube equipment should be met, and this is what led to Hepburn's letter. Stanford Hooper stated the matter frankly in later testimony to Congress. After the war, he said, the Navy was unwilling to grant patent releases to its suppliers because "we were afraid that we were getting in pretty deep." Instead, he and his colleagues tried to devise a scheme "whereby we could buy radio equipment without having to take the patent responsibility; and we suggested that the companies get together and work it out in some way."[8]

That is precisely what "the companies" did, and with such promptness as to suggest that they had been waiting for the right signal. This was hardly remarkable: to do as Hepburn and Hooper requested served their interests no less than the bureau's. Their response, however, went far beyond a mere cross-licensing of tube patents—a straightforward matter that their attorneys could have attended to in a couple of weeks. What emerged from the negotiations was an elaborate protocol designed to govern the future exploitation of continuous wave radio. What began as a plea for the safety of ships at sea ended as a treaty for the allocation of corporate territory. In the process, decisions were taken that had profound consequences for the future of American communications.

[7] Julius Weinberger, research engineer for RCA, stated flatly in March 1920 that "tubes are not available to the mariner at present," and recommended that, in designing direction-finding equipment, RCA produce "a simple set of adjuncts to the regular ship's crystal receiver" rather than the three-step amplifier that the Bureau of Standards had proposed. See Clark Radio Collection, Cl. 5, 1920, Box 65, Weinberger to A. N. Goldsmith, 18 March 1920. Tubes were, however, available to operators who really wanted them. See E. J. Quinby, *Ida was a Tramp* (Hicksville, N.Y., 1975), pp. 43, 48-49, 68-70, and compare *Popular Radio* 2 (October 1922), 143-44. I owe my information on the availability of Japanese tubes to the kindness of Alan Douglas.

[8] "Commission on Communications," *Hearings* before the Committee on Interstate Commerce, United States Senate, 71st Congress, 1st Session, on S.6, A Bill for the Regulation of the Transmission of Intelligence by Wire or Wireless (Washington, D.C., 1930) [hereafter *FCC Hearings*], testimony of Stanford Hooper, p. 315.

Negotiations between GE and the Telephone Company began in early 1920. It proved easy to agree in principle that some cross-licensing of patents was desirable. The tough questions involved how far to go beyond that. How far should the agreements also try to delimit the fields of activity in which each firm should have exclusive rights—exclusive rights, that is, to use its own patents *and* those of the other firm? This was in truth the heart of the matter. RCA and General Electric had already signed a comprehensive cross-licensing agreement, extending to RCA the right to use in radio any of the patents that GE owned or was licensed to use. Negotiations with AT&T were, therefore, carried on in the first instance by GE, but it was understood that extension agreements would be signed by GE with RCA and by AT&T with Western Electric. The desired outcome was, therefore, a situation in which GE and RCA would be free from the competition of AT&T and Western Electric in certain fields, and AT&T and Western Electric free from the competition of GE and RCA in others. When GE negotiated a cross-licensing agreement with AT&T, it intended to stipulate certain fields that it (and by extension RCA) would promise not to invade but would cede to AT&T; and it received in return assurance of other fields that GE and RCA could securely occupy, free from fear of invasion by AT&T and Western Electric. These stipulations and undertakings were written into the terms of the patent licenses.[9]

Now, it is clear that cross-licensing agreements could easily have been negotiated that did not contain these exclusive features. Such agreements would have been tantamount to a literal pooling of radio patents; they would have eliminated any risk of litigation over tube patents in future; and they would have been free of any taint of illegality under the antitrust laws. They would also, of course, have fully satisfied the Navy's request and solved the problem of tube supply about which the Navy had expressed concern. But, for the corporations, that was not the essence of the matter. The essential goal was agreement on the allocation of corporate territory—fields of activity in which each firm could enjoy rights

[9] This statement is not beyond dispute. Thayer, president of AT&T, stated that "As the contract now stands, I have understood it as neither expressing nor implying any obligation on one party to keep out of the field of another, provided getting into that field did not involve the infringement of patent rights of the other." On this David Sarnoff commented, "If the above statement does represent the intention of the parties to the contract, I do not see the purpose or force in each party having ceded to the other, exclusive rights in certain fields under its own patents." Thayer to Gifford, 10 July 1922, and Sarnoff to Harbord, 6 February 1923, both reprinted in Gleason Archer, *Big Business and Radio* (New York, 1939), pp. 74-75.

of exclusive occupancy.[10] Whether such agreements implied restraint of trade was, of course, an important question, to which Young, Thayer, and lawyers on both sides were sensitive. This explains why they took pains to inform the Department of Justice of what they were up to—though all they could elicit from the attorney general was the noncommittal and unarguable comment that it looked like "a good business arrangement."[11] And it explains, too, why it was important to have it on the historical record that the Navy had initiated the process. Neither Young nor Thayer was a novice in antitrust matters; each knew that it was prudent to move carefully.[12]

What kind of compact would the courts uphold? What kind would expose the corporations to antitrust indictment—and, perhaps, to civil suits for triple damages? There was no way to be sure. Where patents and antitrust law overlapped, neither statutes nor the common law nor legal precedent spoke with a certain voice. Every patent, by its very nature, conveyed monopoly rights: the patent laws rewarded an inventor by granting a temporary monopoly that provided insulation from competitive exploitation of the patented art. The essence of a patent was its exclusionary power, and no court in 1920 had ever held that the antitrust laws required a patent-holder to forfeit that exclusionary power the instant it afforded some degree of monopoly in the market. So much at least seemed clear. But there were questions arising now out of the very scale of the affair in contemplation. What if the consolidation of patents and the allocation of exclusive rights were carried to such a length as effectively to eliminate competition and exclude the entry of new firms? Was there a point at which the privileges granted by patents could be abused? What were the implications when patents were used to carve up an industry into exclusive corporate empires?[13]

[10] But note again that this is a question of disputed interpretation; compare Alexanderson Papers, folder 27, "Radio Trial Brief" (1926), p. 7: " . . . the cross licenses (in many cases at least) do not constitute agreements that the parties will not enter into certain fields. They merely convey patent licenses of limited character. The licensees are entirely free to enter the fields not covered by the licenses if they can do so without infringing the patents." This is, of course, a later reading of the agreements, designed to reinforce RCA's defense against antitrust charges.

[11] Case and Case, *Young*, p. 218, citing Young Papers, Box 95, memorandum of meeting with Attorney General Mitchell Palmer, 12 May 1920.

[12] Indeed, Young's initial responsibility when recruited as a vice-president of GE in 1913 had been to monitor compliance with the consent decree of 1911, which had relieved the corporation from the threat of prosecution for restraint of trade in the sale of electrical equipment. Few executives can have been more sensitive, or better informed, on antitrust matters than he.

[13] Compare the remarks of the assistant attorney general in charge of the

* * *

The job of defining exclusive fields began simply enough. Swope's 1918 memorandum clearly intended that, as a general principle, the Telephone Company should confine itself to wired communications and the proposed new company to wireless; but even at that early stage some hint of the complications that lay ahead could be seen in Swope's proviso that the Telephone Company should also have rights in "wireless used in conjunction with wire communication, in the sense that signals pass automatically from one to another."[14] Presumably this referred to short radio links in the wired telephone network, as for instance from the California coast to Catalina Island. Potentially, however, it could have much wider application—for example, to transatlantic radiotelephony. It seemed at first glance a simple enough matter to distinguish between wired and wireless communications and allocate the one field to AT&T and Western Electric, the other to General Electric and RCA, but the more one thought about it, the more blurred the distinctions became. A. G. Davis saw some of the complications when on 5 January 1920— the very day on which the Navy sent its letter to the companies—he gave Owen Young some rough notes to guide him in the negotiations that lay ahead. It was plain what the general rule should be: "Each party free to work in its field under patents of all." But then the complications began: "Each party free to work in other fields for purposes of its own field— exception only where fields collide, as in radio telephony, where must define limits."[15] Radiotelephony was indeed a field in which both groups would want to stake claims.

Davis thought that instances "where fields collide" would be exceptions, but this was not to prove true. Broadcasting was to breed examples every day. Broadcasting was clearly "wireless," but it was also a form

---

prosecution of RCA on antitrust charges in 1930-32: "The case presents a conflict between the anti-trust laws enacted to prevent monopoly and the type of monopoly created by the government through the sale of patents." (Case and Case, *Young*, p. 593, citing the *Herald Tribune*, editorial of 23 November 1932) See also "RCA'S Television: Off to a Big Lead," *Fortune* (September 1948), 194. For a useful analytic survey of the economics of the patent system, see F. M. Scherer, *Industrial Market Structure and Economic Performance*, 2nd ed. (Chicago, 1980), chap. 16, pp. 439-58.

[14] Young Papers, Box 95, Swope to Young, 28 February 1920, enclosing memorandum of 29 May 1918.

[15] Young Papers, Box 95, A. G. Davis to Young, 5 January 1920, enclosing memorandum headed "General Principles." The date of this memorandum makes it clear that negotiations between GE and AT&T were planned before the Navy's letter arrived.

of telephony; and broadcast stations would depend on the wired telephone system for their remote pickups and their network connections. But all this was still in the future when negotiations began, and it was not the kind of thing Davis had in mind. He was concerned that RCA should be able to use wire lines for remote control of its telegraphy stations, just as the Telephone Company executives were concerned about radio links in their phone system. Broadcasting played no important role in the negotiations and was mentioned only in passing in the final document.

Even so, agreement did not prove easy, and the closer the contract got to final form the more complex it became. AT&T spokesmen said later that they would have preferred the exchange of patents to be nonexclusive; if so, they had the support of legal counsel, sensitive to the risk of antitrust action.[16] RCA and GE, however, preferred the definition of exclusive fields to be as specific and concrete as possible; in particular they insisted on exclusive rights to use the patents of the participating companies in radiotelegraphy. But there was no matching disposition to grant to AT&T and Western Electric exclusive rights in radiotelephony, for to do so would have precluded RCA from using voice communications in overseas and coastal radio. The clauses dealing with radiotelephony, in consequence, developed into a tangled thicket of qualifications and provisos that were later to prove sources of acute conflict.

By 12 April 1920 Alexanderson at least was satisfied that GE and RCA had got as good a deal as could be expected. As he interpreted the draft agreement that had been reached by that date, RCA was to have exclusive rights to use the pooled patents in transoceanic telegraphy and ship and aircraft radio. It would also have nonexclusive rights to transoceanic telephony, in cooperation with the Telephone Company, but no rights in radiotelephony over land unless the Telephone Company was unable to supply the service. To the Telephone Company there were to be ceded exclusive rights to "all land radio telephony for toll purpose, or equivalent," while GE would enjoy, as one of its exclusive fields, "broadcasting service, and sale of amateur apparatus, particularly vacuum tubes." The whole agreement, he believed, aimed at "a natural division of fields of activity"; and the only caution he felt it necessary to add was that it should be recognized that, in giving up land telephony, GE and RCA were giving up a large field and were entitled to expect full cooperation from the Telephone Company in return—specifically, in providing connections to RCA's shore stations.[17]

[16] W. S. Gifford to Owen D. Young, 20 April 1925, as reprinted in Archer, *Big Business*, p. 209; compare Case and Case, *Young*, p. 211.

[17] Alexanderson Papers, Alexanderson to A. G. Davis, 12 April 1920.

Alexanderson's comments are revealing. On the one hand he recognized that the Telephone Company was to have exclusive rights to "land radio telephony." On the other, he thought that GE and RCA would have, as one of their exclusive fields, broadcasting service. And between these two provisions he saw neither conflict nor contradiction. Clearly, in his mind, land radio telephony "for toll purposes or equivalent" was something quite different from "broadcasting." His image of radiotelephony was of point-to-point communication for a fee—the analogue of telephony by wires. Broadcasting was something else: it would be a public service, made available without charge, expected to pay for itself, if at all, through revenue from the sale of receivers.[18] Originally both parties to the contract seem to have accepted some such view; the Telephone Company would have exclusive rights to the manufacture of radiotelephone transmitters (including broadcast transmitters) while GE would have exclusive rights to manufacture receivers. Both parties changed their interpretations later, after rights to manufacture transmitters and receivers became very valuable.

A. G. Davis, who did most of the day-to-day negotiating on behalf of GE, was less easily satisfied than Alexanderson. He thought that rights which should be exclusive to GE had been watered down in successive drafts, and he was beginning to worry about broadcasting, partly in response to pressure from David Sarnoff, commercial manager of RCA, who was again advancing his notion of the "radio music box," or radio receiver in the home. The draft agreement granted to GE the right to "establish and maintain . . . stations for transmitting or broadcasting news, music, and entertainment" and to "make, use, sell, and lease wireless telephone receiving apparatus for the reception of such news, music and entertainment."[19] But this was a nonexclusive license, contrary to Alexanderson's opinion; it was not clear how it was to be reconciled with the Telephone Company's exclusive privileges in land radiotelephony for toll; and the right to establish broadcast transmitting stations was not accompanied by the right to manufacture the transmitters themselves. On this last point the license agreement seemed specific: it ceded to GE the right to manufacture radio receivers, but added immediately, "it is agreed that the General Company has no license to equip wireless telephone receiving apparatus . . . with transmitting apparatus, or to sell, lease, or otherwise dispose of transmitting apparatus for use in connection with receiving apparatus sold under this paragraph."[20] If this clause

[18] Compare the attitudes of Young and Sarnoff, as characterized in Case and Case, *Young*, p. 264.

[19] FTC *Report* (1923), pp. 30-39; Case and Case, *Young*, p. 213.

[20] FTC *Report* (1923), p. 134.

meant what it seemed to mean, GE was prohibited from manufacturing radiotelephone transmitters—at least if they were to be used in conjunction with the receivers that it manufactured.

How important were broadcasting rights anyway? The fact was that nobody knew; and most did not greatly care. Of the GE/RCA group, only Sarnoff, it seems, believed that broadcasting might develop into a field with commercial potential. The possibility that GE might develop a "wired wireless" system of transmitting information and entertainment over electric power lines, or that AT&T might try the same thing over the telephone lines, was a source of as much concern. Broadcasting was not, in any case, regarded as a matter of such importance that Young was willing to see the negotiations prolonged in a possibly futile effort to eliminate ambiguities. He had other concerns that made early agreement with the Telephone Company expedient. One of these was the need for a traffic agreement; another was the need for more capital.

* * *

Traffic agreements occupied much of the time and attention of RCA's chief executives during the first few years of the corporation's existence. Most of them were negotiated with foreign governments and telegraph bureaus and were intended to ensure that RCA was designated as the exclusive U.S. agent for the transmission and reception of radio traffic between the United States and the various foreign countries involved. These we shall discuss later in connection with Westinghouse's abortive attempt to break into long-distance radio. A traffic agreement with AT&T, if it could be worked out, had a different purpose. To understand this, we have to know something of RCA's relations with the two domestic wire telegraph companies in the United States: Western Union and Postal Telegraph.

RCA began handling radio traffic between the United States and the rest of the world in March 1920. It soon became apparent that there was a serious lack of balance on the transatlantic circuit between eastbound and westbound traffic. The word-count of messages transmitted to Europe was very much lower than the word-count of messages received, even though the price per word was the same.[21] The reason was

---

[21] For example, in December 1920 144,224 paid words were transmitted eastbound on the British circuit as compared with 173,004 westbound; on the German service the corresponding figures were 80,025 and 233,014, and on the Scandinavian service 8,193 and 22,104. As a contributing factor, W. A. Winterbottom, RCA's traffic manager, suggested that American equipment was more efficient and permitted a higher average speed of reception than in Europe while

not far to seek. RCA had no network of telegraph offices in the United States where messages to be sent to Europe by radio could be handed in. The only such offices were in New York City and Washington, D.C. If you lived anywhere other than these two cities and wanted to send a telegram to a European country, you would take or telephone your message to your local Western Union office (or to one of the much less numerous Postal Telegraph offices).[22] Western Union, however, would not accept such a message for transmission to Europe "via radio." You were not permitted to specify the mode of transmission to be employed. The reason for this was simple: Western Union held a financial interest in the submarine cables and preferred to channel traffic through them. The only way in which you could send a message to Europe by radio, if for some reason you insisted on using that mode, was to file it with a domestic telegraph company as a normal land message addressed to RCA at 64 Broad Street in New York City, from where (if you had previously established credit facilities) it would be transmitted to Europe by radio. This was inconvenient, and few people did it. The result was that RCA was continually losing eastbound traffic: virtually the only messages to Europe it handled originated in New York or Washington, where they could be handed in directly at RCA offices. The problem did not exist on the transpacific circuit, because neither of the telegraph companies had a financial stake in the transpacific cable; Western Union, for example, would gladly accept a telegram addressed to Tokyo to be forwarded "via radio" by RCA from San Francisco.[23] And it did not exist for incoming traffic from Europe, which RCA forwarded at normal commercial telegraph rates from New York to the ultimate recipient.

The blunt fact was that RCA depended for its domestic "feed" on companies whose stake was in a rival communications system. This was a unique situation. In Europe the agencies that handled long-distance radio were either government departments that also ran the domestic

---

the circuits were open. Young suggested that relatively little press traffic moved eastward, and that the reason for this, according to the press associations, was simply that a relatively small amount was used by the foreign press. See the testimony of Winterbottom and Young in *Cable Landing Hearings*, pp. 337-56.

[22] Western Union had about 26,000 offices in the United States while Postal Telegraph had a little over 3,000. In 1928 Postal Telegraph did between 15 and 16 percent of the total landline telegraph business in the United States. See *FCC Hearings*, pp. 1240 and 1472, testimony of David Sarnoff and Newcomb Carlton.

[23] Hiram L. Jome, *Economics of the Radio Industry* (Chicago and New York, 1925), p. 157; *Wireless Age* 10 (October 1922), 55. For an extended discussion, see the testimony of W. A. Winterbottom in *Cable Landing Hearings*, pp. 340-41 and compare the testimony of O. D. Young, ibid., p. 1094.

telegraph system or they were private companies that had a service contract with the domestic telegraphs, as Marconi did with the British Post Office. And it was a situation that made it essential for RCA to come to terms with at least one of the domestic wired systems in the United States—unless indeed it proposed to set up its own national network of radiotelegraph offices and organize its own pickup and delivery service. The costs and uncertainties of establishing such an independent domestic communications system were such that it was never seriously contemplated.

Why the domestic telegraph companies, and specifically Western Union, by far the larger of the two, had not been brought into the creation of RCA from the beginning is a question that admits of no easy answer. It is true, of course, that radio was from the start thought of as competitive with wired communications systems rather than complementary to them; nevertheless, the possibility had occurred to some—for example to E. J. Nally in 1919 and to A. G. Davis in the year following.[24] But there is no record that Western Union was ever approached with a view to buying a stock interest in RCA, nor that Newcomb Carlton, president of Western Union, ever contemplated such a step.[25] As regards the narrower issue of a traffic agreement, Carlton in public always presented Western Union as perfectly willing at any time to make suitable arrangements for the handling of RCA's messages.[26] And so indeed he was—for a price. It was a price, however, that RCA could not pay: 20 percent of RCA's gross receipts from overseas business. This, as Young pointed out, "would not only exhaust but perhaps more than exhaust the entire profits . . . arising from overseas transmission."[27]

Revenue from telegraph traffic was literally RCA's lifeblood in these early years; apart from what it might earn from the sale of equipment to others, the corporation's commercial future depended on moving information and getting paid for it. Traffic agreements with foreign governments and corporations ensured that it would compete effectively with

[24] See extracts from Nally's diary, as cited in Clark, "Radio in War and Peace," chap. 12, pp. 72 and 78; Case and Case, *Young*, p. 217.

[25] A merger between Western Union and RCA's telecommunications division was, however, seriously discussed in 1928-1929, as a possible response to the unification of cable and wireless service in Britain. Western Union backed out of the deal, and negotiations began between Young and Sosthenes Behn for a similar merger with IT&T. These plans were later abandoned because of congressional opposition.

[26] See, for example, his testimony in *Cable Landing Hearings*, pp. 180-81.

[27] Young Papers, Copybook 802 (Radio), Young to Newcomb Carlton, 10 January 1921.

the cables for traffic moving into the United States. But the absence of traffic agreements with any of the land-line communications systems in the United States meant that it was gravely, perhaps disastrously, handicapped in the competition for outgoing traffic. Newcomb Carlton knew this, and, knowing it, asked a price that would have effectively expropriated the net earnings on RCA's capital. Owen Young knew it too and, unwilling to pay Western Union's price, sought an alternative arrangement with the Telephone Company.

Such an alternative arrangement would presumably have involved using local telephone operators to accept and deliver RCA's overseas traffic and AT&T's "long-lines" to channel this traffic to and from New York. It would, at best, have been an imperfect substitute for a traffic agreement with Western Union—a Western Union office was, after all, the place where one sent telegrams in those days, and Western Union's familiar delivery boys were the people who delivered them—but it might have sufficed.[28] We can only surmise: not all of Owen Young's persistent persuasiveness could elicit from the Telephone Company terms that RCA would accept. Thayer was perfectly willing to lease his long lines to RCA, for a price; and he was also willing to buy an equity interest in RCA, though far from enthusiastic about the idea. But to Davis this looked as if Thayer was trying to have it both ways: AT&T could come in as a partner, or it could rent its lines, but it could not expect to do both.[29] On this issue the companies could not reach agreement, and when the cross-licensing contract was finally signed on 1 July 1920, shot through as it was with provisos, qualifications, and ambiguities—a lawyer's paradise and a layman's nightmare—it contained no traffic agreement. For that RCA had to look elsewhere.[30]

[28] Compare Young's remarks at the Cable Landing Hearings (p. 337): "The contract is not considered either by the Radio Corporation or the telephone company as providing the best means for the collection and distribution of overseas messages. The contract merely provides facilities by which such messages can be distributed and collected, regardless of the attitudes of the domestic land wire telegraph companies."

[29] Young Papers, Box 95, Davis to Young, 29 May 1920 (cited in Case and Case, Young, p. 214).

[30] RCA was finally able to negotiate a contract with the Postal Telegraph Company by which the latter agreed to receive at its American offices all transatlantic messages designated to be forwarded "via radio" and to transmit them to RCA in New York. Messages from overseas would also be delivered by Postal Telegraph to any point in the U.S. at which it had an office. The agreement was announced by RCA in its Annual Report for 1922. See also Jome, Economics, p. 157; Wireless Age 10 (October 1922), 55; and Fessenden Papers (State Archives of North Carolina), 1140-15.

Nor did the cross-licensing contract contain any provision for AT&T to become a stockholder in RCA, which had been the third of Young's objectives. On that issue, too, Thayer proved slow to convince, and it was not until 4 August that he and Young reached agreement. By that date hopes for a traffic contract had been abandoned; removal of that complication may have made it easier to define how the two companies should be related. Young was, of course, aware of RCA's need for new capital, but for him AT&T's role in RCA also involved a matter almost of principle. He wanted AT&T's participation to follow the pattern set by General Electric: a large investment of technical expertise and financial resources, matched by the allocation of common and preferred stock. In the end he got his way, despite considerable reluctance on Thayer's part. AT&T agreed to contribute to RCA the sum of $2.5 million, payable in bonds of the Southwestern Bell Telephone Company. In return AT&T received 500,000 shares of RCA preferred and 500,000 shares of RCA common. After discounts and commissions, RCA's treasury realized some $2.4 million from the transaction.[31]

This infusion of new capital represented an important enlargement of RCA's resources. It also represented an important phase in the corporation's evolution toward more autonomous status, for with representatives of AT&T now sitting on the board RCA was no longer the unique creation of General Electric that it had formerly been. This development was carried further in February and March of 1921, when the United Fruit Company also became a major stockholder, to the extent of $1 million. United Fruit (through its subsidiaries, Tropical Radio and the Wireless Specialty Apparatus Company) held several useful radio patents, particularly that on the Pickard crystal detector; it owned a chain of spark-equipped radio stations in Central America and the Caribbean that might one day, if modernized, serve as a useful complement to RCA's facilities in that area; and it had a lawsuit against American Marconi that was still pending against RCA and had so far defied efforts at settlement. Not to be overlooked, too, was United Fruit's impressive political clout in Washington when Central American affairs were involved, and the fact that it was planning to test turbo-electric drive on the ships of its fleet—the equipment to be purchased, possibly, from General Electric.[32] Bringing United Fruit into the RCA fold, in short, made good

[31] Young Papers, Copybook 802 (Radio), Young to Thayer, 9 August 1920; ibid., Box 95, Young to Davis, 4 August 1920; Case and Case, *Young*, p. 219.

[32] Case and Case, *Young*, pp. 215 and 217. Young in 1920 was earning an annual salary of $60,000 as vice-president of GE and $15,000 as chairman of RCA; conceivably in some circumstances Young might have felt a conflict of

business sense; and it was not bad politics either. The arrangement followed what was emerging as the standard pattern. United Fruit, through the Wireless Specialty Company, cross-licensed its patents with RCA, GE, and the Telephone Company, bought 200,000 shares of RCA preferred and the same number of RCA common, and agreed to abandon the hangover of litigation against American Marconi. In return GE bought, for cash, 50 percent of the stock of Wireless Specialty, and United Fruit got a seat on RCA's board.[33]

Why corporations like United Fruit and, more significantly, the Telephone Company signed agreements for the cross-licensing of patents with GE and, by extension, with RCA is clear enough. Technical progress and commercial development would have been seriously inconvenienced if something of that nature had not been done. It is less clear why the delimitation of exclusive fields of activity figured so prominently in those agreements. A nonexclusive pooling of patents would have accomplished all that was necessary for the advancement of the radio art. And insistence on exclusivity had its costs. Certainly, if this principle had been absent, the later history of RCA would have been less marred by intercorporate squabbling and public hostility, for it was the exclusive features of the contracts that set the legal staffs of the "Radio Group" and the "Telephone Group" arguing with each other like rival gangs of teenagers hassling over "turf" while at the same time lending all-too-credible color to charges of restraint of trade. Insistence on exclusivity seemed a way of imposing order on the industry; but the semblance of order was spurious when markets and technology were both in continuous and rapid flux. And, from the public relations point of view, the corporations involved were setting themselves up for antitrust prosecution at the hands of the first administration that saw political advantage in issuing indictments.

It is also not self-evident why the Telephone Company and United Fruit became equity shareholders in RCA. Certainly a patent pool, and even cross-licensing agreements with clauses conveying exclusive rights, could have been negotiated without an investment of capital. RCA's profit expectations were not such as to make the acquisition of shares attractive purely as an investment: no one expected RCA common to earn dividends within the foreseeable future, while the 7 percent dividend rate on the

---

interest between these two roles, but the United Fruit deal did not provide those circumstances.

[33] Paul Schubert, *The Electric Word: The Rise of Radio* (New York, 1971), pp. 201 ff. The agreement with United Fruit is reprinted in FTC *Report* (1923), pp. 142-48.

preferred was no higher than the investing corporations could have earned by employing the capital in their own businesses. What, then, induced them to transfer such sizable sums to RCA's treasury? The answer, apparently, is that Owen Young would not negotiate on any other basis. The subscriptions were the price of admission to the syndicate—the membership dues in the RCA "club"—and RCA was not willing to cross-license patents without them. Young was doing what he had so often done in the public utility field: putting together a viable organization, assembling the resources necessary for it to function effectively, and making sure that all concerned parties had a stake in its survival and success. RCA, for him, was not only an instrument for asserting the autonomous influence of the United States in world communications; it was also an institution for creating a community of interests domestically.

This, however, speaks only to the question of RCA's purposes, and Young's specifically. Why the Telephone Company acceded to this conception is less obvious, and Thayer's reluctance to agree to a stock subscription to RCA suggests that he would have been content with a much less intimate relationship than Young had in mind. In the short run, what AT&T expected to get out of the deal was freedom from patent conflicts, particularly over vacuum tubes. In the longer run it wanted cooperation with RCA in the development of radiotelephony—and protection for what it regarded as its natural interests in that field. The press release issued by AT&T on 26 August made the points explicitly. It quoted at length the letter from the Navy's Bureau of Engineering, clearly implying that the initiative in bringing the companies together had come from the federal government. It described the exchange of licenses as enabling the Telephone Company to "supplement its wire system with wireless extensions . . . as between shore and ships at sea." And it presented the ultimate outcome of the agreement as bringing into a harmonious relation "the world-wide wireless system of the Radio Corporation and the universal service of the Bell System," so that eventually telephone service could be extended to ships at sea and to foreign countries. In conclusion it noted almost parenthetically that, to better carry out the purposes of the agreement, AT&T had bought a minority interest in RCA, and W. S. Gifford, vice-president of the Telephone Company, had been made one of the radio company's directors.[34]

The "harmonious relation," if it ever existed, did not long endure. Dissension over the interpretation of the cross-licensing agreements, particularly as they applied to transoceanic radiotelephony and the manu-

---

[34] AT&T Press Release, contained in Young Papers, Copybook 802 (Radio), Young to A. G. Davis (telegram), 26 August 1920.

facture of broadcast equipment, broke out almost immediately. By the fall of 1923 disagreement was so acute and bargaining positions on both sides had so hardened that further direct negotiation was recognized as futile.[35] Litigation or the use of a neutral arbitrator were seen as the only feasible courses of action and, in the hope that publicity might be avoided and problems resolved in months rather than years, an arbitrator was appointed.[36]

Meanwhile, beginning in February 1922, the Telephone Company proceeded to divest itself of the common and preferred stock it had so recently acquired in RCA, and by the early months of 1923 all these securities had been sold to the general public—the first large block of RCA shares to appear on the market since the company was founded.[37] AT&T's two representatives on RCA's board of directors both resigned in June 1922. It was blandly explained in AT&T's Annual Report for 1922 that ownership of stock in RCA had proved unnecessary for cooperation between the companies and therefore the securities had been disposed of "in line with our general policy to hold permanently only the stocks and securities directly related to a national telephone service." And spokesmen for the company claimed that brokers had been advertising AT&T's stock ownership in order to induce the public to invest in RCA—"which tended to create a moral obligation on this company's part which it did not wish to assume."[38]

Such sensitivity to moral obligations on the part of a large corporation

[35] See below, pp. 482-86. The most comprehensive account of the arbitration proceedings is in Archer, *Big Business*, chaps. 5-9; but see also Leonard Reich, "Research, Patents, and the Struggle to Control Radio: A Study of Big Business and the Uses of Industrial Research," *Business History Review*, 51 (Summer 1977), 208-35.

[36] See J. G. Harbord to Young, 21 September 1923, as reprinted in Archer, *Big Business*, pp. 110-11. Harbord estimated that litigation, taking into account the inevitable appeals, would not yield a decision before three to five years had elapsed. It was also true, of course, that arbitration promised to minimize publicity.

[37] See AT&T, Annual Report for 1922; FCC, Walker Report, p. 21. Some RCA stock held by former shareholders in American Marconi may have changed hands earlier. Furthermore, in June 1922 the directors of RCA voted to offer Young 100,000 shares of RCA common at the special price of sixty cents a share; Young took up the offer and sold the shares on the market between 1922 and 1925. (See Case and Case, *Young*, p. 382) With these exceptions, the sale by AT&T was the first offering of RCA shares to the general public.

[38] Eric Barnouw, *A History of Broadcasting in the United States, Vol. 1, A Tower in Babel* (New York 1966), 1: 123,161.

was no doubt admirable, but skeptical observers widely interpreted the move to mean that AT&T was casting off an alliance that had become distasteful. Any inclination to move in this direction must have been reinforced by the knowledge that, during 1922, the Federal Trade Commission had begun to investigate charges that RCA had been set up by GE and others as a "bogus independent" in an attempt to monopolize the manufacture and sale of radio apparatus, and by the passage of House Resolution No. 568 on 3 March 1923, directing the Federal Trade Commission to investigate whether RCA and its affiliated companies were in violation of the antitrust statutes. The Telephone Company had its own recurrent antitrust headaches and no need to acquire others by too close an association with RCA. Divestiture offered the chance to put a little prudent distance between the two corporations. And prudent it was to prove in 1932 when, with RCA and its associated companies facing antitrust prosecution, AT&T was able to make its separate peace with the Justice Department by calling attention to the fact that it had sold its interest in the radio company many years earlier. Clearly Young's conception of RCA as the integrating focus of American interests in radio was not one the Telephone Company found it possible to share.

Some at least of RCA's staff had hoped for productive cooperation with the Telephone Company at the technical level, but not much came of this. The cross-licensing of patents in itself, of course, brought some benefits. General Electric's work on high-power transmitting tubes, for example, was materially assisted by the work of Houskeeper at Western Electric in perfecting the glass-to-metal seal—one of those seemingly minor achievements in technology that make possible the more dramatic advances.[39] Western Electric benefited, too, from being able to use RCA's transmitting facilities in its long-distance radiotelephone tests, as for example in December 1922 when the first "single sideband" speech transmissions were carried out between RCA's Rocky Point station on Long Island and Western Electric's factory at New Southgate in London, England.[40] But at the day-to-day level there was little technical interchange.

[39] RCA's first major success with tube transmitters in transatlantic radiotelegraph service came in the fall of 1922, when a bank of three 50 kw. kenotrons and six 20 kw. pliotrons operated on two of the RCA circuits to Britain and Germany for a sixteen-hour period, replacing the alternators ordinarily used. These were water-cooled tubes with external anodes, constructed with the Houskeeper glass-to-metal seal. By this date Western Electric had developed a 100 kw. tube. See *Wireless Age* 10 (October 1922), 60 and (November 1922), 55; J. W. Stedenfeld, "William Gibbens Housekeeper [sic]," *The Old Timer's Bulletin* 23 (September 1982), 16-17.

[40] George C. Blake, *History of Radio Telegraphy and Telephony* (London,

Young had given his engineers early warning not to expect too much. Meeting with RCA's technical committee in August 1920, he told them that the agreement with the Telephone Company had been reached only with great difficulty, that "at certain points it almost had to be done by force," and that whether it turned out successfully or not depended above all on the spirit in which it was approached. His advice was to go slowly, not to start any specific project that might depend on cooperation with Western Electric, but rather to try to establish a good working basis for cooperation in the future.[41] This was sound advice but, at the technical level, no such basis for cooperation was ever established. By late 1923 it could be stated as a matter of common knowledge that "cooperation in the matter of research between the two companies . . . has never been realized."[42]

This was hardly remarkable: that kind of technical cooperation could have been achieved only by close person-to-person interaction between the individual members of the two organizations, and that never happened. Indeed, RCA even had difficulty establishing technical cooperation with General Electric—as witness the response when Alfred N. Goldsmith, RCA's research chief, wrote to Alexanderson to ask for a complete file of GE's "design books." Alexanderson replied that, to his knowledge, no such books existed; at GE each department kept its own design data and information was exchanged "only through personal understanding between the design engineers." In fact, Alexanderson commented, it had been found next to impossible to exchange such information by correspondence, and "even when the engineers are working in the same plant, but not in the same building, difficulty is experienced."[43]

If the exchange of technical information between GE and RCA was so difficult, how much more difficult must it have been between RCA and the Telephone Company? There was in fact little interest in close cooperation with RCA on the part of AT&T, except in such instances

---

1928), pp. 326-27. Single sideband is a technique by which the carrier frequency and one sideband of speech frequencies are filtered out at low power levels before the signal is amplified and transmitted. The carrier is reinserted in the receiver by means of a low power local oscillator, after which the speech information is detected and amplified in the usual way. Besides the obvious economy in power, the technique also takes up less of the radiofrequency spectrum.

[41] Young Papers, Additional Papers, Box 3, Minutes of RCA Technical Committee, 25 August 1920.

[42] J. G. Harbord to Young, 21 September 1923, as reprinted in Archer, *Big Business*, pp. 110-11.

[43] Clark Radio Collection, Cl. 5 (1920) Box 65, Julius Weinberger to Goldsmith, 29 March 1920 and Alexanderson to Goldsmith, 1 April 1920.

as the radiotelephone tests, where RCA had facilities that the Telephone Company needed. Why was this? The dynamics of policy formation within AT&T have often seemed inscrutable to the outside world, but in this case we seem to be confronted with a consistent corporate attitude, dating back at least to 1907, which reflected a specific conception of what the Telephone Company's true business was and what its attitude to radio should be. Reduced to essentials, that business was the provision of point-to-point communication over a wired network. One implication of this was AT&T's belief that its survival and growth depended on control of the long lines—the long-distance telephone network that tied together the regional Bell telephone companies.

What did the advent of radio mean for this conception? In the first place, and most obviously, radio components made it possible to extend the wired network over greater distances. This was why AT&T had bought the de Forest triode patents: not primarily with any view to their use in radio—that was an afterthought—but as line amplifiers in the wired network. Secondly, and only slightly less obviously, radio made it possible to extend the coverage of the network where wires could not go. This meant in the first instance to offshore islands, secondly to ships at sea and trains and other moving vehicles, and thirdly to other continents. (Submarine cables capable of transoceanic voice transmission did not come into service until 1955-1956.) From this point of view, radiotelephony was seen as essentially complementary to the wired system; and that was a good enough reason for AT&T and Western Electric to stay on the frontier of research in continuous wave technology applied to the transmission of the human voice. With the long-distance radiotelephone tests of 1915 AT&T had staked its claim to that field. With the single sideband tests of 1922 it showed that it had the full technical capability and could implement it commercially when the market was ready.

But there was a third aspect to radio technology that was less welcome to the Telephone Company. It was, perhaps, in the 1920s just a cloud on the horizon, no bigger than a man's hand, but its presence was recognized. This was the possibility that radio might render AT&T's wired network obsolete. This was a threat not only to the capital values that the wired network represented but also to AT&T's strategic position in the communications industry. Seen from this point of view, radio was a development that the Telephone Company could not ignore, but also one it did not exactly welcome. The result was a certain characteristic ambivalence in the Telephone Company's attitude to radio—an ambivalence clearly evident, for example, in the memoirs of Lloyd Espenschied, who, for many years, as a member of the Telephone Company's technical staff, carried the responsibility for monitoring developments in radio and making sure that the company was not caught napping.[44] Toward radio the Telephone Company consistently played a defensive game. It recognized the potential utility of the medium, as a complement to the wired system, and it kept itself ready to exploit that potential. But even more clearly it recognized radio as a threat—a form of technological competition that, almost overnight, could consign AT&T's long lines to obsolescence.[45] In this interpretation of what radio might mean to long-distance overland communication, the engineers and executives of AT&T were not mistaken: they just misread the timing. The real challenge of radio to AT&T's long lines came only in the 1970s, with the emergence of satellite relays and national microwave networks.

Recognition of this characteristically defensive posture toward radio helps to explain the Telephone Company's attitude toward technical cooperation with RCA and toward the cross-licensing agreements. Technical cooperation was desirable to the extent that it advanced AT&T's strategic objectives—specifically, in ship-to-shore and transoceanic telephony. It was undesirable to the extent that it might move radio technology more quickly to the point where it could offer real competition to the overland wired network. Cross-licensing was desirable to the extent that it gave AT&T access to any advances in electronic technology that GE or RCA might make. But it was even more desirable if—as AT&T claimed—it reserved the manufacture of radiotelephone transmitters to AT&T, for that clause, if it could be enforced, kept GE (and later Westinghouse) out of the telephone business.

It is in these terms that the cross-licensing agreements are to be interpreted. When the broadcasting boom exploded, the exclusive clauses in these agreements acquired a new meaning, because the rights they assigned suddenly acquired a new and unexpected value. And for a time even AT&T was seduced into thinking that it could run a broadcast station like an enlarged public telephone booth. But that was a short-lived affair, and when AT&T sold its "flagship station," WEAF, to RCA in 1926, although personal feelings were bruised, the corporate sigh of

[44] See Lloyd Espenschied, "Origin of Radio Broadcasting" (Columbia University Oral History Collection, Radio Pioneers Series; copyright 1980 by The Trustees of Columbia University in the City of New York; quoted by permission). According to Espenschied, his superiors in the Telephone Company "seemed to regard wireless as a sort of red-headed stepchild, not attractive but continually bobbing up and having to be noticed."

[45] See the discussion in Leonard Reich, "Research, Patents, and the Struggle to Control Radio: A Study of Big Business and the Uses of Industrial Research," *Business History Review* 51 (Summer 1977):208-35.

relief was almost audible.[46] The original and continuing significance of the cross-licensing agreements for the Telephone Company was defensive. They were intended to protect an older technology, that of wired telephony, from the competitive threat posed by continuous wave radio.

*　*　*

The story of how the Westinghouse Electric and Manufacturing Company joined RCA is, in its particulars, very different from that of AT&T, but the underlying theme is the same. Continuous wave radio constituted a single technical system, in the sense that it was necessary that its component parts be integrated in quite specific ways. Each element, used as part of the system, had a utilitarian value much greater than when used alone. In fact, in some cases, apart from the rest of the system it had no value at all. In the case of AT&T, rights to the triode vacuum tube were the major issue, and it was primarily to facilitate the development and use of the vacuum tube that AT&T exchanged patent licenses with General Electric and, for a time, held an ownership interest in RCA. In the case of Westinghouse the analogous issue was control over receiving circuits, specifically those embodying the feedback, heterodyne, and superheterodyne principles. Here too technical considerations exerted pressure toward institutional adjustment. The process was, however, complicated by nontechnical factors, particularly Westinghouse's early entry into popular broadcasting and its drive for an independent role in long-distance radio. Imperfect knowledge and errors in judgment also played their part.

Why the engineers and executives of General Electric and RCA failed to appreciate the critical importance of acquiring a suitable receiver at an early date is a question not easily answered. There was no mystery about the technical necessity. A continuous wave transmitter had to be matched by an appropriate receiver. If it was transmitting Morse code telegraphy, it was essentially switching on and off, at coded intervals, an unmodulated carrier wave, and anyone receiving such a transmission with a simple rectifying detector, such as a crystal or a diode vacuum tube, would hear nothing in the earphones but a succession of meaningless clicks and thumps.[47] Somehow or other the dots and dashes of the con-

[46] Archer, *Big Business*, pp. 251-76; Rupert Maclaurin, *Invention and Innovation in the Radio Industry* (New York, 1949), pp. 111-15; William P. Banning, *Commercial Broadcasting Pioneer: The WEAF Experiment, 1922-1926* (Cambridge, Mass., 1946).

[47] The reader is invited to try the experiment personally, by tuning a general coverage receiver, in the AM (amplitude modulated) mode, to any of the Morse

tinuous wave signal had to be converted into sounds audible to the human ear.[48]

The need for a suitable continuous wave receiver would have been much less urgent if RCA had been planning extensive use of radiotelephony, for in that mode the continuous wave was modulated by audiofrequency tones and any simple rectifying detector would serve to demodulate the signal and reproduce the sound in earphones or (after amplification) in a loudspeaker. This is why the simple crystal set was so effective in the early days of broadcasting. But RCA's mode was to be telegraphy; it had a magnificent transmitter for that mode in the Alexanderson alternator; but it did not have a receiver to match.

Due warning had been given. As early as November 1917 E. P. Edwards had circulated a memorandum that in all respects except its spelling of the key word was right on target. "In offering for sale the Alexanderson system," he wrote, ". . . we have not taken into account the receiving end." Representatives of British Marconi had asked whether "hetrodyne" [sic] reception was necessary for use with the alternator, and in reply General Electric had made a significant admission: all long-distance systems employed heterodyne reception and were "virtually dependent" on it, and the Alexanderson system could dispense with it only by the expedient of modulating the transmitted wave. This, however, would require "the employment of approximately twice the amount of power required to transmit the same distance"—in Edwards's judgment a serious handicap and "a very poor alternative." What should be done? Edwards's recommendation was specific: the heterodyne patents were owned by the National Electric Signaling Company, and General Electric should either

---

code telegraphy that can still be heard on the marine or amateur frequencies. Most modern receivers have an upper or lower sideband (USB or LSB) mode, and if this mode is used the test will not work, since a local carrier signal is automatically inserted. Even in the AM mode, the incoming signal may well "beat" against a steady carrier or a stray oscillation in the receiver itself and create an audible tone. But with these qualifications the test is still an instructive one.

[48] Note, however, that this was not necessary if the signals were to be recorded on paper tape electromechanically (for example, by an "inker") or on photographic film. For very high speed telegraphic reception, such as the 200 words per minute that RCA aimed at on its transatlantic circuits, transcription by human operators using headphones and a typewriter was out of the question. Paradoxically, the early coherers, long abandoned for commercial work, would have received a continuous wave signal perfectly well, albeit only at slow sending speeds. For 200 words per minute as the norm in RCA traffic-handling, see FCC Hearings, p. 1143, testimony of O. D. Young.

buy them or get an exclusive license to use them. Otherwise, he feared, the Marconi Company would do so, which would put a completely different complexion on its negotiations with GE.[49]

The National Electric Signaling Company (NESCO), Reginald Fessenden's old firm, had gone into receivership in 1912. At the time Edwards wrote, rights to Fessenden's patents had passed to a successor firm, the International Signal Company, better known under its later title of the International Radio Telegraph Company.[50] General Electric did in fact make an offer to that company for the heterodyne patent, but it was not accepted and Young did not pursue the matter.[51] It had always been the policy of Given and Walker, the Pittsburgh investors who had financed NESCO, that they would sell Fessenden's system complete or not at all, so it is not at all certain that the heterodyne patent, by itself, was for sale at any price. Two matters, however, are certain. First, Edwards was correct in his analysis: continuous wave transmission called for heterodyne reception, and any possible substitutes for that ingenious and elegant concept were distinctly second or third best. And second, when GE and RCA entered the age of continuous wave radio, they did so without the benefit of an adequate receiver.

Alexanderson later recorded the unsettling effects on RCA's engineering staff when the implications of this deficiency were realized. Shortly after the company began operations in 1920 he was called into Young's office and told, "You will have to find another way to receive signals. The Fessenden patent has become the property of a financial group which makes utterly impossible demands. We cannot deal with them." Alexanderson found this disconcerting: by his own admission, "nobody could beat the Fessenden invention."[52] But perhaps the situation was not entirely hopeless. You could not beat the invention, but you might still hope to beat the patent. Ideas as such were not patentable: if a method could be found that achieved the heterodyning effect but differed from the methods Fessenden had described, it might stand up in court. It had to be tried, anyway. Alexanderson got to work on the problem, as did Roy Weagant, the engineer RCA had inherited from American Marconi, and A. N. Goldsmith, the company's consultant at the College of the

[49] Young Papers, Box 72, Edwards to A. G. Davis, 1 November 1917.

[50] Fessenden Papers, 1140-84, G. K. Woodworth to Fessenden, 12 August 1921.

[51] Young Papers, Copybook 802, Young to Senator David A. Reed, 22 January 1923.

[52] Ernst Alexanderson, as quoted in Philip L. Alger, *The Human Side of Engineering* (Schenectady, 1972), pp. 134-35.

City of New York. By August 1920 they had devices they thought workable, in Alexanderson's case a "synchronous detector" that looked promising and had been checked by GE's patent department. RCA's technical committee approved its use, but only after Young had stipulated that the company's legal counsel also pass on it as not infringing the heterodyne patent. It was necessary, he insisted, to move with the greatest care, as there was "a difficult situation approaching."[53]

And indeed there was. By the summer of 1920 GE's engineers and its patent department had reluctantly faced up to the awkward fact that in future the heterodyne principle would be the controlling factor in radio reception. This meant that, while GE and RCA might control the transmitting patents, control of the key receiving patents rested in other hands.[54] Those other hands, furthermore, after May 1920, were no longer merely those of the International Company, which at a pinch might have been bought out and absorbed into RCA. They now included the Westinghouse Company, General Electric's major competitor. And Westinghouse, it was now evident, had acquired those rights and others not for their nuisance value but in order to establish for itself a position in radio equivalent to that which GE had acquired through its creation of RCA. Alexanderson's device, or Weagant's, or Goldsmith's, were ingenious attempts to invent around the problem. At least they might give RCA something that it could use, something that was not a blatant infringement of the Fessenden patent. But, at best, a long and expensive legal battle seemed in prospect.

* * *

Westinghouse had first become involved in radio during the war. As a lamp manufacturer, it had facilities adaptable to vacuum tube production and by 1918 it was turning out tubes at a respectable rate, though not on the scale of GE or Western Electric. This experience, and its tube production facilities, were part of the heritage it carried into the postwar period. Westinghouse had also, however, designed and manufactured low-powered radiotelephone transmitters and receivers for the Signal Corps and the Navy. On the one hand this had sensitized its senior executives to the possibilities of radio; on the other it had given the

[53] Young Papers, Additional Papers, Box 3, Minutes of RCA Technical Committee, 25 August 1920. A. N. Goldsmith also produced a "modulator" designed to circumvent the Fessenden patent. See Elmer E. Bucher, "A History of RCA" (Sarnoff Research Center Archives), p. 245.

[54] Young Papers, Copybook 802, Young to Reed, 22 January 1923.

company a small cadre of engineers experienced in the design and production of radio equipment. Among these, Frank Conrad was conspicuous. He had done much of the design work for the Westinghouse radiotelephone sets; he had served as the company's representative on Hooper's tube development committee; and he was a technically proficient radio amateur.[55]

It may be that these considerations would not have been enough in themselves to induce Westinghouse to continue its work in radio after the war rather than revert to its normal product line of incandescent light bulbs and heavy electrical equipment. There were, however, two other factors at work. One was the traditional rivalry between Westinghouse and General Electric—between Pittsburgh and Schenectady, if you will. Originally a rivalry between the proponents of alternating currents and those who, in the Edison tradition, favored direct current, this had been transformed over the years into a generalized competition for leadership in the electrical manufacturing industry.[56] It is hard to say how much weight this spirit of emulation carried in high-level corporate policy formation, but that it affected the engineering personnel of both organizations is beyond question. And when, in 1919-1920, General Electric made its important and well-publicized moves into radio, committing itself not only to continued production of radio equipment but also to the creation and support of a radio operating company—indeed, what looked as if it might be the nation's only operating company—the lesson was not lost on Westinghouse.

The second factor at work—and here one has a certain feeling of inevitability—was the intervention of the Navy's Bureau of Engineering, and specifically of Lieutenant Commander Hooper. No one had been more committed than Hooper to the idea that American radio, in confronting foreign governments and corporations, had to speak with a single voice. To compete effectively with the cables and to negotiate effectively with foreign authorities, a monopoly was essential. There was only enough traffic for one company; and if two companies competed for business,

[55] Conrad had been in charge of the development of the SCR-69 transmitter and SCR-70 receiver for the Army Signal Corps. He had, incidentally, little formal education, having left school after the seventh grade. Most of his skills seem to have been acquired as a bench hand at Westinghouse; but his intuitive sense for radio design and problem-solving must have had other origins. See Barnouw, *History*, 1: 66-67. For his membership on the Navy's tube development committee, see Hooper Papers, Box 40, manuscript "History of Radio."

[56] Harold C. Passer, *The Electrical Manufacturers, 1875-1900: A Study of Competition, Entrepreneurship, Technical Change, and Economic Growth* (Cambridge, Mass., 1953), pp. 164-75.

foreign governments and corporations, almost always enjoying monopoly power themselves, would find it easy to play one off against the other. This did not mean, however, that Hooper was content to see a single firm dominate radio within the United States. Here, of course, he faced a dilemma—as did American public opinion in general. Monopoly might be essential to maintain American interests abroad, but domestically it was unacceptable. Broadcasting was to pose that problem in particularly acute form, since by 1922 RCA and its affiliated companies controlled the basic patents on vacuum tubes and on almost all other elements of radio technology. But it was not broadcasting that bothered Hooper in 1919-1920. It was the Navy's growing dependence on General Electric as its sole supplier of radio equipment. Western Electric's interest in bidding on government radio contracts was rapidly evaporating. And, because of the distribution of manufacturing facilities and of patents, no other firm was in a position to bid.[57] It was not prices that troubled Hooper so much as what he saw as GE's growing disposition to dictate what kind of radio equipment the Navy should have.[58] A monopoly in radio in America's dealings with the rest of the world Hooper would endorse; indeed he had done much to bring it about. But, where Navy supply was concerned, he wanted competition—or at least an alternative source to General Electric.[59]

Never one to wring his hands about a problem without taking action, Hooper approached Westinghouse. He talked with the company's man in Washington, a Mr. Hyler. Shortly thereafter two Westinghouse lawyers called on him and serious discussions began. In the meantime there had been important developments in Pittsburgh.

The great weakness of the Westinghouse Company was that it had no defensible patent position in radio. During the war this had been of no significance, since the government protected its suppliers against prose-

[57] See Hooper Papers, Box 37. Hooper claimed personal credit for inducing Westinghouse to enter the postwar manufacture of radio equipment, citing as his reasons the facts that Western Electric had stopped bidding on government contracts and Hooper did not like the idea of GE being the sole supplier.

[58] Hooper Papers, Box 3, Hooper to O. D. Young, 17 January 1921.

[59] Compare Clark Radio Collection, Cl. 5, Book 8, "The Navy's Position in the Radio Patent Field," 10 June 1929; and Hooper Papers, Box 3, Hooper to O. D. Young, 2 December 1920: "The opinion of the best minds in radio, both Government and commercial, appears to be that operation for the moment can only be efficient as a monopoly, either Governmental or commercial. This has always been my conviction. As regards sales, manufacturing, and engineering, my conviction is that a combination would be detrimental to the best interests of the public."

cution for infringement. Peacetime manufacture, whether for the government or for civilian buyers, was quite another matter. And if Westinghouse had any thought of spawning an operating company, a defensible set of patents for both transmitting and receiving was vital. In 1919-1920 this meant—for Westinghouse as it had for Marconi—access to continuous wave technology. Westinghouse originally had no such access.[60] That did not mean that, by shrewd trading and fast footwork, access could not be obtained.

Fessenden had parted with his backers, Given and Walker, in 1911, in a flurry of mutual recriminations. By that date the two Pittsburgh businessmen had invested some $2 million in Fessenden's experiments with little but hopes and expectations to show for it. NESCO's successor firm, the International Radio Telegraph Company, did some business in radio manufacture for the government during the war but found itself high and dry when that market disappeared.[61] Its only assets of potential value were the patents on Fessenden's inventions, particularly the heterodyne receiving system and the rotary spark gap. Given bought out Walker's interest in the company during the war and died shortly thereafter, leaving his estate to members of his family. The company's business affairs, to the extent that it had any, were in the hands of its president, S. M. Kintner. Kintner, formerly Fessenden's student, had succeeded him in the chair of Electrical Engineering at Western University (later the University of Pittsburgh); he had been appointed one of NESCO's receivers when the company went into bankruptcy; and in 1920, by a happy coincidence, he was manager of Westinghouse's research department.[62]

Westinghouse was perhaps fortunate to find these assets right under

[60] Westinghouse apparently held a license for the manufacture of audions from de Forest, under the "personal, non-transferable" rights he had retained when selling his patents to AT&T, and had turned this license over to a subsidiary, the Westinghouse Lamp Works. For the later history of this curious permit, see Gleason Archer, *History of Radio to 1926* (New York, 1938), p. 356.

[61] After the bankruptcy of NESCO, title to Fessenden's patents passed to the International Signal Company; on 28 November 1917 the title of that company was changed to the International Radio Telegraph Company. See Fessenden Papers, 1140-84, G. K. Woodworth to Fessenden, 12 August 1921.

[62] Archer, *History*, p. 201; H. P. Davis, "The Early History of Broadcasting in the United States," in Harvard Business School, *The Radio Industry, the Story of Its Development, as told by Leaders of the Industry* (Chicago and New York, 1928), pp. 189-225, at p. 195; Fessenden Papers, 1140-84, Woodworth to Fessenden, 12 August 1921; Helen M. Fessenden, *Fessenden: Builder of Tomorrows* (New York, 1940), pp. 115, 189, and 191.

its nose in Pittsburgh, but acquiring them proved neither straightforward nor inexpensive. A tentative offer to buy out the stockholders in the International Company was rejected. Instead, on 22 May 1920, agreement was reached on the formation of a new company with a title just sufficiently different to be legally distinguishable: The International Radio Telegraph Company. This firm would be jointly owned by Westinghouse and by the stockholders of the original company; it would hold the Fessenden patents; and, on the model of RCA and GE, it would be the radio operating company while Westinghouse did the manufacturing. Westinghouse undertook to contribute $2.5 million in cash to the new organization, and was to receive in exchange 125,000 shares of no par value common stock. The former stockholders in the old company exchanged their shares for 12,500 shares of 7 percent preferred stock in the new one, with a par value of $1.25 million, plus 125,000 shares of no par value common. The intent clearly was that the preferred stock should represent the old company's radio patents and physical assets. It can hardly be denied that its stockholders did well for themselves. On the other hand, Westinghouse now had a toehold in the radio business; it had the nucleus of an operating organization to match General Electric's RCA; and it had not really committed itself very deeply, for its $2.5 million contribution did not have to be provided immediately. The agreement gave Westinghouse two years to make its payment, with the stock to be kept in escrow in the meantime.[63] As events turned out, that transfer was never made.

The story now reaches a stage where a grasp of the precise sequence of events is essential. Kintner, now president of International Radio, left for Europe early in the summer of 1920. His purpose was to negotiate traffic agreements. There was, after all, little point in building transmitting and receiving stations in the United States if they had no stations in other countries with which to exchange traffic. That implied contracts specifying hours of working, rates charged, division of revenue, and so on; and such contracts were usually accompanied by agreements for the interchange of patents, for the good technical reason that, when two stations worked in partnership, it was in the interest of each that the other be able to handle the traffic as rapidly and accurately as possible. These traffic agreements were the ligaments that interconnected national telecommunications entities—Marconi in Britain, for example, with Telefunken in Germany, the Compagnie Générale de T.S.F. in France, and RCA in the United States. They were functional substitutes for the fra-

[63] Details of the transaction may be found in Archer, *History*, p. 193. Compare Schubert, *Electric Word*, p. 182.

ternal compacts that, in the original Marconi conception of a worldwide system, would have linked together the several Marconi companies. Without traffic agreements no company could play a role in the international communications system. It might build up a ship-to-shore network; it might even try to operate a national overland network, as the Federal Company had tried to do before the war and as Emil Simon's Intercity Radio Company was to attempt after the war.[64] But it could not function in international radio.

Meanwhile, in the United States, in anticipation of a successful conclusion to Kintner's mission, further steps were taken to strengthen the position of International Radio. A serious limitation at this date was that neither Westinghouse nor International held rights to a continuous wave transmitter. The nearest approximation to such a device in their arsenal of patents was Fessenden's rotary spark gap. This, like the Marconi timed spark and the Telefunken quenched spark, was about as close as one could get to a true continuous wave generator and still stay within the bounds of spark technology, but it was unmistakably an obsolete device in 1920. Westinghouse had no tube patents and no radiofrequency alternator.

The situation could have been very difficult, indeed impossible, had it not been for the intervention of Stanford Hooper. Hooper's primary concern, as we have seen, was to build up Westinghouse as a supplier of radio equipment to the government. This was hardly feasible if Westinghouse lacked the necessary patents. The Navy held patents: specifically, it held all the Federal Company's arc patents, purchased in 1918, and it held all the Telefunken radio patents, expropriated in 1917 as enemy property and subsequently transferred to Navy custody by the Alien Property Custodian. Some of these might be valuable, though Hooper personally did not think many of them were. The Federal arc patents were another matter: a license under these patents would give Westinghouse and International access to a tested and proven continuous wave transmitter. Hooper offered the whole collection, both Telefunken and Federal, to Westinghouse, under nonexclusive licenses, in return for a license to the Navy under the Fessenden heterodyne patent and a payment of $250,000. The demand for a cash payment was promptly rejected,

---

[64] Emil Simon, formerly one of de Forest's co-workers, had made a small fortune during the war by manufacturing radio equipment for the government. Information on his postwar venture, the Intercity Radio Company, is hard to come by, but the Simon Papers at the Bancroft Library provide useful insights. For the later history of speculation in Intercity securities (not involving Emil Simon personally) see "Annals of Crime," *The New Yorker*, 11 April 1959.

and Hooper did not insist upon it. What he wanted was, for the Navy, a license to the heterodyne—as he later wrote, "the heterodyne was worth a lot more than the arc patents"—and, for Westinghouse, a set of patents that would give it a fighting chance for survival in radio manufacturing.[65]

Hooper was not at this point interested in the creation of a second American radio operating company. He is on record as stating on several occasions his belief that, at least in North Atlantic traffic, there was room for only one such entity.[66] It was only later that he began to dream of a union between Westinghouse and the Federal Company, and then it was radio communications with China and the west coast of South America that he had in mind, not the European theater.[67] Nevertheless, his offer of a license under the Federal arc patents removed what otherwise would have been an insurmountable obstacle to the strategy that Westinghouse and International were following. The offer was formally accepted on 5 August 1920. This meant that, if Kintner returned from Europe with traffic agreements in his briefcase, International Radio would be technically able to implement them. It had a defensible position in continuous wave radio—a transmitter that many thought at least as good as GE's alternator, and a receiving system better than anything at RCA's disposal.

Kintner, however, did not return successful. In fact, he returned without any traffic agreements at all, the closest approximation being a vague promise from Telefunken that, if International were to build a radio station in the United States, Telefunken would give it the same commercial privileges as it gave to RCA.[68] In Britain and France he did not

[65] The quotation is from Hooper Papers, Box 37; compare Clark, "Radio in Peace and War," chap. 12, pp. 85-86. On 19 March 1921 rights under the arc patents were also granted to the Federal Company, from which they had originally been bought, to facilitate that company's development of its concessions for radio communications in China. See Hooper Papers, Box 3, copy of contract dated 19 March 1921 between the Federal Telegraph Company and the secretary of the navy.

[66] See for example Clark Radio Collection, Cl. 100, v. 2, Hooper's comment on Clark's RCA chapter, "page 86 about line 20," and Hooper Papers, Box 37; Clark, "Radio in Peace and War," chap. 12, p. 87.

[67] Hooper Papers, Box 4, memorandum for National Radio Conference: "If the Federal Company cannot stand on its feet alone, I am in favor of asking the G.E. Co. to divorce the Westinghouse Company from the affiliated organization and getting the Westinghouse and Federal together on the Pacific Coast and in the Far East and West Coast of South America, as a rival to the Radio Corporation." See also Hooper Papers, Box 3, reply to Young's letter to Secretary of the Navy Denby dated 12/22/21.

[68] Archer, History, p. 196.

get even that. His failure was not for want of trying. The fact of the matter was that he had been anticipated by E. J. Nally, who had spent the previous six months negotiating traffic agreements across most of Europe for RCA. Nally, of course, had a large advantage. A traffic agreement with Marconi had been part of the deal by which American Marconi was purchased and RCA created, and that served as the entering wedge. Even so, Nally had found it no easy task. The French government and the Compagnie Générale were particularly hard to convince. They were quite content to pass traffic to the Navy's stations in the United States, as they had done during the war, and it took some argument—for which Hooper took the credit—to convince them that that was no longer permissible under American law and they would have to deal with RCA in future.[69] The Germans, too, had been reluctant to commit themselves.[70] By the end of 1920, however, both Telefunken and the Compagnie Générale had signed and these, with Marconi in Britain, were the critical ones. Others followed: traffic agreements were signed with the Japanese government in 1920, with Poland in 1921, with Sweden the year after that. After difficult negotiations, the so-called AEFG consortium, providing for joint action by Britain, France, Germany, and the United States in developing radio communications with South America, became a reality in 1921. And in 1922 RCA and the Federal Company of California made common cause, through the newly created Federal Telegraph Company of Delaware, to construct radio stations in China; a traffic agreement with the Chinese government followed in 1923.[71]

Haggling over the details of these agreements was to be expected, as was a reluctance on the part of the Europeans to tie themselves to an American firm whose technical and administrative competence was unknown. Nevertheless, RCA was offering them something that most of them had never had before: direct telegraphic connection with North America, and at rates per word lower, at first, than the cables.[72] There

[69] Hooper Papers, Box 3, Hooper to Young, 20 September 1921: "One of the biggest rocks under R.C.'s foundation was the arrangement in Paris last year and I reserve [deserve?] absolute credit for putting that across, as the chances were against it when I arrived."

[70] Young Papers, Copybook 802, Young to Bullard, 13 July 1920.

[71] The AEFG negotiations are described in Archer, *History*, pp. 227-39 and, more soberly, in Case and Case, *Young*, pp. 237-45. For George H. Clark's critique of Archer's treatment, see Clark Radio Collection, Cl. 5, Book 8, comments on Archer's *History*. The development of radio links with China still awaits adequate attention from historians, but see Schubert, *Electric Word*, pp. 257-65.

[72] Young Papers, Additional Papers, Box 3, report of RCA technical committee,

were also the advantages of patent exchange, for traffic agreements were typically accompanied by provision for cross-licensing. Each firm or government bureau was guaranteed the exclusive right to use the other firm's patents in its territory. RCA, in other words, was assured of access to any advances in radio technology made in other countries by the organizations with which it signed cross-licensing agreements; and they likewise were guaranteed access to any advances made by RCA and its affiliated companies.[73]

What had been created, in short, was an international cartel, held together by the cross-licensing of patents and by traffic agreements. The problem for Westinghouse and International Radio lay in the exclusive features of these contracts, which served as effective barriers to the entry of new firms. In this sense the international traffic agreements that RCA signed in 1919-1920 were precisely analogous to the cross-licensing agreements signed domestically with AT&T and United Fruit. They served to delimit territory and to close the market to outsiders.

* * *

At this point Westinghouse, barred from international radio by the cross-licensing agreements, seemed to have few options left; the sensible course of action was surely to call it quits as far as radio was concerned. It had not, after all, spent very much on the venture, beyond its executives' time, for of the $2.5 million due International only some $300,000 had so far been paid.[74] Little was at stake beyond corporate *amour-propre*. Owen Young certainly thought the time had come for a sensible trade, and on 2 September 1920 he phoned his old friend, Guy Tripp, president of Westinghouse, and offered to take International Radio off his hands. The terms were reasonable: 700,000 shares of RCA common and the same number of RCA preferred. In return RCA would get most of In-

---

25 August 1920. In 1920 the cable rate for regular traffic between New York and Britain was 25 cents per word; RCA's initial rate was 17 cents.

[73] The agreements are conveniently reprinted in the FTC *Report* (1923); their main provisions are summarized in Maclaurin, *Invention*, pp. 108-109; and certain of the more important are again reprinted in *FCC Hearings*, pp. 1187-94 and 1220-32. Cross-licensing was extended to the Philips Company in Holland in 1925, with Philips being granted exclusive rights to RCA patents in Holland, Czechoslovakia, Denmark, Estonia, Finland, Latvia, Lithuania, Norway, Sweden, and Switzerland. See *FTC Report* (1923), pp. 51-59 and Maclaurin, *Invention*, p. 107.

[74] Case and Case, *Young*, p. 221.

ternational's assets, including specifically the $2.2 million still due from Westinghouse, and a general cross-licensing of patents would follow.[75]

Tripp refused the offer, which was surprising enough. But even more surprising, Westinghouse moved aggressively ahead to increase its investment in radio and strengthen still further its portfolio of radio patents—again in receiver circuitry. The new acquisitions were significant: Edwin Armstrong's feedback and superheterodyne patents. Before the war Armstrong, then a graduate student in Pupin's Columbia University laboratory, had made a name for himself in radio by inventing the feedback circuit, the basis of all vacuum tube oscillators and therefore of all tube transmitters and most receivers. He had a patent on that circuit (No. 1,113,149) that was potentially of great value, although its validity was challenged by competing claims filed by de Forest, Langmuir, and Meissner of the Telefunken Company. During the war Armstrong had served in the U.S. Army in Europe and there, according to his later account, he had been forcibly struck by the difficulty of building vacuum tube amplifiers suitable for use in intercepting enemy radio signals on the short wavelengths the Germans were then using for field communications.[76] These signals were typically weak and required amplification before they could be detected. The difficulty was not so much one of securing enough gain: that could always be done by adding more amplifying stages in the receiver ahead of the detector. And at long wavelengths that would have been enough. But at shorter wavelengths, with the triode tubes then in use, high-gain radiofrequency amplifiers were unstable: they would break into self-oscillation when, as a result of stray coupling or unwanted feedback, some of the output signal leaked back into the input. Armstrong's solution was to convert all incoming signals down to a fixed intermediate frequency by heterodyning them against a tunable local oscillator in the receiver—hence "superheterodyne"—and then obtain the needed sensitivity and selectivity by amplifying and fil-

[75] Ibid. Young had earlier dictated two letters to Tripp which he did not send. Knowledge of what was said during the telephone conversation rests on Young's handwritten notes on the back of one of those letters. Young and Tripp had known each other earlier, when both worked for Stone & Webster, the public utility management and consulting firm.

[76] Archer, *History*, p. 143; Bucher, "History of RCA," p. 247. For a different version of the origin of the superheterodyne, linking it to attempts to locate enemy aircraft by the spark discharges from their engines, see Armstrong, "Vagaries and Elusiveness of Invention," *Electrical Engineering* (April 1943), 150, quoted by Maclaurin, *Invention*, p. 123. See also Lawrence Lessing, *Man of High Fidelity: Edwin Howard Armstrong* (New York, 1956), especially chaps. 6, 7, and 8.

tering the signal at that intermediate frequency, where stability was more easily secured.

It was an elegant solution to the problem and proved basic to all later advances in receiver design. True, its invention was also claimed by a French designer named Lucien Levy, who could adduce considerable circumstantial evidence to show that Armstrong had learned the idea from him; but that was a matter the courts would decide.[77] In the United States in the summer of 1920 Edwin Armstrong was accepted as the inventor, and on 8 June the patent (No. 1,342,885) was issued to him. He had earlier, while the patent was pending, granted a number of licenses to the superheterodyne to help finance the legal costs of defending his feedback patent, and one of these had been acquired by the International Company on 12 May 1920.[78] David Sarnoff had witnessed a demonstration in February and been greatly impressed, but General Electric had not followed the matter up with vigor.[79] On 5 October Westinghouse took an option on the feedback patent, on the superheterodyne, and also on a number of Pupin's patents on tuned radiofrequency circuits. And on 4 November the option was exercised. Armstrong received $335,000 in cash, the amount to be increased by $200,000 if he won his interference proceedings on the feedback patent against de Forest.

General Electric could have acquired licenses, perhaps even exclusive licenses, to the feedback and superheterodyne patents long before November, if it had chosen to do so. Armstrong was far from rich; his legal expenses were heavy; he was in no position to reject a generous offer

---

[77] Emil Simon, visiting Paris in 1919, was invited to visit the Ecole Militaire, where he met Lucien Levy, a civilian radio engineer. Levy had invented a superheterodyne amplifier and offered Simon the U.S. rights for $5,000. Simon told him that Armstrong claimed to have made the invention and had filed a patent application on it, to which Levy replied that Armstrong had stolen the invention from him when attached to the laboratory as a U.S. Signal Corps captain. Simon was skeptical and refused to buy the rights. Several years later AT&T bought Levy's U.S. patent application for $20,000. Of the nine claims in Armstrong's patent, filed 8 February 1919 and issued 8 June 1920, all were lost in later interference proceedings in the U.S. Patent Office; one claim went to Alexanderson of GE, one to Kendall of Bell Laboratories, and the remaining seven to Levy, backed by AT&T. See Simon Papers (Bancroft Library) and Haraden Pratt Papers (Bancroft Library), Lloyd Espenschied to Pratt, 20 September 1954; Espenschied to B. F. Miessner, 23 March 1963; E.J. Simon to Pratt, 20 April 1963; Pratt to "Pat," 29 March 1962; and Pratt to R. P. Multhauf, 19 March 1963. Compare, for a different version, Lessing, *Man of High Fidelity*, pp. 92-93.

[78] Archer, *Big Business*, p. 91.

[79] Bucher, "History of RCA," p. 247.

and might well have welcomed the support of GE in his bitter and long-continued litigation against de Forest.[80] Instead of which, by the end of 1920 GE found itself confronting a Westinghouse Company securely in possession of a group of patents fundamental to the future development of radio technology. And this was an organization that, less than a year before, had not held a single radio patent of any significance. One has to admire the speed with which Westinghouse had moved, and the confidence with which it had picked out, among the thousands of radio patents then available, the ones that were to prove controlling in the long run. A question can be raised, indeed, as to whether GE and RCA, even at this early date, were not afflicted by a certain technological complacency where radio was concerned. Alexanderson, Weagant, and Goldsmith made up a first-class engineering team, and they were backed up by Langmuir and the GE Research Laboratory. But this very quality may have encouraged a kind of casualness toward innovations originating outside the GE/RCA community—the "not invented here" syndrome.[81] Designing and constructing the new alternator-equipped transmitting stations, in any case, took up much of their time and energy. Top management—Young, Nally, Davis—was preoccupied with RCA's "foreign relations" during these critical months, particularly the all-important matter of traffic agreements. And Sarnoff as commercial manager was primarily concerned with the development of traffic.[82] Perhaps what it comes down

[80] Litigation over the feedback patent was not finally decided by the Supreme Court until 1934.

[81] This judgment may be unduly harsh. An alternative interpretation would be that RCA gambled on de Forest in the interference proceedings. Since AT&T had purchased de Forest's patents and since RCA was cross-licensed with AT&T, a victory for de Forest would give RCA rights under the oscillator patent. This, however, was quite a gamble; and such an interpretation does not explain RCA's failure to acquire rights under the Fessenden heterodyne patent. For a contemporary analysis see Clark Radio Collection, Cl. 5, Box 66, Ira Adams to A. N. Goldsmith, 5 June 1920.

[82] Sarnoff often acted as chairman of RCA's technical committee in Nally's absence, but the primary duties of this committee appear to have been drawing up plans for rebuilding the high-power stations and other contruction projects. See Bucher, "History of RCA," p. 244. It had been suggested in March 1921 that both Sarnoff and A. G. Davis be made vice-presidents of RCA. Rice objected to the appointment of Davis, presumably because of its effect on General Electric. Hence no action was taken and Sarnoff's election was postponed. Young urged that he be elected to the office without further delay on the ground that it was "almost necessary to the organization." See Young Papers, Copybook 802, Young to John W. Griggs, 23 March 1921. The best available biography of Sarnoff, from an author who worked with him, is Carl Dreher, Sarnoff: An American Success (New York, 1970).

to is the fact that, in acquiring licenses to such novelties as the super-heterodyne, RCA and GE thought that there was no urgency. Westing-house knew that there was.

But on what commercial prospects was Westinghouse basing this pro-gram of patent acquisition? Not, obviously, on hopes for a position in telecommunications: those had disappeared with Kintner's abortive mis-sion to Europe. And not, presumably, merely the desire to build up a little more equity and a stronger bargaining position, secure in the knowl-edge that RCA would eventually have to buy International out. The answer was broadcasting.

*   *   *

Precisely when broadcasting began, and who was the first to "broad-cast" radio signals deliberately, are questions one would gladly be spared from answering. There have been many contenders for the honor. Prob-ably Fessenden has as good a claim as any, with his broadcast of speech and music on Christmas Eve, 1906, to an audience of previously alerted amateurs and shipboard operators.[83] But if Fessenden was the first to broadcast for purposes of entertainment, it was not by a wide margin. In September 1907 a number of Lee de Forest's radiotelephone sets were installed on ships of the Navy's "Great White Fleet" on its goodwill cruise around the world, and, although the equipment saw little official service on the voyage, legend has it that one H. J. Meneratti, chief elec-trician's mate on the U.S.S. *Ohio*, earned his place in history by playing phonograph records over the air to amuse his buddies, thus becoming the world's first disc jockey.[84] In January 1910 de Forest certainly broad-cast from the Metropolitan Opera House in New York, unfortunately on the same frequency as a busy United Wireless spark station in the same city.[85] And even earlier he had set up a transmitter in "Telhar-monium Hall," at Thirty-second Street and Broadway, from which he broadcast synthetic music from the Cahill brothers' wonderful telhar-monium machine.[86] Charles "Doc" Herrold, in San Jose, California, is known to have transmitted regular entertainment programs between 1909 and 1917, using a low-powered arc transmitter; and Harold J. Power

[83] H. M. Fessenden, *Fessenden*, pp. 153-54.

[84] L. S. Howeth, *History of Communications-Electronics in the United States Navy* (Washington, D.C., 1963), p. 171.

[85] Thorn Mayes, "DeForest Radio Telephone Companies, 1907-1920" (Mimeo, 1982); Georgette Carneal, *A Conqueror of Space: An Authorized Biography of Lee De Forest* (New York, 1930), pp. 231-32.

[86] Society of Wireless Pioneers, *Sparks Journal* 5 (1982), 6; Carneal, *De Forest*, pp. 207-209.

and his friends were similarly occupied at the Tufts College amateur station near Boston.[87] Even the Navy has a special claim to priority—ironically, for broadcasting the first speeches by politicians, with Congressman Fitzgerald of Ohio and Senator Frelinghausen of New Jersey, supported by the Marine Band, addressing an invisible audience from the Anacostia station on Saturday nights.[88] And there may well be others with claims to recognition at least as good as those usually mentioned.

All these feats predated Frank Conrad's transmissions in Pittsburgh in 1921, which are conventionally taken as representing the birth of broadcasting in the United States. Whether this conventional view is acceptable or not depends partly on semantics, partly on an understanding of what was distinctive and novel about Conrad's work. The word "broadcasting," as applied to radio, seems to have originated with the Navy, where it meant simply transmitting a message to several receiving stations without requiring any of them to acknowledge receipt; the purpose of the practice was to ensure that the ships receiving the message did not disclose their positions to enemy direction-finding stations.[89] In that sense broadcasting was almost as old as radio itself. But Conrad in 1921, and those like Fessenden and de Forest who preceded him, were clearly doing more than that.

The bare facts are not in dispute. Conrad, like all other amateur operators, had been required to suspend operations for the duration of the war emergency. He got permission to put his amateur station back on the air in April 1920, and promptly did so, using the call sign of 8XK. It was a sophisticated amateur station, as one would expect, for Conrad was a skilled engineer; he had been experimenting with radiotelephony for several years; and he had carried much of the responsibility for the radiotelephone design and development work at Westinghouse during the war. He had also been, as already noted, a member of Hooper's tube development board. In short, he probably knew as much about the practical use of vacuum tubes in radiotelephony as anyone in the country at the time. It was hardly surprising, therefore, that when Conrad resumed his ham radio hobby he set up his station for radiotelephony. This was something of a novelty in amateur radio circles, where spark telegraphy

[87] Jane Morgan, *Electronics in the West: The First Fifty Years* (Palo Alto, Calif., 1967), pp. 29-31; Barnouw, *History*, 1:34-36.

[88] Hooper Papers, Box 40, manuscript "History of Radio." Hooper gives no date for these broadcasts but states explicitly that they antedated Conrad's broadcasts in Pittsburgh. Navy broadcasts were suspended in 1922 on orders from the secretary of the navy. See Hooper Papers, Box 4, Hooper to E. H. Quinlan, 27 November 1922.

[89] Clark Radio Collection, Cl. 100, v. 1, p. 64.

was still the norm, and his transmissions attracted much attention in the Pittsburgh area.

So far there is nothing particularly remarkable about the story. Conrad had greater expertise and better facilities than most others who shared the same hobby, but it was predictable that, once they received permission to resume operation, progressive amateurs would start experimenting with continuous wave transmitters and that they would try voice transmissions as well as the more traditional Morse code telegraphy.[90] Many of them, after all, had been radio operators in the armed forces during the war and had learned about tube transmitters and radiotelephony there. In those days licensed amateurs did not wait for manufacturers to introduce them to new techniques; if they wanted to try something, they built the equipment themselves.

Something new happened, however, when Conrad began transmitting music, played on phonograph records, from his amateur station; when, tired of answering individual inquiries, he announced that he would be transmitting on a regular schedule; when the local merchant who provided him with his records began supplying them free in return for mention of his name on the air; and when in September 1920 an enterprising Pittsburgh department store began advertising and selling simple receivers with which Conrad's broadcasts could be heard. And even more clearly was something novel taking place when the Westinghouse Company, responding to local enthusiasm and newspaper publicity, decided to erect a station at its plant in East Pittsburgh and operate it every night with an advertised program, looking to the sale of receivers and goodwill for the Westinghouse name to justify the small expense. That station, licensed on 27 October 1920 as KDKA, was the immediate offspring of the ham radio station Frank Conrad had put in operation in his garage in April of the same year. But between April and October a major innovation had taken place.[91]

But what kind of innovation was it? Not, certainly, a technical one. KDKA, with its six 50 watt tubes, was indeed using "state of the art" technology, but nothing that had not been familiar to radio engineers

[90] Barnouw, *History*, 1:61-62, mentions particularly the activities of Professor Earle M. Terry with amateur station 9XM at the University of Wisconsin; William E. Scripps, publisher of the Detroit *News*, with station 8MK; Charles Herrold in San Jose; Harold Power of Medford; Fred Christian (6ADZ) in Hollywood; and several others.

[91] The events leading up to the establishment of KDKA have often been described. Convenient sources are Davis in Harvard Business School, *Radio Industry*, pp. 189-225; Archer, *History*, pp. 199-210; and Barnouw, *History*, 1: 64-74.

for several years. It was no technical breakthrough that created the broad-casting industry almost overnight. What made the KDKA experiment significant—and the experience of station 8MK in the offices of the Detroit *News* was very similar—was its disclosure that a market existed and that it could be reached with a relatively small investment. That market was, initially, the community of radio amateurs, individuals who knew how to string up a wire antenna and tune a crystal set and were delighted to share those skills with their friends, families, and neighbors. But beyond those amateurs was a vast potential audience with an apparently insatiable appetite for news and music whose existence had previously been almost totally unsuspected.

Radio broadcasting, it has been said, captured the popular imagination. It brought news to a news-hungry public, with an oral directness and immediacy that the printed newspaper lacked. And it provided music that was, seemingly, free, with a quality of reproduction at least as good as the phonograph and with none of the nuisance of winding a crank and changing a record every few minutes—or of being restricted to your private collection of recordings.[92] But radio broadcasting also opened up opportunities for profit, and if all the ways in which those profits could be reaped were not immediately apparent, one at least was. To receive radio broadcasts you had to have a receiver. At first it could be very simple: a crystal, a coil wound (often) on an empty Quaker Oats container, and a pair of headphones. That was enough to get started. But as the number of stations increased, so did the need for selectivity and fine tuning. As the habit of "listening in" caught on, so did the desire to hear the weaker or more distant stations, and that meant a requirement for sensitivity. As radio became part of family life, with utilization no longer confined to those conventionally recognized as technically sophisticated, ease of operation became an important consideration, and after a while the tricky "cat's whisker" and crystal were no longer acceptable. Consumers moved up first to regenerative receivers and then, when the squeals of a poorly adjusted "regen" became intolerable, to the more sophisticated "neutrodyne" or "superhet."

It looked, in short, as if money might be made from this newly discovered market, particularly by a corporation with a strong patent position in receiver circuitry. A conspicuous feature of this social innovation, indeed, was the way in which almost from the beginning it was integrated into the price system and the market economy (in contrast to the experience in other countries).[93] Consider, to underline the point, what Frank

[92] Schubert, *Electric Word*, pp. 213-14.

[93] Compare, for example, Asa Briggs, *The History of Broadcasting in the United Kingdom, Vol. 1, The Birth of Broadcasting* (London, 1961).

Conrad did with his amateur license. He played music; he advertised the firm that lent him records, and he engaged in one-way transmissions to listeners he could not identify—that is, he broadcast. For any one of these activities he would, today, have his license suspended by the Federal Communications Commission.[94] This is because there has come into existence an elaborate code of regulations designed precisely to insulate the operation of amateur radio stations from the commercial market. But it was not so in Conrad's day. And that is why it was so easy for amateur stations to make the transition to commercial broadcasting. There was no clear boundary, no perimeter beyond which the enterprising amateur might not go. There was, certainly, the technical challenge, the sheer pleasure of exercising a new skill. But there was also, for some, the knowledge of a market opportunity and the freedom to respond to it.[95]

In Conrad's case what made the difference was the intervention of his employer, the Westinghouse Company. H. P. Davis, vice-president of Westinghouse, saw the newspaper publicity about Conrad's broadcasts and took the initiative in establishing a station explicitly intended for broadcasting at the company's plant in East Pittsburgh. The move has to be seen in the context of the situation in which Westinghouse found itself at that time. It had gone to some expense to acquire the Fessenden and Armstrong-Pupin patents. It had committed itself to purchasing a major interest in International Radio. But it had failed completely to break into telecommunications. The exclusive traffic agreements that served as RCA's defensive ramparts remained intact. The blunt fact of the matter was that, in the early fall of 1920, Westinghouse had failed to establish itself in the only field of radio communications that then offered prospects of commercial profitability. Other options had been explored: demonstrations of radiotelephony for the Fall River Line and for the New York, New Haven and Hartford Railroad, for example. But without success.[96]

In the circumstances, one may admire the promptness and energy with which Westinghouse moved to exploit the first hints of a commercial market for broadcasting and yet consider some reaction of that kind

[94] One-way transmissions are permitted under certain specified conditions— for example, for experimental purposes, for emergency communications, and for the transmission of code practice and official bulletins consisting solely of subject matter having direct interest to the amateur radio service as such. See Federal Communications Commission, *Rules and Regulations*, Pt. 97, Amateur Radio Service, paras. 97.89 and 97.91 (Washington, D.C., various dates).

[95] For a partial list of amateur stations that made the transition to commercial broadcasting, see Barnouw, *History*, 1:82.

[96] Ibid., p. 66.

highly predictable. H. P. Davis later made the connection explicit. International Radio, he said, owned a handful of ship-to-shore stations and Westinghouse had already given some thought to equipping them for radiotelephony and instituting a regular broadcast news service—the intended audience being ships at sea. Lack of interest on the part of the shipping lines killed that idea. Westinghouse, however, continued to search for a suitable market for the technical capability it knew it possessed. "A large sum of money," said Davis, "expended for control of the International Radio Telegraph Company emphasized in our minds the necessity for developing our new acquisition into a service which would broaden, popularize, and commercialize radio to a greater extent than existed at that time, in order to earn some return on this investment as well as keep the radio organization together."[97] This was the primary corporate motive for the move into broadcasting. Westinghouse had invested funds and prestige in radio technology; broadcasting held out the prospect of recouping the investment.

* * *

Word of what Westinghouse was up to got a mixed reception in Schenectady and New York. There were those, of course, who dismissed broadcasting as a fad that would soon pass: President Rice of General Electric, for example, thought that a solidly based corporation like GE should stay out of such frivolity.[98] Young and Nally still thought of RCA as first and foremost an overseas communications company. But there were others, at somewhat lower levels in the organization, who felt with some justice that GE had once again been caught napping by its more agile Pittsburgh rival. W. C. White of the GE Research Laboratory later recalled that he and his co-workers were "amazed at our blindness. . . . We had everything except the idea."[99] And at AT&T, too, pressures mounted for a move into broadcasting.

The fact of the matter was, however, that the idea had been around for some time. Sarnoff had proposed construction and sale of his "radio music box" while still assistant chief engineer of American Marconi in 1916. It is not clear what he intended to use as a transmitter, for American

[97] Davis in Harvard Business School, *Radio Industry*, p. 193.

[98] Dreher, *Sarnoff*, pp. 60-61.

[99] William C. White, "Reminiscences" (Columbia University Oral History Collection, Radio Pioneers Series; copyright 1980 by The Trustees of Columbia University in the City of New York, quoted by permission), cited in Barnouw, *History*, 1:73-74.

Marconi had no continuous wave generator, but in other respects his memorandum to Nally was quite explicit in its vision of a network of local low-powered broadcasting stations transmitting news and music to simple receivers in the home.[100] But Nally showed little interest. To expect a man whose whole life had revolved around telegraphy to respond to such a vision was perhaps asking too much; and for American Marconi to have pioneered in entertainment broadcasting would have required a drastic wrenching of corporate traditions. In 1916, in that organization, the idea was out of time and out of context. But Sarnoff did not abandon it; in January 1920 he presented the idea to Young and in March, in response to a request for an estimate of probable business, he predicted sales of one million sets in the first three years at a selling price of $75 per set.[101] On 17 June, at a meeting of the RCA technical committee, funds were appropriated to get the project under way. By this time the "radio music box" phrase—guaranteed to set any engineer's teeth on edge—had been dropped and the set was referred to as the Radiola. Engineers at GE had estimated that it would cost about $2,000 to build a prototype, and that it could be done in four to six weeks. The consensus of the RCA committee was that this was impossible, but it was decided to go ahead anyway. There was, as it happened, $2,000 available. This money had originally been appropriated for the Weagant tube project, GE's attempt to invent around the de Forest triode patents; with the signing of the cross-licensing agreement with AT&T this had become unnecessary and in May 1920 the project had been dropped.[102] The minutes of the committee leave the impression that, in the absence of this small windfall of research money, Sarnoff's project might again have been shelved.[103]

[100] Sarnoff's memorandum, authenticated by Bucher, is reprinted in Archer, *History*, pp. 112-13. According to Archer, Sarnoff had earlier transmitted music from the Marconi station in Wanamaker's department store in New York City. But how music was transmitted from a station equipped for spark telegraphy is not explained.

[101] Barnouw, *History*, 1:79; Archer, *History*, p. 189. The reader should be aware that these early memoranda and forecasts are important elements in the Sarnoff legend, and that the documentation for them is not as secure as might be wished. Barnouw relies heavily on Archer and on official biographies of Sarnoff; Archer in turn relies on Elmer Bucher, formerly an engineer with RCA and one of Sarnoff's protégés.

[102] Clark Radio Collection, Cl. 5, Box 65, RCA Research Department, report for month of May 1920.

[103] Clark Radio Collection, Cl. 5, Box 66, minutes of meeting of RCA technical committee, 17 June 1920.

It can hardly be said, therefore, that RCA and its member companies were caught totally unprepared by the advent of broadcasting. But the matter had never been regarded as pressing, and the sum authorized for Sarnoff's project is a fair indication of the importance attached to it in corporate planning. The events of late 1920 changed all that. It was not inevitable that RCA, GE, or the Telephone Company would choose to follow the example set by Westinghouse and enter broadcasting directly, in the sense of setting up their own stations and organizing their own programs. It was certain, however, that as the demand for broadcasting increased, so would the demand for equipment and particularly the demand for vacuum tubes. This demand could be met only by RCA, in its role as sales agent for GE and the Telephone Company (apart, that is, from tubes sold by de Forest personally and by firms producing tubes with licenses from de Forest or with no license at all). And equipment using tubes, whether receivers or transmitters, could be manufactured only by the member companies of RCA, or by such other firms as they might choose to license. The cross-licensing agreement had, among other clauses, allocated to GE the right to manufacture broadcast receivers and to the Telephone Company the right—the exclusive right, its executives believed—to manufacture radiotelephone transmitters. Broadcasting confronted both companies with an urgent and exponentially growing demand for which neither was prepared. And that demand was of a novel type: it was a demand, not from shipowners, telegraph companies, or government departments, but from local entrepreneurs and homeowners. Any attempt to enforce exclusive rights, to the detriment of consumers or potential competitors, anything that could be construed as deliberate restriction of output, was certain to have grave social and political consequences.

＊　　＊　　＊

What Westinghouse had done was demonstrate that a latent demand for broadcasting existed and that it could be served by relatively unsophisticated facilities and a modest investment of capital. This did not mean, however, that Westinghouse, or any other company outside the RCA group, was in a position to serve that market legally. Any transmitter Westinghouse might manufacture was certain to use vacuum tubes and would necessarily infringe patents held by the RCA consortium. So would every receiver beyond the simplest crystal set. By the late fall of 1920 Westinghouse had four different receivers ready for the consumer market;

not one could be sold without inviting legal action.[104] Just as RCA's traffic agreements had kept Westinghouse out of international radio, so now the armory of patents held by RCA and its member companies threatened to exclude it from the domestic market. The Fessenden and Armstrong patents held by Westinghouse were indeed valuable assets. But every regenerative receiver needed at least one vacuum tube; every heterodyne set needed a local oscillator.

As had been true somewhat earlier with AT&T, the pressures for integration could not be ignored. Westinghouse needed access to patents that the RCA group controlled; and similarly RCA, GE, and the Telephone Company needed access to the patents that Westinghouse had acquired. A patent war—as Young had reassuringly told the RCA technical committee in August—was really not probable.[105] There were simpler ways. It was, after all, a question of trading.

In September 1920, after Kintner's abortive mission to Europe and before the establishment of KDKA in Pittsburgh, Owen Young had made Westinghouse what he considered a fair offer: RCA would absorb International Radio; Westinghouse would get 700,000 shares of RCA common and the same number of RCA preferred; and a general exchange of patent licenses in the radio field would be arranged. The offer was refused. On 5 October Westinghouse acquired the Armstrong feedback and superheterodyne patents. On the tenth of that month Young reported to his board that he had raised his offer to one million shares each of RCA common and preferred. There was no immediate response from Westinghouse.

What held up agreement was neither the price Young offered nor the intrinsic desirability of a patent exchange, but rather the tricky question of how the business of radio manufacturing should be divided up. If Westinghouse were to join the group, it would be as a manufacturer of radio equipment for RCA to sell, as was true of General Electric and the Telephone Company. What proportion of RCA's business should go to Westinghouse and what to the other firms? In the case of the Telephone Company the issue had been handled—at least on paper—by specifying fields of activity: in other words, the type of equipment determined which firm would manufacture it.[106] This would not work with Westinghouse, which had every intention of manufacturing broadcast receivers—GE's

[104] Clark Radio Collection, Cl. 14, abstract of manuscript history of radio; Schubert, *Electric Word*, p. 206.

[105] Young Papers, Additional Papers, Box 3, report of RCA technical committee, 25 August 1920.

[106] Case and Case, *Young*, p. 224.

assigned field in the compact with AT&T—and, if it could get away with it, transmitters also.[107] The alternative was a system of quotas: with the exception of those devices explicitly reserved to the Telephone Company, RCA's requirements for equipment covered by patents would be met by GE and Westinghouse, with the business divided between the two in stated proportions.

This, of course, was treading on dangerous ground. In the first place, any such prorating system would greatly increase RCA's vulnerability to antitrust action. This had been of some concern during the negotiations with AT&T; it was a much more sensitive matter when dealing with Westinghouse, traditional competitor of General Electric in the electrical industry. And in this case there was no timely letter from the Navy to rationalize the affair. Secondly, it posed tricky problems in determining the proportions in which output and sales should be divided, for there was little historical record to provide guidance and no obvious basis for setting percentages. A. G. Davis of GE was concerned enough about this issue to question the need to bring Westinghouse into the combine at all: GE and AT&T could get along without the heterodyne and feedback patents, he thought, even if that meant that RCA had to make do with second-best apparatus.[108] And thirdly, any such system was bound to introduce serious inflexibility into RCA's procurement and sales. This might have been of little importance if the company had remained primarily a telecommunications firm. But now it was entering a new industry; it was going to be selling to a different public; and it was certain to be facing a kind of competition it had not met before—competition from a score of small firms hungry for business, not overly scrupulous in their respect for patent rights, and prompt to respond to the shifts of consumer preferences. RCA had already given some evidence of slow corporate reflexes; the arrangement proposed with Westinghouse did not promise to make them faster. How was RCA likely to fare in the fast-paced market for broadcast receivers when it had no manufacturing or design facilities of its own, when reports from retailers and salesmen had to percolate slowly back to the engineers at GE and Westinghouse to have any effect? Sarnoff in particular had reservations on this score; sales experience in the next few years would fully justify them.

But these doubts and uncertainties did not override the pressures for integration. For GE and RCA the Fessenden and Armstrong patents were as near to indispensable as any patents could be. And International Radio, backed by Westinghouse, was a potential source of competition better

[107] Archer, *Big Business*, pp. 27, 98.
[108] Davis to Rice, 8 December 1920, quoted in Case and Case, *Young*, p. 224.

removed. Westinghouse, for its part, despite all its fast footwork in the preceding months, had worked itself into an almost impossible situation. Whatever its future role in radio might be, whether in broadcasting or in telecommunications or in manufacturing, it had to get licenses under the GE-AT&T patents. Consolidation of property rights was essential if the technology was to be effectively used.

Young left most of the negotiating to the two corporate presidents—Edwin Rice for GE and Edwin M. Herr for Westinghouse—and between them they worked out a solution to the only real difficulty that remained: the prorating of output. With the exception of types of equipment explicitly reserved for Western Electric, AT&T's manufacturing subsidiary, RCA would order 60 percent of its requirements of radio equipment from General Electric and 40 percent from Westinghouse.[109] These percentages were based roughly on the two companies' shares of the electrical business of the country.[110] The other clauses were in accordance with Young's offer of the previous October. The International Company was purchased outright. Westinghouse was issued one million shares each of RCA common and preferred. RCA received the $2.5 million that Westinghouse had agreed to subscribe to International. Seats on RCA's board of directors went to representatives of Westinghouse and International. And an agreement was drafted providing for the cross-licensing of radio patents. Young for RCA and Tripp for Westinghouse signed a preliminary agreement on 25 March 1921 and the final contracts were signed on 30 June.

[109] These percentages referred only to equipment covered by patents. For other devices RCA was free to buy from any supplier. To maintain amicable relations with United Fruit, Young urged that its subsidiary, the Wireless Specialty Apparatus Company, be encouraged to compete for RCA's business. See Young Papers, Copybook 802, Young to E. P. Edwards, 27 January 1922.

[110] FCC Hearings, p. 1311, testimony of David Sarnoff. This does not imply that, between them, they handled 100 percent; merely that their respective shares were in the ratio of 3 to 2.

TEN

RCA in Transition

M IF Owen Young had a grand design for the structure of RCA and was not merely seizing opportunities as they came along, the entry of Westinghouse marked its realization.[1] It might seem that he paid a high price for the achievement, and the question can certainly be raised whether key patents could not have been cross-licensed without making Westinghouse a major stockholder in RCA and without purchasing the International Company.[2] This, however, was the model that had been followed with General Electric, AT&T, and United Fruit, and Westinghouse was to be no exception. Young wanted more than an integration of technology, which a simple cross-licensing of patents would have achieved; he wanted also an integration of interests.

For a time it looked as if the structure he had built would hold together. And certainly RCA in the year 1921 was an impressive-looking organization, a formidable alliance of all the major corporations in the country with a stake in radio.[3] In its primary task (as seen by its founders) of

[1] It is worth noting that Young's latest biographers are skeptical as to whether Westinghouse was one of the original "building blocks" that Young envisaged in setting out to create RCA and at one point toy with the image of the avid book collector "needing one last volume to complete a valuable set." See Josephine Young Case and Everett Needham Case, *Owen D. Young and American Enterprise: A Biography* (Boston, 1982), pp. 220, 225.

[2] For an informed discussion of the terms of the transaction, and in particular of the curiously large quantity of RCA preferred stock transferred to Westinghouse, see Case and Case, *Young*, p. 225. Preferred stock had previously been limited to the value of the cash and other tangible assets involved in the trade.

[3] Ownership rights in the spring of 1921 were divided as follows (common and preferred stock had equal voting rights, share for share):

long-distance radiotelegraphy it was making a creditable showing. The effect on toll charges had been felt quickly. For more than thirty years the cable companies had kept their rate for regular traffic between New York and London at 25 cents a word. When RCA opened for business on 1 March 1920 its corresponding rate was 17 cents. This was raised to 18 cents on 1 January 1921, and in April 1923 both RCA and the cable companies settled on a rate of 20 cents.[4] Statistics of total traffic handled showed a healthy upward trend, from 7 million paid words in 1920 to 17.4 million in 1921 and 22.5 million in the year following.[5] In 1923 radio circuits were in operation between the United States and Britain, Germany, France, Poland, Italy, Norway, and Japan, with service to Holland, Sweden, Brazil, and Argentina expected to begin shortly. For most of these countries this was the first direct telegraphic service with the United States they had ever had. In 1923 RCA handled about 30 percent of the transatlantic traffic in competition with some seventeen cables, and on the transpacific circuit, where only a single cable linked the United States with the Orient, an impressive 50 percent.[6] Equally important, it was providing a service that was quick and reliable—something that the Marconi Company had never achieved before the war and that the overloaded cables had been unable to furnish after the war. Young was confident that by 1924 RCA—a "lusty infant," he called it,

	Common	Preferred	Total	%
GE	1,876,000	620,000	2,985,626	30.1
Westinghouse	1,000,000	1,000,000	2,000,000	20.6
AT&T	500,000	500,000	1,000,000	10.3
United Fruit	200,000	200,000	400,000	4.1
Individuals	1,667,174	1,635,174	3,302,348	34.9

See Gleason Archer, *Big Business and Radio* (New York, 1939), p. 8; Eric Barnouw, *A History of Broadcasting in the United States, Vol. 1, A Tower in Babel* (New York, 1966), 1: 73; Case and Case, *Young*, p. 225; FTC *Report* (1923), pp. 20, 22. These figures refer to the spring of 1921, before AT&T began to sell off its stock holdings. Statistics for shares held by individuals (mostly former stockholders in American Marconi) have been derived as residuals. Note that the FTC *Report* gives 5,734,000 as the total number of shares of common stock outstanding, which is at variance with the figures in the above table.

[4] FTC *Report* (1923), p. 36.

[5] Sarnoff, in Harvard Business School, *The Radio Industry, the Story of Its Development, as told by Leaders of the Industry* (Chicago and New York, 1928), p. 104.

[6] Hooper Papers, Box 5, Hooper to Captain Samuel W. Bryant, 2 February 1923; Clark Radio Collection, Cl. 5, Press Release, 23 November 1922.

with "real money in it"—would be able to meet the dividend payments on its preferred stock that would then become payable.[7]

This structure, however, had been put together on the basis of certain particular assumptions about the functions that RCA was intended to perform, the representatives of the public interest to which it would be primarily answerable, and the expectations that it would have to meet. By 1923 all these parameters had begun to shift, and large cracks were appearing within RCA itself, between RCA and the agencies that had earlier been its strongest supporters, and between RCA and important segments of public opinion.

* * *

Internal tensions first became evident in relations with the Telephone Company. President Thayer of AT&T had never much liked the idea of his organization holding an equity interest in RCA, and broadcasting brought this antipathy out into the open. The decision to sell off AT&T's stock interest in RCA during 1921 and 1922 presaged the problems that lay ahead, and by 1922 a polarization between the "Radio Group" (RCA, GE, Westinghouse, and United Fruit) and the "Telephone Group" (AT&T and Western Electric) was explicitly recognized by everyone involved. The points of contention were many, but the major issues all centered on broadcasting. Did AT&T and Western Electric, its manufacturing subsidiary, have the exclusive right to make broadcast transmitters, as their reading of the cross-licensing agreements indicated? Was the manufacture of broadcast receivers an exclusive field for GE and Westinghouse, or could Western Electric also enter that lucrative market? Who had the right to use the group's patents to operate broadcast stations for profit by selling time "on the air"? Was that "public telephony for toll" or something different? Could stations operated by GE or Westinghouse use the Telephone Company's wires for broadcast station pickups and network connections, or did that represent invasion of AT&T's exclusive field of wired telephony?

Each of these issues stemmed from the rise of broadcasting and the explosive growth in the demand for vacuum tube transmitters and receivers—developments that had been, if not completely ignored, given no more than incidental attention when the original cross-licensing agreements were drafted. Disagreement over how those compacts should be interpreted in the new context of radio broadcasting rapidly developed into open hostility between the personnel of the two groups and of the

[7] Young Papers, Van Hornesville, N.Y., Young to N. Dean Jay, Morgan, Harjes and Co., 15 January 1923; Young to Thomas W. Finley, 9 December 1922.

broadcast stations that they operated. AT&T, acting on its belief that radiotelephony for toll (and therefore broadcasting) was a Telephone Company prerogative, tried to restrict use of its phone lines to its own stations and those it had directly licensed under its patents, forcing stations supported by Westinghouse and GE to rely on telegraph lines and shortwave radio, which were much less satisfactory, for their hookups and network connections. GE and Westinghouse, in turn, denied the Telephone Company's claim that it alone had the right to manufacture broadcast transmitters and license their use, while insisting that they alone had the right to manufacture receivers or license others to manufacture them. And, belatedly reasoning that radiotelephony for toll and broadcasting were quite different activities, they argued that the Telephone Company had no right to enter commercial broadcasting in the first place.[8]

Attempts to mediate these and other disputes informally during 1922 proved abortive and produced nothing but a hardening of attitudes on both sides. In the closing days of 1923 formal arbitration was decided on, as provided for in the original agreement, both sides agreeing in advance that there would be no appeal from the decision. Roland W. Boyden, a highly respected Boston lawyer, well-known for his work on the Reparations Commission, agreed to serve as arbitrator. A draft of his decision became available in November 1924. It quickly became clear, when its terms were analyzed, that if it were permitted to become final the consequences would be unfortunate for all the parties involved. To be sure, on almost all the disputed points the arbitrator proposed to rule in favor of the Radio Group: the exclusive right to sell receiving sets; the right to the use of pickup lines; the right to collect tolls for broadcasting. And, most remarkably, he proposed to deny to the Telephone Company the right to use any of the Radio Group's patents in broadcast transmitters. But matters were not as simple as they at first appeared.

This draft decision, if it had ever been made final, would obviously have been a major setback for the Telephone Company, effectively eliminating it from any active role in the broadcast industry. But, appearances to the contrary, it would not have been entirely favorable to the Radio Group. In the first place, the decision appeared to state that, although members of the Radio Group were free to operate broadcast stations themselves, they were not licensed to sell broadcast transmitters to others,

[8] These disputes and the attempts to resolve them are the substance of Gleason Archer's *Big Business and Radio*, to which the reader is referred for fuller information. For a more analytic treatment, see Leonard Reich, "Research, Patents, and the Struggle to Control Radio: A Study of Big Business and the Uses of Industrial Research," *Business History Review* 51 (Summer 1977), 208-35.

if those transmitters used any of the patents of the Telephone Group (as they were certain to do). And an analogous restriction would apply to the Telephone Group. It was a question, therefore, whether any of the companies would be legally free to sell transmitters to outside buyers. This would hardly be a popular position for RCA to defend in public. Furthermore, the draft decision raised the disconcerting possibility that the cross-licensing agreement, if interpreted as the arbitrator wished to interpret it, was illegal under the Sherman Antitrust Act. The Telephone Company, therefore, had one last ditch defense open to it: it could reject the arbitrator's decision outright, on the grounds that it could not knowingly be party to an illegal contract. This would be an awkward argument for RCA to challenge.

The situation had its seriocomic aspects. In drawing up the original contract each party had deployed the best legal talent at its disposal, and great pains had been taken to protect what were regarded as essential interests. Yet concerning the prizes now at stake the contract spoke in language so ambiguous that the lawyers on each side—in some cases the very individuals who had helped to draft the document in the first place—could not agree on what the original intentions had been. Informal mediation having failed, the cumbersome mechanism of formal arbitration had been invoked. But now, in March 1925, with the arbitrator on the verge of handing down his formal report, both sides exerted every effort to prevent that report from being made. RCA appeared to have won a major victory. But the victory had been too complete, for the Telephone Group, finding the arbitrator's proposed settlement unacceptable, was now prepared to abandon the cross-licensing agreement completely rather than accept its terms; and it was RCA that found itself on the defensive.

If any doubts remained on this score they were removed on 17 March 1925 when the Telephone Company released—not to the public but to GE and RCA—a legal opinion prepared at its request by John W. Davis, a distinguished lawyer of national reputation, a former solicitor general and erstwhile Democratic candidate for the presidency. Davis held that the cross-licensing agreement, or at least certain clauses in it, as interpreted by the arbitrator, violated the Sherman Act. If the contract were regarded as divisible, he argued, the Telephone Company was released from any obligation to respect those clauses that were illegal—for example, those that restrained it from manufacturing broadcast receivers. If it was taken as indivisible, then the whole document was illegal and the Telephone Company was under no obligation to honor any part of it.[9]

---

[9] The essential paragraphs in Davis's opinion are reproduced in Archer, *Big Business*, pp. 193-98.

It was, of course, only an opinion, no matter how reputable its author. The danger was that the Telephone Company, relying on this opinion, would refuse to accept the arbitrator's ruling, and the conflict might end up in the law courts, where it could hardly avoid the publicity that so far it had escaped. None of the parties in the dispute wanted that to happen: their own private system of justice might not be working very well, but it was preferable to the alternative.[10]

The tedious and protracted negotiations by which this impasse was eventually resolved have been fully described elsewhere, and we may summarize them briefly here.[11] It was clear to the participants by the end of 1924 that some further trading of territory could not be avoided. Each side had, or thought it had, rights that the other side needed, or thought it did. And over the whole proceeding hung the possibility of unwelcome publicity. There had to be a settlement of some kind, some redefinition of the economic territory that each corporate group should occupy. Until one could be worked out, Young for RCA and Gifford for the Telephone Company agreed verbally that each side would proceed as it had before the arbitrator had made his preliminary judgment known, and arrangements were made to delay, first temporarily and then indefinitely, the handing-down of a final decision. Responsibility for hammering out acceptable terms was delegated to Sarnoff for the Radio Group and Edgar S. Bloom for the Telephone Company, and by early summer they had reached tentative agreement on the major issues. AT&T would receive exclusive patent rights in the field of public service telephony but would withdraw from broadcasting; while GE, Westinghouse and RCA would enjoy exclusive rights in the fields of wireless telegraphy, entertainment broadcasting, and the manufacture of radio tubes for public sale. AT&T also received the exclusive right to provide wired telephone service for radio, an important offset to its loss of rights in broadcasting.

The only remaining point of difficulty was how large a role, if any, the Telephone Company should play in the manufacture and sale of broadcast receivers. This was a more tricky issue than appeared on the surface. The market for receivers was large and potentially lucrative, and the Radio Group was reluctant to cede any part of it to Western Electric; on the other hand, Telephone Company executives thought it important that their group should retain a share of the business at least large enough to keep their engineers abreast of developments in receiver design. A compromise was finally worked out which provided that the Telephone

[10] As Erik Barnouw has pointed out, details of these negotiations were not revealed until Sarnoff made his records available to Gleason Archer for the writing of *Big Business and Radio*. See Barnouw, *History*, 1: 184.

[11] For a full account, see Archer, *Big Business*.

Company might manufacture each year up to $3 million worth of receivers and $2 million of tubes and appliances free of royalties, with a royalty of 50 percent to be paid to RCA on any sales above those figures. On 7 July 1926 a set of contracts was signed embodying the terms of agreement, and the prolonged conflict finally came to an end. Ironically, the Telephone Company never made use of its hard-won right to manufacture broadcast receivers for sale. In 1926 it began its withdrawal from active involvement in broadcasting by selling its "flagship" station, WEAF, to RCA for $1 million. Its nationwide system of "long lines," however, was essential to network broadcasting and, as the networks expanded, AT&T came to derive a substantial and steadily increasing income from leasing this wired system. But it was out of the entertainment business and that particular experiment with "radio telephony for toll purposes" was at an end.

The RCA that emerged from these rewritten compacts of 1926 was significantly different from the RCA that had signed the original contracts of 1920. By 1926 marine and international radiotelegraphy was only a part of its business, and by no means the most profitable part. It was deeply involved in domestic broadcasting, as a distributor of receivers, transmitters, and components, as a licenser of competing manufacturers, and, with the formation of the National Broadcasting Company in 1926, as owner of broadcast stations and a network.[12] Still formally a sales agent for General Electric and Westinghouse, without manufacturing facilities of its own, it was reaching out for a more independent role, one that would enable it to cope more effectively with the competition it faced in these new markets. The steady accretion of power to David Sarnoff, culminating in his appointment as president in 1930, reflected these changes. Sarnoff had pointed the way toward broadcasting; he had conducted the tough negotiations during the last stage of the fight with AT&T; and he had a clear sense of the kind of corporation he wanted RCA to become. That did not involve depending on GE and Westinghouse to do its manufacturing for it.

＊　＊　＊

As RCA moved toward a larger role in broadcasting and the domestic consumer market for radio, its relations with the federal government shifted. Increasingly it found that it had to deal with the Department of State when foreign concessions and contracts were involved, with the

---

[12] Barnouw, History, 1:185-86. NBC, incorporated in September 1926, was owned 50 percent by RCA, 30 percent by GE, and 20 percent by Westinghouse.

Department of Commerce on questions of frequency allocations, and with the Department of Justice, the Federal Trade Commission, and the Congress on antitrust and licensing issues. Relations with the Navy Department, on the other hand, became attenuated, partly because the Navy's role in radio shrank with the rise of broadcasting, partly because industry representatives deliberately shifted their attention toward the new centers of power that were emerging in Washington.

Great pains had originally been taken to ensure that the alliance between RCA and the Navy Department would remain firm. Admiral Bullard occupied a seat as government representative on RCA's board of directors and faithfully attended almost all its meetings until he was posted to the Far East in July 1921 (after the agreements with Westinghouse had been negotiated). Hooper showered Young with advice and admonitions. And most, though not all, of RCA's major policy decisions were informally cleared with the Navy before being implemented. By the fall of 1921, however, strains were already evident.

The entry of Westinghouse into RCA caused the first major difficulties in the relationship. In this matter Hooper believed that the Navy had been deliberately kept in the dark, and he came near to charging Young with having acted in bad faith. Young had been careful to sound Hooper out initially, admitting that he was "plenty worried" about the prospect of Westinghouse entering international radio in competition with RCA. He thought that Guy Tripp might be willing to bring Westinghouse into RCA instead; but, if he did, it would be called a monopoly. Which would Hooper rather see?

Hooper's response was that RCA should take Westinghouse in. There was not enough international radio traffic to support two companies; and a monopoly in the United States was necessary to combat monopolies in other countries. He made only two stipulations: first, he wanted to see actual and not just apparent competition in the sale of radio equipment to the government; and second, the government should be officially consulted before any final decision was made. He received—or thought he did—Young's assurance that both these conditions would be met.[13]

What exactly Hooper had in mind when he stipulated that the government should be officially consulted before a final agreement was reached is not clear: presumably something more than a letter to the Navy De-

[13] Hooper Papers, Box 3, memorandum of conference, 21 December 1921, with Senator Root; Clark, "Radio in Peace and War," chap. 12, pp. 88-89. According to Clark, this conversation took place in Albany during a break in "the international Radio Conference" (presumably the Preliminary Conference on Electric Communications, held in Washington, D.C., 8-15 November 1920).

partment. In the event, not even that was forthcoming. Rumors that the alliance with Westinghouse had been consummated began to appear in the newspapers in September 1921, but Hooper refused to believe them. "A notice recently appeared in the press that the deal with the Westinghouse has been closed," he wrote to Young. "Of course, I realize that this is incorrect, at least in part, as I have had your repeated assurances that nothing would be done until the government had had opportunity to pass officially on the matter."[14] In early November he wrote again in terms that reflected concern and frustration: "We are rather hurt that you appear to have deserted those who tried so hard to do a patriotic duty in getting the Radio Situation started right, by going abroad and doing a lot of things we are unaware of, which may or may not have the Government's approval, also closing up the Westinghouse deal. . . . You asked me to trust you for four months, and I gladly stated I would, but it is getting to be a hard job."[15] And later in that month, in an official "Memorandum regarding Mr. Owen D. Young," he put his doubts on record. Young, he reported, had given Hooper his "personal word" that he would take the matter up formally with the government before entering into the combination with Westinghouse, but he had not done so. "The point is that although he is an excellent man we must take his word for nothing. . . . We may find it desirable to encourage a monopoly in long distance international communications but if we do so we must protect the American public from a natural resulting monopoly in radio research, radio patents, ship radio service, sale of radio apparatus, etc., and the time to protect the public is to get a definite official written statement out of the Radio Corporation and the G.E. Co. now before we hand them a monopoly of high power communication on a silver platter."[16]

Young later apologized for failing to keep Hooper informed. The agreements with Westinghouse had been signed, he said, while he was away on a visit to Japan; he had not expected the business to be acted on until his return, and he had certainly intended to write the letter that had been stipulated. This explanation, if not wholly convincing, smoothed over the immediate resentment, but the episode was not forgotten. The Federal Trade Commission, in its Report on the Radio Industry in 1923—RCA's

[14] Hooper Papers, Box 3, Hooper to Young, 20 September 1921.

[15] Hooper Papers, Box 3, Hooper to Young, 5 November 1921.

[16] Hooper Papers, Box 3, "Memorandum Regarding Mr. Owen D. Young," 17 November 1921. In a "Memorandum of Conference with Secretary Hoover" of the same date, Hooper charged that "on three separate occasions" Young had promised to consult officially with the government on the Westinghouse matter but failed to live up to his promises.

first brush with the antitrust laws—called attention to the fact that every one of the cross-licensing agreements signed by RCA had been shown to and approved by representatives of the Navy Department, except the agreement with Westinghouse.[17] For Hooper it was one more piece of evidence to suggest that RCA was not developing into the 100 percent American radio company, answerable to the Navy, that he had originally envisaged.

There was more to the incident than hurt feelings. The immediate problem was a breakdown in communication between Young and Hooper personally—whether intentional or due to oversight it is impossible to say. But underlying that was a structural shift in the relationship between RCA and the Navy Department. Talk of "monopoly" made this apparent. No group had stated more forcefully the need for a single organization to represent American interests in radio than had the Navy. But now Hooper was beginning to worry about what that meant for the domestic market. How was monopoly in the international field to be reconciled with competition at home? It was all very well for him to talk about getting an "official written statement" out of GE and RCA before the Navy condoned the union with Westinghouse, but what did he imagine such a statement should contain? And how were its provisions to be enforced? On these matters he was silent. If RCA and its member companies were really concerned about antitrust matters, it was toward Congress, the Department of Justice, and the Federal Trade Commission that they should have been looking, not the Navy Department. In that arena the Navy was virtually irrelevant, except insofar as its actions in the past could be invoked to justify the industrial structure of the present.

In external affairs also, RCA under Young's leadership was not comporting itself as the Navy had hoped and expected. Hooper's policy in this area—and since the departure of Josephus Daniels there was little apart from Hooper's convictions that could be called a Navy policy—had consistently been summed up in the phrase, "A Monroe Doctrine of Radio." What this meant in practice was perhaps not as self-evident as Hooper thought it was. For him the phrase had several levels of meaning. In the first instance, it meant the exclusion of foreign interests from the ownership or operation of radio stations within the United States. Secondly, it meant that in South and Central America (and, by implication, in Canada, although that country was seldom mentioned), external radio communications should be controlled by United States interests—possibly in cooperation with local companies but certainly not in cooperation with Europeans. And thirdly, and most generally, it implied that through-

[17] FTC *Report* (1923), p. 24.

out the world American radio interests—as represented, for example, by RCA—should as far as possible function autonomously rather than in cooperation with other nations. In all of these sentiments Hooper was reflecting the prewar conviction that Britain had dominated world communications through its ownership of the submarine cables and of the sources of gutta-percha, and the determination that, in the new age of international radio, that pattern should not be allowed to repeat itself. (See above, Chapter Five.) The United States was to be the new hub of the world's telecommunications system.

This vision of how world communications should evolve cast both RCA and the Navy Department in particular roles: the one as corporate instrument of the national purpose, the other as governmental monitor of performance. And it was because RCA did not, and perhaps could not, conform to that vision that its relations with the Navy became increasingly distant. The rebuilding and modernization of the world's radio communications necessarily involved RCA in cooperative, rather than confrontational, relationships with the corporations and government bureaus that were its counterparts in other countries. The network of traffic agreements was only the most obvious indication that what was being put together was an international system. And in the construction of that system Young's strategy of cooperation and partnership—his characteristic searching for the deal that would give all parties what they felt they had to have—proved more effective than the xenophobic nationalism that inspired Hooper's vision.

The contrast between the two philosophies became very evident in connection with the so-called AEFG consortium, hammered out by Young in a series of difficult negotiations with representatives of British, French, and German radio interests in the fall of 1921. Under this agreement, RCA, British Marconi, the French Compagnie Générale de Télégraphie sans Fil, and the German Gesellschaft für Drahtlose Telegraphie, m.b.H., granted all their radio communication rights in the South American republics to a board of trustees, which was to hold them on behalf of the four parties in equal shares. RCA, in addition to its normal quota of trustees, was to name the chairman, who was to be a prominent American not connected with RCA, and the chairman could break a tie or veto any action taken by the majority of the trustees that in his opinion was unfair to the minority. In effect this meant that no action could be taken without American approval. Under the direction of the trustees, national companies were to be formed in each Latin American country to build radio stations and conduct its international communication services. Each station so built was to be under the direct control of an operating committee with four members, one representing each country. (No provision was made, however, for representation of the host countries, either on the operating committees or on the board of trustees.)[18]

In Young's view, and in that of later commentators, the formation of the AEFG consortium was a major accomplishment: an agreement for international economic cooperation arrived at without government pressure or intervention; an outstanding contribution to the peaceful reconstruction of the world economy; and—a point of some significance—the first international agreement to be reached after World War I in which the German delegation functioned as an equal participant.[19] Equally important, it was functionally effective: it got the job done. It got radio stations built in Latin America and it obviated the duplication of facilities, rate-cutting, and waste of radio channels that competitive expansion would have implied. For Young, indeed, there was no real alternative to international cooperation. The Germans and the French had already secured concessions in Argentina. British Marconi was already established in Brazil. How could one pretend that these countries did not have legitimate interests in that area? The problem was not one of keeping the Europeans out; it was one of making sure that the United States was not left behind.

Hooper saw the matter differently. For him there could be no true "Monroe Doctrine of Radio" in Latin American unless the United States had its own stations there. Traffic agreements he could accept, but (as he wrote to Young) "exclusive contracts, such as your company appears to be making abroad, can only result in an international pool, which,

[18] Young Papers, Copybook 802, Young to James R. Sheffield, 7 December 1921.

[19] Case and Case, *Young*, pp. 237-43; Ida Tarbell, *Owen D. Young: A New Type of Industrial Leader* (New York, 1932), pp. 135-37. Compare Michael J. Hogan, *Informal Entente: The Private Structure of Cooperation in Anglo-American Economic Diplomacy, 1918-1928* (Columbia, Mo., 1977), pp. 140-46. Archer (*History*, p. 242) states that Young insisted that Germany be included in the arrangements for the consortium in order to offset the anticipated opposition of Britain and France, and portrays the negotiations as a conflict between British and American imperialism. George H. Clark will have nothing to do with this interpretation, on the ground that German radio interests were already established in Latin America and could not be left out. "To ascribe either to the British, French, or Germans, the idea of excluding from the September conference (1921) one of the companies which already had a real position in South America and all of them previously agreed among themselves on mutual cooperation in the organization of international radio communications with South America, is contrary to truth as well as to logic in every respect." (Clark Radio Collection, "Radioana," Cl. 5, book 8, 871-975)

directly or indirectly, will result in submerging the interests of each country with the others in the group."[20] "What we want," he insisted, "is a Monroe Doctrine of Radio in South America, and the sooner we get some stations erected there the more liable we are to have it."[21] And it was in line with these convictions that Acting Secretary of the Navy Roosevelt advised the secretary of state in 1923 that the AEFG trusteeship "cannot be considered a bona fide American interest deserving the assistance and protection of this Government."[22]

In this case the need for prompt action if European interests were not to preempt the field carried the day for Young. Both the State Department and the Commerce Department favored RCA's position, and Navy reservations were in the end overruled. But when Young proposed the same kind of arrangement for China—an international consortium to take over all existing concessions, with RCA playing a major role in providing equipment and arranging finance—Hooper dug in his heels, and this time he had the support of the State Department. He wanted to see an American radio company operating in China independent of the cooperative ties with other nations that Young had in mind, and both the Navy and the Department of State were counting on the Federal Telegraph Company of California to play that role. Federal had signed a contract with the Chinese government in 1921 providing for the construction of a network of high-powered stations in China for communication with its stations in California. Unfortunately, this concession conflicted with concessions previously granted to British Marconi (through the Chinese National Wireless Company), to the Mitsui Bussan Kaisha of Japan, and to the Great Northern Telegraph Company, a Danish firm with a concession granted in 1896 which, in the company's opinion and in that of certain Chinese officials, gave it a complete monopoly of all the external communications of China, whether by cable or by radio. The concession granted to Federal, therefore, immediately drew fire from the British, Japanese, and Danish governments, and the State Department found itself defending Federal's contract, rejecting claims to exclusive privileges as a violation of United States treaty rights, and asserting an Open Door policy for radio communications with China.[23]

---

[20] Hooper Papers, Box 3, Hooper's reply to Young's letter to Secretary of the Navy Denby, 22 December 1921.

[21] Hooper Papers, Box 4, Hooper to Young, 3 October 1922.

[22] Hooper Papers, Box 5, copy, Acting Secretary of the Navy Roosevelt to the Secretary of State, 19 December 1923.

[23] Hogan, *Informal Entente*, chap. 7, pp. 129-58, especially p. 146. See also Young Papers, Copybook 802, Young to Sheffield, 7 December 1921 and Young to Hon. Elihu Root, 12 December 1921; Hooper Papers, Box 4, memorandum

The fact of the matter was, however, that Federal had neither the technology nor the manufacturing capacity nor the financial resources to exploit its concession unaided. It badly needed the support of a major American corporation with good banking connections. Young's plan was for all the parties involved to transfer their concessions to a board of trustees, with the government of China to select an eminent Chinese citizen to act as chairman; and he proposed that RCA and Federal should jointly form the American component of the group. In this plan he had, in principle, the support of British Marconi, of the British Foreign Office, of the French and Japanese governments, and eventually even of the Federal Company. The Navy and the State Department, however, were hard to convince, despite warnings that, if the United States stayed out, the British, French, and Japanese companies might well form an independent consortium, strong enough to shut RCA and Federal out of the transpacific radio business.[24] International cooperation of this kind was not what Hooper thought of as a proper policy for American radio, while the State Department continued to insist on an Open Door for American radio companies and execution of Federal's original concession.

Eventually these reservations were overcome, or at least muted. RCA and Federal joined to form the new Federal Telegraph Company of Delaware (with RCA owning some 70 percent of the voting stock) and this organization, with the State Department's reluctant blessing, joined with the British, French, and Japanese in an international consortium, with Federal to specialize in China-United States traffic and the British, French, and Japanese interests to handle traffic in Southeast Asia and between that area and Europe. All this was not enough, however, to get the radio stations built, in the face of growing Japanese opposition and the convoluted obstructionism of the Chinese authorities. In 1927 RCA settled for second best, building its western "Radio Central" in the Philippines and relaying traffic from there by radio to Hong Kong.[25]

In all these negotiations—both the cross-licensing agreements at home and the traffic agreements and consortia formed abroad—Young and his fellow executives believed that they were being true to the original conception of what RCA should be and what functions it should perform. They had brought about that integration of American interests in radio that the Navy had called for. They had broken the logjam of radio patents that threatened to hold up development. They had asserted an autono-

to National Radio Conference; Box 3, memorandum on China dated 17 December 1921; and memorandum of conference with Secretary of Commerce Hoover, 8 December 1921.

[24] Hogan, *Informal Entente*, p. 149.

[25] Paul Schubert, *The Electric Word: The Rise of Radio* (New York, 1971) pp. 257-65; Archer, *History*, p. 328.

mous role for the United States in the reconstruction of the world's communications system. If the test was performance, and if performance was measured in terms of the goals set for the corporation when it was created, RCA had little to apologize for.

Hooper's objections to what had been done made little sense to Young. "It seems to me," he wrote to Hooper, "that we have a Monroe Doctrine of radio in South America, because the activities in that field of the European nationals, although cooperative, is [sic] subject in the last analysis to the control and direction of the Americans."[26] Surely the real test was a pragmatic one. The Argentine station would be in operation in 1923. Plans for the stations in Brazil were far advanced and they should be functioning in 1924. As for the rest of the world, "with American stations in Sweden, in Poland, and in Europe, and with communications already established with England, France, Germany, and Norway, and others rapidly coming, with the improvement of service in the Pacific, and the increased volume of communications with Japan, with cooperation with the Federal Company in China established, it seems to me that we have in a three year period largely established our right to the slogan of 'World Wire Wireless' and realized on the vision which you and Admiral Bullard put before us."[27]

But Hooper would not be reassured. He recalled the emotions with which he and Bullard had first approached General Electric. "I felt a great patriotic inspiration at the time and felt that I was doing something great for the nation. But my disillusion since then has been enough to make a sad old man out of a patriotic, enthusiastic lad."[28] He understood why Young felt he had to act as he did: he had to keep in mind the interests of the stockholders. But was it not possible that he was interpreting those interests too narrowly? The stockholders, he felt sure, were "good average American citizens" who would be glad to give up part of their dividends if the result was something in which they could take patriotic pride. "Most Americans," wrote Hooper in a memorable phrase, "are charitable at heart provided they get a reasonable dividend."[29]

* * *

It was not in the interests of RCA to be at odds with any branch of the federal government. The problem lay in the growing uncertainty as

[26] Hooper Papers, Box 4, and Young Papers, Copybook 802, Young to Hooper, 9 October 1922.
[27] Ibid.
[28] Hooper Papers, Box 4, Hooper to Young, 3 June 1922.
[29] Hooper Papers, Box 4, Hooper to Young, 3 October 1922.

to who, if anyone, was responsible for radio. The Navy based its claims to jurisdiction on the report of an Interdepartmental Board of Wireless Telegraphy appointed by Theodore Roosevelt in 1904 in an attempt to put a stop to squabbling over bureaucratic territory. This had given primary responsibility for radio to the Navy Department by recommending that it establish a coastal radio system, that private stations be restricted so as not to interfere with naval radio, and that other federal agencies obtain the service and equipment they needed from the Navy. This preeminent position had again been acknowledged in the Radio Act of 1912, which allocated to the Navy's stations preferred channels in the radiofrequency spectrum, protected them from interference from private companies, and banished the amateurs to frequencies above 1,500 kHz, which were then thought to be of no commercial or military value.

It was under powers granted by this act that the Navy assumed control of American radio in April 1917 and retained it for the duration of the war emergency. Josephus Daniels's persistent attempts to induce Congress to perpetuate exclusive control of radio by the Navy in peacetime represented the high point of the Navy's aspirations, and the defeat of those attempts foreshadowed the shift of authority to other hands. By the early twenties the Navy's reliance on Roosevelt's 1904 edict was coming to seem an anachronistic and inadequate defense against the aspirations of other federal departments. International radio involved issues of commerce and diplomacy that the Navy Department was in no position to resolve. The growth of civil aviation and a domestic air mail service brought with it the need for land radio stations to provide weather and navigational information, raising once again the specter of Post Office control of radio on the European model, which the Navy had thought had been safely laid to rest. The rise of broadcasting raised new questions of frequency allocation, the control of interference, and the licensing of patents that the Navy had neither the personnel nor the constitutional authority to resolve. And broadcasting also brought on the scene a host of hungry entrepreneurs with little taste for government regulation of any kind, but a special aversion to control of radio by the Navy. And yet if the Navy, in this changed context, had no power to oversee radio, who did?

Owen Young, responding to the Navy's insistence that RCA should secure government approval for its policies, argued that, although acceptable in principle, this was almost impossible in practice. What department could grant approval? None, because no such exclusive jurisdiction over radio had been authorized by Congress. "My own experience," he told Secretary of the Navy Denby, ". . . has been discouraging. Several different departments of the Government are attempting to deal with it.

. . . In the endeavor to cooperate with one, it frequently happens that you antagonize another. In other words, there is not today . . . any uniform policy of the Government, and there is no department of the Government authorized to formulate or exercise a policy."[30] Hooper's response was to reassert the Navy's claim to primacy. If Young had heard conflicting views from other departments, "these views must have been expressed as the views of individuals without full knowledge of the interdepartmental agreement and inter-relations. The Navy Department and the War Department are in full accord as regards the Radio Situation, and neither Department acts as regards foreign relations without full accord by the Department of State. . . . If your company chose to go elsewhere for advice without consulting this Department, and somewhat contrary to the opinions of officers in this Department, you are reaping the results of your actions."[31] But Hooper was well aware that, if RCA did not like the advice it got from the Navy, it would go elsewhere. He noted particularly Young's personal friendship with Secretary of Commerce Herbert Hoover, and RCA's growing tendency to go to the Department of Commerce, rather than the Navy, for guidance and approval.[32]

Hooper was not misreading the omens. The major bureaucratic challenge to the Navy's hegemony over radio was to come from Hoover's Department of Commerce. Determined to bring broadcasting, frequency allocation, and the licensing of stations under his jurisdiction, Hoover skillfully mobilized the support of the broadcasting industry, the engineering profession, and the radio amateurs behind the development of a national radio policy under the aegis of the Department of Commerce. If the substance of such a policy was left conveniently vague, it was at least clear that it would be coordinated by the secretary of commerce, not by the Navy or the Post Office. Opposition to the centralization of authority over broadcasting that this might imply, reinforced by suspicion of Hoover's own political aspirations, led Congress to follow a different route: that of the "independent" regulatory commission—first the Federal Radio Commission of 1927 and then the Federal Communications Commission in 1934. This is the regulatory structure under which the industry functions today; we cannot explore its evolution here, except to note its

[30] Hooper Papers, Box 3, Young to Hon. Edward Denby, 22 December 1921.

[31] Hooper Papers, Box 3, Hooper's reply to Young's letter to Secretary of the Navy Denby, 22 December 1921.

[32] Hooper Papers, Box 3, memorandum of conference, 1 December 1921, with Dr. Stratton of the Bureau of Standards; ibid., 17 December 1921, memorandum of conference with Secretary Hoover, Thursday, 8 December 1921.

origins in the decline of Navy control and the interagency struggles of the 1920s.[33]

From RCA's point of view more was involved than the usual jostling for political and departmental power in Washington. A shift in political alliances had to follow the shift in the corporation's role and functions. The Navy had played an essential part in the creation of RCA and in the consolidation of corporate interests and integration of technological subsystems that marked the entry of AT&T, United Fruit, and Westinghouse. It had provided the bureaucratic protection; it had defined the corporate mission; it had given the new organization its identity as an instrument of national purpose. But now new uses were developing for radio technology and new demands were making themselves felt for radio equipment and services. As domestic markets bulked larger in RCA's affairs while international communications shrank in relative importance, the Navy's support came to seem, from RCA's point of view, more and more dispensable. The risks that RCA now faced were not the kind that the Navy could do much to reduce. The expectations that RCA was now called upon to meet came not from the Navy but from broadcast listeners—forty million of them, it was estimated, by 1928—and from the firms that made up the new radio industry.

RCA's failure to meet these expectations vividly illustrated the change in its circumstances and the difficulty it experienced in reorienting its sense of corporate purpose. The firm had come into existence in a flush of patriotic fervor, and this was still reflected in the rhetoric that its spokesmen used. But by 1922-1923 RCA was, for millions of Americans, "the radio trust"; it was no longer seen as an instrument of the national will but as an exploitative monopoly. This shift in the public image of the corporation reflected a shift in its social role and in the markets it served. And it carried with it first the threat and then the reality of prosecution as a conspiracy in restraint of trade under the antitrust laws: the Federal Trade Commission inquiry of 1922-1923 and the indictment under the Sherman Act in 1930. Reaction to political attack and antitrust indictment, in the form of the consent decree of 1932, created an RCA quite different in its structure and orientation from that which had existed in the 1920s.

Of all business leaders of his day, Owen Young, with his wide experience as expert counsel in the public utility industry and as vice-president for policy in General Electric, should have been most sensitive to public

[33] For an excellent recent analysis, see Philip T. Rosen, *The Modern Stentors: Radio Broadcasters and the Federal Government, 1920-1934* (Westport, Conn., 1980), especially pp. 3-76.

charges of monopoly, to the threat of political attack, and to the risk of antitrust indictment. And the evidence of his correspondence and memoranda suggests that he was well aware of these hazards. RCA, as it existed after the entry of Westinghouse, controlled, directly or through its affiliated companies, every American patent of importance in the field of continuous wave radio technology. If any had escaped the net, if there existed outside the control of RCA any residual or personal rights to important radio devices or circuits, it was either through oversight or because licenses had been granted (as for example by de Forest and Armstrong) before the patents themselves had been purchased. With minor and unimportant exceptions, RCA in 1921 and the corporations associated with it controlled continuous wave technology in the United States as it had evolved up to that date. And, beyond this, because it was backed by the formidable scientific and engineering resources of Western Electric, General Electric, and Westinghouse, not to mention the foreign firms such as Marconi, Phillips, and Telefunken with which it had signed patent agreements, this group appeared likely to control developments in the future also.

This consolidation of rights to continuous wave radio technology had been the controlling principle of RCA's creation. Young and his associates were under no illusions about this. And as long as the corporation's primary orientation was toward the outside world, as long as its primary market was international radiotelegraphy and its primary responsibility was to function as the designated instrument of American radio policy, objections and criticisms were muted. RCA had its mandate from the federal government. Without the intervention of the federal government it would never have come into existence. No apologies were offered because none was called for. But broadcasting changed all that.

The Federal Trade Commission inquiry reflected the changed climate of expectations. It had been initiated in 1922 in response to the filing of a complaint to the effect that General Electric and others had set up RCA as a "bogus independent" with the intent of acquiring a monopoly in the manufacture and sale of radio apparatus. Inquiry into that charge was almost complete when the commission received broader instructions from Congress in the form of House Resolution 548, calling for an investigation of the ownership of patents in the radio industry, its pricing practices, and the existence of contracts, leases, or agreements that might tend to convey exclusive rights or privileges in the reception or transmission of messages by radio. Receipt of this resolution led the commission's staff to undertake a more comprehensive inquiry. It had two major foci: the cross-licensing agreements, and RCA's policies in the sale of vacuum tubes.

The commission, it should be noted, was not charged with the task of determining whether any of the facts it might uncover in its investigation might constitute a violation of the antitrust laws. Its responsibility was purely investigative—to provide such facts as might aid the House to determine whether the antitrust statutes had been violated, and whether further legislation might be advisable.[34] Within the limits of this mandate, the commission's investigators performed a creditable balancing act. They had no difficulty in showing that, in radio communications between the United States and foreign countries, RCA did have a monopoly and did refuse to sell or lease apparatus to potential competitors in that field. Nor was it hard to demonstrate that RCA and its affiliated companies did have substantial control of the radio art through their ownership of patents, that this concentration of ownership of patents had been a primary motive for the creation and expansion of RCA, and that it was very difficult if not impossible for any firm to function in the radio industry without licenses under RCA's patents. If, in certain passages, the language of the report seemed somewhat pejorative, this was a characteristic hard to avoid when discussing monopoly and concentrations of economic power. In general the tone was one of professional neutrality. The report itself showed little trace of demagoguery. Its strength lay not in any recommendations for action, for it contained none, but in the way in which it laid out for public inspection the complex network of agreements and contracts through which the radio industry had been reconstructed. Here, reprinted *in extenso,* were all the traffic agreements, all the cross-licensing agreements, all the sales agreements that had gone into the ordering of the American radio industry since the war. And if, to anyone with the patience to read the report and its exhibits, RCA might in the end seem to squat somewhat ominously at the center of this complex network, the image was perhaps not inappropriate.

Against charges that it held a monopoly in external radio communications RCA had ready defenses. It was much more vulnerable to charges of discriminatory practices and suppression of competition internally. Allegations of this nature had led the FTC to undertake the inquiry initially—Section 5 of the Federal Trade Commission Act had specifically outlawed "unfair methods of competition"—and they had generated much of the political heat responsible for House Resolution 548. Here RCA's control of the key vacuum tube and receiver patents placed it in a difficult position. Should it license these patents to others, thus generating competition in markets that it intended to exploit itself? If so, what royalties

[34] FTC *Report* (1923), pp. 7, 10.

should be charged? Given the nature of the market for broadcast receivers, decisions on these issues inevitably had a political dimension.

There was no clear road through these thickets. The complaints filed with the FTC had alleged, among other things, that RCA practiced discrimination in the sale of tubes, urging the company's jobbers not to sell tubes for use in receivers made by other manufacturers and penalizing those who did so. By the end of 1922 there were some two hundred of these independent manufacturers, and they had the larger share of the market: of the $60 million spent on the purchase of radio receivers in that year, only $11 million went to RCA.[35] This was dismal sales performance for a firm that, in terms of its patent portfolio and the engineering resources at its disposal, should have been able to dominate the market. In some cases these independent manufacturers undoubtedly believed that they were not violating RCA patents—for example, those who used the popular "neutrodyne" circuit patented by L. A. Hazeltine of Stevens Institute.[36] Others knew they were infringing but thought they could get away with it. All these sets, however, required vacuum tubes and it was RCA's contention that, apart from sales to amateurs, it was the only legal supplier of tubes. This was the reason for the pressure RCA put on its distributors, dropping those who ordered only tubes (since most of these ended up in receivers made by other firms) and favoring those who carried and pushed the entire RCA line. It was an understandable policy, particularly at a time when the manufacturers were straining to keep up with the demand for tubes and receivers; but it was a highly unpopular one.

It was also of dubious legality. The basic Fleming diode patent expired in 1922, and after that date considerable numbers of new firms entered tube production, despite the fact that the triode patent still had six years to run. Among these was Lee de Forest, relying on the residual rights that allowed him to sell tubes to amateurs. RCA knew very well that de Forest was selling tubes to people who could be called "amateurs" only in the broadest sense of the word, and asked him, in accordance with his original agreement with AT&T in 1917, to get from purchasers an agreement that any tubes they bought from him would not be used for commercial radio communication. When de Forest's company refused, RCA brought suit—and lost, the judge holding that to use the 1917 covenant in this way would be to use it for a purpose for which it was

[35] Barnouw, *History*, 1: 115, citing *Broadcasting*, 1939 Yearbook, p. 11.

[36] Rupert Maclaurin, *Invention and Innovation in the Radio Industry* (New York, 1949) pp. 127-29.

never intended. The result was a flurry of damage suits against RCA which were settled only at considerable expense.[37]

With receiver circuits, loudspeakers, and other components, it was much the same story. Even when what RCA took to be its legal rights could be enforced—which was less often than the company's lawyers tended to believe—this could be done only at considerable cost in terms of public reputation, relations with the rest of the industry, and heightened risk of antitrust indictment. Increasingly, therefore, RCA moved toward a policy of general licensing, accepting the existence of competition in fields that at one time it thought it could control and relying on low license fees and the threat of legal action to keep its competitors in line. The true strength of its position lay not in the arsenal of patents inherited from the past but in the fact that General Electric, Western Electric, and Westinghouse were the leading centers of industrial research in electronics. Through this process the technology of continuous wave radio became generally available to the radio manufacturing industry, despite the fact that the original intention had been to centralize it in RCA and its affiliated firms. For this development the pressure of public opinion, the unexpected weakness of RCA's legal position in certain key instances, and the omnipresent threat of punitive action under the antitrust laws were responsible.

The FTC Report had laid out the corporate structure of RCA in detail. There was no mystery to the cross-licensing contracts and stock ownership that linked GE, AT&T, United Fruit, and Westinghouse to the Radio Corporation, nor to the way in which these contracts allocated exclusive rights and fields of activity. But the report itself had little to say about these matters: its main emphasis was on trade practices. In this respect it reflected the thrust of the Federal Trade Commission Act of 1914 and of the Clayton Act of the same year—indeed, of Wilsonian Progressivism in general—rather than the Sherman Act's suspicion of concentrated economic power as such. In the short run this worked to RCA's benefit: the FTC's formal antitrust complaint, filed in 1924, was dropped in 1928, and the gradual relaxing of RCA's licensing policies quieted some, though by no means all, of the company's most vocal critics. This did not mean, however, that on antitrust issues RCA could breathe freely. The out-of-court settlement of a civil antitrust suit brought by Fessenden in 1926 demonstrated RCA's vulnerability.[38]

[37] Ibid., pp. 129-31.

[38] Ibid., p. 135. Fessenden sued for alleged violation of the Clayton Act in RCA's use of patents originally issued to him. RCA and its affiliated companies settled out of court by payment of $500,000 in damages.

To the outside world RCA presented the image of a stable and well-integrated structure. None of the internal dissension that in 1922-1923 was already setting the "Radio Group" at odds with the "Telephone Group" had shown in the FTC Report. The arbitration proceedings of the mid-twenties were successfully kept out of the newspapers—no compliment to the investigative reporters of the day. There was little to suggest the presence of internal tensions. In fact, however, RCA's structure and its relations with the other members of the Radio Group were changing very rapidly in this period, and for reasons independent of the threat of antitrust action.

The major force behind this evolution was Sarnoff's drive for what he called "unification." In a sense "disintegration," were it not for its unfortunate connotations, would have been a more appropriate term, for what Sarnoff was after was a loosening of the ties that, since its creation, had bound RCA closely to its corporate parents. The particular end in view, for Sarnoff, was a state of affairs in which RCA would possess its own manufacturing facilities, so that manufacturing and selling could be united in the same organization—hence "unification." And the principal motive for this, apart from Sarnoff's personal ambitions, was the hope of improving RCA's lackluster performance in the market for broadcast receivers.

A long step in this direction was taken in 1929, when RCA purchased the Victor Talking Machine Company. Victor, a world-famous and highly respected name in the phonograph business, had suffered badly during the twenties from its initial failure to take the competitive threat of radio seriously. By mid-1928, however, its executives had come to believe that entry into the manufacture of radios, and particularly of radio-phonograph combinations, was essential and, encouraged by their bankers, they opened negotiations for the sale of the company to RCA. From RCA's point of view Victor was an attractive acquisition, partly for its name, trademark, and portfolio of contracts with well-known recording artists, but principally for its large manufacturing facilities. GE and Westinghouse jointly advanced $32 million, on the credit of RCA, to finance the purchase, with $22.5 million of this intended to retire the Victor preferred stock and the remainder to modernize the Victor factories.[39] Approval for an exchange of shares with RCA was obtained without much difficulty from the Victor stockholders, and on 26 December 1929 the RCA-Victor Company was incorporated as a subsidiary of RCA.

RCA now for the first time controlled its own manufacturing plant.

[39] Archer, *Big Business*, p. 345.

But Sarnoff wanted more than this. He wanted RCA to take over the radio manufacturing business of GE and Westinghouse, so that these companies would withdraw completely from that field and leave RCA in possession. In October 1929, while still executive vice-president, he had persuaded RCA's board to approve in principle the unification of the manufacturing, engineering, and selling of all the radio devices manufactured by GE and Westinghouse and sold by RCA. This did not mean, however, that the boards of GE and Westinghouse would approve the policy, nor that acceptable terms for the swapping of assets could be arranged; and it was toward these further steps that Sarnoff bent his efforts after his election as president of RCA in January 1930. Backed by Owen Young, he met remarkably little resistance. Swope, president of GE, and Andrew W. Robertson, his opposite number in Westinghouse, bargained hard about particulars but they did not oppose the principle. This was testimony, perhaps, less to Sarnoff's powers of persuasion than to the logic of the situation, for the unwieldy arrangement that tied RCA to the design and manufacturing facilities of GE and Westinghouse, while denying these firms the right to enter the market directly, was clearly benefiting no one but the competition.

Terms of the agreement were presented to and approved by RCA's board of directors on 4 April 1930. RCA received manufacturing licenses under the patents of GE and Westinghouse in the fields of radio equipment, phonographs, and moving pictures; manufacturing facilities and real estate, including particularly Victor's plant in Camden, New Jersey, GE's Harrison Tube Plant, and Westinghouse's Lamp Works in Indianapolis; the royalties that RCA had previously been collecting and transferring to the other two firms; their stockholdings in a number of RCA subsidiaries, including RCA Victor and NBC; forgiveness of the $32 million loan that had been made for the purchase of the Victor Company; and a number of other considerations of less importance. In return for this largesse RCA transferred to GE and Westinghouse a little over 6.5 million shares of its common stock (selling at this time at about $40 per share) but no cash or other assets. Comparisons are difficult, but it may well be true, as one historian has claimed, that the transaction represented one of the largest transfers of assets in industrial history up to that time. It also marked, or seemed to, the final success of Sarnoff's drive for "unification." At the same time, because of their holdings of RCA's stock, GE and Westinghouse retained a large commitment to the firm's success.

The agreements to carry out the transfer were signed on 23 April 1930. On 13 May the Department of Justice filed an antitrust suit against RCA and its associated companies alleging violations of the Sherman Act.[40] The timing struck some observers as too close to be coincidental, and Young for one believed that the indictment had been filed to prevent the unification agreements from going into effect. Certainly he and Sarnoff had underestimated the impression they would make on the public: what from inside RCA looked like a sensible reallocation of functions to the outside world seemed like the ultimate centralization of power in the "radio trust," with perpetual control vested in the giants of the electrical industry, General Electric and Westinghouse.[41] But, although announcement of the unification agreements may have affected the timing of the government's action, the indictment had clearly been in preparation for some time. RCA and the constellation of corporations linked to RCA had from the beginning offered a tempting target for antitrust lawyers, and a documentary record adequate to support a plausible indictment under the Sherman Act had been fully laid out by earlier inquiries. Nevertheless, the spring of 1930, with the financial community still jittery after the market shocks of the previous fall, seemed a strange time for a Republican administration to launch a major attack on big business, and it is probably true that political considerations had their influence. It was highly desirable for the Hoover administration to present itself as able and willing to stand up to the large corporations and defend the rights of the consumer and the small businessman. Owen Young was already being mentioned in knowledgeable circles as a likely Democratic candidate for the presidency and if, by attacking the corporations with which he had been so closely identified, his reputation could be tarnished a little, there might be partisan advantage in that too.

The FTC's investigation of 1922-1923 and its formal complaint filed in 1924 focussed on RCA's business practices, particularly its licensing procedures.[42] The 1930 indictment by the Department of Justice, in con-

---

[40] Petition in Equity No. 793, U.S. District Court, Delaware, in *United States of America v. Radio Corporation of America et al.*, (1930). Defendants named in the original petition were AT&T, Western Electric, RCA, General Electric, Westinghouse, RCA Photophone, RCA Radiotron, RCA Victor, the General Motors Radio Corporation, and General Motors itself (which had, with RCA, formed General Motors Radio in 1929 to exploit the market for automobile radios). In the early part of 1932 an amended and supplemented petition was filed which named as additional defendants International General Electric, Westinghouse Electric International Company, National Broadcasting Company, and RCA Communications.

[41] See, for example, the remarks of Senator Clarence Dill, as summarized in Case and Case, *Young*, p. 497.

[42] The FTC complaints were dropped in 1928. Carl Dreher (*Sarnoff: An Amer-*

trast, was aimed at RCA's structure. The department had two major objectives: first, to compel RCA to sever its corporate ties with GE and Westinghouse; and second, to eliminate all exclusive features from the cross-licensing agreements. Neither of these features was new: they dated from the corporation's earliest years and had been matters of public knowledge at least since publication of the FTC Report. What led the Justice Department to base an antitrust indictment on them in 1930 was its recent success in a case involving gasoline cracking patents in the oil industry.[43] In that case the lower courts had held that exclusive agreements based on patent pooling and patent licensing could be in violation of the Sherman Act. This was novel doctrine, and of course the decision was appealed. At the time the proceedings against RCA were initiated the appeal had not yet reached the Supreme Court. When it did, the decision was reversed. But in 1930 the attorney general had reason to believe that he had new grounds for an attack on RCA and its associated corporations.

Young put the essential issues succinctly when urging his friend Charles Neave to lead RCA's defense team. The suit was based, he wrote, on the theory that the original integration of patents, carried out during the formation of RCA, violated the Sherman Act because it tended to suppress competition. Further, the subsequent agreements on patent licenses, along with the stock interest held by GE and others in RCA, likewise suppressed competition because they extended the monopolistic effect of the patents beyond their expiration dates. RCA and its associated companies held that the original setup and the cross-licensing agreements were legal; the Justice Department held that they were not.[44]

The situation was a very dangerous one for RCA for a reason that may not be immediately obvious. The Radio Act of 1927, in specifying the powers and responsibilities of the new Federal Radio Commission, had laid it down in Section 13 that no licenses for radio transmission were to be issued to any individual or corporation that had been finally adjudged guilty by a federal court of unlawfully monopolizing or at-

---

ican Success [New York, 1970], pp. 134-35) refers to antitrust proceedings against RCA initiated by the Department of Justice in 1924, but this seems to be an error.

[43] Case and Case, *Young*, p. 499.

[44] This represents a drastic summarization of a highly complex issue, but it is hoped that it will suffice for a lay interpretation. For more professionally phrased statements of the legal issues, see Young to Neave (radiogram), 16 May 1930, reprinted in Case and Case, *Young*, p. 500, and FCC, *Walker Report*, Vol. 14 of Exhibits, "Report on Bell System Policies and Practices in Radio Broadcasting," 1 December 1936, pp. 566-68.

tempting to monopolize radio communication through control of the manufacture or sale of apparatus, through exclusive traffic arrangements, or by any other means.[45] RCA had already had one close encounter with this clause of the statute as a result of its unsuccessful attempt to force the De Forest Company out of commercial tube manufacture. The company's receiver in that case (it had subsequently gone bankrupt) had charged that RCA's practices violated the Clayton Act; the U.S. District Court had agreed, and the decision had been upheld on appeal. RCA had escaped punitive action in that case only by the delicate argument that the monopolistic practices referred to involved equipment only and not "communication," but it had been a close shave, with two of the five commissioners voting to void RCA's licenses.[46] Final conviction on antitrust charges, in short, would very probably put RCA out of the communications business, both for broadcasting and for long-distance traffic. It could not function without station licenses.

In view of the uncertainties, RCA and its associated companies had a strong interest in preventing the case from coming to trial. No mere change in current corporate behavior, however, could be an acceptable response to the government's charges. The complaints were levied at the very structure of the organization, and the attorney general showed no inclination to dismiss them, or agree to a consent decree, in the absence of firm assurances that changes in that structure would be made.

A consent decree, the professorial cliché has it, preserves the competitors but not necessarily competition. The parties against whom the complaint has been lodged admit no guilt, but promise that in future they will not behave as they have in the past. In the case of RCA and its affiliated companies the question was less behavioral than it was existential: RCA, GE, Westinghouse, AT&T, and Western Electric agreed that in future they would not bear the same relation to each other as they had in the past. In the case of AT&T and Western Electric this presented no great problem. Since 1922 they had held no stock interest in RCA, so no question of divestiture could arise; and their executives insisted that, if they had had their way, the patent licensing agreements would have been nonexclusive from the start. To make assurance doubly sure, on 18 December 1931 the Telephone Company gave GE written notice of cancellation of the 1926 license agreement, and a new contract approved by the Justice Department was executed on 1 July 1932. That done, AT&T and Western Electric were essentially out of the conflict.[47]

[45] The Act is reprinted in Appendix B, pp. 300-15, of Barnouw, *History*, 1.

[46] Barnouw, *History*, 1: 256-57; Federal Radio Commission, *Fifth Annual Report, 1931* (Washington, D.C., 1931), pp. 9, 74-75.

[47] For reactions to the new agreement by Telephone Company personnel, see

RCA, GE, and Westinghouse found no such easy relief, but even they, once it was clear that the Justice Department would not drop the suit, seem to have accepted a consent decree as the least undesirable outcome possible. There was a sense, indeed, in which the changes the Justice Department wanted were no more than an extreme form of Sarnoff's "unification" program. That program had not contemplated more vigorous competition among the three firms: as Sarnoff saw it, GE and Westinghouse would withdraw from radio manufacturing and leave that field to RCA. Nor had it contemplated requiring GE and Westinghouse to dispose of their stock interest in RCA and withdraw from participation in direction and management. In these respects what the Justice Department wanted went beyond what Sarnoff had thought he could get. He was not inclined, however, to oppose the general thrust of the department's program, provided that the interests of RCA were protected.

The date for the trial had originally been set for 15 November 1932. The problem was to work out, before that date, such a rearrangement of assets and liabilities among the three firms as would enable each of them to function in the radio industry, offer at least the prospect of new and vigorous competition in the future, and meet the Justice Department's stipulations for divestiture and nonexclusive pooling of patents. Eighteen months of hard bargaining followed. Not until 21 November 1932 (trial having been postponed), were the terms of a consent decree acceptable to the three corporations and to the Justice Department delivered to the presiding judge.

In every major respect the department got what it had sought from the beginning. All the cross-licensing agreements were rewritten so as to purge them of their exclusive features; GE and Westinghouse agreed to dispose of all their stock in RCA within three years and to refrain from holding or acquiring such stock in future; they were no longer to have seats on RCA's board of directors (though Young and Robertson, by special dispensation, were allowed to stay on for a further five months); and, after a two-and-half-year period to permit it to reorganize its resources, RCA was to be subject to the open competition of its former

FCC, *Walker Report*, pp. 264-65. F. B. Jewett, president of Bell Laboratories, concluded that "while a casual reading of the agreement by one not thoroughly conversant with all the factors may appear to establish the basis for an enlarged free development in most of the fields, this is not actually the case." He saw no need for any of the companies to fear new competition in fields where they had already attained a "commanding position." G. E. Folk, AT&T's general patent attorney, concurred; in fact, he said, if Jewett's interpretation were not correct, AT&T would never have signed the agreement.

parents.[48] It is easy to understand, indeed, why the radio decree is commonly regarded as one of the great achievements of the Sherman Act, comparable to the meat-packing decree of 1920 and the dissolution of Standard Oil in 1911.[49]

But if the outcome was a victory for the Justice Department, it was by no means a defeat for RCA. Not part of the consent decree itself, but an essential element in the negotiations leading up to it, was a series of transfers between the three corporations that in several ways left RCA in a stronger position than it had occupied before—to the extent that any corporation's position could be called strong in 1931-1932. It was the task of working out these trades, rather than acceptance of the Justice Department's stipulations, that had prolonged and complicated the negotiations.

All parties seem to have agreed, almost from the beginning, that in the final outcome RCA would have to be left a viable organization, with a fighting chance for survival in complete independence of its parents. In principle this was not something to be taken for granted: the Justice Department could have demanded the dissolution of the company. One vital but easily overlooked hurdle in the negotiations may well have been passed when the department was induced to admit that it saw nothing illegal in the way RCA had been set up in the first place. That granted, the problem became one of ensuring that, when GE and Westinghouse divested themselves of ownership, RCA could still function. And here the Radio Corporation's very vulnerability became, for Sarnoff, an important bargaining asset.[50]

Already, in his successful drive for "unification," Sarnoff had established the principle that RCA should be an integrated unit, with its own manufacturing, design, and research facilities; now he contended that it had to be financially strong if it were to survive. This implied, first, that RCA should retain all the assets that had been allocated to it in the unification agreement of 1930; and second, that its outstanding debts to GE and Westinghouse should be drastically reduced. He got what he

[48] The provisions of the consent decree are conveniently summarized in FCC, *Walker Report*, Vol. 14 of Exhibits, pp. 64-68; Case and Case, *Young*, p. 592; Barnouw, *History*, 1:267; and Archer, *Big Business*, pp. 378-79.

[49] Later critical analysis, however, has not entirely validated the euphoria that greeted these antitrust decisions at the time. See, for example, on the meat-packing decision, Robert M. Aduddell and Louis P. Cain, "Public Policy Toward 'The Greatest Trust in the World,' " *Business History Review* 55 (Summer 1981), 217-42; and by the same authors, "The Consent Decree in the Meatpacking Industry, 1920-1956," ibid. 55 (Autumn 1981), 359-78.

[50] Compare Case and Case, *Young*, p. 593.

wanted. RCA's unfunded debt to its two parent corporations in November 1932 was approximately $18 million, two-thirds of it to GE.[51] Of this, approximately half was cancelled outright. To offset the rest, RCA transferred to GE $1,587,000 in debentures plus the RCA Building on Lexington Avenue in New York City (valued for the purpose at $4,745,000); Westinghouse received $2,668,000 in debentures. This was a very substantial writing-down of corporate obligations. On the other hand, it saddled RCA for the first time with a burden of fixed-interest debt, and in the disastrous market conditions of 1932 (RCA had a net loss of over $1 million in that year) this was not something to be taken lightly. But there were offsetting considerations. RCA retained all the gains it had won in the unification agreements. It was protected from GE and Westinghouse competition for an initial two and a half years. It had unrestricted rights—albeit nonexclusive rights—to all GE, Westinghouse, and AT&T patents. And above all, perhaps, as Sarnoff saw the world, he was at last free to run his own corporate empire as he wished. In that sense the Department of Justice had won for him prizes he could never have won for himself.

* * *

The RCA that emerged from the consent degree of 1932 was a very different organization, in both structure and function, from that which GE and American Marconi had created in 1920. What can be said in general terms about the processes that brought about this transformation?

There were, in the first place, the internal dynamics of the organization itself. As with any organization, RCA served as a vehicle for the hopes and ambitions of its members. In this case it was profoundly affected by the drive for power of the man who became its president in 1930. But Sarnoff's hunger for greater freedom of action, his striving to get out from under the protective oversight of General Electric and Westinghouse, both reinforced and were reinforced by market imperatives. In the market for radio equipment that developed after the rise of broadcasting, RCA was seriously handicapped by its dependence on design and production facilities that it did not itself control. Smaller firms, even under the competitive handicap of having to pay royalties to RCA for licenses under that firm's patents, consistently showed themselves more responsive to consumer preferences and quicker to introduce the innovations in design and construction that consumers demanded. RCA's satellitic relation to GE and Westinghouse may not have been the only

[51] For the exact figures, see Archer, *Big Business*, p. 376.

factor responsible for its low market share, but it was perceived at the time to be a large part of the explanation, and this lent force to the drive for autonomy.

Secondly, there were dramatic changes on the demand side of the market. RCA had been created to serve and develop the market for intercontinental and marine radiotelegraphy. Its establishment was one element in a positive strategy of technological management, tailored to that market. Within three years of its founding, however, it was facing a new and quite different market: that for broadcasting equipment and services. And from this market still others developed: for equipment for talking pictures, for phonographs and recordings. These markets developed because the technology of continuous wave radio proved to be more versatile than had been anticipated. RCA was not well prepared to cope with these new developments, either in its structure or in the attitudes of most of its members, and for much of the 1920s, instead of aggressively managing the application and diffusion of radio technology, it found itself reacting to threats and challenges impinging on it from the outside. Only toward the end of the decade does one sense a return to more confident and aggressive strategies. The creation of broadcasting networks implied the reimposition of more centralized control, which the original proliferation of independently owned local broadcast stations had threatened. And success in persuading the Federal Radio Commission to grant the preferred "clear channels" to the powerful network-affiliated stations worked in the same direction.

These demand-side shifts caused acute strains within the structure of RCA, cast the corporation in a new social role, and exposed it to a new set of expectations. Internal conflict between the Radio Group and the Telephone Group reflected the inadequacy both of the original cross-licensing agreements and of the procedures for resolving conflicts that had been incorporated in them. Sarnoff's call for "unification" was essentially a strategy designed to break RCA loose from GE and Westinghouse and give the corporation greater autonomy and mobility in an increasingly competitive market. Involvement in that market, a market in which RCA sold to homeowners and competed with small business, exposed the corporation to new political risks, and in the face of those risks RCA again found itself trapped in a defensive, reactive mode. RCA was now seen not as the defender of American national interests against a potentially hostile outside world, but as itself a threat to American values. Resentment against RCA's licensing policies and fears of the concentrated economic power that it represented—particularly ominous when it involved control over information—resulted in antitrust indictment and, with the consent decree of 1932, abandonment of exclusive licensing and separation of RCA from its corporate parents.

Institutional considerations such as these explain much of the history of RCA, and of the radio industry in general, in the 1920s. There were internal drives for greater autonomy. There were external changes in markets, public expectations, and the political environment. And out of the interaction of these forces emerged a new structure for RCA and for the industry. Underlying these processes, however, was something more simple and fundamental. This was the unfolding of continuous wave radio technology—the progressive opening-up of new uses, new markets, new possibilities for further development. At the heart of this process lay that most versatile of electronic devices, the vacuum tube. Without cheap, efficient and reliable vacuum tubes there would have been no broadcast industry, no cheap receivers in the home, no low-cost broadcast transmitters—and no talking pictures or high-quality sound recordings either. In that sense it was the unexpected versatility of the vacuum tube, the seemingly inexhaustible potential for further development that was implicit in the device, and the unpredictable course of that development, that kept the radio industry in constant flux through the 1920s and later. Arcs and alternators had opened the age of continuous wave radio. They were the devices that made spark obsolete and made the transmission of speech and music possible. But their potential for development was soon exhausted. Not so with the vacuum tube.

During the 1920s these potentials were being explored. It was not just a matter of new types of tubes—the screen grid, the pentode, the water-cooled power tube, and so on—nor of new circuit arrangements, important though these were. Much more profound in its long-run implications was the fact that, after a long pause, the frontier of development once again began to move into new regions of the electromagnetic spectrum.

Vacuum tubes made this possible. Arcs and alternators functioned in the very low frequency range—Carnarvon with a wavelength of 14,000 meters, Tuckerton with 16,800: these were typical frequencies—and although it was probably not impossible to get them to radiate at much higher frequencies, it would have been very difficult to do so. Furthermore, there seemed little point in trying. Everybody knew that, for long-distance work, you had to use long wavelengths. An image used by Alexanderson was very persuasive. Stand on the shore and watch the waves come in: the little ripples never travel very far, but the great long swells may have traversed the ocean.[52] And experience seemed to support

[52] Ernst Alexanderson in IRE *Proceedings* 9 (April 1921), 83-90, quoted in

the generalization. From this many things followed: the need for very high transmitting antennas (a substantial fraction of a wavelength); for very high power, so that the ground wave, despite large losses from absorption, could radiate over great distances; and, since large antennas and powerful transmitters were costly, for large and heavily capitalized organizations to run them. Reinforcing this was the fact that, in the very low frequency range of the spectrum, each clear channel took up a large fraction of the available "space." Hence there could never be more than a score or so stations functioning on those frequencies; once they were built and operating, it was virtually impossible for a new station to find a channel. In these circumstances "squatter's rights" to the spectrum were what counted, and international conferences could do little more than endorse and register the allocation of frequencies that had already taken place.

The discovery of shortwave long-distance propagation in December 1921 was largely the work of American amateurs, but the Marconi Company and RCA were not far behind.[53] The consequences for radio technology were far-reaching. In the first place, it relieved congestion at the very low frequencies and opened up the radio spectrum to new entrants. Secondly, it sharply reduced the costs of building and operating a telecommunications system: with a directional antenna, a good superhet-

James E. Brittain, ed., *Turning Points in American Electrical History* (New York, 1977), pp. 207-18.

[53] Clinton B. DeSoto, *Two Hundred Meters and Down: The Story of Amateur Radio* (West Hartford, Conn., 1936), esp pp. 70-78; John Clarricoats, *World at Their Fingertips* (London, 1967), pp. 62-72. As might be expected, priority in the discovery of long-distance shortwave propagation is hotly disputed. It appears well established, for example, that H. J. Round of British Marconi, using a 100 meter wavelength, conducted successful long-distance tests in 1920, one year before the amateur transatlantic tests, and that Frank Conrad of Westinghouse began experimenting with transmissions on a wavelength of about 100 meters in 1921. See Pratt Papers, Armstrong to Pratt, 19 January 1953, and Pratt to Editor, *Communications and Electronics*, 16 June 1953. British Marconi began test transmissions on 32 meters from Poldhu in 1924, and signals were well received in Montreal, Buenos Aires, and Sydney. During 1923 RCA constructed and operated a shortwave transmitter at Belfast, Maine, using wavelengths of 120, 90, 60, and 40 meters, but the results of these tests were overshadowed by the surprising success of the Poldhu experiments. What radio engineers considered truly remarkable about these tests was that such distances could be covered on the short waves during daylight. For an interesting exchange between Edwin Armstrong, C. B. Jolliffe, and H. H. Beverage on the relative contributions of the Marconi Company and RCA, see *FM and Television News*, July, August, and September 1948.

erodyne receiver, some empirical knowledge of radio propagation, and access to the short waves, an amateur could now do with 1 kilowatt what RCA and Marconi had been trying to do with 200. And thirdly, it reminded radio engineers of a fact they had come close to forgetting: that the radio spectrum was open-ended, up to the frequencies of infrared light.

This was perhaps the most important consequence of all, for the exploration of ever-higher frequencies and the design of equipment for those frequencies was to be the central theme of radio research for the next half century. As that exploration proceeded, new frequencies became available for use, each new segment a massive increment to the resource base. Without these new frequencies, television, FM broadcasting, radar, microwave networks, and satellite relays would have been impossible. In that sense the opening-up of the short waves in 1921-1923 was fundamental to all later development.[54] Underlying that accomplishment, however, was the vacuum tube oscillator. And underlying that was a concept that had once been no more than a dream and an ideal: continuous wave radio.

[54] Including the invention of the transistor, since it was the inherent limitations, in terms of electron transit time, of even the smallest vacuum tubes at ultrahigh frequencies that impelled the search for a solid-state amplifier and oscillator.

ELEVEN

Epilogue

⎍〰〰〰 IN 1925 Ernst Alexanderson, then at the height of his reputation, addressed the American Institute of Electric Engineers on "New Fields in Radio Signalling." Radio technology, he said, had gone through a period of rapid change and was at that moment catching its breath before starting out on new developments. It was experiencing "one of those pauses . . . which occurs in every engineering development," when technique had caught up with commercial requirements and was enjoying a breathing spell before embarking on new efforts. Such pauses in the advance of technology were necessary, he argued, because in its initial stages development was always carried on at a loss, and if it were not for these breathing spells, when the innovations of the previous phase began to pay off commercially, the sources of financial support would soon dry up.[1]

To some of the audience the idea that radio technology was enjoying a breathing spell of any kind in 1925 may well have seemed odd. Any engineer involved in vacuum tube development, or in the design of broadcast transmitters and receivers, or in the exploration of shortwave ionospheric propagation could pardonably have believed that he was struggling up a steep technological incline, not resting on a plateau. Nevertheless, when Alexanderson said that in 1925 an era of technological development in radio was drawing to a close, he was in one sense perfectly correct. By that year continuous wave radio had come to be accepted as the norm. The technical challenge for radio engineers now was to explore its full potential.

[1] E.F.W. Alexanderson, "New Fields for Radio Signalling," *GE Review* 27 (April 1925), 266-70. For a general analysis of technological plateaux, see Devendra Sahal, *Patterns of Technological Innovation* (New York, 1981).

This was a radical change. Between 1912 and 1925 radio technology had gone through a revolution. It was not just a matter of new hardware. Ideas had changed too. Radio was now conceptualized in terms not of spark discharges but of continuous waves of constant frequency. This was what men like Fessenden and a few others had dreamed of, long before they knew how to accomplish it. The period between 1912 and 1925 was the time in which the dream became reality. There were still thousands of spark transmitters in existence but everyone who knew anything about radio recognized that they were obsolete, the inconvenient residue of a technology now dead. The future lay with the continuous wave.

Three innovations had made this possible: the alternator, the arc, and the vacuum tube. In the case of two of these, the alternator and the arc, 1925 did indeed mark the end of an era. Technically, the Alexanderson alternator had been perfected by 1919. The design was frozen at that date and there were no major improvements or modifications thereafter. Economically, the process of diffusion began in 1917, when the prototype 50 kilowatt unit was installed at New Brunswick, and ended in 1924, when the last two 200 kilowatt units were shipped to Brazil, only to be ignominiously returned to the United States for scrapping three years later. In all, twenty of the big 200 kilowatt machines were manufactured. Two were sold to British Marconi and installed at Carnarvon in 1921. Two were sent to Poland for the Warsaw station that went on the air in 1923. Two equipped Sweden's station near Varberg, opened in 1924. Two, as mentioned, were intended for Brazil. And the rest went to RCA: two each at New Brunswick, Marion, Bolinas, Tuckerton, Kahuku, and Radio Central.[2] Few major innovations have undergone so little further development after first introduction. Few have been diffused so rapidly.

Diffusion of the arc transmitter began earlier but ended at about the same time. By 1917 Federal arcs were standard equipment at all U.S. Navy high-powered radio stations and several smaller units had been installed on battleships. When in 1918 the Navy designed the Lafayette station to serve as the European terminal for military and diplomatic radio traffic, it was taken for granted that a Federal arc would be installed.

[2] Thorn L. Mayes, "The Alexanderson 200-kw. High-Frequency Alternator Transmitters," Society of Wireless Pioneers, *Historical Papers*, "Ports O'Call," 4 (n.d.), 37. Alternators were also manufactured in Japan from General Electric blueprints by the Shibaura Company. A 400 kilowatt installation (presumably a pair of 200 kilowatt machines) was completed in 1922 at Haranomachi station and was used for communications with the United States. (Personal correspondence, H. Kaji to the author, August 1980).

That, however, was the last arc station the Navy built. Arcs were to have been installed at the Monroe, North Carolina, station, intended for Caribbean and Latin American communications, but plans for that installation were cancelled in 1920. No new Navy arc stations were built thereafter, and existing installations were phased out as suitable tube transmitters became available. As for civilian use, the revived Federal Company after 1919 intended to use arcs for its proposed China stations, but these plans never reached fruition. Apart from marine installations, diffusion of the Federal arc effectively ended with the end of the First World War. Minor technical improvement continued thereafter, but there were no major new developments in arc transmitter design and no large new units were built.

In other parts of the world installation of arc transmitters petered out not long after. C. J. de Groot's station in Java, which went on the air in 1923, seems to have been the last of the high-powered arcs.[3] When in 1923 C. F. Elwell published his monograph on the Poulsen arc generator (part engineering treatise, part sales promotion literature), he was able to list, throughout the world, no fewer than seventy-eight arc-equipped stations of more than 25 kilowatts input power.[4] That, however, was the end of the road. Elwell's book summarized a mature technology but generated no new orders. By the 1920s the market for arc transmitters had disappeared, as had the market for alternators. With tube technology now available, engineers had no interest in further work with alternators and arcs, and operating companies had no interest in purchasing them. When, after 1923, the short waves were opened up for long-distance transmission, tube equipment monopolized the field.

When, therefore, Alexanderson suggested in 1925 that radio technology was enjoying a breathing spell before attempting further advances, he was using the metaphor in a particular sense. There was no pause in tube development in the mid-twenties. There was no pause in the exploration of new modes of radio propagation and new ranges of the spectrum. And there was no pause in the application of radio technology to new uses. In one sense only was the metaphor appropriate: deployment of arc and alternator transmitters was, by 1925, complete. These devices

[3] Kaye Weedon, "PKX-Bandung: The Story of de Groot's Mountain Gorge Antenna and Giant Arc Transmitter at Malabar, Java, 1917-1927" (Unpublished slide lecture); Pratt Papers, Federal Telegraph Company file, Box 1, report by R. A. Lavender, 22 December 1923; "Radiotelegraphy in the Dutch East Indies," Radio Review 2 (November 1921), 574-82; "The High Power Station at Malabar, Java," IRE Proceedings 12 (December 1924), 693-722.

[4] C. F. Elwell, The Poulsen Arc Generator (New York, 1923).

had led the way in establishing continuous wave transmission as the standard technology of radio. That revolution was now over.

* * *

A technological shift of this magnitude is, in effect, a discontinuity in history, and in any field of inquiry the analysis of discontinuities presents difficulties. A detailed account of what actually happened, such as we have given in the preceding chapters, can provide the empirical groundwork, but in itself it leaves one dissatisfied. We want to know why the discontinuity happened and what its significance was, and a narrative history does not answer those questions. Each of the new continuous wave generators had its own history. Each was "managed" in a different way, and with different consequences, by the business and governmental institutions affected. There are a few gaps in the evidence but at the level of particular events we know a lot about what happened. This does not absolve us, however, from raising more general questions. What can be said in general about the way this technological shift occurred and the way it was handled? What light does it throw on the general problem of understanding how technologies change and how they influence, and are influenced by, human thought and action?

To ask such questions is to raise issues of philosophy, ideology, and method, and in all three respects we do not start with a clean slate. The history of technology has its greatly respected pioneers: men like Lewis Mumford, A. P. Usher, and Lynn White, Jr., in the United States; T. S. Ashton and L.T.C. Rolt in Britain; Marc Bloch in Belgium. These were, however, individual scholars, each with his own point of view and method, and none of them set the stamp of his philosophy or method on the field as a whole. General interest in the history of technology as a field for academic research and teaching began, in the United States, in the 1950s and 1960s. In the early years it was motivated largely by persistent misgivings in schools of engineering about the next generation of engineers.[5] Suggestions that engineers were being too narrowly trained, that they emerged from their schooling with the mind-set of technicians rather than professionals, resulted in curricular reforms that required engineering students to take a certain proportion of their courses in the humanities and social sciences. When these requirements proved unpopular and the courses selected not always well-suited to the needs of the students, there developed within a few engineering schools courses in the history of

[5] Eugene S. Ferguson, "Toward a Discipline of the History of Technology," *Technology and Culture* (January 1974), 13-30.

technology—a move that, as Eugene Ferguson puts it, was at first "un-noticed by historians generally and quietly tolerated by a preoccupied faculty of engineering."[6] In response partly to these developments, partly to a more general groundswell of interest, there came into existence a professional association of historians of technology and, not surprisingly, a concern over how the history of technology should be written and whether the new generation of scholars working in the field could do any better than their predecessors. What were the appropriate historical methods? Was a "discipline" of the history of technology possible?

The fact that much of the support for teaching and writing in the history of technology at this time came from the engineering community had certain implications for the way the subject was approached. At first, for example, it reinforced an implicit assumption that technological improvement always improved social welfare, or at least that the burden of proof was on those who said it did not. And it encouraged a belief that what retarded such improvement, and hence limited the contribution that technology could make to welfare, was the failure of human institutions and human personalities to adjust quickly enough. Technology, that is, was seen as the dynamic, constructive force and society as the confining shell that constrained its beneficent effects.[7] These assumptions, of course, were seldom stated explicitly and would hardly have withstood critical scrutiny if they had been. But they underlay much of the more popular writing that was done, and some of the teaching, too.

During the 1960s interest in the history of technology as part of the education of engineering students intersected with a second set of concerns. On the one hand there was, after Sputnik, a general anxiety over whether the nation had lost the technological "edge" that, many believed, had served it well in the past. On the other, there was growing concern over the effects of technological change on the environment and on the nature of warfare. These concerns gave to the study of technology an urgency and immediacy it had lacked before. They also brought a shift of emphasis. Technology now appeared as a threat, not a blessing, as a daimonic force in human affairs, often injurious to the natural environment and, in the form of nuclear weapons, potentially terminal for life

---

[6] Ibid., p. 15.

[7] See George H. Daniels, "The Big Questions in the History of Technology," *Technology and Culture* 11 (January 1970), 1-21. Daniels cites the writings of Roger Burlingame as representative of this philosophy but recognizes that even Burlingame "at times took a rather dim view of the technology which he invested with such active force." For a general survey of trends in the historiography of technology, see John Staudenmeier, *Design and Ambience: Historians and Technology, 1958-77* (Ann Arbor, Mich., 1980).

on the planet. The central question now seemed to be not how the rate of society's adjustment to technological change could be accelerated but how—indeed, whether—society could bring its technology under control.

Teaching and writing in the history of technology have not been insulated from these shifts in public attitudes. There are two areas in particular in which the field reflects the contexts in which it has developed. In the first place there has been a continuing interplay between two contrasting approaches, often labelled the externalist and the internalist. The internalist approach, which comes naturally to historians with a strong engineering or technical background, tries to identify the inner logic by which a particular technology develops: the genealogy or line of descent, as it were, through which a particular device or system has evolved. The externalist approach, on the other hand, which appeals more to the humanists and social scientists working in the field, concentrates on the interrelations between a technological system and the context (social, economic, political, and so on) within which it functions. The emphasis is on how particular elements in that context influence the course of technological change, and on how technology influences the course of political, economic, and social development.

The second polarization that has emerged in the historiography of technology is between those who view technological development as essentially a controlled or managed process and those who see it as following its own autonomous logic. David Noble, for example, depicts modern Americans as confronting "a world in which everything changes, yet nothing moves."[8] Continuous revolutionary change in the forces of production—that is, in technology—seems to leave undisturbed a static social structure. What has happened, Noble asks, to the classic Marxian dialectic, in which changes in the forces of production, generated within a social order, nevertheless undermine that order? What has neutralized that great dynamic of history? Why is technology, at least in America, no longer a truly revolutionary force? He finds part of the answer in the way in which the modern corporation has learned to manage technological change for its own ends, largely by co-opting professional engineers, the prime generators of scientific technology, into corporate management and identifying the ideology of the engineer (the design and operation of rational systems) with the ideology of the corporation itself. Noble sets himself apart from those who think of modern technology as if it had a life of its own, an immanent logic that transcends the desires

---

[8] David F. Noble, *America by Design: Science, Technology, and the Rise of Corporate Capitalism* (Oxford, 1977), p. xvii. Quotations from this book by permission of the copyright holder, Alfred A. Knopf, Inc.

and intentions of its creators. To him such views are part of the mystification of history. Technology is not a disembodied force: it is part of social existence and in the modern world it is thoroughly and continuously "managed." Hence its very limited ability to effect social transformation.

The school of thought usually regarded as of opposite polarity to Noble's is represented by the work of Jacques Ellul, particularly as it has been expounded and elaborated by Langdon Winner.[9] For Ellul, technology (if this is the proper translation of his term, *la technique*) is very far from being controlled or managed by human beings in the contemporary world. On the contrary, it is technology that controls us—or, perhaps better expressed, has swallowed us up. It is the environment in which we live, as a fish swims in water, and for the most part we take it for granted, until something goes wrong.

At one level of analysis, the notion that technology can control human beings is, as Winner points out, "patently bizarre," and George Kateb among others has poured scorn on the idea that somehow machines can develop a volition of their own and exchange roles with their creators.[10] There is, however, more to Ellul's position than that; his arguments raise questions about the interaction between technology and human choice that are not easily dismissed.

The crux of the matter is that, for Ellul, *technique* is much more than machinery. The essence of technique is method and organization, and in that sense the idea that technique has come to dominate life in the late twentieth century becomes much more plausible. Obsession with technique, with the search for the most efficient method, has indeed become central to our lives, both in their public and in their most private aspects. Ellul's thesis, stripped to its bare essentials, is that we have become concerned with techniques—means of achieving ends—to the exclusion of concern for the ends themselves. This is what enables him to say that technique has become autonomous, that it has fashioned an "omnivorous world which obeys its own laws."[11] It is now "a reality in itself, self-

[9] Jacques Ellul, *The Technological Society*, trans. John Wilkinson (New York, 1964); Ellul, *The Technological System*, trans. Joachim Neugroschel (New York, 1980); Langdon Winner, *Autonomous Technology: Technics-out-of-Control as a Theme in Political Thought* (Cambridge, Mass.: MIT Press, 1977). All quotations from *The Technological Society* by permission of the copyright holder, Alfred A. Knopf, Inc.

[10] Winner, *Autonomous Technology*, p. 13; George Kateb, *Utopia and its Enemies* (New York, 1963), p. 109.

[11] Ellul, *The Technological Society*, p. 14.

sufficient, with its own special laws and its own determinations."[12] It "maps its own route, it is a prime and not a secondary factor, it must be regarded as an 'organism' tending toward closure and self-determination; it is an end in itself."[13]

Whether this is an acceptable interpretation is a question not to be decided without detailed attention to the arguments and evidence that Ellul himself presents; it cannot be settled here. At the very least, however, it deserves to be taken seriously. Most of us, probably, if asked to name the institutions and practices that keep technology "under control," would mention the legal system, the political process, certain durable cultural values, and above all our confidence that, in the last analysis, we are free to choose, to accept technology or reject it as we wish. This is exactly what Ellul calls into question. He asks us to consider whether the legal system, the political process, and our cultural values, far from being restraints on technology, are not in fact extensions of technique, dominated by the search for rational efficiency that is the essence of the technological mentality. Our confidence in voluntarism he holds to be spurious. We may believe that we make free choices but in fact we do not. Technology forces the choice. "Man is absolutely not the agent of choice. He is an apparatus registering the effects. . . . Man decides only in favor of what gives the maximum efficiency."[14] Technology, as Ellul sees it, permeates society and culture; it fashions society in terms of its necessities; our choices are themselves part of the technological process and reflect the same concern with method, efficiency, and organization as do all other elements of that process.

It is easy to fault Ellul for his methodology. He defines technique so that it includes virtually every aspect of modern life, and concludes that technique is a self-determining system, that it is autonomous and follows its own laws. From such a definition it is hard to see how any other conclusion could emerge. Technology, the ensemble of techniques, becomes coterminous with society as a whole, including its values.[15] This

[12] Ibid. p. 9.

[13] Ellul, *The Technological System*, p. 125.

[14] Ibid. p. 239.

[15] Ellul's position shifted somewhat between publication of *The Technological Society* in 1954 and the appearance of *The Technological System* in 1980, perhaps in response to criticism. In the later book technology is depicted as a system that lives in and off society and is "grafted upon it." It is said to use society as an "underpinning." The criticism in the text is still valid, however, for Ellul insists that all elements of life are bound up with technology and that the "totalization" of technology has produced a "veritable integration" of all the human, social, economic, political, and other factors (*The Technological System*, pp. 203-204).

is why Ellul's theses are impervious to empirical testing; what he offers is not a theory but a vision, to be accepted or rejected according to the new insights it makes possible.

How can this kind of vision be reconciled with the point of view of a thinker like David Noble? One's first impulse is to deny that any reconciliation is possible. Technology is either subject to management or it is not; the course of technical change is either "by design"—somebody's design, if not ours—or it is not. To Noble, talk about autonomous technology is sheer mystification, and dangerous mystification at that. To Ellul, when we talk of managing technology we deceive ourselves, and our only hope of understanding the human condition is to escape from that self-deception.

Nevertheless, it is not impossible that there may exist a frame of reference comprehensive enough to accommodate both points of view. Exploring what such a frame of reference might be will bring us back to the debate between the internalist and externalist schools of thought, to the ideas concerning the nature of technology put forward in the Prologue, and eventually to our analysis of the innovation that was continuous wave radio. Let us bear in mind, however, that there are limits to what empiricism can accomplish; differences in ideology are not to be resolved by appeals to evidence.

The essence of the approach suggested in our Prologue was the conception of technology as a kind of knowledge—the kind that deals with our capacity to manipulate the natural environment for human ends. Knowledge is organized information. We argued that invention is best regarded as a process by which information comes to be organized in new configurations or gestalts; and we suggested that a useful strategy for studying such processes was to analyze the way in which flows of information previously separate are from time to time brought together, intentionally or by chance. The points of intersection or confluence of these information flows would probably prove to be, we suggested, the social locations where new combinations emerged. And we argued that the areas of overlap, where different social subsystems met and intermingled, were locations where distinct communications networks interconnected and therefore where the probability of confluence was high. Of special significance, therefore, were the individuals who functioned as translators at these locations and the institutions that developed in the areas of overlap, functioning partly as elements in one subsystem (for example, the economy), partly as elements in others (for example, the

For pointed comments on Ellul's use of systems theory, see the review by Steven L. Del Sesto in *Technology and Culture* 23 (January 1982), 81-85.

governmental and/or the technological systems). The way in which this model has shaped our interpretation of the emergence of continuous wave radio and the formation of RCA hardly needs to be underlined.

Noble, Ellul, and Winner have little to tell us about the emergence of technological novelty. They are concerned primarily with the nature and functioning of technological systems. Similarly, the distinction between the externalist and internalist approaches to the history of technology is more a difference of opinion over the forces that guide the direction of technological change than over the process of invention itself. Nevertheless, a conception of technology as knowledge and of invention as a process of combining and recombining information has implications for both these issues.

We are inclined to think of systems of knowledge as static. We tend to see them as bodies of information to be learned, like an encyclopedia that sums up what is known, or claims to. In fact this is seldom if ever the case. Any system of knowledge exhibits at least two dynamic tendencies. There is a straining toward internal consistency, toward the elimination of contradictions. And there is a straining toward extension, toward exploring how far the knowledge can be carried and what its limits may be. Only when each of these tendencies has fully played itself out can that body of knowledge be said to have reached stasis. Such a final state is theoretically conceivable but it must be very rare, if only because, as it is extended and elaborated, any body of knowledge will encounter anomalies that it cannot explain, phenomena with which it cannot deal. This is, of course, the essence of Thomas Kuhn's theory of scientific revolution. But the principle is of wider applicability. It holds for legal codes, for design principles, for a parent's hard-won knowledge of how to deal with an unhappy adolescent, as much as it does for scientific theories.

One way of expressing this point is to say that every paradigm has its limits. "Normal practice" in any field of activity consists in working within the accepted paradigm and takes the form of attempts to eliminate contradictions and explore the limits of applicability. This is familiar doctrine. Its relevance to the history of technology lies in the fact that this is the kind of work that an "internalist" approach necessarily highlights. The emphasis in this approach is on the internal dynamics of development of a technological system, and these internal dynamics are precisely the drive to consistency and extension that are the normal characteristics of any body of knowledge.

Edward Constant's theory of presumptive anomalies, discussed briefly in our Prologue, refers to situations in which the limits of extension are perceived prospectively by individuals working within an accepted body of technological knowledge and practice. When that occurs, there begins a movement to shift the acquisition of knowledge and the elaboration of practice onto a new track, one that represents a change of direction, the perception of the problem in different terms. As Constant emphasizes, the heart of the matter is a cognitive change, and his use of the adjective is significant. One comes to *know* in a different way, to see the situation differently, as when turning a kaleidoscope causes all the pieces to fall suddenly into a new pattern. This gestalt shift is essential to the kind of technological discontinuity that Constant refers to.

Technological discontinuities of this kind are hard to handle within a purely internalist frame of reference. They resist incorporation into an evolutionary scheme of the classic Darwinian type (although more recent models of evolution may be more hospitable). To explain the shift to a new technological paradigm one has to invoke exogenous forces. In Constant's model these are always injections of new knowledge or insights from science. This may well be, however, an unnecessary limit on the generality of his analysis.[16]

In terms of the model we have been developing here, Constant depicts a situation in which the inflow of new information from science precipitates the start of a process that ends up as the radical restructuring of a body of technological knowledge and practice. But, as we have indicated, the technological system participates in exchanges of information and resources not merely with science but also with government and the economy, to name merely the two other social subsystems we have explicitly specified. It may indeed be the case that in certain historical contexts infusions of knowledge from science played a uniquely significant role, but there is no reason in principle why this should be so. Science is not the only information-generating sector of society. Technology responds to signals not only from science but also from government and the economy. Constant himself has vividly described the partly nonrational motivation of those few "*provocateurs*" who initiate the shift away from an accepted paradigm. Insights from science may indeed give such "fanatics" a clue as to the new directions to follow. But their dedication to the cause may receive no less support from their anticipation that accepted practice will shortly meet failure in the marketplace. And analysis of the form such failure will take may furnish information as to what the new system must be capable of doing.

For reasons such as these, Constant's original emphasis on the uniquely important role of scientific information should probably be relaxed some-

[16] Compare Rachel Lauden, "Models of Scientific and Technological Change," *Technology and Culture* 23 (January 1982), 78-80.

what. Presumptive anomalies can arise for other reasons. Similarly, a question can be raised as to whether it is useful to distinguish so sharply between the kind of discontinuity that occurs in a paradigm shift and the more gradual and incremental modes of technological change. Are the social processes at work so different in the two cases? The analogy to Kuhn's theory of scientific revolution is seductively attractive, but technological systems are market-oriented as scientific theories are not. Their interaction with the economic system therefore differs. Both the rate and the direction of technological change are influenced by market signals in a way that is not true of science. And this is true both of the incremental changes characteristic of the development of any normal technology and of the larger discontinuities that, in Constant's terminology, are paradigm shifts.

A technological system, at any point in time, can be thought of as confronting several possible lines of development. These are the vectors, or trajectories if you will, along which that system can move; they represent the system's potential for change. Some of them represent no great divergence from the path the system has followed in the recent past. Others represent acute changes in direction. What determines which vector—or vectors, since several may be followed simultaneously—the system will follow?

An "internalist" answer to this question will run in terms of the internal logic of the system itself, or what we have referred to as its drive to eliminate contradictions and explore the limits of applicability. An "externalist" answer will, in contrast, emphasize the signals to which the system responds—the information generated in other sectors of society that, in effect, "steers" its development. The answers seem incompatible, but in fact they are not, for they are answers to different questions.

It is essential to distinguish between the variables that determine the set of technological vectors that a system faces at a given time, and those that determine the rate at which it moves along one or more of these vectors.[17] The former—the possibilities that are technically open for further development—are internally determined; they follow from the content and structure of the system as a body of knowledge. But in explaining the latter—the selection of vectors that are in fact followed and the rate at which the system moves along those vectors—a much larger role must be accorded to external, socioeconomic factors.

Consider the history of radio in these terms. If we take our stand, imaginatively, in the year 1912 and credit ourselves with good infor-

[17] A vector that is ignored or abandoned is of course one along which the rate of movement is zero.

mation about the state of radio technology in that year and its development in the recent past, we could specify a number of possible lines of development that it might follow in the near future. We could speak knowledgeably, for example, about the advanced forms of spark transmitter then becoming available. We would know of the new continuous wave generators. We would certainly have heard of Fessenden's heterodyne principle. And we would be aware, as a theoretical possibility, of that elusive goal of many experimenters, radio telephony. But we could not, merely from our knowledge of the technology itself in that year, predict which of the possible lines of development the radio art would in fact follow in the years ahead. To attempt that kind of forecast we would have to introduce information extraneous to radio technology itself. We would note the Navy's new interest in the Federal Company's arcs. We would wonder whether General Electric would continue building radio alternators now that NESCO was out of the picture, and to whom they would be sold. If we were very well informed, we might even speculate about what AT&T intended to do in radio, now that it owned the rights to de Forest's audion. And of course we would wonder whether spark radio was really on the way out, and whether those people would prove correct who had been saying for years past that spark was a dead end and the future belonged to the continuous wave. If these and similar speculations were to make any sense, if reasonable individuals were to give them a hearing, they would have to be grounded on information going far beyond radio technology itself. In particular they would require information on probable markets, investments of capital, rates of return, and public policy.

What is true of these particular devices and these particular corporate and government interests is true of the shift from spark to continuous wave transmission in the large. From at least the year 1900 it was recognized by some radio engineers and scientists, though not by all, that radio waves could be generated not only by spark discharges but also by high frequency alternating currents. From that point on, the continuous wave represented one of the possible vectors along which radio technology could travel. As understood by persons versed in the art at the time, it was a pronounced departure from conventional practice. Continuous wave transmitters and receivers were not incremental improvements on the equipment used for spark. To follow the vector of continuous wave radio called for a marked deviation from the path conventional wireless technology was following. The rate at which radio technology moved along that new vector was a function not only of the technical characteristics of continuous wave radio but also of the resources that could be attracted to that line of development. And these in turn were a function of the entrepreneurial abilities of men like Fessenden, Elwell, and de Forest, of corporate and government policy, and in the last analysis of information on the probable market performance of continuous wave equipment in competition with other modes.

A clear distinction between a set of feasible vectors of development facing a technological system at a given time, and the rate at which it moves along one or more of those vectors, should lay to rest any idea that internalist and externalist approaches are necessarily in conflict. They are not; they merely refer to different aspects or phases of the process. We have to recognize the selective influence of the socioeconomic environment on the set of possibilities that a technology confronts. This influence is made effective primarily through the allocation of scarce resources among alternative technological vectors. It is a matter of selective investment. Complications arise from the fact that these investment decisions are typically based on imperfect information and commit resources to a future that is imperfectly foreseen. Hence the highly speculative character of investment when a technological system passes through a sharp change in vectors, and hence too the emphasis on capital gains rather than income from operations in the motives of those making the investments: they are in the business of creating new capital values. But, these complications aside, the matter is clear in principle. The content and internal structure of a technological system do not uniquely determine the direction in which nor the rate at which it will be developed. These are determined by the interaction of the technology with its socioeconomic context, and in particular by the information and resources provided by the economic system and in some cases by government. Selective investment is the central process at work: it shapes the probability distribution of outcomes.

Let us turn to the second of our two polarities. On the one hand we have those who think of technology in the modern world as having been brought largely under social control. Its potential for bringing about radical social change has been for the most part neutralized and it now serves as one of the methods by which social order is maintained. The large bureaucratically organized corporation is viewed as the principal agent of technological management and its influence over the engineering profession and engineering education is seen as one of the main channels through which such management is exercised. This does not, of course, rule out cooperative action with the machinery of government. Indeed, in some versions of this theory, the distinction between governmental and corporate bureaucracies becomes largely irrelevant. Both are agents for the management of technology in the interests of social order, and their staffs of salaried experts are members of the same "technocracy."

Ellul's analysis moves on a different plane. He is concerned not with methods of social control but with the leading characteristics of modern culture—a culture that, in his view, has come to be pervaded by an obsession with technique, to the neglect of that concern with values and the ends of action that was once the hallmark of civilized behavior. To speak of the "management" of technology in the modern world makes little sense to people of Ellul's way of thinking, if by "management" is meant the application of a scale of values that is not itself determined by the obsession with technique.

Nevertheless, Ellul's vision and Noble's have more in common than their contrasting vocabularies and imagery might suggest, and once again the conception of technology as a form of knowledge or organized information can help to suggest a possible synthesis. If we take this conception seriously, the question of management or control appears in a somewhat different guise. In what sense can an individual, an organization, or a social class be said to be in control of, or to be managing, a body of information? Clearly, if they can exclude others from access to the information or if they have decision-making power over its use. In that sense information is no different from other forms of property: the test of ownership is the power to exclude others and to decide on disposal. This is what scholars like Harold Innis had in mind when, analyzing the history of communications, they wrote in terms of monopolies of knowledge and the degree to which different media—papyrus, clay tablets, the printed book, radio, and so on—lent themselves to the formation of such monopolies.[18] Some technologies—and the word in this context has to be used in the plural—are more easily managed than others.

Ellul's vision is of a world in which our thoughts, feelings, and actions are controlled by the large-scale technological systems in which we participate. He is not greatly concerned with particular technologies. He asks us to consider what it means to live in a world that has become obsessed with efficiency—method or "technique"—to the neglect of the ends that efficient means are supposed to serve. The management of technological systems is, for him, not problematic; it is taken for granted.

[18] Harold A. Innis, *Empire and Communications* (Oxford, 1950); "The Newspaper in Economic Development," in *Political Economy in the Modern State*, ed. Innis (Toronto, 1946), pp. 1-34; *The Press: A Neglected Factor in the Economic History of the Twentieth Century* (Oxford, 1949). See also William Christian, ed., *The Idea File of Harold Adams Innis* (Toronto, 1980) and, for a recent appraisal of Innis's thought, Daniel J. Czitrom, *Media and the American Mind: From Marx to McLuhan* (Chapel Hill, N.C., 1982), esp. pp. 147-82.

His interest is in what happens to us when the management of technological systems becomes the prime order of business for humanity, when
the relationship between tool and toolmaker is reversed and we become
the instruments of the technological systems that we have created.

To manage a technological system means to manage the information
on which it is based and that in turn means to control access to the
information and the uses to which it is put. Those who speak of a
technological system as being autonomous or "out of control" do not
mean that it is not being managed in that sense; they mean that those
who are doing the managing are following the system's own logic—they
are in that sense the instruments of the system's dynamics. Decisions
about access and use are being made in the interests of maintaining and
expanding the system, not in terms of criteria exogenous to the system.
A society that has reached that stage of development—one whose functioning has come to depend on the efficient operation and integration of
large-scale technological systems—is indeed a society functioning "by
design." That design, however, is set by the requirements of orderly
functioning, which is precisely what is meant by Ellul's "technique."

Langdon Winner approaches the issue by way of his concept of "reverse
adaptation." The conventional and idealized view is that, in rational
behavior, means are adapted to ends; technological rationality in particular implies the use of the most efficient means possible to achieve ends
that are "given." Beyond a certain stage of technological development,
however, according to Winner, we face precisely the reverse relationship;
ends are adapted to the means available, which are technological systems;
and rationality (if the word still has meaning in this context) implies
finding ends that will keep those means employed and smoothly functioning. This stage is reached when we find ourselves dealing, not with
individual tools or machines, but with large technological systems or
networks—"systems characterized by large size, concentration, extension, and the complex interconnection of a great number of artificial and
human parts."[19] Such systems have become typical of twentieth-century
technology and pervade all technologically advanced societies. They represent, Winner tells us, a quantum jump over the power and performance
of the smaller, more segmented systems of previous eras, and they impose
on individuals and societies imperatives that reflect their requirements as
systems.

Systems of the kind Winner is referring to—one thinks, for example,
of a nation's defense establishment, or an electric power grid, or of the
network of supply and communications associated with space explora-

[19] Winner, *Autonomous Technology*, p. 238.

tion—are characterized not only by large size but also by considerable internal complexity and specialization and by complex interactions with other systems. Control and planning become indispensable if the system is to avoid what Winner calls apraxia—the failure of the system as a whole because of the failure of a single component or a single set of interdependencies. But control and planning take on a special meaning in this context: it is outcomes that have to be controlled and planned in the interests of the instrumentalities. The system requires for its own regular and continued functioning that tasks be clearly specified—or created, if not already present—and that outcomes be predictable. Hence "reverse adaptation": it is now ends that must adjust to the requirements of means, not vice versa.

Winner does not use the term, but what he is referring to can be thought of as a kind of "system imperialism": the tendency of any large and complex system to reach out to control more and more of its environment in the attempt to reduce uncertainty and improve the odds for its survival and expansion. On a certain scale the phenomenon has long been familiar to students of bureaucracy, public or private. But Winner asks us to think of it as the dominant characteristic of the modern age. Underlying it is the logic of modern technology—the sheer brute efficiency of large-scale systems *if* the requirements of those systems are met.

There are, of course, alternative ways of regarding the bureaucracies that pervade the modern world, and some may be more impressed by their tendency to trip over their own feet than by their vaunted efficiency. That is not our present concern. Our interest is in the strategies they use to "manage" technology, and the question at issue is whether these strategies reflect a technological imperative of some kind or leave some room for the application of other criteria of choice—that is, for deliberate management. We raise these issues in the hope that they will throw light on our central theme: the introduction of continuous wave radio.

Winner suggests five strategies by which large-scale systems attempt to control their environments. They will, for example, try to control the markets for their inputs and outputs—markets that conventionally are thought of as providing an economic discipline that the organization must respect. They will exert themselves to control or influence the political processes that are in theory supposed to regulate their behavior. They will seek new "missions" when the function that was their original reason for being has been fulfilled. They will discover or create crises to justify their further expansion. And in general they will propagate or manipulate the needs they purport to serve. All these are modes of "reverse adaptation" in Winner's sense. They are techniques for eliminating or minimizing uncertainty. And they involve converting into instrumentalities of the system what had previously been thought of as external constraints. What before were ends to be served are transformed into means to be used.

There is certainly no difficulty in finding examples to illustrate Winner's five strategies. A glance at any newspaper, or at the *Congressional Record*, will provide a ready supply. And many other writers have discussed at length the ways in which recognition of such processes makes necessary important revisions in, for example, our theories of the competitive market and of democratic political processes. What Winner contributes—and here he is clearly in the Ellul tradition—is his insistence that these particular processes are expressions of a single underlying tendency. From this perspective the distinction between public and private, governmental and corporate, becomes very hazy and uncertain. What we see are societies made up of complex organizations, intimately connected with each other in networks of interdependence, some perhaps expanding while others contract, but each striving to control and stabilize its environment. These organizations are seen not just as the carriers of technology but as its expression and embodiment. They are the institutional representations of technique in Ellul's sense.

* * *

Let us descend from the somewhat dizzying level of these generalizations to the concrete particulars of continuous wave radio. There is, of course, no question of testing Ellul's propositions, nor those of Winner or Noble, if only because of their strong ideological component. But they can be used to illuminate the narrative.

In some respects, if one wanted to find a case study to illustrate what is involved in the management of technology, it would be hard to find a better one than this. It is early enough in time that the role of government is relatively easy to unravel—that is, before heavy government investment in industry by way of the military establishment became the norm. It is set in a period when corporations proceeded with more candor and impunity than they do today. In the case of the General Electric Company and the Navy Department we have remarkably rich collections of personal and organizational records to work from. And we are dealing with a technology that in its time was certainly revolutionary. The result is a sequence of events that is, so to speak, relatively transparent to analysis.

We have three main areas of concern. We are interested, first, in the process of invention and innovation—that is, in the emergence of technological novelty. Within that area our particular concern is to see whether an analysis in terms of information flows is feasible and useful. Second, we are interested in the behavior of large organizations, governmental and private, and particularly in the strategies they use to manage and control new technologies. And thirdly we would like to know whether, at a highly empirical level, we can discern evidence of the kind of reverse adaptation that Winner writes about: whether, that is, there are signs of a reversal of the normal means–end relationship, so that goals and purposes are found or created to suit the requirements of a technological system rather than vice versa.

There have been three main lines of continuity in our narrative. One starts with Fessenden, leads through his work with the National Electric Signaling Company, his collaboration with Alexanderson in the production of GE's first radiofrequency alternators, GE's decision to continue development of the device after NESCO became inactive, the negotiations with the Marconi interests, the Navy's intervention in those negotiations, and finally the formation of RCA. The second begins with Elwell at Palo Alto, his decision to stay in radio after the McCarty radiotelephone tests, his importation of the first Poulsen arcs from Denmark, the formation of Federal Telegraph and Telephone, its construction of a commercial radiotelegraph network in the western United States and to Hawaii, the sale of the first arcs to the Navy, the Navy's adoption of arc transmitters for its high-powered chain, and the Navy's decision in 1918 to purchase the Federal Company's patents to prevent their being acquired by Marconi. And the third concerns de Forest, his struggle to develop a radio system that would rival Marconi's, his search for a sensitive detector, the invention of the triode audion, his work for the Federal Company, the discovery of the audion amplifier and oscillator, and their purchase by AT&T. These three themes converge with the creation of RCA as an instrument of national communications policy, its conversion into an agency for the integration of ownership rights in continuous wave radio, and its later transformation in the course of its attempts to manage the diffusion of vacuum tube technology during the emergence of popular broadcasting.

It is easy to tell this story in terms of the personalities involved. And it is not difficult to organize it around the devices—the alternator, the arc, the vacuum tube—that made continuous wave radio possible. In the preceding chapters we have tried to do both: to show how the evolution of particular pieces of technological hardware interacted with the life histories of particular individuals and the fortunes of particular organizations. Let us try now to see how the elements of the story might fall into place when seen from a different perspective. If we single out these devices as crucial, what new configurations of knowledge did they represent? What bodies of scientific and technological information went into their creation? If we emphasize the activities of certain individuals as particularly important, what knowledge and what sources of information did they contribute? If, with the advent of the continuous wave, radio technology did begin to move along a sharply different vector, what technological traditions went into defining that vector and what economic or other interests influenced the allocation of resources to its exploitation?

Whether Reginald Fessenden can be called a scientist is a matter of semantics. If diplomas, certificates, and degrees are the criteria, he was not. What he knew about electricity he learned from his own reading and experimentation, his discussions with Kennelly, and his work for Edison and the Stanley Company. Whether his methods are properly called scientific is also a matter of definition. He was certainly capable of systematic experimentation and he insisted on precise measurement. But the dispassionate search for truth for its own sake, the exploration of nature's uniformities for the mere joy of discovering them, had little place in his motivation. Fessenden was interested in knowledge because it was useful. What he found out from his tests and experiments he wanted to put to use, and promptly.

Nevertheless, Fessenden was among the first, if not the first, to spot the presumptive anomaly that radio technology was facing and to identify the vector along which it would in future have to move. What part did science play in these events? To point out that he was very well versed in the scientific literature is only the beginning of an answer. Also relevant is the fact that he had been teaching electrical theory, with particular reference to Hertzian waves, at Purdue and Pittsburgh and that, through membership in the AIEE and participation in its programs, he was in touch with the most advanced knowledge of alternating current theory and practice available in the country at the time. No amount of argument will make Fessenden into a pure scientist. But, to the extent that work on high frequency oscillations had a theoretical component in 1898, Fessenden knew of it and understood it.

And that, after all, is precisely the point. Was Fessenden an engineer or a scientist? The question has no meaning: he was something of both— a highly theoretical engineer and a highly pragmatic scientist. When he presented his paper to the AIEE in 1898, were he and the other members discussing scientific questions or engineering ones? The answer is the same: the discussion revolved around issues that to an engineer must have seemed very theoretical and to a scientist very practical. Fessenden was functioning in both worlds; he was part of both communications networks. In this he was not alone: Pupin and Steinmetz, both present when Fessenden read his paper, could be characterized in the same way. And this was typical of the locations where the most advanced thinking on Hertzian waves, wireless telegraphy, and high frequency oscillations was to be found at the turn of the century. They were sites where theoretical speculation about Hertzian waves met and mingled with the ideas of experimenters working on wireless communication and engineers working on alternating currents.

But was it his knowledge of science and his acquaintance with scientists that alerted Fessenden to the inadequacies of spark radio? Only indirectly. Recall that his initial work in radio was concerned with improving detectors, not transmitters. This was the focus of his research at Pittsburgh and of his first radio patents. He wanted a device more sensitive than a coherer, and one that would give a quantitative indication of signal strength when coupled to a tuned circuit. Anyone working with detectors of that type could hardly fail to become aware of the way a spark signal splattered its emissions all across the radiofrequency spectrum. Work with coherers and Morse inkers would not make that point so dramatically: the coherer would "trigger" and the inker would make its mark on the tape no matter how broad the signal. Given a tuned detector that would selectively measure signal strength at a large number of frequencies, however, it would be clear that a spark signal dissipated its power in unnecessary radiation; a true continuous wave signal would concentrate the available energy on a single frequency. This was the point that Pupin made so forcefully after hearing Fessenden's presentation to the AIEE in 1898. With a spark discharge you could never exploit the full advantages of tuned circuits, either in transmitters or receivers.

Was this a scientific insight? In a sense it was. It rested on a knowledge of the theory of resonance, Rayleigh's work on acoustical resonance and Lodge's on electromagnetic resonance in particular. But in another sense it was an eminently practical observation, the full force of which would be obvious only to someone who had actually worked with wireless detectors and "syntonic" circuits. Practical experience, in short, disclosed—to some observers, not to all—the limitations of spark-generated signals and pointed to continuous wave radiation as a way of overcoming those limitations. When that practical lesson sank in, the scientific knowledge was there to rationalize it. But the initial stimulus came not from science but from work with the actual generation and detection of radio signals.

And, of course, not everyone grasped the significance of what was learned and not everyone saw the full implications. The early history of wireless is replete with horror stories of carefully arranged demonstrations degenerating into debacles because of deliberate or accidental interference between stations. Experimenters were slow to appreciate the necessity for selectivity and precise tuning. The reason is clear: spark transmitters, in the nature of the case, could not be tuned precisely to a single frequency; and there was little point in trying to build highly selective receivers when a single powerful spark transmitter could in effect wipe out whole regions of the spectrum. Greater freedom from interference, economical use of the available wavelengths, and efficient use of transmitter power all depended on making the shift to the continuous wave.

Fessenden was one of the first to grasp what that really meant for radio technology. Having grasped it, he accepted its implications. He had to have a transmitter that would generate true continuous sine wave alternating currents at radio frequencies. Even Pupin, whose insight into the logic of the matter was no less clear than Fessenden's, ventured no further than to suggest a very large increase in spark frequency, approximating more closely to a continuous wave but still remaining within the spark tradition. To use the vocabulary suggested earlier, this represented movement along the same technological vector. It was certainly a feasible option, a route easier to follow and more predictable in its outcome than the search for a true continuous wave transmitter. Marconi followed it, with his synchronized spark and disk discharger, and Telefunken with Max Wien's quenched spark. So indeed did Fessenden himself. By 1912 his big rotary spark was the most powerful transmitter available in the United States, and the Navy would certainly have adopted either that machine or the Telefunken quenched spark if Elwell had not shown up with his arc. But Fessenden knew very well that the rotary spark was a compromise—a good compromise in engineering terms, and perhaps a commercially profitable one, but not the transmitter he had to have if true continuous wave radio was to be a reality. That depended on the alternator.

Designing a radiofrequency alternator was in principle a simple matter. All you had to do was find a way to prevent the machine from disintegrating at extremely high rotor speeds. That, however, is rather like saying that to achieve controlled nuclear fusion is simple. So it is, in principle: all you have to do is fulfill certain known requirements for a sustained reaction.[20] What we intend to convey, when we speak of matters that are simple in principle but remarkably difficult in practice, is that we are facing a pure problem of technology. What is to be finally accomplished is clearly specified, and the theory of the matter is known. The problem is to bridge the gap between theory and a functioning physical device.

[20] Joan Lisa Bromberg, *Fusion: Science, Politics, and the Invention of a New Energy Source* (Cambridge, Mass., 1982).

This is a matter of design, and that is why some people are inclined to think of design as the essence of technology.

Designers of European radio alternators, such as von Arco and Goldschmidt, tackled the problem by running their alternators at relatively low rotor speeds and then multiplying the frequency of the resulting alternating current by various ingenious schemes. Alexanderson chose not to take that approach: his alternators generated the output frequency directly. One implication was that he faced all the problems of high rotor speeds in their most extreme form. This ruled out any thought of a revolving armature carrying coils of wire, as in conventional power alternators. Any such device would have disintegrated from centrifugal force. The characteristic feature of an Alexanderson alternator was the knife-edged steel rotor with slots cut in its circumference. (See Pl. 3) Serious problems of balance, alignment, lubrication, and speed control remained, of course. But a rotor of that type was not likely to disintegrate.

It was an elegant solution, and testimony to Alexanderson's skill as a designer. It was also, however, testimony to the accumulated technical lore and expertise of General Electric's Schenectady shops. When in 1901 Fessenden, by then totally immersed in wireless communication, took his ideas for a radiofrequency alternator to Steinmetz, he was making contact with a different engineering tradition: the tradition of power engineering. Alexanderson himself was always clear on the matter: his accomplishments in radio, he used to say, were merely the application of known principles of power engineering to a new field, that of wireless communication. This was excessively modest but the point was sound. The formation of the Fessenden-Alexanderson partnership marked not merely the collaboration of two talented individuals but also the merging of two bodies of technical knowledge, the confluence of two streams of technical information: one coming from power engineering, the other from communications. The result was a remarkably productive partnership and a remarkable spurt of creative engineering.

Thanks to the research of James Brittain, we are now thoroughly informed about the nature and style of this partnership and about the development of the Alexanderson alternator itself.[21] It was a device el-

[21] James E. Brittain, "C. P. Steinmetz and E.F.W. Alexanderson: Creative Engineering in a Corporate Setting," IEEE *Proceedings* 64 (September 1976), 1414-17; "Power Electronics at General Electric: 1900-1941," *Advances in Electronics and Electron Physics*, Vol. 50 (New York, 1980), pp. 411-17; "The Alexanderson Alternator: An Encounter between Radio Physics and Electrical Power Engineering" (Paper presented at the joint meetings of the Society for the History of Technology and the History of Science Society, Philadelphia, 31 October 1982); "E.F.W. Alexanderson (1878-1975): The Remarkable Career of an Engineer-Inventor," *The Bent of Tau Beta Pi* 67 (Summer 1976), 7-11.

egant both in concept and in physical embodiment, delicately balanced, as precise in its construction and smooth in its operation as a Swiss watch. But it was also a massive piece of machinery, generating pure sine wave oscillations of current at radio frequencies with a stability that no comparable device of its day could match. It was, in short, a beautiful piece of machinery, a masterpiece of the art. We are missing something important if we fail to sense the pride that it inspired in Alexanderson and his co-workers. Negotiations with the Marconi interests would never have been so difficult and tortuous if they had not been carried on against a background of distrust. Alexanderson and his colleagues were, at bottom, quite unconvinced that the Marconi people appreciated or would properly use the device they had built. It is possible, too, that Alexanderson's insistence that GE continue developing the device, after orders from NESCO ceased, was grounded partly in an almost aesthetic sense of the elegance of the machines he was building. And one wonders whether GE would have grasped as eagerly as it did at the prospect held out by the Navy of the formation of an American radio company if it had not promised GE the opportunity to oversee the deployment of its alternators directly.

Because of this affective component, and because it did the job it was designed to do so well, it is easy to become enthusiastic about the Alexanderson alternator. Its engineering quality was certainly very high, and this was to be expected, for it was designed and manufactured by a firm that had already brought the manufacture of high-speed rotating electrical machinery to the level of a fine art. In that sense the alternator represented a mature technology, as did (for example) the clipper ship in the age of sail or the Lockheed Constellation in the age of propeller-driven aircraft. Creations like these strike one as beautiful; there is an economy of design that appeals immediately to the aesthetic sense. But they are products at the end of a technological vector: the future holds little in the way of further development for any of them. The Alexanderson alternator was a fine machine but it had no offspring. The idea of generating radio waves by running specially designed alternators at very high speed was carried no further. Instead, the industry turned to the vacuum tube. And the early vacuum tubes, at the start of their technological vector, were neither very efficient nor very beautiful. They were curious-looking glass bottles with wires sticking out in odd directions and they were not at all impressive—except to those would could see their future potential.

The alternator represented the confluence of two technological traditions: electric power engineering and wireless communications. The arc transmitter represented a similar confluence, but here the tradition that intersected with the new art of wireless had its origin in electrical illumination. The arc light as a source of illumination had nothing at all to do with communications. We have sketched in an earlier chapter the curious sequence of events that brought the two conceptions together: attempts to eliminate the undesirable noise that an arc light emitted when fed with imperfectly smoothed direct current; experiments with the arc as an amplifier of voice currents; and Duddell's serendipitous discovery that an arc could be made to oscillate by shunting a tuned circuit across the discharge. Intersecting with that line of development we have Elihu Thomson's magnetic blowout in the United States and finally Poulsen's idea of operating the arc in a hydrocarbon vapor. With Poulsen we have for the first time the four essential elements of an arc radio transmitter: the arc flame itself, the resonant circuit in parallel, the magnetic field, and the hydrocarbon vapor atmosphere.

There could hardly be a better example of how the process of invention can extend over considerable periods of time and require, not a single breakthrough, but the progressive putting-together of known elements in new combinations. The arc as a generator of undamped electrical oscillations was not something invented at a particular point in time; it was a device that emerged from an extended process and it changed and evolved while going through that process. Clearly the process had both scientific and technological components. On the one hand there was the interest in eliminating what originally seemed to be an imperfection in the arc itself—its tendency to generate noise—and then in putting that phenomenon to practical use. On the other hand there was the scientific interest in exploring what went on inside the arc plasma itself and why the arc seemed to defy Ohm's Law. Modifications in the arc circuit produced a device that would oscillate continuously as long as it was provided with power. Further modifications raised the frequency out of the audio range and into the radio spectrum.

In all this one nowhere receives the impression of a deliberate, directed search for a new form of wireless transmitter, far less of the perception of a presumptive anomaly in spark technology. There is much tinkering with an intriguing device. There is much scientific interest in its behavior. But it is not until we reach Valdemar Poulsen, in whose hands all four essential elements come together, that we find a clear recognition that what has been invented is a device capable of transmitting information over distance without wires. And, as far as we can tell, Poulsen was the first to realize that this device, because it generated undamped oscillations, made possible radiotelephony and not merely radiotelegraphy.

Among users of arc technology, Elwell was the first to show the almost ideological rejection of spark that we find in Fessenden. In fact, Elwell was the more extreme of the two, for after the early and abortive tests of the McCarty system, he never had anything at all to do with spark, whereas Fessenden, as we have seen, was willing to compromise. Scientific insights into the limitations of spark seem to have had little to do with this early and total commitment to the continuous wave. More relevant is the fact that Elwell's point of entry into radio was by way of telephony, and his conviction, on the basis of practical tests, that spark and telephony could not be reconciled. There is some historical irony in this. Elwell clearly understood that for wireless telephony he needed a continuous wave transmitter, and it was that need that took him to Denmark. Demonstrations of wireless telephony, using the transmitter that he brought back from Denmark, were critical to his success in raising capital and interesting the San Francisco business community. But the Federal Telegraph Company never seriously involved itself in wireless telephony and never offered a commercial wireless telephone service. The arc transmitters that Elwell and later Fuller built were used exclusively for telegraphy. The point is an interesting one. The hope of developing a system of wireless telephony that could either challenge AT&T in its own field, but without the capital investment required to build a duplicate wired network, or else present such a competitive threat that AT&T would have to buy it, was the *ignis fatuus* that many early wireless experimenters followed, including Fessenden and de Forest. Elwell was no exception. Telephony was what lured him into radio and what inspired his early entrepreneurial successes. Above all, it was what converted him to the continuous wave and specifically to the Poulsen arc. But it was not the economic niche that, in the end, the Federal Company's arcs occupied.

Scientific theorizing and experimentation had played an important part in the development of the Poulsen arc in Europe, but for Elwell science and his friendships with scientists played a rather different role. He did of course receive at Stanford good training as an electrical engineer, but probably as important was the fact that he won the trust of the Stanford faculty and access to their intellectual (and in some cases financial) capital. It is striking how directly and unhesitatingly he headed for Denmark once his decision to develop a radio telephone system was taken. This suggests that he was confident of his information, that he knew of the claims made for the Poulsen system, that he knew indeed how to get in touch with Poulsen personally. Access to trusted sources of information is clearly indicated at this point, and it is probable that advice from friends on the Stanford faculty was of critical importance. Not to be overlooked either is his early experience with electric furnaces and the design of heavy-duty transformers, work that was channeled to him through the good offices of a Stanford professor and in which he received good advice from the same source. It was in that assignment that he first became acquainted with arcs and acquired confidence in working with very large currents. At Stanford, and through the Stanford faculty, Elwell found himself at the point of confluence of a variety of flows of information; his reaction to them did much to shape his future career.

On the evidence of his autobiography and his achievements, Elwell seems to have been a talented engineer, supremely self-confident. He was not a theorist, however, and not a speculative thinker. Science and scientists were for him useful resources on which he could draw, and he knew how to do that very effectively. But it was not scientific insight that steered him toward continuous wave radio in the first place; rather it was his sense of the commercial opportunity latent in radio telephony and his knowledge that spark transmitters could not do the job. As chief engineer of Federal Telegraph, his value to the company rested not on his scientific knowledge but on his driving energy and practical engineering know-how.

The history of the Federal Company would have been very different if Elwell had not persuaded the Navy to test his 30 kilowatt arc against Fessenden's rotary spark at Arlington in 1912. True, the company had already faced the need for higher power output because of the unreliable performance of the 30 kilowatt arcs on the San Francisco—Honolulu circuit. But it was the impression Elwell's transmitter made on the Navy, and the Navy's subsequent adoption of arcs, first for Darien and then for all the stations of its high-powered chain, that made a new approach to arc transmitter design essential. The pressure was from the new market that Elwell had opened up; the limitation was the highly empirical way in which he had tackled the "scaling-up" of Poulsen arcs up to that point. Elwell's abrupt departure from the Federal Company shortly after the first Navy contract was signed may have been partly coincidental; the ostensible reason was not a difference of opinion over design practices but the company's refusal to back Elwell's ambitious plans for transpacific expansion. But one cannot help wondering whether perhaps he knew that his usefulness to the company was waning and that a new approach was called for.

Fuller's appointment as chief engineer marked adoption of that new approach. Its main characteristics were a theoretical specification of the principal variables affecting arc transmitter output, careful experimentation to determine quantitatively the effect of varying one or more of the variables, and the derivation of a series of empirical formulas to guide transmitter design thereafter. Work along these lines laid the basis for the remarkably rapid increases in transmitter output power that the company was able to achieve between 1912 and 1919; and it was the fruits of that work—the knowledge it had created—that the Navy tried to keep out of the hands of the Marconi Company by purchasing Federal's patents in 1918.

Two points seem worth noting. First, when Fuller joined the Federal Company he brought with him techniques of engineering analysis that Elwell had not possessed, or at least had not used. This was partly a matter of personality, but partly it reflected the fact that Fuller was a younger man and that university instruction in electrical theory had made considerable advances since Elwell left Stanford. Second, Fuller came directly from employment with NESCO in Brooklyn, where he had been working on the design of small arcs for use as local oscillators in heterodyne receivers. This, on top of his graduate work at Cornell, had taught him that the arc would yield to controlled experimentation and to the use of "modelling" (both in the sense of theoretical models and in the sense of using small arcs to test the effect of changing parameters). Fessenden, of course, had left NESCO by this time, but some of the Fessenden tradition remained, and this was part of the baggage that Fuller brought with him when he moved to Palo Alto.

The significance of Fuller's work can be gauged not only by the rapid strides made in arc transmitter performance after 1912 but also by the lengths to which the Navy was prepared to go to keep the Federal arc in American hands. To be sure, the Navy's purchase of Federal's patents and property was a hasty and ill-considered act, and some at least of Federal's officers and directors did not emerge from the episode with entirely clean hands. Nevertheless, the fact that it happened at all is testimony to the distance that radio had travelled along that particular technological vector since Elwell brought his first tiny arc back from Denmark ten years before. What was at stake was not the Poulsen arc itself but the Elwell-Fuller developments, which by 1918 had carried arc transmitter technology far beyond the European level. Around these developments the Federal Company had been built, as had the Navy's long-distance communications network.

And the rate of development was indeed remarkable: from 100 watts input in 1909 to 30 kilowatts in 1913, 200 kilowatts in 1915, 500 kilowatts in 1917, and 1,000 kilowatts in 1919. By way of comparison, Fessenden gave his first alternator order to General Electric in 1901; the 50 kilowatt alternator, the first capable of long-distance service, was tested at New Brunswick in 1917, and the 200 kilowatt alternator, roughly equivalent to a 500 kilowatt arc, in 1919. The alternator, to be sure, was a more difficult product to manufacture, but on the other hand the Federal Company in Palo Alto had neither the physical facilities nor the engineering resources available to General Electric in Schenectady. Part of the explanation for the longer development time lies in GE's hesitancy to undertake construction of higher-powered machines after Fessenden left NESCO in 1912—that is, after the assured market had disappeared. But much of the credit must also be given to Federal's single-minded dedication to the arc, the device on which its fortunes totally depended, and to the guaranteed market that after 1913 was provided by Navy orders.

This dedication to a single device, evident to some extent in Fessenden and very strongly in Elwell, is not visible at all in de Forest, despite all the mythology that has grown up around the audion. In fact it is hard to depict de Forest as a convert to continuous wave radio in any sense until after 1907 when, with the formation of the Wireless Telephone Company, he began working with small Poulsen arcs. Up to that point his main objective had been to develop a system of radiotelegraphy that would rival Marconi's without too blatantly infringing Marconi patents. In that endeavor, despite much borrowing and adaptation, he remained within the spark tradition.

De Forest was of course a scientist. Did he not have a doctoral degree from Yale to prove it? It is difficult, however, to point to any specific connections between his scientific training and his later accomplishments, and his style as an inventor was hardly what we normally associate with scientific method. His experimental techniques were based on intuition and a willingness to try anything rather than on a firm grasp of the underlying theory. His technical writing shows a disconcerting tendency to slip into poetic imagery at exactly those points where precision is most called for. What he derived from his training at Yale was a good grounding in mathematics and electrical theory, although as we have seen he had pointed criticisms to make of the instruction he received in both these fields. From the research he carried out for his doctoral dissertation he learned all he needed to know about electrical resonance. And from his general training he became familiar with the rich resources of information available in the scientific periodicals and the proceedings of scientific societies—an asset of which he made good use in later years.

At two critical points in his career de Forest was able to make notable advances in radio technology because he found himself so situated that he could integrate previously separate bodies of knowledge and had the intelligence and imagination to take advantage of the fact. The reference is, of course, to the invention of the triode audion detector in 1906 and the triode amplifier and oscillator in 1912. We are not concerned here with questions of priority; our interest is in the flows of information that came together to make these inventions possible.

In the case of the triode detector we start with the "glow tube" that de Forest used in his Yale experiments, an unsatisfactory detector for any commercial purpose but embodying nevertheless the idea of detecting Hertzian waves by their effect on ionized gas in a partially evacuated glass tube. The later flame detectors, though no closer to commercial utility, represented the same line of thought. Added to this was de Forest's growing conviction that what he was really looking for was some kind of relay: a device by which feeble antenna currents could be used to control much larger currents, the variations of which could be heard in an earphone or used to control a recording device. This relay concept is clearly evident in the second of de Forest's flame detectors; the flame passes over two pairs of electrodes, one in the antenna circuit, the other in the earphone circuit.

It should be underlined that the conductivity of heated gases was an entirely respectable field of scientific research at the time. De Forest's flame detectors may look somewhat outlandish to us, but there was a logic to them with some basis in physical science. Between these exposed flame detectors and the later glass-enclosed audions there is both an intellectual connection and a connection in the underlying imagery. As far as de Forest was concerned, the audion, as he conceived it, functioned by virtue of the gas it contained. The filament was there to heat the gas, and the electric current that passed from cathode to anode was carried by gas molecules.

The idea of the detector as a relay and the idea of detecting Hertzian waves by their effect on heated gas were two of the four essential elements that went into the audion detector as de Forest invented it. The third was the idea of enclosing the electrodes in a partially evacuated glass tube and heating the residual gas by means of an incandescent filament. We have noted work along these lines by several European scientists. As far as de Forest was concerned, however, it seems safe to conclude that he got the idea from the Fleming valve, although de Forest himself denied this. The Fleming valve itself represented a different line of intellectual descent, one that starts with the incandescent lamp, investigation of the blackening of the inside of the lamp bulb, and research on the Edison effect. The carriers of this flow of information were Edison himself, William Preece of the British Post Office, and John Ambrose Fleming. Fleming is, for our purposes, the key individual, not so much because his research added greatly to knowledge of the Edison effect, but rather because he was the first to identify the process unambiguously as rectification and to suggest that the current carriers were not gas molecules but electrons. To Fleming also goes the credit for first suggesting use of the effect for the detection of wireless waves. We have noted that Fleming himself stood at the confluence of three flows of information: the research done on the Edison effect; the operating experience of the Marconi Company; and the then-novel theory of the electron.

The fourth essential element in the triode audion was the control grid, inserted between heated cathode filament and cold anode plate. This was the vital step that de Forest took and Fleming did not. Similar use of control elements in cathode ray tubes has been suggested as one source of the idea. We have noted, however, de Forest's insistence that a de Forest audion was not just a Fleming valve with a control grid. Physically, that it precisely what it was. Intellectually, however, and in terms of the imagery underlying the device, de Forest's assertion has merit. This is where his conception of the detector as a relay came into play. Antenna currents coupled to the control grid controlled much larger currents flowing from cathode to anode. Fleming, in contrast, was thinking in terms of rectification—Was that not how he had interpreted the Edison effect?—and in terms of the one-way flow of fluids through a valve. A more striking illustration of the importance of visual imagery in the process of invention would be hard to find.[22] And the consequences were important. What would it have meant for later history if the Marconi Company had controlled both the patents on the Fleming valve and those on the triode?

Development of the audion as an amplifier and oscillator resulted from demands being placed on the device that it had not originally been designed to meet. De Forest had described one of his early audions as a "device for amplifying feeble electric currents," but it was not in fact used for that purpose and probably could not have been. As a detector it met serious competition from the simpler and more robust crystal detectors and had little market success. This is why there is that strange hiatus in the development of the audion between 1906 and 1912, a period when the potentials of the device remained largely unexplored and de Forest himself seemed to have lost interest in it.

There were two exceptions to the general neglect: Hammond's laboratory outside Gloucester and the Federal Company's laboratory in Palo Alto. At both these locations the audion was called upon to perform functions other than detection. At Gloucester it was to serve as amplifier and oscillator in systems designed for the remote guidance of small craft and torpedoes. At Palo Alto it was to serve as amplifier to assist in the reception of high-speed telegraphy and later as oscillator in heterodyne receivers. Another way of phrasing this is to say that in both locations the men working with audions were confronted with a different set of

[22] Eugene S. Ferguson, "The Mind's Eye: Nonverbal Thought in Technology," *Science* 197 (26 August 1977), 827-35.

market signals. They were given information on desired end-uses that had not been present, or had not been brought into prominence, before.

Lowenstein brought to his radio guidance research the knowledge of the audion he had acquired while working with de Forest in New York. He added to that his knowledge of the highly specific and stringent demands of the system Hammond was working on and in particular the need for stable single-frequency continuous wave signals and for sensitive high-gain amplifiers. De Forest in Palo Alto possessed the same knowledge of the audion and knew in addition of the special problems of the Federal Company in transcribing high-speed telegraphy from wire recorders. Both men also knew of AT&T's urgent need for a "repeater" if it was to make good on Carty's pledge of transcontinental telephone service by the end of 1914.

The strategic role of Hammond as "translator" at this juncture deserves emphasis. Not only did he finance Lowenstein's work, specify the direction it should take, warn him of the progress being made in Germany, and urge him to demonstrate his amplifier to the Telephone Company. He also called Alexanderson's attention to the audion and supplied him with the first specimen to be examined in the General Electric Research Laboratory. This was a critical development. Irving Langmuir's research had already taken him into the theory of pure electron emission from incandescent filaments in high vacua, and he was able to see at once the possibilities latent in de Forest's gassy audion. Hammond also wrote about the audion to Beach Thomson, president of Federal Telegraph, and although as far as we know the letter was never acknowledged, we have de Forest's word for it that in all probability this was the event that initiated vacuum tube research in the Palo Alto laboratory. Significantly, it was the audion's capacities as an oscillator, not as an amplifier or detector, that Hammond emphasized.

Telephone technology played a role at this point that it had not played during the development of the audion as a detector. De Forest at Palo Alto had two men working with him, one of whom—Van Etten—was an experienced telephone technician. Lowenstein and Hammond were also working with telephone circuits, as we know from their correspondence. This does not necessarily mean that either group was thinking of the Telephone Company as the primary market, although the $1 million prize that AT&T was rumored to be offering for a workable line amplifier was certainly an incentive. It does mean, however, that they were able to draw upon the considerable stock of technical information that had been accumulated by the telephone industry. This was particularly significant in de Forest's case. Van Etten supplied the kind of hardheaded practical experience in telephone circuitry that de Forest lacked; he designed the input and output transformers for the amplifier; and he was the one whose skepticism de Forest and Logwood, the third member of the team, had to overcome.

Lastly, there was the intersection of de Forest's audion amplifier circuit with the theory of thermionic emission. This took place at two locations: first at Schenectady, through the mediation of Hammond and Alexanderson, and second at AT&T headquarters in New York, through the mediation of John S. Stone. Langmuir at Schenectady and Arnold in New York looked at de Forest's audion and saw in it something he had not seen: the possibility of converting it into a true high-vacuum device, capable of handling large currents and large voltage differentials, operating not by gaseous conduction but by electron emission. What is truly remarkable is not that they were able to achieve this insight, for both were first-class physical chemists and thoroughly conversant with electron theory, while de Forest was not. The point that needs to be underlined is that neither had investigated the audion before, despite the fact that its existence and some at least of its functional characteristics had been generally known since 1906. This is particularly striking in the case of Arnold and the other Telephone Company engineers, who were under considerable pressure from higher echelons in the organization to develop an audiofrequency amplifier and had been conducting a systematic search for one. The plain truth of the matter was that they had never thought of looking to the wireless industry for relevant information. That was a different body of knowledge, and the unexamined assumption was that it had nothing to offer the technology of a wired telephone system. The confluence of the two streams of information had not yet occurred. Information about the audion was not secret; it was just separate, or was thought of by telephone engineers as separate, which amounted to the same thing. In a sense the Telephone Company was paying the price for abandoning the tentative explorations into wireless communications that had been initiated by the Boston office before 1907. It is hard to believe that Lowenstein would have received such a chilly reception, or that de Forest's audion amplifier would have taken AT&T's research staff so much by surprise, if those explorations had been continued. The creative breakthrough occurred belatedly, when previously discrete bodies of information about the audion, about the technology of wired telephony, and about the theory of electron flow were finally merged. Once it happened, of course, it all seemed very simple and obvious.

The discovery of the audion oscillator seems to have been a clear example of serendipity. The people at Federal Telegraph knew about heterodyne receivers: Fuller had been experimenting with small arcs as local oscillators for just such receivers before joining Federal's staff. And they understood that the low "grunting" noise made in earphones by a radiotelegraph signal recorded at high speed and then played back at a much lower speed could be converted into a higher tone, easier to copy, if it could be heterodyned against a local oscillator. But the development of such an oscillator had very low priority for de Forest's team. They concentrated on the amplifier and when their amplifier broke into self-oscillation they interpreted it as a problem, as indeed it was. Clear and explicit recognition of a circuit as an oscillator circuit, producing an audio note that could be controlled by varying the circuit constants, did not come until August 1912. Even then there is nothing in the laboratory notebooks to indicate a realization by de Forest that he had, potentially, not merely an audio oscillator, but also a generator of radio waves—that is, an entirely new form of continuous wave transmitter. The matter is controversial and any conclusion is to some extent a personal judgment, but the evidence strongly suggests that de Forest did not think of his oscillator circuit as potentially a continuous wave radio transmitter until he was questioned on the point by Stone in New York in October 1912, just before the AT&T demonstrations. Lowenstein and Hammond had taken that essential step earlier: Hammond's letter to Beach Thomson of January 1912 explicitly referred to the oscillating audion as an alternative to the arc and the alternator—that is, as a radio transmitter.

That an audio amplifier would break into self-oscillation as a result of stray coupling of the output signal back into the input was certain. One would have to take precautions to prevent it from happening. That someone working with such a circuit would realize sooner or later that it was generating continuous high frequency alternating currents was as inevitable as such things can be. Once that step was taken it was bound to be realized that there was no reason in principle why radiofrequency oscillations, and not merely audiofrequency ones, could not be generated. And with that mental step one would have a radio transmitter, as well as an oscillator eminently suited for converting Fessenden's heterodyne receiver at last into a practically useful device. The vacuum tube oscillator, however, was not the result of any search for a continuous wave transmitter. It was the byproduct of the search for an audiofrequency amplifier.

For the vacuum tube, as for the alternator and the arc, invention was not an act, capable of being located precisely in time and space, but a process extending over time in which information from several sources came to be combined in new ways. To understand the process it is essential to understand the previously separate flows of information and stocks of knowledge that came together to produce something new. It is characteristic of this process that the understanding of the thing that is being created, the invention that is emerging, changes as the process advances and new information is brought to bear. The identification of exactly when the invention was made and of who made it becomes largely arbitrary. One ends up thinking of the invention as, so to speak, plastic or malleable, and of the process by which the invention is made as essentially social and cooperative.

We do not usually write about invention this way. We like to dramatize discoveries; we like to think of the "eureka moment" when light dawns and all is suddenly clear and the invention is made (forgetting that analysts like Koestler and Usher who have written of eureka moments and acts of insight have laid at least equal stress on the long period of gestation and scene-setting that precedes such moments, and the long period of critical revision and gradual improvement that follows it).[23] Also, we are inclined to think of invention as an act rather than a process because of the bias built into our patent laws. If property rights in a new discovery are to be secured, it is important to be able to establish priority in time, and there are famous examples (the telephone, the vacuum tube oscillator) where a few hours or days have made all the difference. So it becomes important, in patent applications and later litigation, to think of invention as an act that can be dated in time and attributed to identifiable individuals with precision.

This bias, however, should not be allowed to corrupt our historical interpretations nor our understanding of how invention and discovery happen. Examples such as that of the vacuum tube argue convincingly for an interpretation that views invention as a process with considerable duration in time, one to which many individuals contribute in a substantial way, and in which the conception of the thing invented or discovered changes. It may be possible, in certain instances, to identify in that process a particular point at which everything "falls into place"; but to call that *the* moment of invention is an arbitrary decision on our part, and if our patent laws and popular romances encourage us to think in this way, that is our misfortune. Such moments indicate only that the last essential interconnection has been made, the last necessary bit of information has been integrated. If we think of a new invention or discovery as a new configuration of information, we should recognize that these new configurations emerge only after a substantial period of preparation, in which disparate and ill-related pieces of information are jumbled around in the mind, and that their emergence is followed by another lengthy period of review and revision. The latter period itself is often

[23] Arthur Koestler, *The Act of Creation* (New York, 1967); Abbott Payson Usher, *A History of Mechanical Inventions*, 2nd ed. (Cambridge, Mass., 1954).

hardly distinguishable from the period of preparation leading up to a new discovery.

* * *

Already we are verging on the issues of control and management that authors like Noble, Ellul and Winner have raised. Let us leave aside for the moment the complex problems concerning the autonomy of technique. At a commonsense level, a technology can be said to be managed if the outcomes are in conformity with the expectations and intentions of those who developed it. The closeness of fit between outcomes and initial conception—between how the technology is conceived *ex ante* and how it is perceived *ex post*—is the basic issue.

Seen from this point of view, the three devices that have concerned us present interesting contrasts. At one level of analysis they were equivalent devices. Functionally, they were alternative techniques for generating high frequency alternating currents. That is to say, they were alternative ways of proceeding along the same technological vector: that of continuous wave radio. At a lower level of abstraction, however, they were very different. They had different technical genealogies; they represented different configurations of information; and they were managed in different ways and with different degrees of success.

In the case of the alternator and the arc, expectations and outcomes matched very closely. And it is no coincidence that with these two devices the process of invention was linear, or very close to it, in the sense that it involved few abrupt changes in direction, few unpredicted developments, few surprises. This is by no means to deny that some of the early expectations were disappointed: neither the alternator nor the arc, for example, proved of much importance for radiotelephony, although both could be and were used for that purpose. And it is not to deny the very challenging problems of design and manufacture that had to be surmounted en route to the final perfected product. Alexanderson's work on the alternator was a virtuoso performance on any scale of engineering excellence, and such a machine could have been manufactured at few other places in the world at that time. Similarly, Elwell knew exactly what he wanted when he first went to Denmark, and he brought back exactly what he went for; but there were many headaches before really high powered arcs could be built, and it took the more highly developed analytic skills of Fuller to carry development beyond the 30 kilowatt level. All this is true; and in these particulars lies much of the interest. But it is also true that the hopes and expectations of those who initiated the process of invention were closely matched by the characteristics of the devices that resulted. The big 200 kilowatt alternators with which RCA equipped its stations would have impressed and pleased Fessenden but they would not have surprised him. The 1,000 kilowatt arc Fuller designed for the Lafayette station must have struck Elwell as a fine achievement; but, if he looked at it, he would have seen a larger and more sophisticated version—which is to say, a version with greater information content—of the tiny 100 watt arc he had brought from Denmark.

The vacuum tube was not like that. De Forest fumbled his way to creativity, and in that respect he may more accurately typify the process of invention than does Elwell's work with the arc or Alexanderson's with the alternator. There is, significantly, no single moment in the long process leading up to the triode vacuum tube that one can reasonably point to as *the* moment of invention. The one episode that looked like a classic "eureka moment"—the gas flame experiments in de Forest's Chicago apartment—turned out to be misleading. Even the crucial step that differentiated de Forest's device from Fleming's—the insertion of the control grid—seems to have been taken without, at the time, any consciousness of radical change. It was just one more variation to try. More fundamentally, the device that finally emerged—the "hard" vacuum tube that dominated radio technology up to the invention of the transistor in 1946— was significantly different from de Forest's audion. It had uses and functions that had been no part of de Forest's original objectives. And it operated according to scientific principles that he had great difficulty in accepting and that were certainly not the principles by which he thought the audion worked when he invented it. The invention changed drastically during the process of invention. What finally emerged was a device infinitely more versatile, more powerful, and more full of potential for the future than the narrowly defined device that had been the original goal of de Forest's efforts.

So far we have been comparing the conception of a device with its realization. That is technological management in one sense, and in that sense the alternator and the arc represented managed technology while the vacuum tube did not. The development, production, and deployment of the alternator was a process managed by General Electric; for the arc the analogous process, from importation of the technology through development through deployment, was managed by Federal Telegraph. The only possible qualification is, in both cases, the intervention of governmental authority through the Navy Department; whether this is to be thought of as an exception or as part of the management process is a question to which we shall return shortly. No corresponding statement can be made with reference to the vacuum tube. Corporate management of vacuum tube technology begins with acquisition of the patent rights by AT&T. That was, to be sure, the start of a long and important process of development and refinement, but it was a long way from initial conception.

There is, however, another sense in which one can talk about the management of technology. We have been discussing control of the technical process from conception to realization. But management of technology may also refer to its consequences, and here the question of whether the outcome matches the expectations appears in a different guise. There are technological advances the consequences of which merely fulfill expectations, to a greater or lesser degree; there are others that transcend those expectations, generating outcomes that are different in kind, not merely in degree, from the ideas that inspired them. Technologies of this second type appear to defy management from the start; they generate consequences that are unanticipated. Institutions react to them, rather than controlling them, and management, if it occurs at all, is management with a timelag. For historians of technology, among others, the distinction is important. Most technological improvement is mundane, routine stuff, a matter of incremental improvements, of performing a function a little faster or a little more precisely or with a little less labor. In the aggregate, the cumulative effects of that kind of technological change are very large; but the effects of each increment can be predicted and managed—they are the stuff of which econometric models are made. A few technological developments are not of that nature. They are truly revolutionary in the sense that they introduce a large, irreversible discontinuity in development that was neither predicted nor planned. To say this is not to assert the indispensability of any particular innovation. As they used to say, there is always more than one way to skin a cat; and in its effect on such massive aggregates as the national income, no single invention makes much difference in the short run.[24] Some technological developments, however, do open doors that before were closed and make evident possibilities that before were not seen. One such was the triode vacuum tube.

[24] Any residual beliefs in the "indispensability" of any single invention has been dissipated by the work of Robert Fogel. See Robert W. Fogel, *Railroads and American Economic Growth: Essays in Econometric History* (Baltimore, 1964), and "A Quantitative Approach to the Study of Railroads in American Economic Growth," *Journal of Economic History* 22 (June 1962), 163-97; and compare Paul A. David, "Transport Innovation and Economic Growth: Professor Fogel on and off the Rails," in *Technical Choice, Innovation and Economic Growth: Essays in American and British Experience in the Nineteenth Century*, ed. David (New York, 1975).

The alternator and the arc were no minor contributions to the technical development of radio. The arcs proved that a commercial radiotelegraph service could survive in competition with the wired telegraph and telephone systems over land, and in competition with the submarine cables across many miles of ocean. As important, they provided the U.S. Navy with a radio network connecting all its major bases and assuring reliable communications with units of the fleet years before any other naval power had that capability. The alternator provided indispensable communications facilities between the United States and Europe during World War I and, after the war, gave RCA the equipment it needed to equip its long distance stations in the United States and establish new stations in foreign countries. Between them, the alternator and the arc proved that high-powered continuous wave radio communication not only was possible but was, in competition with the older technology of spark, the more efficient mode. In the shift from spark to continuous wave, it was the alternator and the arc, not the vacuum tube, that broke away from the old conventions and showed what the new technology could do.

In doing that, the alternator and the arc magnificently fulfilled the expectations of those who had built them and the hopes of those who, years earlier, had predicted that spark technology was facing a dead end and would have to yield to the continuous wave. The institutional repercussions were important also. Command over continuous wave technology made it possible for the United States to rid itself of foreign ownership of radio facilities on American soil and to create a corporation charged with the responsibility of representing and advancing American national interests in telecommunications. This too had been part of the expectations of those who, at some risk to their careers, had backed continuous wave radio when it was still a new and controversial technique.

But the vacuum tube had consequences of a different order entirely. Nobody planned to make the audion into a continuous wave transmitter in the first place. And nobody planned to make it the basis for a new industry—public broadcasting—not to mention a myriad of other later uses, such as talking pictures, television, and radar. These were all unplanned consequences, part of the unanticipated fallout from a device whose versatility seemed to increase from year to year. Why the difference? Why, with the vacuum tube, did it seem as if human ingenuity, like the sorcerer's apprentice, had unleashed forms of magic that could be neither ignored nor controlled, while with the alternator and the arc everything had been so managed, so tidy, so neatly arranged?

One could say that the alternator and the arc each represented the end of technological vectors, while the vacuum tube was the beginning of one. But that does no more than define the problem in different terms. What made the vacuum tube so hard to manage was, on the one hand, the ease with which it could be duplicated, once the tricks of the trade were known, and on the other the surging mass demand for tubes once the broadcasting boom got under way. Receiving tubes were not hard to imitate: anyone with access to facilities for making incandescent bulbs, together with a Gaede or Langmuir vacuum pump, could produce tubes that, if not as good as those marketed by RCA, were still acceptable. (Transmitting tubes were a different story.) This was not the kind of device around which it was easy to build a "monopoly of knowledge," and RCA was never in fact able to suppress or control the trade in bootleg tubes. This ease of imitation thrust RCA and its corporate allies back on legal defenses—that is, on enforcement of their patent rights. Formidable on paper, these defenses proved fragile in practice in view of the public outcry against the "radio trust" and the risk of political intervention. This vulnerability to punitive political action and popular protest itself reflected the characteristics of the vacuum tube, for it had made radio entertainment an article of mass consumption.

The same disruptive effect of the vacuum tube is evident in its effect on the intercorporate treaties that had brought AT&T, Westinghouse, and United Fruit, as well as GE, into the RCA fold. No compacts could have been more carefully drawn up, with more precise attention to every conceivable eventuality, than these. They were masterpieces of legal draftsmanship, documentary testaments to a joint corporate commitment to manage the new art of continuous wave radio in every detail. Broadcasting and the vacuum tube reduced them to a shambles. Designed to minimize intercorporate conflict, the documents became themselves sources of acrimonious dispute, their enforceability and indeed their intent cast into question by the rise of a "field" that was partly wired communication, partly wireless, partly public service telephony for toll, and partly something entirely new. These corporate compacts had been drawn up by men who had for their frame of reference older and more familiar technologies: either wired telephony or long-distance point-to-point radio. Vacuum tube technology applied to broadcasting presented a radically different set of problems. If radio had remained the preserve of alternators and arcs the agreements might have proved very durable— monuments to successful technological management. As it was, with the diffusion of the vacuum tube, they became within a few years no more than embarrassing encumbrances. Hopes for management of the broadcasting industry hinged thereafter (until the 1980s) on control of the networks and of the wire connections that made networks possible.

If the test of technological management is the goodness of fit between intentions and outcomes, we can hardly describe the history of the vacuum tube in the 1920s as an example of success. And the most important lesson to be learned is probably how misleading it can be to talk of technology in the singular instead of technologies in the plural, particularly when management is at issue. Some technologies lend themselves easily to management by centralized bureaucracies; others do not. The vacuum tube, in the early phases of its diffusion, was a technology that did not. The reasons lie partly in its physical characteristics, partly in the nature of the demand that it encountered and helped to create. The vacuum tube transformed the radio industry because it made possible public broadcasting to receivers in the home. Radio became domesticated. Its products became consumer goods, its messages became advertising instead of "traffic." In the process the firms that had seemed likely to control the industry acquired completely new constituencies: entrepreneurs hoping to set up local broadcasting stations; firms wishing to manufacture components and complete receivers; listeners wishing to buy receivers and tubes and to hear programs. These new constituencies represented important markets and the exploitation of these markets transformed the structure and functioning of the industry. But since they had expectations to be met and the political muscle to make those expectations effective, the environment in which the industry functioned was also transformed.

RCA existed as the institutional embodiment of the attempt to control and manage continuous wave radio technology, an attempt in which both private and governmental organizations had joined. It had been created to function in one world—the world of alternators and transoceanic radiotelegraphy—as the chosen instrument of national telecommunications policy. In that world the relevant sites of power had been other large corporations, like Marconi and Telefunken, engaged in the same business and linked to RCA by traffic agreements and patent pools, and government agencies like the British Post Office and the U.S. Navy that were directly concerned with communications policy. That world still existed in the late 1920s, and RCA still functioned in it, with growing prestige and efficiency. But alongside it there had grown up the hurlyburly world of commercial broadcasting, a world in which the rules of the game and the names of the players were quite different. That world had been made possible by the vacuum tube.

By both tests, then—the match between conception and realization, and the match between intentions and outcomes—the alternator and the arc may be classed as successfully managed technologies, while the vacuum tube, in its early phases, can not. But what of the broader and more complex issues raised by Ellul and Winner? Continuous wave radio and the institutions that developed around it clearly came to constitute a technological system. Did this system develop according to its own autonomous logic? Was it in that sense out of control? This is a different and more difficult question. As we have already argued, the issue raised by Ellul and Winner is not whether technological systems are managed by human agents or not, but rather whether that management reflects the imperatives of the system or, conceivably, the imposition of some other criteria of choice. Are means being adapted to ends, or are ends being created or molded or manipulated in the interests of the available means?

Throughout our historical narratives we have been dealing with men and organizations dedicated to finding better methods for achieving objectives—that is, to technique. For every thousand hours spent by these men and organizations in evaluating means for achieving ends, perhaps less than one was spent in discussing whether the ends were worth achieving or whether other ends better deserved the energies that were being expended. (The fact that Owen Young occasionally did exactly that is one of the qualities that endear him to historians and made him valued by his contemporaries.) If that is what Ellul means by the domination of technique, we may concede the argument to him, for what it is worth, without further ado. There is little need for detailed case studies to prove the point. The world of affairs is not staffed by philosophers.

There is, however, one episode in which it appears that the logic of the technological system was not allowed to play itself out to the full, where a different set of criteria was brought to bear, and where explicit attention was paid to the ends of action and not just to the means. We refer, of course, to the intervention by representatives of the Navy Department in the negotiations between General Electric and the British and American Marconi companies for the sale of alternators. Here explicit political arguments were deployed. As a result, the organization that controlled access to the new technology was induced to abandon its original objectives and adopt a strategy that was markedly different. Surely this is a case where the technology was very far from being out of control or "autonomous."

It may be conceded immediately that this episode fully supports our argument that the alternator represented managed technology. It is not even necessary to believe that the Navy's intervention represented any real interruption in the process. There was, as we have seen, an effective informal communications network already in place linking Edwards at GE with Hooper in the Bureau of Steam Engineering. At a more formal level, if Young's letter to Roosevelt of 29 March 1919 cannot quite be interpreted as an invitation to the Navy to intervene in the negotiations, it was at the very least a clear signal that, if the Navy intended to do anything, now was the time to do it. The conventional interpretation of these events depicts GE as being taken by surprise by the Navy's action. This reading of the story should be treated with considerable skepticism. Officers at the Navy Department in Washington knew what GE was doing, and responsible executives at GE knew what action to expect from the Navy. Similar skepticism is indicated with regard to two other examples of Navy intervention: the purchase of the Federal Company's patents, and the Navy's letter to the tube manufacturing companies calling for cooperative action to eliminate patent disputes over tubes. In each of these cases it is difficult to take seriously an interpretation that portrays the Navy's action as catching the corporations unprepared. The image that should be in our minds is, more likely, that of a play in which all the actors know their lines very well and need no help from the prompter.

If this is the case, we must consider the entry of political considerations, through the Navy Department, not as an interruption in the process of technological management but as an integral part of the process. In terms of the model suggested earlier, the development of continuous wave radio was taking place in one of those areas of overlap where the technical, economic, and political realms met and interacted. If the technology was to be effectively managed, information from all three sources had to be integrated into the decision-making process. The Navy's function was to serve as a channel for information from the political system. Information from that source commanded attention because it was, or could be, backed up by effective sanctions and because it was stated in a vocabulary (national security and the national interest) that tended to halt debate.

The Navy purchased the Federal Company's patents because it believed, on slender evidence, that those patents or rights under them would otherwise be sold to the Marconi companies. It induced GE to cancel negotiations for the sale of alternators to the Marconi companies because it believed that, if the sale went through, it would fix in British hands the power to dominate world communications. It appealed to GE and Western Electric to settle their patent disputes in order that, in the interests of safety at sea, marine operators could be supplied with vacuum tubes. In each case the Navy's action cut through a very complex situation in which there was room for considerable disagreement over the appropriate ends to be pursued. Once the action was taken, discussion of the appropriate ends effectively ceased. Invocation of the national interest, as interpreted by the Navy, was sufficient to short-circuit debate over the ends to be served. To repeat: there is no implication that the Navy's formula for action was imposed on reluctant cooperators. Quite the contrary. Suspension of debate over ends meant a clearing of the air, a reduction of uncertainty, a welcome freedom to take action. Men and organizations could now go ahead and do what they did best: concentrate on designing and carrying out the best methods for attaining ends that were now taken as given.

It is worth noting how, in these cases, the Navy was able to induce an atmosphere of crisis. Whether it was the Federal patents, or GE's alternators, or the vacuum tube patents, the requirement was not just for action but for action at once. And in each case there was just enough evidence to make the invocation of crisis believable. With the alternator and the arc, the "enemy at the gates" was the Marconi Company and behind that British imperialism and British "domination" of international communications. What "domination" meant with reference to communications, and whether the British government or British-based corporations actually exercised any power that could be called "domination" were questions seldom asked. Examples of cable censorship during the war were of doubtful relevance; titular ownership of the transatlantic cables was of little significance when, as Lloyd George pointed out, most of the cables were under long-term lease to American corporations; and allegations that Marconi was on the verge of achieving a monopoly of long-distance radio were hardly convincing to anyone who had seen the French station at Tuckerton or the German one at Sayville or the Federal Company's installation in Hawaii. But none of that mattered very much: "domination" was a cue word that evoked a predictable response. And there were others: when Admiral Bullard took Owen Young to one side and confided to him a "state secret" received from the president in Paris, it made an impression. That "state secret" began its life as a worried and preoccupied chief executive asking his personal physician if he would be good enough to tell the new director of naval communications to keep an eye on American interests in radio. There was nothing secret about it; Wilson, his doctor, and Bullard, at the time, had only the vaguest idea of what was at stake. But it made little difference.

There is no intention of attributing to these individuals any greater measure of deviousness than is the common lot of humanity. They were, by and large, quite ordinary people trying to do what seemed to them the obviously sensible thing to do in the light of the information available. They were free in principle to act in other ways, to make other choices. What prevented them was the thought that, in the circumstances, these other acts and other choices would not have been quite as sensible. And that perhaps is the central point. Hegel once wrote that history proceeds behind our backs. There are alternatives in front of us, among which we try to make the most reasonable choice we can. But there are also processes going on of which we are not aware, or perhaps not aware until some time has passed. What was in fact happening as a result of these initiatives by the Navy was that impediments to the development and diffusion of continuous wave technology were being eliminated. This is true, paradoxically, even of the attempts to block transfers of the technology to the Marconi companies. There is ample evidence that few people knowledgeable about radio between, say, 1912 and 1919 had any confidence in the ability of the Marconi companies to adopt and aggressively develop new technology. This may have been a misperception: Godfrey Isaacs was nobody's fool and knew very well that the company he headed had, in a technical sense, been caught sleeping at the switch. The behavior and achievements of the Marconi Company after 1919 do not suggest technological conservatism. But in 1919 the belief was there, and it counted for something.

Throughout the actions and statements of Navy personnel in this period there runs the constant theme of hostility to the British role in international communications and distrust of British intentions. The explicit intent of Navy intervention in the sequence of events that led up to the formation of RCA was to create a distinctively American institution and advance distinctively American interests. As far as overt intentions were concerned, that was the objective sought, and control of the relevant technology was the means, not vice versa. Now nationalism and xenophobia were no strangers to American life in 1919. It would have been strange if they had not played some role in the evolution of postwar radio policy. There was a sense that communications lay too close to the heart of national sovereignty to be entrusted to the hands of foreigners. A nation that did not control its own external communications, it was felt, had not yet attained full status as a nation. Sentiments of this kind, never fully analyzed, underlay much of the hostility to the Marconi Company, and not merely in the Navy Department. The vigor with which such sentiments were exploited, however, suggests that nationalism and distrust of the foreigner were also serving as means. Appeals to patriotism could be counted on to generate a predictable response. And the belief that one was responding to such an appeal served to rationalize actions that otherwise might be difficult to explain, such as refusal to sell equipment to a valued customer. In this sense use of the symbols and rhetoric of nationalism was part of the technique used to advance the technology of continuous wave radio. There was a constant interplay between means and ends, technology serving now as an instrument of nationalism, nationalism now as an instrument of the technology.

What difference did it make? Did the Navy's actions and GE's response really change the outcome in an important way, or merely clothe that outcome in a different institutional form? Did the technology of continuous wave radio follow a historical logic of its own? Suppose that events had followed a different course, that RCA had never been created: in what important ways would the outcome have differed? Suppose that Wilson, his mind preoccupied with other things, had said nothing to Cary Grayson about radio as they took their morning drive through the streets of Paris. Suppose that Admiral Bullard, with no preliminary briefing in Paris, had on his arrival in Washington listened with tolerant skepticism to Hooper's calls for immediate action and promised, perhaps, that he would "look into it." Or suppose that the directors of General Electric had proved immune to Bullard's eloquence and, putting their patriotism in their back pockets, had said flatly that GE was not a communications company, that it was poor business for a firm to compete with its own customers, and that they intended to sell alternators to Marconi, no matter what the Navy said about it. How would the long-run outcome have differed?

To suggest answers to these questions is to engage in hypothetical history and no great confidence can be placed in the results. But one can, without straining the imagination, conceive of a different but nevertheless plausible alternative scenario. In this hypothetical alternative world, if the approach to General Electric had been handled differently, or if GE's reaction had been other than it was, GE would have contracted to sell its alternators to Marconi, there would have been no GE-backed drive for "an American radio company," and Josephus Daniels's campaign for government ownership would have confronted the demand of the American Marconi Company that, once the war emergency was officially over, the rights of private property should be respected and its stations returned to private hands. There is no reason to suppose that Congress would have been any more receptive to the idea of government ownership than it in fact was, nor that Wilson could or would have given Daniels's campaign the decisive leadership that just might have made it politically acceptable. In the circumstances the end of the state of war emergency would have seen American Marconi back in business and GE free to sell and ship its alternators wherever it wished.

And it would have sold them to Marconi. There were, it is true, some final details still impeding the signing of a contract when Bullard and Hooper intervened, but nothing that could not have been handled. And the order in prospect was a large one: in May 1919 the Marconi negotiators had been ready to place a firm order for twenty-four alternators, fourteen for the American company and at least ten for its British parent. It is worth pausing for a moment to consider what an investment on this scale would have implied. If we allow two alternators for each station, which was later RCA's standard practice, this first order would have provided American Marconi with seven major long-distance stations and the British company with five. Acquisition of alternators in these numbers would have done more than merely reequip the Marconi network as it had existed in 1914. What was planned was an enlargement of that network into a truly worldwide system. The stations that RCA later equipped in Europe, in Poland and Sweden, would have been equipped instead by Marconi. Included too would have been expansion into Latin America and across the Pacific to Japan and China. Marconi would also, of course, have bid strongly for the Imperial Chain contracts—that prospect explains why the British part of the contract was for "at least" ten machines. And all this would have been in addition to the basic transatlantic circuits, which formed the major traffic artery of the system.

It was an impressive design and we can understand how easily American observers saw behind it the specter of British "domination" of world communications, how they carried forward into the age of radio the images formed in the age of telegraph cables, when it had been all too easy to visualize the world's submarine cables as a giant spider's web, with Britain controlling the central hub. A radio communications system was really not like that. You could not cut a radio link as you could cut a cable, although you might jam it. Interception of messages passed by radio did not require physical possession of the medium, as it did with cables. Radio had implications for the control and dissemination of information that the cables did not. Nevertheless, in both systems there was a drive for completeness, for integration, and it was this drive that was so easily identified with the dynamics of imperialism.

People like Hooper and Bullard were not in error when they saw in the Marconi design a thrust toward worldwide integration. The men behind the Marconi companies knew quite well that, although their component companies might be separate legal entities, each subject to the laws of a particular political jurisdiction, nevertheless they were integral parts of a communications system that transcended national boundaries. The technical function of this system was to pass information quickly and accurately from one point on the surface of the globe to another, and it performed this function most efficiently when the system was fully integrated, when it was complete in its coverage, and when there were no barriers, technical or political, to the movement of messages. It required no sophisticated grasp of information theory to understand this: the matter was clear to anyone who "handled traffic," whether by cable or by radio or by any other means. As these communications systems grew, their component parts became increasingly interdependent and the systems became more completely integrated. This was a tendency that transcended national boundaries and the ambitions of any nation-state.

Intervention by the U.S. Navy, or by anyone else, could not change the logic of development of the world's telecommunications system. It could, however, accelerate or retard the process, and it could change its institutional form. That was the true historical significance of the formation of RCA. Creation of that company meant that after 1919 the United States would be integrated into the global telecommunications network in a different way, through a national organization that was expected to respond to American needs and aspirations more directly than a unit of the transnational Marconi organization would have done. A new center of decision-making was created, and it is no coincidence that its first order of business was to build the stations that would interconnect it with other elements of the world network and to negotiate the traffic agreements that would govern its relations with other units in the system. In the absence of this new organization, reconstruction of the radio links that connected the United States to the outside world would have taken place under the auspices of the Marconi system.

And it could have been handled smoothly enough. The organizational structure already existed; the funds were available; so were the experienced operating personnel. And, with new access to the design and manufacturing expertise of General Electric, Marconi stations could have moved long-distance traffic with an efficiency and confidence that had never been possible in the age of spark. There would, of course, have been friction with the Navy Department, and the Department of State would have shown increasing concern as the question of American radio penetration of China and Latin America came to the fore. But problems of that kind would have been nothing new to the Marconi companies; indeed, they were precisely analogous to the constant frictions that developed with the Post Office and the Foreign Office in Britain. Such differences were to be expected when a transnational organization had to deal with the executive agencies of a national state. They were part of the costs of doing business in a world still afflicted with nationalism.

Regardless of the Navy's intervention, therefore, the development of the world's telecommunications system would probably have proceeded much as it did. The system had in that sense an inherent logic of growth, a drive for extension and integration as a system, irrespective of particular institutional forms. This was the groundswell of change. In that limited sense a kind of technological autonomy has to be admitted. The limits to that autonomy become evident, however, when we recall that continuous wave radio technology provided much more than a new and more efficient instrumentality for telecommunications. It also made possible public broadcasting, and that was a system of a rather different nature. In that field the development of radio technology encountered a cultural phenomenon that it had not itself created: a mass market for information and entertainment the existence of which had been barely suspected. What made public broadcasting such a remarkable social innovation was the fact that in it two independent streams of development converged: on the supply side, the technological process that had led to that most versatile of devices, the vacuum tube; on the demand side, the cultural processes, generated in a print-dominated age, that had created by the second decade of the twentieth century a hunger for news and entertainment delivered orally in the home. Broadcasting did not result from the inexorable unfolding of a technological imperative latent in radio technology; it burst on the scene when continuous wave radio finally found a mass market.

Contract for Establishment of High Power Radio Service

CONTRACT of two parts made and concluded this ———— day of May, 1919, by and between ———— represented by the President of said corporation, party of the first part (hereinafter called the Company) and the United States represented by the Secretary of the Navy, party of the second part (hereinafter called the Government).

WHEREAS, the Government, particularly in view of the prime importance to the Government, the people of the United States, and the Navy at all times of reliable means of world-wide radio communication, desires to encourage the establishment of the best available means of such communication, comprehending a chain of wireless stations covering as far as necessary the important countries of the world, capable of successfully competing with all other radio systems; and

WHEREAS, it is an essential part of the value to the Government of a system of international radio communications that the American part of it be wholly controlled by American interests, and that other portions of it be controlled as far as possible by American interests; and

WHEREAS, the General Electric Company has devised and built apparatus and systems particularly well adapted for said purpose and has recently been solicited by foreign interests, private and otherwise, to allow such apparatus and systems to be used by such interests under circumstances that might not sufficiently advance the interests of the Government and the people of the United States; and

WHEREAS, the Government owns certain letters patent pertaining to radio communication, including those relating to radio communication by the arc system of generation known as the Poulsen patents and Fuller improvements thereon, the patents on the Atlantic Communication Company's system and other patents that were acquired in connection with and incidental to the prosecution of the present war; and

WHEREAS, the General Electric Company, desiring to serve the Government to the best of its ability with respect to the establishment and maintenance of such a system of world-wide radio communication as is desired by the Government has caused to be organized the present Company and has transferred to it the full right to its experience and letters patent for long wave radio communications as hereinafter defined for the purpose of enabling it to make and carry out the present contract; and

WHEREAS, the possibility of financing and carrying on the undertaking contemplated is largely dependent upon the encouragement, cooperation, and assistance of the Government in every reasonable and proper way, and specifically, but not exclusively, as hereinafter stated; and

WHEREAS, the Government has for some years past used every effort to have formed a purely American company for erecting and operating such a system of world-wide radio communication as that above mentioned; and

WHEREAS, the General Electric Company has claims against the Government for unlicensed use of various radio inventions controlled by it, and has been requested by the Government to present such claims for discussion and settlement, and the present Company is in a position to grant to the Government proper releases of such claims; and

WHEREAS, it is desirable in the Government's interests, to have the use in its field of the inventions of the Company and of the General Electric Company for radio purposes and desirable in the interests of the people and for the development of the business interests of the United States that the Company should acquire to the extent hereinafter indicated the radio inventions of the Government and of the General Electric Company in order that it may develop and use the best possible system of radio communication.

NOW, THEREFORE, it is hereby agreed by and between the Company and the Government, the latter acting in pursuance of all authority conferred by law, as follows:

1. For the purposes of this contract long wave communication and transmission are defined as radio communication and transmission by wave lengths suitable, as the art from time to time advances, for communication for distances of one thousand (1,000) statute miles or more. Long waves are defined as waves suitable for such communication. Short waves are defined as waves too short to be suitable for such communication. Nevertheless the Government may at any time treat as short waves hereunder waves of such lengths below six thousand meters as the Government may at that time exclusively appropriate for any communication of less than one thousand miles range.

The expression Government field as used herein means:

A. All land, ship, and air-ship stations owned by the Government in so far as the same are used for transmitting and/or receiving Government messages as distinguished from commercial and other messages at any wave length. Government messages shall include in addition to ordinary Government business, free news service furnished by the Government for the public benefit.

B. All land stations owned by the Government in so far as the same are used for transmitting and/or receiving any messages to and from ships by short waves.

C. All ship and air-ship stations owned by the Government on Government-owned ships or air-ships in so far as the same are used for transmitting and/or receiving Government messages and incidental commercial messages at any wave length.

D. All land stations owned by the Government in so far as the same are used for transmitting and/or receiving to and from ships by long waves Government messages and incidental commercial messages.

The expression Company field as used herein means:

a. All land stations in so far as the same are used for transmitting and/or receiving by long waves commercial and other messages as distinguished from Government messages.

b. All land stations not owned by the Government in so far as the same are used for sending and receiving messages by short waves, except that while the Government pursues its present policy of owning all shore stations transmitting to and receiving from ships by short waves, known as Coastal Stations, such Coastal Stations shall not be included in the Company's field.

c. All ship and air-ship stations transmitting and/or receiving messages by any wave length except Government owned stations on Government owned ships and air-ships.

2. Independently of the construction of the stations referred to in Article 7 hereof, and in consideration of the grant by the Government in Article 3 hereof, the Company hereby grants to the Government without royalty, except such royalty as may be payable to others but not to the General Electric Company by virtue of contracts under which the Company has or shall have acquired particular patents, a full, free, non-transferable license restricted to the Government's field and exclusive in that field to manufacture, cause to be manufactured and use inventions covered by and claimed in any and all United States letters patent that the Company now owns and under which it has or may hereafter acquire during the terms of this agreement the right to grant such license, and all letters patent that may be issued on applications for letters patent that the

Company now owns, to the full extent of the terms for which such letters patent are or may be granted respectively, so far as they relate to or are applicable to radio. The Company also agrees that the General Electric Company will make a grant to the Government in the language of this article.

3. Independently of the construction of the stations and in consideration of the grant and agreement in Article 2 hereof and in further consideration of Article 6 hereof, the Government hereby sells, assigns and transfers to the said Company, its successors and assigns, all United States patents, inventions, applications for patents and rights or licenses under or in connection with patents, which the Government now owns in so far as the same relate to or are applicable to radio, the same to be held and enjoyed by the Company, its successors and assigns as full and entirely as the same would have been held by the Government had this assignment and sale not been made, the Government reserving, however, to itself:

a. A full, free, non-transferable license restricted to the Government's field and exclusive in that field to manufacture, caused to be manufactured and use the inventions of said patents and applications to the full end of the terms for which such letters patent are or may be granted, respectively.

b. The right to grant non-exclusive licenses to others for short wave work.

And the Government agrees to perform every reasonable lawful act necessary or desirable to confirm in the Company the rights herein granted and the full, free, and undisturbed possession of the same, and in particular the Government agrees that the Secretary of the Navy will not grant for long wave work except for its own use without the written consent of the Company the authorization referred to on the third page of a certain contract, dated the 15th day of May, 1918, by and between the Federal Telegraph Company, the Poulsen Wireless Corporation, and the United States of America.

4. The Government will not utilize the inventions covered by the licenses herein granted, agreed to be granted and reserved or the inventions covered by any patents referred to in this agreement or any apparatus or devices heretofore or hereafter purchased by the Company or of the General Electric Company, except in the Government field. Without impairing or waiving the restrictions of the licenses or enlarging the reservations of the Government, the Company authorizes the Government to use the long wave stations which the Government at the time has built for Government purposes, in general commercial communication without royalty, until January 1st, 1921 and thereafter to utilize them in general commercial communication on payment to the Company of a royalty

equal to 5% of the gross receipts from the business under this article 4, within thirty days of the expiration of each Government fiscal year but only from and to such places at which the Company shall not at the time be maintaining a commercial service direct or through land lines and/or short cables to and from one of its stations, or where the Company shall fail to handle a message at a station under its control.

5. The provisions of Articles 2, 3 and 4 hereof shall apply as far as permissible under the law to any patents, patent applications, licenses or rights under or to take patents acquired by either party or by the General Electric Company during the term of this agreement, and all licenses and rights to be granted under such patents shall run for the life of the patents respectively.

The Company agrees that the General Electric Company will continue its present practice of requiring those of its employees likely to make inventions along this line of work to assign such inventions to it; the Company also agrees to pursue a similar practice.

The admission of validity implied in the acceptance of licenses and assignments hereunder is limited to the field and term for which such licenses exist.

Each party releases the other from any and all claims arising from past infringements by the other on any radio patents which such party owns.

The long wave rights acquired by the Company from the Government hereunder are not assignable except that they may with the approval of the Navy Department be assigned to a successor corporation.

6. The Company agrees for itself and for the General Electric Company that neither it nor the General Electric Company will make, sell, or dispose of for use in the United States except for its or their own use any radio apparatus designed for transmission over distances exceeding one thousand (1,000) miles without the consent in writing of the Navy Department, which consent shall not be refused provided the purchaser is a loyal American citizen or an American corporation and offers sufficient security that he or it will remain such and will remain free from foreign domination; the Company further agrees that neither it nor the General Electric Company will make any working agreements with reference to radio with any foreign company or Government or for the sale of long wave apparatus without the written consent of the Navy Department except as the General Electric Company may be required so to do by existing contracts.

7. The Company will promptly and diligently endeavor to procure such patent rights and licenses, in addition to those it now owns, as may be necessary or desirable for the purposes contemplated, and the Government will, so far as is lawful and proper, assist in so doing, being kept

informed of the terms on which such rights are secured. If these rights can be secured to its satisfaction, the Company will undertake, making every reasonable effort in the undertaking, to establish, maintain, and operate in an efficient and business like manner a chain of wireless stations intended to constitute a high grade international system of communications.

The Company represents that its charter contains such lawful restrictions as are possible to insure that the beneficial ownership and the voting power of at least eighty per cent (80%) of the voting capital stock is at all times in citizens and/or corporations of the United States, and that it provides that its officers and employees whose duties and functions are to determine and control its policy shall be citizens of the United States; the Company agrees that only citizens of the United States shall be employed in the United States as either sending operators or receiving operators in its various radio stations, provided that the free use of radio telephony shall not be limited or restricted except as may be provided for by general legislation, as for example in time of war.

8. The Government shall cooperate in every reasonable and proper way with the Company in such undertaking and in the securing of proper concessions for the establishment of stations in foreign countries and in this country, and will use its best efforts in all proper and legitimate ways to foster and promote the proposed operations.

9. The Company agrees that it will at all times give to the Government's messages on Government business, up to ten per cent of the Company's gross business in each month precedence over other messages at rates not exceeding 75% of its commercial rates for messages transmitted under comparable conditions and that it will take all reasonable precautions to the end that all its stations both in the United States and abroad shall in case of war be available, so far as the Government takes over radio in war time, exclusively for the Government's messages and so maintained as to be of the greatest possible utility to the Government in case of war as follows:

A. By maintaining its stations, as far as is reasonably possible, in good working order and using due diligence to replace the equipment to conform to the advances of the art.

B. By employing in charge of each station, so far as allowed by the laws of the countries in which the stations are located, a citizen of the United States.

C. By encouraging its operators to become and remain Naval Reservists.

10. The Government agrees fully and permanently to license the proposed operations and to provide for and permanently assign to the Com-

pany ample and suitable wave lengths for the proposed operations in accordance with the then existing state of the art.

11. The Government agrees that no departmental regulations shall be made discriminating against the Company in the Company's field; the parties agree to conduct their operations in their respective fields in accordance with the contract and with full regard for the rights of each other. The Government agrees strictly to enforce proper and reasonable regulations designed to prevent radio interference with the Company's messages in its field and to limit as far as possible the license of radio apparatus, systems, or devices which cause harmful interference from the point of view of the then existing state of the art, so that such interference shall not occur.

12. The Government agrees so far as reasonably possible to furnish at cost to the Government at all times and at places convenient to it the testimony of its officers and employees and copies (certified if desired) of its records as far as may be needed to establish facts material to any patent litigation involving the interests of the Company in its field.

13. A representative of the Government, to be designated by the Secretary of the Navy or by the President, shall have the right of discussion and presentation of the Government's views and interests concerning all matters coming before the Company's Board of Directors that may affect the Government's interests, but such representative shall not have the right of voting upon any questions to be decided by the Board of Directors; this representative shall be given due notice of the meetings of the Board of Directors of the Company as shall be required to be given the Directors. The Company consents that in case of war the Government may if it so elects take over any and all of its stations and other property for the conduct of Naval and Military operations on such terms of compensation as may be agreed upon between the parties or as may be prescribed by such act of the Congress as may be applicable thereto with just compensation for the use of the patents and other property during the period for which the Governments retains possession. On the passing of the emergency the Government shall restore the property and franchises in as good condition as when taken over.

14. The Company shall be entitled to use the entire plant and equipment to be installed as herein stipulated in the Company's field conformably to present and future law without hindrance or control by the Government except as in this agreement stipulated.

15. It is agreed between the parties hereto that the United States long wave radio patents, patent rights, and good will now owned by the Company, including those acquired from the Government hereunder, are, in the Company's field, of the value of Five Million Dollars ($5,000,000)

plus a royalty of five per cent (5%) on the gross receipts from business in the Company's field. The Government may at any time take over the said patents, rights and good will for long wave work on the basis of this value, provided that at the same time it also takes over at their fair value at the time subsequently acquired long wave radio patents, rights, and good will of the Company and the tangible property of the Company, all for long wave work.

16. The Government recognizes and admits that in acquiring the apparatus and devices useful for radio purposes, manufactured and being manufactured by the General Electric Company, whether for the station at New Brunswick, New Jersey, or otherwise, and in acquiring other apparatus and devices useful for radio work manufactured and to be manufactured by the General Electric Company and in remodeling and altering the new Brunswick station and other stations in such manner as to embody and utilize inventions heretofore owned by the Company or acquired by it hereunder, it acquires and has no license to use the same, except in the Government's field; the Government therefore agrees that it will not use such apparatus, devices, or stations, except in its field or permit others to use them in the Company's field. The Government agrees not to sell material, apparatus, or devices useful for radio work embodying an invention or a then existing patent which the Company then owns, unless in such a manner as not to convey to the purchaser or another any right to use such apparatus under such patent.

17. Each party agrees from time to time as may be desired by the other party to execute such further assignments, licenses and other papers as are necessary and desirable to carry out the intent hereof.

18. This agreement shall run until January 1, 1945. Licenses or assignments that ought to have been granted or executed up to that time, or later, on inventions acquired up to that time shall be granted, executed, and delivered after said date and shall run for the life of the patents respectively. On the expiration of this agreement the Company shall continue to be free to operate and shall at all times so long as it continues its operation be afforded fair and reasonable facilities and a proper number of wave lengths in accordance with the then existing laws.

19. This agreement shall be binding upon and inure to the benefit of this Company, the General Electric Company and any Company conducting radio business in the United States or owning or beneficially interested in United States radio patents, provided that such Company is controlled either by this Company or by the General Electric Company.

20. No member of or delegate to Congress, officer of the Navy, or any person holding any office or appointment under the Navy Department, is or shall be entitled to any share or part of this agreement or to any

benefit arising therefrom, but this stipulation, so far as it relates to members of or delegates to Congress, shall not be construed to extend to this agreement, it being made with an incorporated Company.

IN WITNESS WHEREOF, the United States of America, represented by and acting through the Secretary of the Navy who exercises the power of the executive department of the Government for this purpose, and all other powers him hereto enabling, has caused these presents to be executed, and the Secretary of State and the Secretary of Commerce have approved the same, and the Company has caused these presents to be executed by its President and its corporate seal to be affixed thereto by its Secretary acting under authority conferred by its Board of Directors.

SOURCE: Manuscript Division, Library of Congress, Hooper Papers, Box 2, file May-August 1919.

Index

Library of Congress Cataloging in Publication Data

Aitken, Hugh G. J.
  The continuous wave.

  Includes index.
  1. Radio—United States—History. I. Title.
TK6548.U6A65 1985   384.6'0973   84-22265
ISBN 0-691-08376-2 (alk. paper)
ISBN 0-691-02390-5 (pbk.)

Hugh G. J. Aitken is George D. Olds Professor of Economics and American Studies at Amherst College. He is author of several books, including *Taylorism at Watertown Arsenal* and *Syntony and Spark: The Origins of Radio*.